Surface Membrane Receptors

Interface Between Cells and Their Environment

NATO ADVANCED STUDY INSTITUTES SERIES

A series of edited volumes comprising multifaceted studies of contemporary scientific issues by some of the best scientific minds in the world, assembled in cooperation with NATO Scientific Affairs Division.

Series A: Life Sciences

Recent Volumes in this Series

Volume 2 – Nematode Vectors of Plant Viruses
edited by F. Lamberti, C.E. Taylor, and J.W. Seinhorst

Volume 3 – Genetic Manipulations with Plant Material
edited by Lucien Ledoux

Volume 4 – Phloem Transport
edited by S. Aronoff, J. Dainty, P. R. Gorham, L. M. Srivastava, and C. A. Swanson

Volume 5 – Tumor Virus–Host Cell Interaction
edited by Alan Kolber

Volume 6 – Metabolic Compartmentation and Neurotransmission: Relation to Brain Structure and Function
edited by Soll Berl, D.D. Clarke, and Diana Schneider

Volume 7 – The Hepatobiliary System: Fundamental and Pathological Mechanisms
edited by W. Taylor

Volume 8 – Meat Animals: Growth and Productivity
edited by D. Lister, D. N. Rhodes, V. R. Fowler, and M. F. Fuller

Volume 9 – Eukaryotic Cell Function and Growth: Regulation by Intracellular Cyclic Nucleotides
edited by Jacques E. Dumont, Barry L. Brown, and Nicholas J. Marshall

Volume 10 – Specificity in Plant Diseases
edited by R. K. S. Wood and A. Graniti

Volume 11 – Surface Membrane Receptors: Interface Between Cells and Their Environment
edited by Ralph A. Bradshaw, William A. Frazier, Ronald C. Merrell, David I. Gottlieb, and Ruth A. Hogue-Angeletti

The series is published by an international board of publishers in conjunction with NATO Scientific Affairs Division

A	**Life Sciences**	Plenum Publishing Corporation
B	**Physics**	New York and London
C	**Mathematical and Physical Sciences**	D. Reidel Publishing Company Dordrecht and Boston
D	**Behavioral and Social Sciences**	Sijthoff International Publishing Company Leiden
E	**Applied Sciences**	Noordhoff International Publishing Leiden

Surface Membrane Receptors

Interface Between Cells and Their Environment

Edited by

Ralph A. Bradshaw, William A. Frazier,
Ronald C. Merrell, and David I. Gottlieb

Washington University School of Medicine
St. Louis, Missouri

and

Ruth A. Hogue-Angeletti

University of Pennsylvania School of Medicine
Philadelphia, Pennsylvania

PLENUM PRESS • NEW YORK AND LONDON
Published in cooperation with NATO Scientific Affairs Division

Library of Congress Cataloging in Publication Data

Main entry under title:

Surface membrane receptors.

(NATO advanced study institutes series: Series A, Life Sciences; v. 11)
"Lectures presented at the NATO advanced Study Institute . . . held in Bellagio, Italy, September 13-21, 1975."
Includes index.
Hormonal receptors–Congresses. 2. Cell membranes–Congresses. 3. Plasma membranes–Congresses. 4. Cellular recognition–Congresses. I. Bradshaw, Ralph A., 1941- II. Nato Advanced Study Institute. III. Series. [DNLM: 1. Cell membrane–Physiology–Congresses. QH601 N11s 1975]
QP187.S865 574.8'76 76-25821
ISBN 0-306-35611-2

Lectures presented at the NATO Advanced Study Institute entitled
Surface Membrane Receptors: Interface Between Cells and Their Environment,
held in Bellagio, Italy, September 13-21, 1975

A Division of Plenum Publishing Corporation
227 West 17th Street, New York, N.Y. 10011

Printed in the United States of America

Preface

The NATO Advanced Study Institute entitled "Surface Membrane Receptors: Interface Between Cells and Environment" was held in Bellagio, Italy September 13-21, 1975. This meeting was an attempt to bring together in an international and interdisciplinary forum scientists who are studying recognitive phenomona which take place at the surface membrane of cells. While an attempt was made to restrict the subject areas covered at the meeting to those experimental systems which have been biochemically characterized to some extent, it will also be noted that some contributions to this volume represent a preliminary identification of interesting regulatory substances which might reasonably be expected to act at the cell surface. This book is divided into four sections reflecting the subject areas covered during the course of the meeting. The first section entitled "Membrane Structure and Receptor Function" is intended as an overview of the role of membrane structure in determining the regulatory properties, physical state, structure and location of cell surface receptors. It should be noted that the plasma membrane itself provided the unifying theme for the intentionally diverse contributions to this volume. The following three sections represent an arbitrary division into three levels of structural complexity of the things in their external environment with which cells must specifically interact. These include interactions with other cells (Section II), macromolecular substances (Section III), such as hormones, growth factors and deleterious macromolecules such as bacterial toxins, and finally small molecules (Section IV) such as neurotransmitters, chemoattractants and hormonal substances. While each of these areas has its own unique problems and to some extent, unique methodologies, it is hoped that the scope of this volume will stimulate in its readers, as it did in those of us present at the Institute, an appreciation of the utility of interdisciplinary communication and interdisciplinary approaches in addressing the problems of membrane receptor structure, function and regulation.

The editors wish to thank the Scientific Advisory Committee of NATO, Merck Sharp and Dohme, The Kroc Foundation and The National Science Foundation for providing the financial support which made this meeting possible. Also, we wish to thank Mr. G. B. Mallone and his staff of METAVIAGGIO, Como, Italy for their extensive assistance in making the preparations for this meeting. Finally, we thank Mrs. Dorothy Hogue for her patient and untiring assistance in the editing and final preparation of the manuscript.

Ralph A. Bradshaw
William A. Frazier
Ronald C. Merrell
David I. Gottlieb
St. Louis, Mo.

Ruth A. Hogue-Angeletti
Philadelphia, Pa.

June, 1976

Contents

SECTION I

MEMBRANE STRUCTURE AND RECEPTOR FUNCTION

SECTION II

CELL-CELL INTERACTIONS

Simple Organisms

Aggregation

SECTION IV

CELL-SMALL MOLECULE INTERACTIONS

Section I

Membrane Structure and Receptor Function

THE FLUID MOSAIC MODEL OF MEMBRANE STRUCTURE: SOME APPLICATIONS TO LIGAND-RECEPTOR AND CELL-CELL INTERACTIONS

S. J. SINGER

Department of Biology, University of California at
San Diego, La Jolla, California 92093 (USA)

The general principles governing the molecular organization of membranes have become clarified in recent years, and we are in the midst of a period when increasingly detailed experimental information about the structures of membranes and their molecular constituents is being obtained at an explosive rate. This structural information must be relevant to an understanding of membrane functions, and in this paper, we attempt briefly to show how these structural insights may illuminate some aspects of the two main problem areas that are the foci of this symposium: ligand-receptor and cell-cell interactions.

The Fluid Mosaic Model of Membrane Structure

The fluid mosaic model of membrane structure (37, 41) is currently widely accepted as describing the general features of the organization of the lipids and proteins of functional membranes. The model has been considered in some detail in the references cited above and elsewhere (38, 39), and will therefore be discussed here only to that extent and in those directions required for the problems addressed.

The proteins of membranes. A first consideration is to recognize that there are two distinctive classes of proteins associated with membranes: peripheral and integral proteins (37, 38). A protein may be categorized as peripheral by a set of operational criteria (38) that reflect its relatively weak binding to the membrane and its close similarity to ordinary soluble proteins. On the other hand an integral protein is characterized by the fact that hydrophobic bond-breaking agents are required to dissociate it from the membrane;

by its binding affinity for lipids and non-ionic detergents; and by its insolubility in the monomeric state in aqueous solutions in the absence of detergents. These operational criteria suggest a very different kind of structural association of peripheral and integral proteins with the membrane (38): integral proteins are assumed to interact directly with the membrane lipids, while peripheral proteins are thought to be attached to the membrane peripheral to the lipid bilayer, generally at sites on specific integral proteins where the latter protrude from the surface of the lipid bilayer (see below).

Integral proteins. The majority of membrane-associated proteins fall in the category of integral, and include most membrane enzymes and receptors. The hypothesis was advanced some years ago (22, 45) that the distinctive feature of an integral protein molecule is its amphipathic structure. It was proposed that such molecules were more-or-less globular, but were differentiated into hydrophilic and hydrophobic segments, with the hydrophilic segment(s), containing essentially all the ionic residues of the protein, protruding from the membrane into the aqueous phase; and the hydrophobic segment embedded in the non-polar interior of the lipid bilayer. Evidence has since been obtained that several integral proteins of membranes do indeed exhibit such amphipathic structures (38).

The basis for this amphipathic structure is thermodynamic, and stems from the facts that ionic and highly polar groups (such as saccharide moieties) are in a substantially lower free energy state in contact with water than with a non-polar environment, while non-polar residues are in a lower free energy state in contact with a non-polar than an aqueous environment (37).

There are a number of ways one can conceive an amphipathic protein to be situated in a membrane subject to these thermodynamic restrictions. A protein may consist of only one polypeptide chain or alternatively of several polypeptide chain subunits, and these subunits may be only part-way embedded, or may entirely span the membrane with hydrophilic segments protruding from both sides of the membrane (Fig. 1). It was suggested (37, 38) that those proteins involved in the facilitated or active transport of hydrophilic small molecules and ions are all similar in structure, consisting of two or more subunits spanning the thickness of the membrane. Such a structure may generate a central water-filled channel lined with ionic residues and one or more specific active sites for binding the ligand that is to be transported. In this model, the translocation event in transport is taken to be a quaternary rearrangement of the subunits (Fig. 2), which might require an additional source of energy beyond that derived from the binding of the ligand itself. A similar proposal had been made earlier by Jardetzky (19). There is now good evidence that several proteins involved in membrane transport have subunit structures similar to that of Fig. 1 right (16, 21).

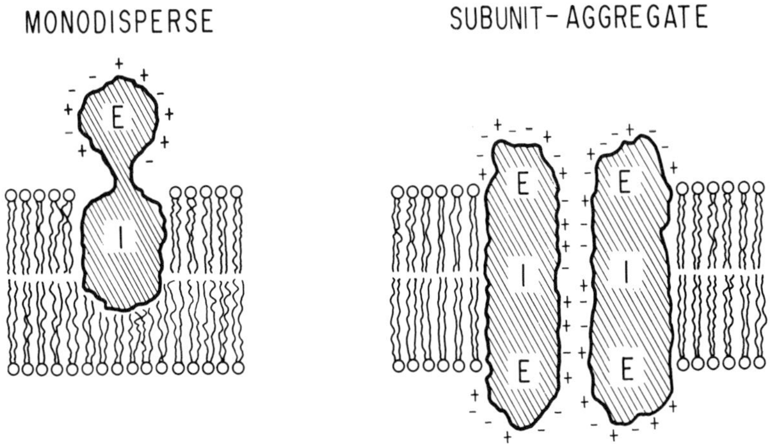

Fig. 1. A schematic representation of two of the ways in which integral proteins might be associated with the membrane. The dumbbell shape of the monodisperse protein is meant to suggest a structural basis for the fact that the integral protein cytochrome b_5 has a limited region of its amino acid sequence that is especially sensitive to proteolysis. E and I refer to exterior and interior regions of the protein, respectively. From Singer (38); further details are given in that article.

<u>Peripheral proteins</u>. The existence of the peripheral proteins has not yet been as widely appreciated as has that of integral proteins. They are the minority of membrane-associated proteins. We have discussed at length (38) a number of different membrane proteins we regard as peripheral, including as the best-studied examples cytochrome c of the inner membrane of mitochondria, and the protein complex known as spectrin on the cytoplasmic surface of erythrocyte membranes. As suggested in a preceding section, a peripheral protein should generally become non-covalently bound to a specific site on a particular integral protein, and subserve its functions by virtue of such binding. Thus, there is good evidence that cytochrome c is specifically attached to sites on two different integral proteins, cytochrome c reductase and cytochrome oxidase, and mediates electron transport between them. Released from the mitochondrial membrane by 3 M KCl, however, cytochrome c is a water-soluble monomeric protein, exhibiting no unusual properties that would explain its membrane-binding other than by the attachment to integral proteins just discussed.

It follows that there is a considerable analogy between recognized peripheral protein such as cytochrome c, and externally added ligands such as antibodies, lectins, and polypeptide hormones. The latter proteins are also ordinary water soluble proteins that become

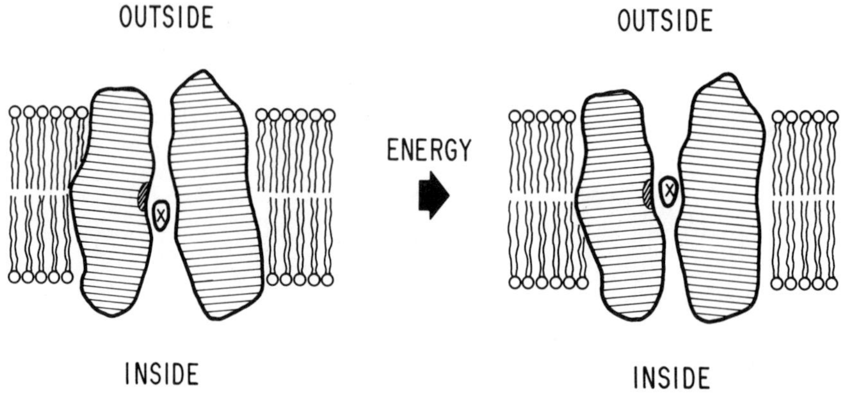

Fig. 2. A schematic representation for a proposed mechanism for active transport. A subunit aggregate protein that spans the membrane with a water-filled central channel (Fig. 1) can undergo an energy-driven quaternary rearrangement that translocates a hydrophilic or ionic ligand X from one side of the membrane to the other (37, 38).

attached to membranes by binding to specific sites on integral proteins (antigens, oligosaccharides, and hormone receptors, respectively) which are embedded in and protrude from the membrane. The analogy may be more than merely formal, as is explored in the subsequent discussion of ligand-receptor intersections.

Discussion of the spectrin complex is deferred to the section dealing with the control of mobility in membranes.

Fig. 3. (NEXT PAGE) A highly schematic representation of the fluid mosaic model of membrane structure, with associated peripheral proteins. The view is of the cytoplasmic face of a membrane, with some integral proteins (I) spanning the thickness of the membrane and others only part way embedded in the fluid lipid bilayer (as seen in cross-section). A particular kind of peripheral protein (P) is shown as specifically bound (non-covalently) to a site on the I protein where the latter protrudes from the membrane into the cytoplasm. In view A), the peripheral protein P is in a monomeric state but in B) it has become self-aggregated: (The aggregation is here depicted as one-dimensional, but it could be more complex.) By virtue of the attachment of P to I, the aggregation of P necessarily ties down the I molecules which are no longer individually mobile in the fluid membrane. The conversation from state A) to state B) provides a simple mechanism for the reversible immobilization of integral proteins in either limited or extensive areas of a fluid membrane. [From Ref. 29].

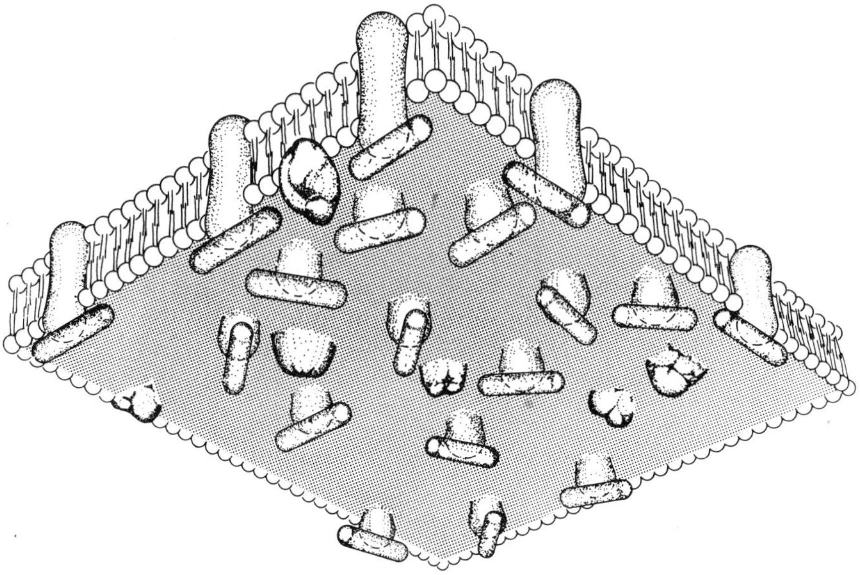

Fig. 3A.

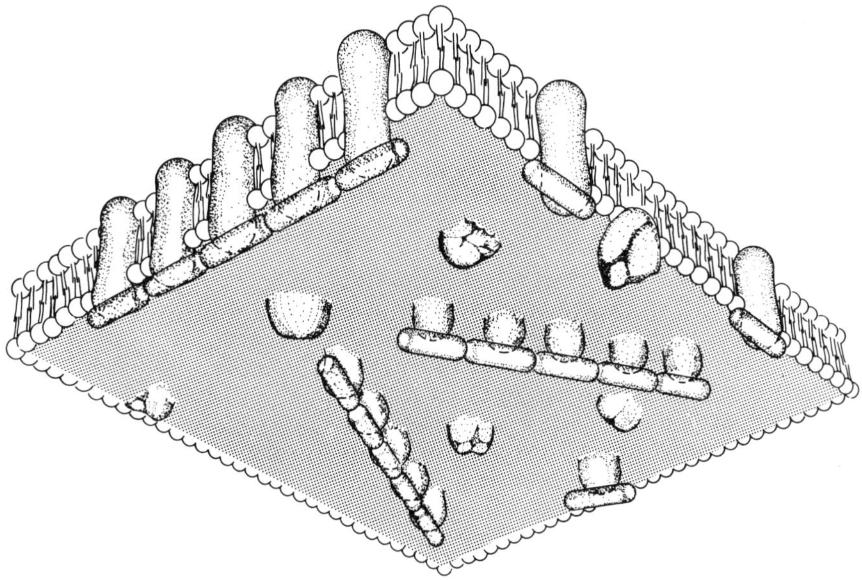

Fig. 3B.

Mosaic structures in membranes. The evidence is conclusive that the bulk of the phospholipids of functional membranes is arranged as a bilayer although some fraction of these lipids may interact more strongly with integral proteins in the membrane. At least a part of the bilayer lipid is in a fluid state under physiological conditions. The intercalation of globular integral proteins into bilayer lipid results in a mosaic structure (13). In principle, however, two different kinds of mosaic might be allowed (41) with either a protein or a lipid matrix to the mosaic. In the former case, sterochemically-defined protein-protein interactions in the plane of the membrane could produce an extended two-dimensional lattice, with pockets of fluid lipid bilayer intercalated into the interstices of the protein lattice. It was suggested that such a mosaic might consist of only one, or a small number of different proteins, if repeating protein-protein interactions were required to generate the structure. Also, such a mosaic would be a relatively rigid structure, with little or no rotational or translational mobility of the proteins and lipids in the plane of the membrane. The purple membrane of halobacteria, containing the single protein bacteriorhodopsin (16, 27) seems to satisfy these predictions for a protein matrix mosaic structure. Synaptic structures and gap junctions in membranes may be other examples, but in general, protein matrix mosaics are relatively rare, and even where they exist, they appear to be present as part of a membrane of the opposite kind, with a lipid matrix.

A mosaic membrane with a lipid matrix is what we have called the fluid mosaic model of membrane structure (Fig. 3A). It predicts that individual membrane proteins would be randomly distributed at long range in the plane of the membrane, and embedded in an appropriately fluid lipid phase, would be capable of rapid rotational and translational diffusion in the plane of the membrane. An analysis of the evidence led us to conclude (41) that most membranes were fluid mosaic structures under physiological conditions, and a large body of evidence obtained since that time has supported this view (12).

The fluidity of membranes and its control. What purposes are served by having a fluid mosaic structure for a membrane, with its components capable of lateral diffusion in the plane of the membrane? We can be certain that even if we do not know all facets of the answer to this question, it is an important question to ask because it appears that the fluidity of membranes is under homeostatic control. This is indicated, for example, by some recent experiments of Ingram (18) with the membranes of *E. coli*. He found that if growth is carried out at a constant temperature in the presence of a series of alcohols at different non-lethal concentrations, the bacteria exhibit a lag phase before undergoing exponential growth. During the lag phase, the average fatty acid composition of the lipids of the bacterial membrane undergoes changes such as to compensate for the increased fluidity of the membranes that results from the uptake of the alcohol into the lipid bilayer. Apparently only after this compensatory change occurs can the bacteria resume their normal growth rates in the continued presence of the alcohol.

Besides suggesting that the fluidity of the membrane is metabolically controlled, these results imply that those integral protein enzymes of the bacterial membrane that are involved in lipid biosynthetic reactions somehow sense changes in the fluidity of the membrane, i. e., their activities are affected by the membrane fluidity. If these enzymes are effected, then other unrelated but structurally similar enzymes may also be. There are a number of possible explanations for such enzymic rate sensitivity to membrane fluidity: for example, an enzymic reaction may involve as a rate limiting step the translational diffusion in the membrane of the integral enzyme and its membrane-bound substrate to form the enzyme-substrate complex; alternatively, a membrane enzyme may be a subunit aggregate (Fig. 1) undergoing a rate-limiting allosteric change (Fig. 2) in the course of its catalytic function, such that its rate is altered by a change in the lipid viscosity.

The lateral diffusion of integral proteins in a membrane may be critically important to a number of different membrane phenomena, as is discussed in some detail elsewhere (40), and in a later section. If this is true, it may also be important to control the mobility of such components. One means for such control can involve metabolic changes in the composition of the membrane lipids to make the matrix of the membrane more or less fluid, as was observed in the experiments of Ingram (18) just mentioned. Another mechanism of control, however, may involve peripheral proteins. In our original paper on the fluid mosaic model (41), we suggested that where such restrictions on the mobility of integral proteins appear to exist, "some agent extrinsic to the membrane (either inside or outside the cell) interacts multiply with specific integral proteins" and thereby inhibits the mobility of these proteins. Actomyosin-like contractile proteins peripherally attached to the cytoplasmic surface of plasma membranes may serve as these "agents", and by a metabolically-controlled aggregation-disaggregation equilibrium (Fig. 3A, B), may not only define the shape of the cell but also reversibly restrict the mobilities of integral proteins to which they are attached. Thus, the spectrin complex bound to the cytoplasmic surface of the erythrocyte membrane (34, 35), and an actin-smooth muscle myosin-like complex peripherally attached to the cytoplasmic surface of fibroblast plasma membranes (29), may function in a manner related to the scheme represented in Fig. 3.

In other cases, peripheral components bound at the outer surface of a plasma membrane may serve similarly to restrict the mobility of integral protein components in the membrane. It has been suggested (D. Mazia, personal communication), for example, that the vitellin layer attached to the exterior surfaces of invertebrate oocytes may be an example of such a system. And as will be discussed in a following section, externally added antibodies and lectins directed to cell surface receptors can exercise related functions.

The asymmetry of membranes. Another important aspect of membrane structure that has emerged in recent years is the asymmetry with which the proteins and phospholipids are distributed in the two halves of the membrane bilayer. This is most clearly demonstrated in the membrane of human erythrocytes, the most intensively studied membrane to date. The proteins of the membrane have specific orientations in the plane perpendicular in the membrane (42), and among the phospholipids, phosphatidylcholine and sphingomyelin are predominantly localized to the outer half layer, and phosphatidylethanolamine and phosphatidylserine in the inner half layer (44), of the membrane bilayer. The maintenance of this asymmetry is most probably attributable to the very slow rates with which the proteins and phospholipids rotate from one surface to the other (flip rates) in the intact membrane. These slow flip rates can be explained (41) by the same thermodynamic arguments we advanced in predicting that integral proteins are amphipathic molecules; very large free energies of activation are required to force the ionic and highly polar groups of the proteins and phospholipids through the non-polar interior of the membrane to reach the other side.

It is very likely that an asymmetry of both proteins and phospholipids exists in membranes other than erythrocytes. This is clear for the glycoproteins of many cell surfaces (25, 26) whose carbohydrate moieties appear to be exclusively oriented to the outer surface of the plasma membrane; but is still not established for the phospholipids. In what follows, however, we will assume that such asymmetry of distributions is a general features of membranes.

One important consequence of this asymmetry is embodied in the bilayer couple hypothesis (33). This hypothesis proposes that the two halves of an asymmetric membrane may respond differently to a particular perturbation, such as the introduction of a drug, or a change in temperature, or some metabolic process. If one half of a closed membrane expands or contracts relative to the other half, the membrane will be placed under stress, and among other possible consequences, may respond by a change in curvature. Conversely, a change in membrane shape or curvature is taken to reflect a change in the relative surface areas of the two halves of the membrane bilayer (34). We suggest later on that such stresses set up in a plasma membrane, behaving as a bilayer couple, may be important to the transmission of signals across the membrane.

Molecular Structure and Fluidity in Some Membrane Functions

This brief discussion of some of the structural features of membranes, as we presently conceive them, is by way of introduction to a consideration of their relevance to the membrane functions that are the subject of this volume. As a first application of these concepts, let us turn to a consideration of surface phenomena of lymphocytes, which have received a great deal of intensive investigation because of their crucial role in the immune response, but which in principle are no doubt broadly relevant in cell biology.

Clustering of lymphocyte surface receptors by antibodies and lectins. At about the time that the fluid mosaic model appeared, a remarkable series of experiments were reported (24, 43) on the effects of anti-Ig antibodies on the IgM-like molecules (antigen receptors) in the cell surfaces of B-type mouse lymphocytes. If fluorescent labeled antibodies to mouse Ig were added to these cells at 4°, the antibodies bound to the receptors, and a relatively uniform distribution of fluorescence was observed which eventually formed small patches all around the cell periphery ("patching"). If, however, these cells were brought to 37°, the fluorescence rapidly (in about 2 min) migrated into one region of the surface ("capping"); if incubated further at 37°, the fluorescence was removed from the cell surface by pinocytosis of the capped regions into the cell interior (10, 11). Neither capping nor endocytosis occurred with univalent Fab fragments of the antibodies, indicating that simple binding alone was not enough; a cross-linking of the IgM-like receptors was important to these processes. The capping and endocytosis, but not the patching, was inhibited by NaN_3, showing that they required energy.

Capping and endocytosis were subsequently found also to be inhibited by appropriate amounts of concanavalin A (Con A) bound to the cell surface (47). This inhibition by Con A could be reversed by colchicine. While the precise molecular mechanisms involved in these effects of Con A are not entirely certain, what does appear clear is that Con A binding to a relatively small fraction of its receptors on a B lymphocyte somehow severely restricts the mobility of the Ig receptors (and perhaps other integral proteins) in the membrane; and that colchicine reverses this effect. The latter observation suggests that microtubules are in some manner implicated, but no direct electron microscopic demonstration of microtubule involvement has been forthcoming. Whatever the details, some immobilization mechanism involving peripheral components on the cytoplasmic surface of the membrane, such as is schematically represented in Fig. 3, is likely to be operative when Con A binds to the external surface of the membrane and restricts the mobility of other receptors.

It was also found that if the anti-Ig antibodies were removed from cells after their Ig receptors had been cleared from their

Fig. 4. (NEXT PAGE) A scheme for the mechanism of endocytosis following upon the binding of a ligand L to its specific receptor R in a fluid membrane. R molecules may or may not span the thickness of the membrane. The binding of L to R molecules that are initially randomly and uniformly dispersed in the membrane (panel a) leads to a clustering of receptors at 37° (panel c). This clustering can be induced either by a cross-linking of multivalent L molecules with multivalent R molecules, or, as shown, by monovalent L and monovalent R (see text for details). The clustering of R molecules produces some stress (not explicitly shown in the figure) in the bilayer-couple membrane which may induce some critical permeability change locally in the membrane, which, it is suggested, activates an actomyosin-like peripheral protein system attached to the cytoplasmic face of the membrane. A contractile process then produces the endocytosis of the regions of the membrane so affected. This is proposed as a general mechanism which is not only involved in down-regulation of hormone receptors (see text) but also in the uptake of specific proteins by cells, and in phagocytosis (39, 40).

---

surface membranes, and the cells were cultured at 37° in the absence of the antibodies, within 6-8 hours the Ig receptors reappeared on the surfaces.

These were the first of many experiments that have since been carried out with multivalent antigens, with antibodies to a variety of surface receptors, and with multivalent mitogens (lectins), using a wide range of eukaryotic cells, that have demonstrated the patching, capping, and endocytosis to be very general phenomena (39). An important characteristic of these experiments was the independent capping and endocytosis of a particular membrane receptor by its specific ligand. Thus, IgM receptors on lymphocytes could be capped by anti-Ig antibodies without affecting the distribution of the H-2 histocompatibility antigen, and conversely, anti-H-2 antibodies could cap the H-2 antigen without altering the distribution of the IgM receptors.

The molecular mechanisms involved in these phenomena may be analyzed in terms of the structural concepts discussed in the first part of this chapter. A schematic representation of what may be going on is given in Fig. 4. A ligand L, upon binding to its specific receptor R (an integral protein in the plasma membrane of the cell) causes a clustering (patching, or patching followed by capping) of the originally uniformly dispersed R molecules if the lipid is sufficiently fluid to allow it. The ligand L may be multivalent and induce the clustering by virtue of its cross-linking of multivalent R molecules, but in principle, this is not necessary. As we suggested in our original paper (41), in some cases the binding of a

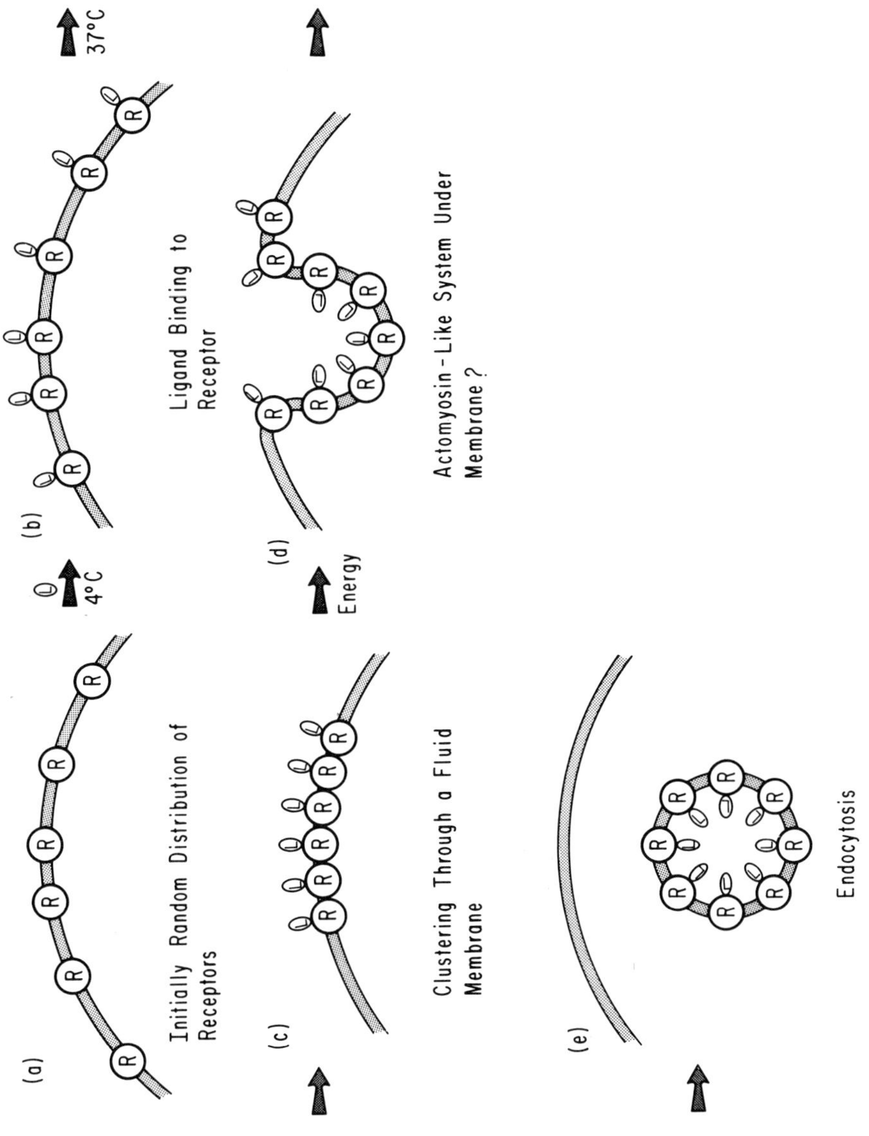

Fig. 4. Absorption of Proteins Across Membranes

univalent ligand L to its receptor R may so alter its conformation as to render it specifically "sticky" for other R molecules (either with or without L bound to them.) The ligand L would thereby function as a peripheral protein restricting the free diffusion of individual integral R molecules in the plane of the membrane.

We have proposed (39) that some such clustering process is obligatory for endocytosis to occur following the binding of some ligand L to a cell surface. Since clustering occurs through the diffusion of receptors in the plane of the membrane, it follows that such ligand-induced endocytosis requires that the membrane be fluid.

What happens immediately following upon clustering is a matter of speculation at present. We suggest that the bilayer couple behavior of membranes is implicated, i.e., for any of a variety of possible reasons, certain kinds of clustering of receptors affect the surface area of the outer half of the membrane bilayer differently from that of the inner half. The membrane is thereby placed under stress; perhaps the fluidity of the clustered regions of the membrane is thereby altered (2), and the rates of some critical ion translocations (Ca^{++}, for example, see Rasmussen *et al.*, 31) are significantly changed, as discussed in an earlier section. In turn, as a result of such local changes in permeability of the membrane, local activation of an actomyosin-like peripheral protein aggregation is produced on the cytoplasmic surface of the membrane. The effect of this aggregation is proposed to be an increase in the surface area of the inner half of the membrane relative to the outer half, leading to the local invagination of the membrane and its subsequent endocytosis (Fig. 4) (35, 36).

This proposal has by no means been experimentally verified in its entirety at this time, but it has the merits of coherence and of experimental predictability. We have already shown (36), for example, that certain compounds which cause intact human erythrocytes to assume a crenated shape, inhibit the capping and endocytosis of lymphocyte IgM-like receptors that are induced by anti-Ig antibodies. The anionic amphipathic compounds 2,4 dinitrophenol (DNP) and 2,4,6 trinitrophenol (TNP) are erythrocyte crenators in the range of 1-10 mM concentration. According to the bilayer couple hypothesis, we have attributed this shape change (33) to the preferential binding of these compounds to the lipid in the outer half layer of the bilayer membrane, leading to an expansion of the outer half relative to the inner, and to the evaginations or crenations observed. The suggestion is that DNP and TNP similarly intercalate preferentially into the outer half layer of the *lymphocyte* plasma membrane, expanding the surface area of the outer half relative to the inner half. In this manner, these compounds counteract the *opposite* area changes required for capping and endocytosis by anti-Ig antibodies to occur, and therefore inhibit these processes.

We have discussed these lymphocyte surface phenomena at some length because it seems clear that they have general import for a variety of cellular phenomena, as will next be considered.

Ligand-receptor interactions. There are three topics we will discuss under this heading: the nature of receptors, the phenomenon of cooperativity, and the down-regulation of hormone receptors.

i) The nature of receptors. Molecules that are the receptors for polypeptide and catecholamine hormones, lectins, and other ligands are in the main likely to be integral proteins in the membrane, with a hydrophilic segment of the protein protruding from the surface of the membrane into the aqueous phase. Very little is known about these receptors at present. Whether they exist in the membrane as monomeric or oligomeric structures, and whether they are part-way embedded or span the entire thickness of the membrane, are important matters about which we have little or no direct evidence. As mentioned earlier, glycoproteins appear to have their covalently-bound oligosaccharide chains oriented exclusively to the outer surface of the plasma membrane of eukaryotic cells (25, 26); it is possible that all integral proteins that have a hydrophilic segment at the outer surface are glycoproteins, but it is too early to make such a generalization. If this were true, however, all receptor proteins would be glycoproteins. In this context, it has been claimed that the insulin receptor is a glycoprotein (7).

An interesting exception to the generalization that receptors are proteins is the case of the receptor for cholera toxin, which appears to be a ganglioside glycolipid, GM1 (see van Heyningen, elsewhere in this volume). That the binding of a hormone to a protein receptor, or of cholera toxin to a lipid receptor, can produce similar effects (for example, in stimulating adenylate cyclase) suggests that the membrane events following either type of binding are similar. This may mean either that: a) the ganglioside is non-covalently bound to an integral protein in the membrane, which subsequent to the toxin-binding behaves in the same way as a receptor protein to which a hormone is directly bound; or b) that ligand binding at either protein or lipid may have similar structural sequellae, such as clustering, bilayer couple stresses, etc., as discussed above. In this connection, it is of considerable interest that the binding of cholera toxin to lymphocyte membranes does induce a capping of its receptors (5, 32).

In the remainder of this section, we want to follow up on our original suggestion (41) that membrane fluidity and the ligand-induced clustering of receptors in the plane of the membrane might play a critical role in the mechanisms of hormone or other cell activation processes. This suggestion has been taken up and amplified by others (7, 30) since.

ii) Negative cooperativity. In several papers in this volume, the phenomenon of negative cooperativity in the binding of hormones to their specific receptors is discussed. This phenomenon appears to be widespread but not universal in such systems (8). Before the fluid mosaic model became widely accepted, one hypothesis to explain cooperative phenomena in membranes (4) took as the model of membrane structure the repeating lipoprotein subunit model (3). These authors proposed that such lipoprotein units might exist in two conformational states, and that transitions from one to the other state were cooperative among these units. This proposal is untenable as such since it is now clear that members are not generally organized as a repeating lipoprotein-subunit structure. Another proposal is that of Levitzki (23) who suggests that the receptor molecules are organized as stable specific oligomeric clusters in the membrane. The specific binding of hormone to one receptor in a cluster, it is proposed, can induce changes in the conformations of the other receptors in the cluster, and thereby elicit negative cooperativity. In this model, the matrix of the membrane and in particular its fluidity, plays no direct role in the phenomenon.

We suggested a quite different possibility (41) in the light of the fluid mosaic model: "It is possible...that a particular integral protein can exist in either of two conformational states, one of which is favored by ligand binding; in its normal unbound conformation, the integral protein is monomolecularly dispersed within the membrane, but in the conformation promoted by ligand binding, its aggregation is thermodynamically favored. The binding of a ligand molecule at one integral protein site, followed by diffusion of the non-liganded protein molecules to it, might then lead to an aggregation and simultaneous change in conformation of the aggregated protein within the membrane...." The remaining unbound receptor sites in such a cluster would then bind the ligand with a lower affinity than that of the first site, to produce the negative cooperativity. The main point is that a dynamic clustering is suggested by this mechanism, in contrast to the static clustering of Levitzki (23).

While we suggest that many, if not all, cases of negative cooperativity in hormone-receptor interactions result from dynamic clustering processes, it does not necessarily follow that all such receptor clustering must result in negative cooperativity. Whether negative cooperativity follows upon clustering would depend on many variables, such as the orientation of the binding site on the receptor molecule, the conformational changes induced on binding of the hormone, etc.

One interesting corollary of this dynamic model of negative cooperativity is that if the mechanism is to be physiologically relevant, the rate of dissociation of the hormone from the first receptor must not be substantially greater than the rate of the diffusional collision of the individual receptor molecules in the plane of the membrane; otherwise, clustering would only rarely occur.

Indeed, the rates of dissociation of insulin-receptor and epinephrine-receptor bonds in their high affinity states have half-times of the order of an hour or so at 37° (see De Meyts; Limbird and Lefkowitz, elsewhere in this volume), whereas average diffusional collision times in a membrane at 37° must be of the order of several seconds or less (depending on the concentration of receptors in the membrane) (12).

The experimental findings to date concerning negative cooperativity in hormone-receptor interactions can at least equally well be explained by the dynamic as by static models discussed above. For example, there is a marked effect of temperature on the negative cooperativity of insulin binding to its receptors, with low temperatures favoring the high affinity state and high temperatures the low affinity state (9). This would follow if the low affinity state arose only through the diffusional clustering of hormone-bound and unbound receptors, a process which would be inhibited if the viscosity of the membrane lipids became too large at low temperatures. Similarly, Con A has been found to eliminate the negative cooperativity of of insulin-receptor interactions without affecting the high affinity binding of the insulin to its receptor (9). Since the binding of Con A to lymphocytes appears to restrict the mobility of Ig receptors in the membrane (see previous section), if the insulin receptors were similarly restricted, the system would be locked into the high affinity state. Colchicine should reverse this effect, however. Therefore, while static mechanisms of negative cooperativity are by no means ruled out, the concept of a dynamic clustering leading to negative cooperativity is entirely consistent with the results just mentioned.

iii) Down-regulation of receptors. There are other terms that could be used to describe this phenomenon, including subsensitivity, or desensitization. The characteristics of the specific effect we refer to are discussed in the chapters by Limbird and Lefkowitz, elsewhere in this volume. In the presence of suitable concentrations of a hormone, the number of receptor binding sites for that hormone on an intact cell decreases over a period of time, without change in the original affinity of the remaining receptor sites for the hormone. In some cases, different hormones can induce similar losses of receptors on the same cell, each specific for its own receptor. After removal of insulin from liver cells that have lost a large fraction of their receptor sites, and upon return of the cells to culture at 37°, the original insulin binding activity is gradually recovered on the cell surfaces.

All of these results are remarkably parallel to those which occur on the binding of anti-Ig antibodies to the IgM-like receptors on B lymphocytes, as has been discussed in detail in the previous section. In particular, we would attribute the down-regulation to the clustering and endocytosis mechanism schematically represented in

Fig. 4. The binding of the hormone to its receptor produces a specific clustering of its receptors which then leads to the endocytosis of the clustered regions in the membrane. (That insulin receptors are indeed mobile in fat cell and liver cell membranes, and can be endocytosed upon insulin binding, is suggested in experiments using ferritin-conjugated insulin (20)). Experimental tests of this proposal are evident. For example, down-regulation should be blocked by suitable concentrations of NaN_3 and 2,4,6 trinitrophenol: agents, that, as discussed earlier, generally inhibit capping and endocytosis. Furthermore, direct demonstration of endocytotic vesicles, lined on their inner surfaces with insulin, in down-regulated cells should be possible using ferritin-labelled antibodies to insulin (28).

An attractive feature of the proposals presented in this section is that the apparently unrelated phenomena of negative cooperativity and down-regulation in hormone-receptor interactions are thereby connected to one another, both being successive consequences of the clustering of receptors in a fluid membrane that is induced upon specific binding of the hormone. On the other hand, it does not follow from such a mechanism that every hormone-receptor system that exhibits down-regulation must also show negative cooperativity (since clustering of receptors need not always lead to negative cooperativity), nor conversely, that every system showing negative cooperativity must also exhibit down-regulation (since clustering of receptors, particularly if limited to the formation of small patches, need not always lead to activation of the contractile mechanisms involved in endocytosis).

It should be made clear at this point that the clustering of its specific receptors that we propose to be induced by insulin, epinephrine, and related hormones is of a type not yet definitively demonstrated experimentally. Ligand-induced clustering of receptors that is observed when antibodies bind to surface antigens on cells, or when lectins bind to glycoproteins, are most likely the result of a multivalent ligand-multivalent receptor cross-linking reaction, producing a two-dimensional aggregate in the plane of the membrane. On the other hand, if insulin induces clustering of its receptors, it is presumably doing this as a univalent ligand. De Meyts et al. (9) have argued that at concentrations below 10^{-7} M insulin binds to its receptor as a monomer. It is still conceivable, however, that once bound to its receptor, insulin may aggregate with itself in some specific way, thereby indirectly generating a multivalent ligand. (The clustering and endocytosis observed upon binding ferritin-conjugated insulin to membranes (20) would be important evidence for the ability of univalent insulin to induce clustering if it could be proved that only one insulin molecule was attached to each ferritin, and that the very large ferritin molecule was not otherwise promoting the clustering).

In this regard, the system involving the interaction of catecholamines with the β-adrenergic receptor in frog erythrocyte membranes is particularly interesting. Agonists such as *l*-isoproterenol and *l*-epinephrine not only show negative cooperativity of binding and down-regulation of receptors, but also stimulate adenylate cyclase in the membrane (see Limbird and Lefkowitz, elsewhere in this volume). If these agonists do indeed induce a clustering of their receptors, they must be doing this as univalent ligands, by the mechanism we have previously suggested (41). Again, an experimental prediction of this mechanism is that the down-regulation (by endocytosis) of β-adrenergic receptors by agonists should be inhibited by erythrocyte crenators such as 2,4,6 trinitrophenol. Furthermore, when agonists down-regulate the β-adrenergic receptors, and thereby produce a corresponding decrease in epinephrine-stimulated adenylate cyclase activity, there is no decrease in the basal activity nor in the fluoride-stimulated activity of the cyclase. If down-regulation is due to a clearing of β-adrenergic receptors from the membrane by clustering and endocytosis, these results suggest that adenylate cyclase molecules are not simultaneously cleared from the membrane with the receptors. In turn, this suggests that the β-adrenergic receptor molecules and adenylate cyclase molecules do not form a stoichiometric complex in the membrane, for if they did, they might be expected to be simultaneously endocytosed. On the other hand, if the receptor and adenylate cyclase molecules are physically independent of one another in the membrane, even after the hormone binds to its receptor, then the activation of adenylate cyclase by the hormone must occur by a mechanism which is quite different from any of those which are under consideration at present (7). One possibility is that when a sufficiently extensive clustering of the receptors is induced by the binding of the hormones, a stress is produced in the entire membrane by the bilayer couple mechanism as discussed in an earlier section; this stress in some manner causes the activity of the adenylate cyclase embedded in the membrane to change. In this connection it is interesting that while antagonists, such as *l*-alprenolol, exhibit negative cooperativity of binding to the β-adrenergic receptors, they neither stimulate adenylate cyclase nor induce down-regulation. This could be rationalized by the mechanisms just discussed if antagonists produce a clustering of receptors which is extensive enough to induce negative cooperativity of binding, but not extensive enough to apply sufficient stress in the bilayer couple membrane either to stimulate the adenylate cyclase, or to activate the mechanisms for endocytosis.

It is clear that at the present time these suggestions are highly speculative, and many other explanations for these phenomena of ligand-receptor interactions can equally well be entertained. Nevertheless, the novelty of these proposals and the distinct experimental tests they suggest I hope justify their elaboration in this discussion.

Cell-Cell interactions. In several of the papers in this volume, problems of specific interactions among cells are discussed. The

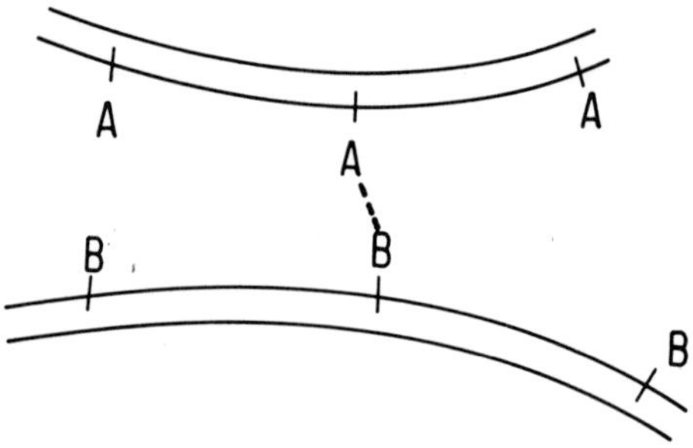

Fig. 5. A schematic view of the interaction of specific membrane-bound components, A and B, leading to cell-cell adhesion. The formation of the first A, B bond (center) makes it more probable that the A, B bond on the left will form, because of the entropy factor discussed in the text. The A, B pair on the right, however, is depicted as too distant from each other to form a bond unless molecules can diffuse laterally in their respective membranes and/or the membranes can change shape.

molecular mechanisms for these and other specific cell-cell interactions appear to fall into two categories: those where integral components of the two apposed plasma membranes bind to one another (direct interaction), and those where a multi-valent peripheral protein, such as a lectin, binds simultaneously to integral components on the two apposed membranes (indirect interaction). For the purposes of the following discussion, we will concentrate on direct interactions, but the matters raised are equally relevant, and easily generalized, to the indirect case.

There are no doubt many factors that contribute to the overall binding affinity, or adhesiveness, of one cell for another, including the surface charge properties of the cells, the degree of exposure of the surface molecules that form the specific bonds between the cells, and the surface density of these interacting molecules. There are three factors, however, that are especially emphasized here, because they are not often taken into account in this connection. Some of these have already been pointed out elsewhere (39, 41).

i) Enhancement of affinity due to multivalent binding. Clearly, the more single bonds formed between two cell surfaces, the stronger will be the overall binding of the two cells to one another. But there is another way in which multiple binding can markedly enhance cell-cell adhesiveness, a factor which is well understood in the analogous problems in immunochemistry (6, 17) and is illustrated in Fig. 5. Suppose that we have a number of identical integral proteins A protruding from one membrane, interacting with components B protruding from the other (A and B might equally well be present on both membranes). Let us assume that the first bond formed between A and B is characterized by some particular affinity, given by the equilibrium association constant K_1. A and B molecules in the membranes near the site of the first A-B bond can now form bonds with a much greater affinity constant than K_1, provided that there is no interference by steric factors. This increased affinity is due to the fact that the second A and B molecules to form A-B bonds, and all subsequent ones as well, are already in proximity to one another as a result of the formation of the first A-B bond, i.e., the free energy change for the second reaction is much more negative than for the first because it involves a much smaller entropy decrease than the first. Therefore, the capacity to make multiple bonds at sites where two cell surfaces adhere can increase the overall adhesiveness not only in proportion to the number of bonds formed, but also by virtue of this pronounced enhancement in affinity of all bonds after the first.

ii) Mobility of binding sites. The critical requirement for such multiple bonds to form, however, is that there be no steric interference to their formation after the first bond has formed. It is clear, however, that if the density of A and B sites in the membrane is uniform and low, and if these sites are immobilized in the plane of the membrane, then after the formation of the first A-B bond, the A and B molecules that are sufficiently close to the site of the first bond may not be properly juxtaposed to form any additional bonds beyond the first. On the other hand, if the unbonded A and B sites were sufficiently mobile in the plane of the membrane, then before the first A-B bond could dissociate, the second and further A-B bonds might form through the rapid diffusion of A and B in the two membranes into juxtaposition near the site of the first bond. (Energetically, the loss in entropy due to the clustering of A and B sites within the two membranes would be more than made up for by the enthalpies of formation of the A-B bonds.) Therefore, in cases where there is a random uniform dispersal of small numbers of such intercellular binding sites in the surface membranes of two cells, we suggest that the mobility of these sites in the membranes may be critical to the formation of strong intercellular adhesions involving these sites (Ref. 41, Fig. 7; Ref. 39, Fig. 7).

To illustrate that these effects are indeed operative we reproduce the electron micrograph in Fig. 6, which shows the regions of adhesion between a pigeon erythrocyte and peripheral lymphocyte that were agglutinated with ferritin-conjugated Con A. In two distinct areas of adhesion between the two cells, ferritin particles can be seen to be densely packed between the apposed membranes, whereas elsewhere on the surfaces of the two cells, there is only an occasional ferritin particle seen. Now, the Con A receptors in the membranes of pigeon erythrocytes and lymphocytes are normally uniformly dispersed and mobile under physiological conditions.[1] Clearly, Fig. 6 indicates that these Con A receptors exposed on the

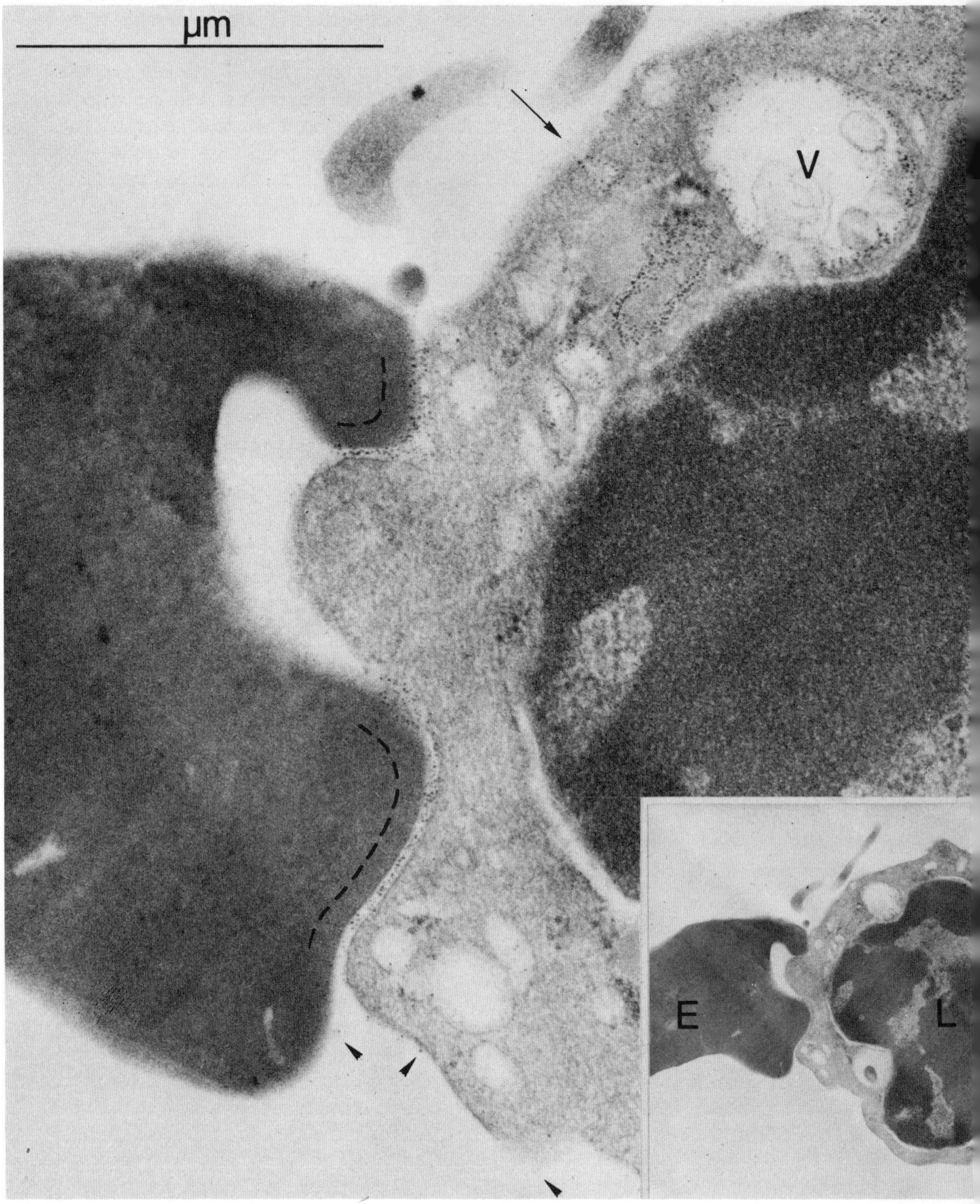

Figure 6. See Insert.

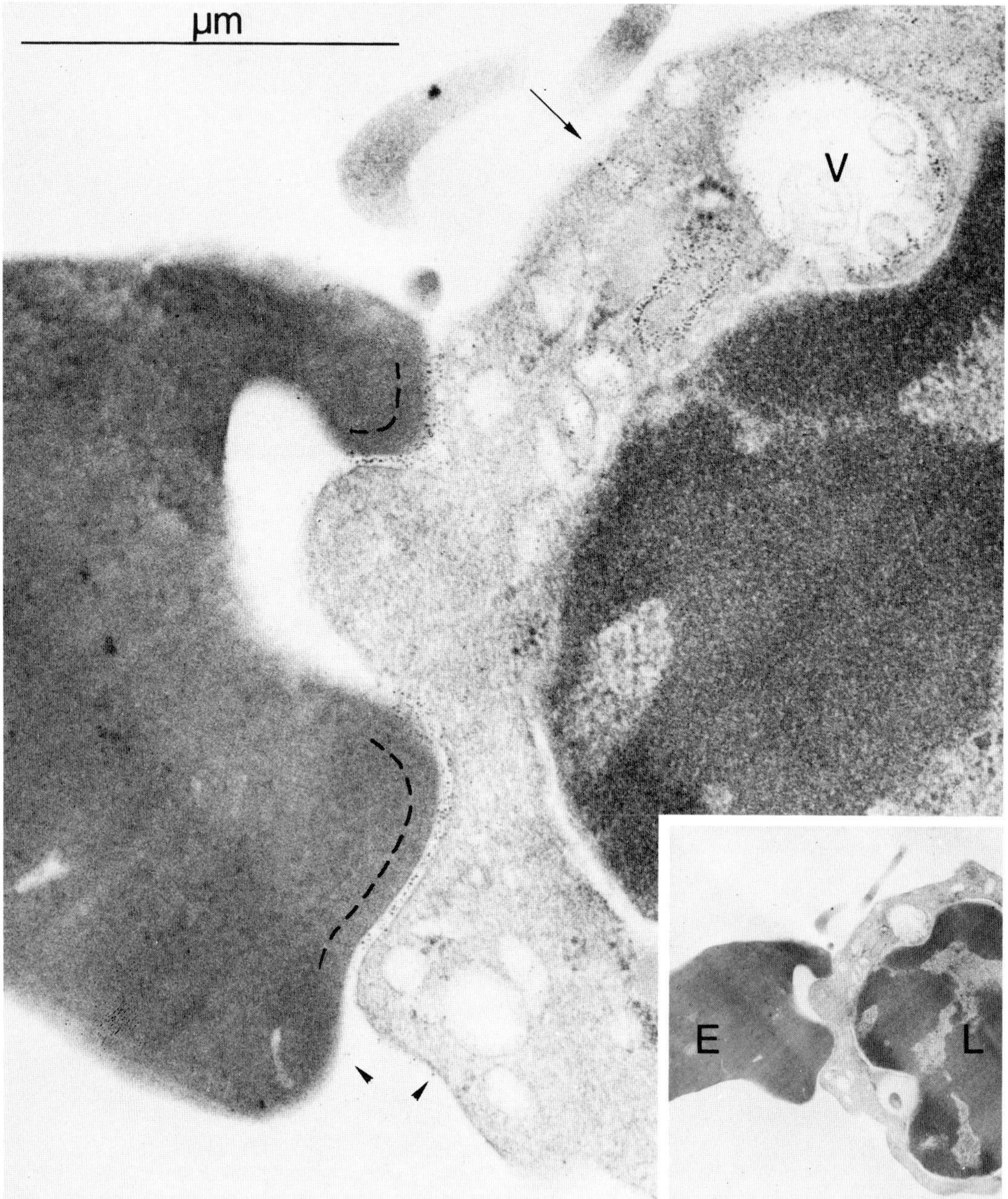

Figure 6

Fig. 6. An electron micrograph of a thin section through a pigeon erythrocyte (E, inset) and a peripheral lymphocyte (L, inset), agglutinated by a ferritin conjugate of concanavalin A, (Fer-Con A). This micrograph demonstrates a number of phenomena discussed in the text. Note first that at the two regions of adherence of the cell (marked by the dashed lines) the intercellular gap is densely lined with Fer-Con A, whereas elsewhere on the surfaces of the two cells, Fer-Con A is only sparsely bound (tail-less arrows). This strongly suggests that the Con A receptors have been mobilized by a process of lateral diffusion into the regions of adherence of the two membranes upon binding and cross-linking by the Fer-Con A. Secondly, note how the normal disk shape of the erythrocyte is distorted to accommodate to these adhesions with the lymphocyte surface. Thirdly, an invagination of the plasma membrane (tailed arrow) and a vesicle (V) in the lymphocyte are both densely lined with Fer-Con A, exactly as expected from the stages d) and e) of the clustering-endocytosis mechanism depicted in Fig. 4. The bar denotes 1 μm.

two cell surfaces have been cross-linked by the ferritin-conjugated Con A after they had diffused into the regions of the membranes of the two cells where contact had been made, and that the regions of adhesion coincide with these regions of clustered receptors.

While it is the early events in cell adhesion that we have been discussing, it should be recognized that such clustering of cross-linked binding sites into regions of cell-cell contact may have direct consequences that are profoundly important to the subsequent interactions of the two cells. For example, the clustering of receptors may produce stresses and permeability changes in the membranes where the cells are bound to one another, as has been discussed in an earlier section. Such permeability changes may in some cases activate some actomyosin-like filament system on the cytoplasmic surfaces of the apposed membranes, as was invoked to explain endocytosis of clustered regions of membranes in isolated cells (Fig. 4). Such systems might indeed form, but endocytosis might be inhibited by cell-cell contact at these clustered sites. Molecular events similar to those just discussed may operate, for example, in the phenomenon of contact inhibition of motility (1). At the edges of isolated motile normal cells, the membrane extends in a very flexible "ruffle". When two motile cells make contact at these ruffled edges, the remarkable electron micrographs of (14) show that there almost instantaneously develop filamentous arrays on the cytoplasmic surfaces of the ruffle membranes where they are in contact. We suggest that when the fluid ruffle membranes make contact, receptors in the two membranes migrate rapidly into clusters forming intercellular bonds; that this clustering produces permeability changes in the apposed membranes that activate an actomyosin-like filament system which is peripherally attached to the membrane, perhaps as suggested in Fig. 3. (The inability of

[1]R. Scheckman and S. J. Singer, unpublished observations.

malignantly transformed cells to show contact inhibition of motility, which correlates with their inability to form such filamentous arrays when their ruffles make contact (15), may then be due to some property of the cells that either prevents these membrane permeability changes from occurring, or from being effective in causing the filaments to be generated.)

Another possible consequence of such clustering-adhesion phenomena might be to introduce polarity into cells that were structurally undifferentiated prior to adhesion. In an isolated cell with a fluid mosaic membrane, the cell surface might initially be everywhere uniform when averaged over a time interval of minutes. But the formation of stable intercellular adhesions through clustered receptors might be the crucial initial step in polarizing a cell, by thus producing a structural and chemical differentiation of one or more regions of its originally uniform cell membrane.

iii) Morphological changes in a fluid membrane. If the cell surfaces of two interacting cells can readily change shape so as to allow a greater total contact area between the cells, the number of intercellular bonds and hence the total adhesiveness of the cell could be optimized. The convoluted shapes often exhibited by interacting cells is no doubt attributable to this factor. The distorted shape of the pigeon erythrocyte shown in Fig. 6 is most likely also an expression of morphological changes to promote adhesion to the lymphocyte.

While such shape changes are fairly well-recognized to be involved in many instances of cell-cell interactions, it may not be widely realized that they may require that the membrane be fluid. According to the bilayer couple hypothesis (33; see earlier section), a change in curvature of a membrane requires a corresponding change in the surface areas of the outer and inner halves of the membrane bilayer. For example, if a region of the membrane of cell I envelopes a protruding region of the membrane of cell II, the cell I membrane in the region of this invagination has a larger area on its inner than its outer half layer. Such local relative area changes might conceivably occur by any of a variety of mechanisms. For example, lipid or protein components might be rapidly removed from, or inserted into, the half layer of the membrane in contact with the cytoplasm of the cell; or lipid or protein components already in the membrane might diffuse out of this local region of the membrane into an adjacent region in one half layer or the other; or neutral species such as free fatty acids or cholesterol might "flip" from one surface to the other. Each of these possible events requires that the membrane be a fluid if it is to occur rapidly enough to produce the area and shape changes required.

The last two sections, on the possible roles of receptor clustering and cell membrane shape changes in cell-cell interactions, have emphasized a requirement for the fluid character of the membranes participating in these interactions. These considerations may be involved for example, in the experiments of Wekerle et al., (46) who studied the adherance of rat T-lymphocytes to monolayers of mouse fibroblasts, which involves the recognition by the rat cell of specific histocompatibility antigens on the surface of the mouse cell. They found that while adherence occurred at 37°, it did not at 4°, and furthermore it did not at 37° in the presence of NaN_3, conditions which inhibit the clustering of receptors in the lymphocyte experiments discussed in an earlier section.

In conclusion, we have attempted in this paper to discuss some general aspects of membrane structure and function as we currently recognize them, and then have applied these considerations to an analysis, deliberately speculative in some respects, of the phenomena of ligand-receptor and cell-cell interactions that are addressed in this volume. The only interest in such an analysis is if it provokes new ideas and new experiments in these complex areas at the borderline between molecular and cell biology.

ACKNOWLEDGEMENTS

Original studies discussed in this paper that were carried out in our laboratory were supported by USPHS grants GM-15971 and AI-06659. We are grateful for discussion with Drs. Jesse Roth, Pierre DeMeyts, and Lee Limbird, and to Dr. De Meyts for sending us preprints of papers in press.

REFERENCES

1. ABERCROMBIE, M., HEAYSMAN, J. E. M. and PEGRUM, S. M., Exp. Cell Res. 67 (1971) 359.
2. BARNETT, R. E., SCOTT, R. E., FURCHT, L. T., and KERSEY, J. H. Nature 249 (1974) 465.
3. BENSON, A. A., J. Amer. Oil Chem. Soc. 43 (1966) 265.
4. CHANGEUX, J. P., THIÉRY, J., TUNG, Y. and KITTEL, C., Proc. Nat. Acad. Sci. USA 57 (1967) 335.
5. CRAIG, S. W. and CUATRECASAS, P., Proc. Nat. Acad. Sci. USA 72 (1975) 3844.
6. CROTHERS, D. M. and METZGER, H., Immunochem. 9 (1972) 341.
7. CUATRECASAS, P., Ann. Rev. Biochem. 43 (1974) 169.
8. DE MEYTS, P., J. Supramol. Struct. (1975) in press.
9. DE MEYTS, P., BIANCO, A. R. and ROTH, J., J. Biol. Chem. (1976) in press.
10. DE PETRIS, S. and RAFF, M., Eur. J. Immunol. 2 (1972) 523.
11. DE PETRIS, S. and RAFF, M., Nature New Biol. 241 (1973) 257.
12. EDIDIN, M., Ann. Rev. Biophys. Bioeng. 3 (1974) 179.
13. GLASER, M., SIMPKINS, H., SINGER, S. J., SHEETZ, M., and CHAN, S. I., Proc. Nat. Acad. Sci. USA 65 (1970) 721.
14. HEAYSMAN, J. E. M. and PEGRUM, S. M., Exp. Cell Res. 78 (1973) 71.

15. HEAYSMAN, J. E. M. and PEGRUM, S. M., Exp. Cell Res. 78 (1973) 479.
16. HENDERSON, R. and UNWIN, P. N. T., Nature 257 (1975) 28.
17. HORNICK, C. L. and KARUSH, F., Israel J. Med. Sci. 5 (1969) 163.
18. INGRAM, L. O., J. Bact. (1976) in press.
19. JARDETZKY, O., Nature 211 (1966) 969.
20. JARRETT, L. and SMITH, R. M., J. Biol. Chem. 249 (1974) 7024.
21. KAYTE, J., J. Biol. Chem. 250 (1975) 7443.
22. LENARD, J. and SINGER, S. J., Proc. Nat. Acad. Sci. USA 56 (1966) 1828.
23. LEVITZKI, A., J. Theor. Biol. 44 (1974) 367.
24. LOOR, F., FORNI, L. and PERNIS, B., Eur. J. Immunol. 2 (1972) 203.
25. NICOLSON, G. L. and SINGER, S. J., Proc. Nat. Acad. Sci. USA 68 (1971) 942.
26. NICOLSON, G. L. and SINGER, S. J., J. Cell. Biol. 60 (1974) 236.
27. OESTERHELT, D. and STOECKENIUS, W., Nature New Biol. 233 (1971) 149.
28. PAINTER, R. G., TOKUYASU, K. T. and SINGER, S. J., Proc. Nat. Acad. Sci. USA 70 (1973) 1649.
29. PAINTER, R. G., SHEETZ, M. and SINGER, S. J., Proc. Nat. Acad. Sci. USA 72 (1975) 1359.
30. PERKINS, J. P., Advances Cyclic Nucleotide Res. 3 (1973) 1.
31. RASMUSSEN, H., GOODMAN, D. B. and TENENHOUSE, A., CRC Crit. Rev. Biochem. 1 (1972) 95.
32. REVESZ, T. and GREAVES, M., Nature 257 (1975) 103.
33. SHEETZ, M. and SINGER, S. J., Proc. Nat. Acad. Sci. USA 71 (1974) 4457.
34. SHEETZ, M. and SINGER, S. J., Proc. Nat. Acad. Sci. USA 73 (1976) in press.
35. SHEETZ, M., PAINTER, R. G., and SINGER, S. J., Cold Spring Harbor Conf. on Cell Motility (1976) in press.
36. SHEETZ, M., PAINTER, R. G. and SINGER, S. J., J. Cell Biol. (1976) in press.
37. SINGER, S. J., Structure and Function of Biological Membranes, (L. I. Rothfield ed.), Academic Press, N. Y. (1971) 145.
38. SINGER, S. J., Ann. Rev. Biochem. 43 (1974) 805.
39. SINGER, S. J., Advances Immunol. 19 (1974) 1.
40. SINGER, S. J., Control Mechanisms in Development, (R. H. Meints and E. Davies eds.) Plenum Press, N. Y. (1975) 181.
41. SINGER, S. J. and NICOLSON, G. L., Science 175 (1972) 720.
42. STECK, T. L., J. Cell Biol. 62 (1974) 1.
43. TAYLOR, R. B., DUFFUS, W. P. H., RAFF, M. C. and DE PETRIS, S. Nature New Biol. 233 (1971) 225.
44. VERKLEIJ, A. J., ZWAAL, R. F. A., ROELOFSEN, B., COMFURIUS, P., KASTELIJN, D., and VAN DEENAN, L. L. M., Biochem. Biophys. Acta 323 (1973) 178.
45. WALLACH, D. F. H. and ZAHLER, P. H., Proc. Nat. Acad. Sci. USA 56 (1966) 1552.
46. WEKERLE, H., LONAI, P. and FELDMAN, M., Proc. Nat. Acad. Sci. USA 69 (1972) 1620.
47. YAHARA, I. and EDELMAN, G. M., Proc. Nat. Acad. Sci. USA 69 (1972) 608.

STRUCTURE AND FUNCTION OF MEMBRANES AND THE ROLE OF THE PLASMA MEMBRANE IN METABOLIC REGULATIONS

Efraim RACKER

Section of Biochemistry, Molecular & Cell Biology
Cornell University, Ithaca, New York 14853 (USA)

Membranes are essential for life as we know it. It is very likely that the origin of life is associated with the formation of vesicular membranes. It is difficult to imagine that the most improbable of events, the creation of life, could have taken place without such a device of organization and concentration. Membranes are structures that envelop cells and intracellular organelles, thereby creating compartments. It was not Caesar, but a membrane who first said: divide and conquer. By protecting the delicate components of living protoplasm against a hostile environment, membranes have made the survival of a genetic apparatus possible and have made life liveable.

Membranes are remarkable structures. On the one hand they are rugged, withstanding severe physical and chemical attacks launched either from natural sources or by clumsy biochemists who aim to resolve them. On the other hand, membranes are endowed with remarkable flexibility and contain delicate metabolic machineries.

<u>Structure of membranes</u> - It is now generally agreed that membranes are not all alike. But the question is, "are all membranes created equal?" Do they have a common architecture, some common structural proteins and are the interactions between their constituents governed by some common laws?

Ten years ago there was so much controversy on the basic organization of membranes that (almost) everybody was confused. That was bad. Now the pendulum has swung the other way and we have reached a state of consensus which we should challenge to avoid the obstruction of creative experimentation by a scaffold of dogmas. Yet it seems certain that the basic structure of membranes is that

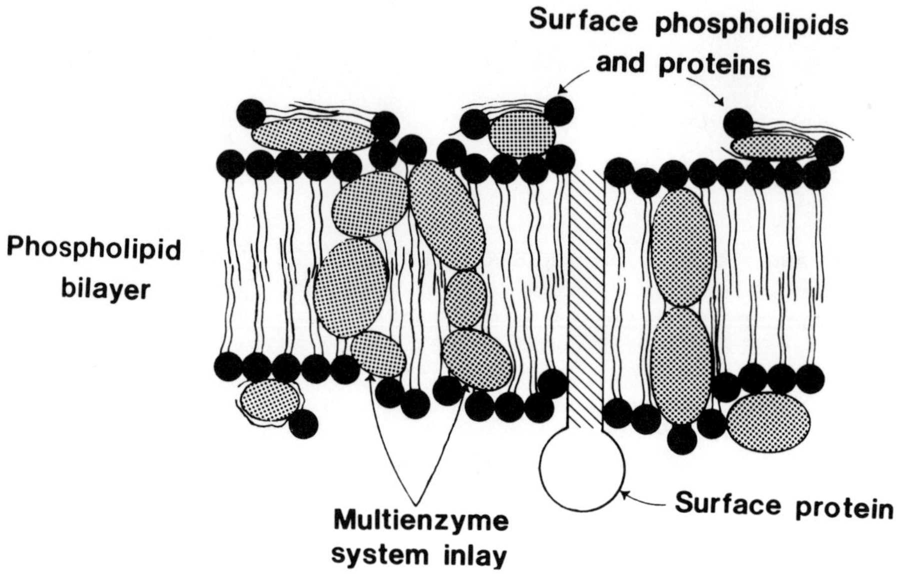

Fig. 1. The structure of biological membranes.

of a phospholipid bilayer with numerous inlays of proteins and overlays of proteins and phospholipids as shown in Figure 1. Since the inlays may range from single proteins to chains of respiratory enzymes and even purple patches (11) which interrupt the continuity of the phospholipid bilayer, we see the emergence of an organization of domains with individualistic properties. Realization of this pattern will protect us from generalizations such as the frequently quoted fluidity of membranes.

Function of membranes

a) Compartmentation and communication - The primary function of membranes is compartmentation. The plasma membrane creates the compartment that protects the protoplasm and allows the cell to retain the ingredients required for life. The membranes of intracellular organelles provide the isolation and protection of machineries that fulfill specialized functions such as those of mitochondria and of nuclei.

But with the privilege of isolation comes the responsibility of communication. Thus the membranes must be endowed with devices that permit the import from the common market of the evironment of food and ions which have appropriate passports of entry. The membranes must also contain devices that allow the export of undesirable metabolites and waste products. We shall discuss later some of their communication channels in greater detail.

b) Secondary functions - Superimposed on the two basic roles of membranes there are numerous secondary functions that were acquired during evolution. They were responsible for the amazing diversity of membranes that have made any generalizations almost futile. For example, the complex machinery required for oxidative phosphorylation dominates the properties of the inner mitochondrial membrane, while the plasma membrane is dominated by the export-import business.

Nature has taken advantage of the relatively sedate properties of membranes and has used them for the anchorage of enzymes for the sake of regulations. Thus we can explain the somewhat unexpected and apparently fortuitous attachment of enzymes to membranes such as the location of hexokinase or monoamine oxidase on the plasma membrane or of glyceraldehyde-3-phosphate dehydrogenase on the outer mitochondrial membrane.

Experimental approaches - There are three approaches to the study of membranes. Experiments with (a) intact cells or (more or less) intact organelles, (b) resolved and reconstituted systems and (c) model membranes. In the order listed they have decreasing significance in terms of cellular physiology and increasing significance in terms of depths of analysis and interpretation. I advocate that we should use all three approaches and trust none. Since we have experimented in these three areas of research, I shall illustrate them with examples. In the context of this volume I shall emphasize the plasma membrane.

(a) Intact cells - Some twenty years ago we became interested in the phenomenon discovered by Warburg (22) that tumor cells have a high aerobic glycolysis. Because of controversies that lasted for decades but evolved more heat than light, this phenomenon has been shoved under the rug with the excuse that it is both secondary and unimportant. This, in spite of the fact that among hundreds of biochemical properties that have been claimed to be universal for malignant tumors, the high aerobic glycolysis is one of few or perhaps the only one that has survived the test of time and experimentation. Although I hasten to concede that the high aerobic glycolysis is indeed likely to be a secondary phenomenon and even not unique for tumors, I know of no malignant cells that do not glycolyze. Moreover, there is convincing evidence that within related tumor lines the rate of aerobic glycolysis is proportional to the degree of malignancy (2, 21). The aerobic formation of lactic acid may be therefore an essential even though not a sufficient feature of malignancy. How can we establish whether this is actually the case?

Our plan was to discover first the cause of the high aerobic glycolysis of tumors and then to search for ways to repair the lesion. In the course of our studies we have become convinced that

TABLE I

Effect of Na^+, K^+, Ouabain on Glycolysis of Ascites Tumor Cells[a]

Additions	μmoles lactate/30 min/mg -dinitrophenol	μmoles lactate/30 min/mg + dinitrophenol
Ascites tumor cells in buffer	0.36	1.24
" + rutamycin	0.43	0.41
" + ouabain	0.12	1.03
" - K^+	0.16	-
" - Na^+	0.15	-
" - K^+, Na^+	0.04	-

[a]Experimental conditions were as described elsewhere (19).

the reason for the high aerobic glycolysis of tumors is an increased ATPase activity. It is rather remarkable that so many modern reviews and textbooks do not mention the fact that "ATPase" is a glycolytic enzyme: without ADP and P_i there is no glycolysis. True, the glycolytic pathway does not care from where the supply of ADP and P_i is coming, but cells contain only catalytic amounts of these cofactors and lactate formation ceases when the excess of ATP generated during glycolysis is not consumed. Indeed this is a very sensible way to do business, adjusting production to demand. Simple calculations convinced us that the biosynthetic processes, which consume ATP, are not responsible for the major supply of ADP and P_i in the tumor cells. Moreover, inhibitors of protein and nucleic acid synthesis have only minor effects on the rate of glycolysis. It appears that there are varieties of processes that can contribute to the ADP-P_i pool. In Ehrlich ascites tumor cells we have established that the plasma membrane ATPase is responsible for the regeneration of these cofactors and for the high aerobic glycolysis (19). In some other cells the mitochondrial ATPase is the chief contributor and in the third group of cells we have not been able to identify the ATPase (20). I shall restrict the discussion today to cells in which the plasma membrane enzyme sustains the aerobic glycolysis.

Our experiments with the Ehrlich ascites tumor illustrate the complexity of the approach with intact cells. Our conclusion that the Na+K+-ATPase is responsible for the high aerobic glycolysis is based on the following experiments. As shown in Table I lactate formation in ascites tumor cells is inhibited by ouabain at concentrations that are required to inhibit Rb^+ uptake into the cells. That this inhibition is in fact caused by a depletion of ADP and P_i rather than by e.g. a depletion in K^+ or inactivation of a glycolytic enzyme, is demonstrated by the high rate of lactate production even in the presence of ouabain when mitochondrial ATPase is activated by 2,4 dinitrophenol (Table I). When either Na^+ or K^+ is deleted from the medium, lactate production is markedly diminished.

Further evidence was obtained by analysis of cells that have been treated with dextran sulfate and made permeable to adenine nucleotides and P_i. In the presence of an excess of P_i and AMP or ADP glycolysis proceeds in dextran sulfate treated cells independent of ATPase activity and is completely resistant to ouabain or other inhibitors of the ATPase (19).

The second problem which we faced was whether the high ATPase activity of the plasma membrane is an expression of a high content of this enzyme in the plasma membrane or of an inefficient operation of the Na^+K^+ pump. An answer to the first question is difficult to obtain because of difficulties associated with the unavailability of an acceptable normal control cell for comparison, the eternal dilemma of cancer research. We therefore concentrated on the second possibility and examined the efficiency of operation of the pump. This is also not a simple matter and we adopted an approach which was used several years ago by Whittam and Ager (23). Since glycolysis is a tightly coupled process, one mole of ADP must be converted to ATP for each mole of lactate formed. Provided the supply of ADP is derived from the hydrolysis of ATP by the plasma membrane Na^+K^+-ATPase, we can calculate how many molecules of K^+ or Rb^+ are taken up for each lactate formed. Thus the Rb^+/lactate ratio is an index of the moles of Rb^+ transported for each ATP hydrolyzed and of the efficiency of pump operation. However, this efficiency will be overestimated when other processes (e.g. in mitochondria) compete with glycolysis for the available ADP; it will be underestimated when ADP generating processes are operative. Since we have established that in Ehrlich ascites tumor cells the major contributor to the ADP pool is the Na^+K^+-ATPase, we are not likely to underestimate the efficiency. An overestimation is more probable because in these cells mitochondrial phosphorylation competes with glycolysis for ADP and P_i (24). In spite of this, the Rb^+/lactate ratio in these cells is quite low indicating that more than one ATP is hydrolyzed for each Rb^+ that is translocated (Table II). In the presence of 2,4 dinitrophenol the Rb^+/lactate ratio is a small fraction of 1.

By a stroke of luck we discovered that some naturally occuring plant products called bioflavonoids such as quercetin, increase the efficiency of operation of several pumps. As shown in Table III quercetin inhibits the mitochondrial ATPase activity without affecting the P/O ratio (6). In this respect it acts like the natural

TABLE II

The Efficiency of the Na+K+ Pump of Ascites Tumor Cells[a] as Measured by the Rb+/Lactate Ratio; Effect of Quercetin

Additions	Rb+/lactate ratio -dinitrophenol	Rb+/lactate ratio + dinitrophenol
Ascites tumor cells	0.8	0.18
" + quercetin (8 μg/ml)	3.8	2.05

[a]Experimental conditions were as described (20).

TABLE III

Effect of Quercetin on ATPase Activity and P:O Ratio of Submitochondrial Particles[a]

Additions	ATPase μmoles/10 min	P/O ratio
Submitochondrial particles	0.83	0.96
" + quercetin(20 μg)	0.28	1.02
" + F_1 inhibitor (2.15 μg)	0.38	1.13

[a]Experimental conditions were as described (6).

mitochondrial ATPase inhibitor, a protein which was isolated by Pullman and Monroy (13). Addition of quercetin to glycolyzing ascites tumor cells markedly increased to Rb+/lactate ratio (Table II). The value of 3.8 must be an overestimate of the efficiency as pointed out above. The value in the presence of dinitrophenol which eliminates oxidative phosphorylation should be a better indication of the true efficiency, since quercetin inhibits also the mitochondrial ATPase, leaving ion translocation as the major contributor to the hydrolysis of ATP. Although we realize that an accurate estimate of the absolute efficiency is difficult by such indirect methods, the comparative data in the presence and absence of quercetin leave little doubt that the efficiency of Rb^+ pumping is greatly increased by the flavonoid, either in the absence or presence of dinitrophenol.

How does quercetin work? Here we have to leave the intact cells which do not lend themselves to studies of molecular mechanisms and return to the problem of quercetin after we have discussed the second approach.

(b) Resolutions and reconstitutions of membranes - Our work on the resolution and reconstitution of the system of oxidative phosphorylation and on its relation to the mitochondrial proton pump has aroused our interest in other ion pumps (15). We have conducted studies on the Ca^{++} pump of sarcoplasmic reticulum because of its simplicity, and on the Na^+K^+ pump because of its relevance to the tumor problem. The strategy of our approach has been identical in each case: we isolate from the membrane which contains the pump, the ATPase protein which can be looked upon as the site of energy transformation. In the case of the mitochondrial protein pump the energy of a proton flux is transformed into ATP which is generated from P_i and ADP. In the case of the Ca^{++} and Na^+K^+ pumps ATP energy is transformed into an ion flux. The basic reactions are the same (Fig. 2). We are interested in obtaining answers to two questions: what is the mechanism of these energy transformations, and how is the process of ATP generation or hydrolysis controlled to allow for the remarkable efficiency observed in nature.

Some answers to these questions have come from experiments on the resolution and reconstitution of the Ca^{++} pump, some from experiments with the Na^+K^+ pump. When I first reconstituted the Ca^{++} pump (14) with a purified preparation of the Ca^{++}- ATPase from sarco-

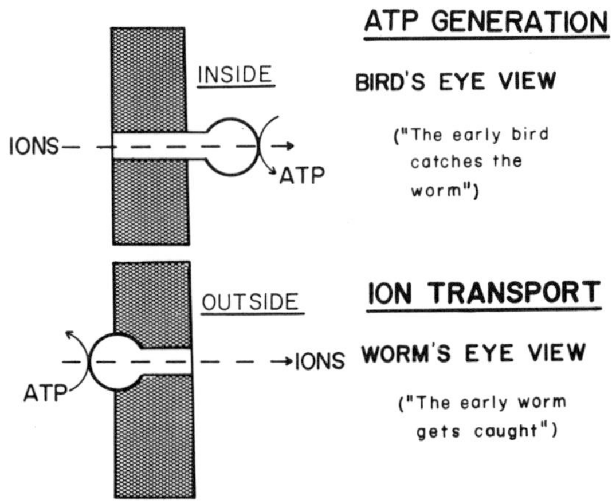

Fig. 2. The converging fields of oxidative phosphorylation and ion pumps.

TABLE IV

Comparison of ATPase Activity and Ca^{++} Translocation

The ATPase activity of the isolated fractions was assayed at 37° (8). The Ca^{++} transport activity and the ATP hydrolysis during this process was measured at 5 min incubation at room temperature as described (18).

	Isolated fractions	Reconstituted vesicles		
Fraction	ATPase μmoles/min/mg protein	Ca^{++} translocation (nmoles/min/mg protein)	ATP hydrolysis (nmoles/min/mg protein)	Ca^{++}/ATP ratio
R_{3a}	5.6	50	160	0.31
R_{3b}	8.0	242	630	0.38
R_{3c}	11.0	365	930	0.39
R_{3d}	8.6	646	708	0.91
R_{3e}	5.0	826	490	1.7

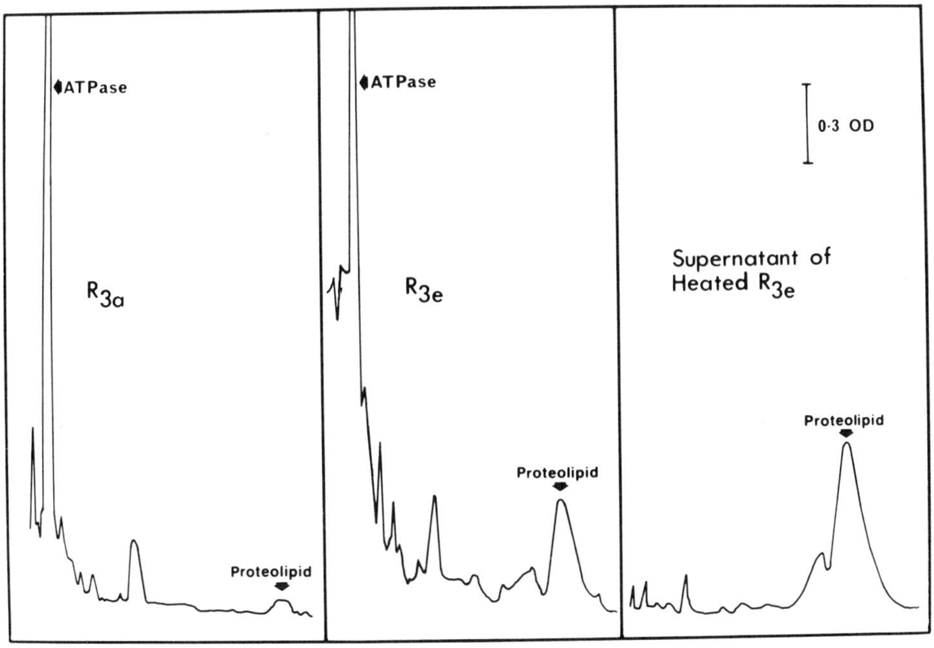

Fig. 3. Acrylamide gel patterns of ATPase preparations containing proteolipid.

plasmic reticulum and crude phospholipids, the rates of Ca^{++} translocations were rapid, but the efficiency was low. A comparison of various ATPase preparations obtained from a published purification procedure (8) revealed striking differences in the efficiency of pumping after reconstitution. As shown in Table IV the Ca^{++}/ATP ratio varied between 3.0 and 1.7. An analysis of the fractions on acrylamide gels (Fig. 3) revealed that the best fractions contained considerable amounts of sarcoplasmic reticulum proteolipid (9) while the poor fractions contained only little. When an active fraction was heated and the denatured ATPase removed, the supernatant still contained the proteolipid. When such a preparation was reconstituted together with an ATPase of low pumping efficiency, there was a marked increase in the Ca^{++}/ATP ratio (Table V). Thus it appears that the proteolipid of sarcoplasmic reticulum is a coupling factor (18). We obtained a clue to its mode of action when we found that the proteolipid, when added in excess, eliminated net ATP-driven Ca^{++} translocation. It also acted as a Ca^{++} ionophore when it was added to Ca^{++} loaded liposomes. I have outlined elsewhere our current interpretation of these findings in terms of the mechanism of Ca^{++} translocation (16).

TABLE V

Stimulation of Ca^{++} Transport by Heat Stable Coupling Factor

(Experimental conditions were as described (18)

	Ca^{++} translocation	ATP hydrolysis	$\frac{Ca^{++}}{ATP}$ ratio
	nmoles/min/mg protein		
Exp. 1			
R_{3a}	50	160	0.31
R_{3a} + 3 μg coupling factor	144	180	0.80
R_{3a} + 6 μg "	180	130	1.40
Exp. 2			
R_{3a}	190	730	0.26
R_{3a} + 4.5 μg coupling factor	370	510	0.72
R_{3a} + 15 μg "	120	1010	0.12

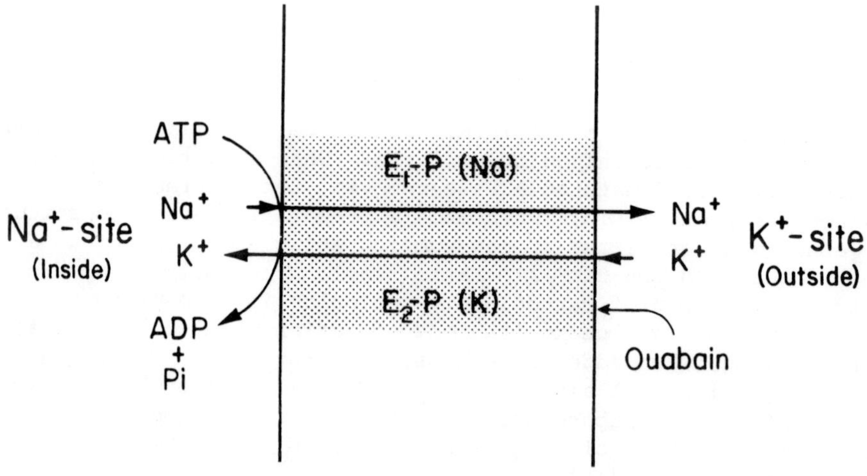

Fig. 4. The orientation of the Na^+K^+ pump.

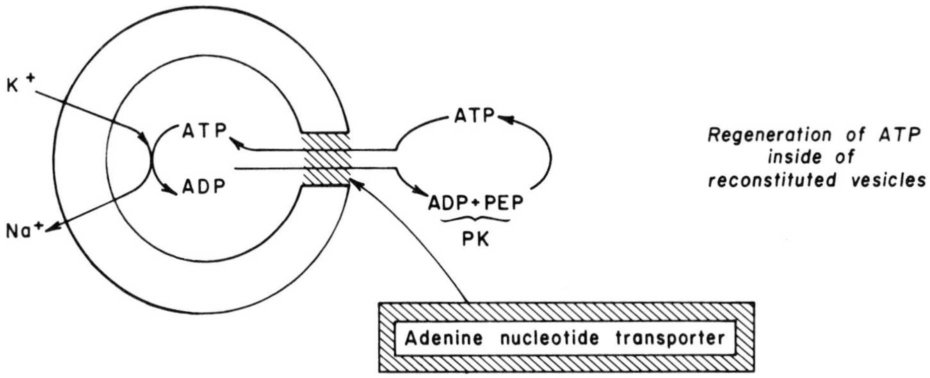

Fig. 5. Scheme of reconstitution of a $Na^{+}K^{+}$ pump with a natural orientation.

The reconstitution of the $Na^{+}K^{+}$ pump posed special problems. As shown in Fig. 4, the ATP-site in intact cells is on the inside of the plasma membrane and is responsible for the outward movement of Na^+. The ouabain-site is on the outside of the plasma membrane. Since liposomes are not permeable to ATP, there are two possibilities: we can try either reconstitution of a pump with a natural orientation which would require incorporation of a nucleotide transport system as illustrated in Fig. 5 or reconstitution of an inverted pump. In several laboratories (3, 4) including our own, reconstitution of such an inverted pump has been achieved. These vesicles catalyze an ATP-dependent translocation of Na^+ from the outside to the inside. As expected (see Fig. 4) the Na^+ uptake is insensitive to ouabain added from the outside but is inhibited by ouabain which is incorporated during the reconstitution procedure. In contrast to the Ca^{++} ATPase which can almost be quanitatively inserted into liposomes, we have not been able to obtain a high yield of reconstitution with the $Na^{+}K^{+}$ ATPase. Less than 20% of the total ATPase added during reconstitution was converted into a ouabain insensitive ATPase. On the other hand, in the presence of ouabain the efficiency of Na^+ translocation was very high. Estimates of the Na^+/ATP ratio were not as accurate as in the case of the Ca^{++} pump because of the lower rates of ion translocation but values of 3 or higher were obtained in several experiments. We do not know whether a proteolipid participates also in Na^+ transport. We will have to learn how to reconstitute a less efficiently operating pump in order to have an opportunity to explore this possibility.

Meanwhile, we have studied the mechanism of action of the $Na^{+}K^{+}$ ATPase and particularly the effect of quercetin. The experimental findings which we shall report elsewhere[1] have been partly confirmatory of others (1, 7, 12) who have studied the formation of phosphoenzyme from either ATP or P_i. In addition we observed that the addition of Mg^{++} prior to the formation of phosphoenzyme from P_i leads to a major change in the conformation of the protein as indicated by the exposure of groups that are reducible on addition of radioactive borohydride.

We have also made numerous attempts to bind reagents that interact with the phosphoenzyme during turnover, catching the aspartyl phosphate group during catalysis. These experiments have thus far been negative. Finally, we have made some progress with our analysis of the mode of action of quercetin. Since the ascites tumor ATPase of the plasma membrane has not as yet been prepared in a highly purified form, we have mainly used for these studies ATPase preparations from frozen sheep kidney (when they are available from Pel Freeze) and more recently from the electric eel.

Quercetin is an effective inhibitor of the hydrolysis of ATP by this enzyme. An analysis of the individual reactions catalyzed by the enzyme revealed that it inhibits the reversible phosphorylation of the enzyme with P_i, but has little or no effect on phosphoenzyme formation from ATP[1]. This may explain why in the ascites tumor cells quercetin inhibits at low concentrations the ATPase activity but not Rb^+ translocation, thereby increasing the efficiency of pump action.

(c) Model systems - Two different types of membrane model systems have been widely used in recent years: liposomes and Mueller-Rudin black membranes. Our own experience with the latter has been rather limited, but it is apparent from the literature (10) that very significant analyses of electrical events can be performed. With the appropriate ionophores many physiological phenomena of excitable membranes have been reproduced with these artificial membranes.

The experiments described under the heading of resolution and reconstitution were performed with liposomes prepared either by sonic oscillation or by solubilization of the phospholipids and the proteins with cholate followed by the slow removal of the detergent by dialysis (5). We are now using five different procedures to incorporate proteins into liposomes and the various advantages and disadvantages of these methods have been discussed elsewhere (17). I should like to stress here that we do not claim that we reconstitute structures that resemble the original membrane. We are reconstituting functional artifacts so that we can manipulate various phospholipids

---

[1] Kuriki, Y., Fisher, L., and Racker, E., in preparation.

and proteins and thereby analyze their role in the physiological process. It would be too much to expect from one or two isolated proteins which are incorporated into an excess of phospholipid bilayers to have exactly the same properties as the original membrane. On the other hand we can learn from these model liposomes what the minimal requirements are, e.g. for the operation of a pump; we can learn something about the asymmetry of the pump components both with respect to protein and phospholipids; finally we can learn something about the assembly of the pump by designing methods of reconstitution that might be operative in vivo.

REFERENCES

1. ALBERS, R. W., KOVAL, G. J. and SIEGEL, G. J. , Mol. Pharmacol. 4 (1968) 324.
2. BURK, D., WOODS, M. and HUNTER, J. , J. Natl. Cancer Inst. 38 (1967) 839.
3. GOLDIN, S. M. and TONG, S. W., J. Biol. Chem. 249 (1974) 5907.
4. HILDEN, S., RHEE, H. M. and HOKIN, L. E. , J. Biol. Chem. 249. (1974) 7432.
5. KAGAWA, Y. and RACKER, E. , J. Biol. Chem. 246 (1971) 5477.
6. LANG, E. R. and RACKER, E., Biochim. Biophys. Acta 333 (1974) 180.
7. LINDENMAYER, G. E., LAUGHTER, A. H. and SCHWARTZ, A., Arch. Biochem. Biophys. 127 (1968) 187.
8. MAC LENNAN, D. H., J. Biol. Chem. 245 (1970) 4508.
9. MAC LENNAN, D. H., YIP, C. C., ILES, G. H. and SEAMAN, P., Cold Spring Harbor Symp., Quant. Biol. 37 (1972) 469.
10. MUELLER, P. , Energy Transducing Mechanisms (E. Racker, ed.) (1975) in press.
11. OESTERHELT, D. and STOECKENIUS, W. , Proc. Nat. Acad. Sci. USA 70 (1973) 2853.
12. POST, R. L., KUME. S. and ROGERS, F. N. , Mechanisms in Bioenergetics (G. F. Azzone et al. eds) Academic Press (1973) 203.
13. PULLMAN, M. E. and MONROY, G. C. , J. Biol. Chem. 238 (1963) 3762.
14. RACKER, E. , J. Biol. Chem. 247 (1972) 8198.
15. RACKER, E., Dynamics of Energy-transducing Membranes (Ernster, Estabrook and Slater, eds.) Elsevier Scientific Publishing Co., Amsterdam, The Netherlands (1974) 269.
16. RACKER, E. , Biochemical Society Transactions (1975) in press.
17. RACKER, E. , Proceedings of 10th FEBS Symp., Paris. (1975).
18. RACKER, E. and EYTAN, E. , J. Biol. Chem. (1975) in press.
19. SCHOLNICK, P., LANG, D. and RACKER, E. , J. Biol. Chem. 248 (1973) 5175.
20. SUOLINNA, E-M., LANG, D. R. and RACKER, E. , J. Natl. Cancer Inst. 53 (1974) 1515.

21. SWEENEY, M. J., ASHMORE, J., MORRIS, H. P. and WEBER, G., Cancer Research 23 (1963) 995.
22. WARBURG, O., Uber den Stoffwechsel der Tumoren, Springer Verlag, Berlin (1926).
23. WHITTAM, R. and AGER, M. E., Biochem. J. 97 (1965) 214.
24. WU, R. and RACKER, E., J. Biol. Chem. 234 (1959) 1036.

Section II

Cell-Cell Interactions

CELLULAR RECOGNITION IN SLIME MOLDS: EVIDENCE FOR ITS MEDIATION BY CELL SURFACE SPECIES-SPECIFIC LECTINS AND COMPLEMENTARY OLIGOSACCHARIDES

Samuel H. BARONDES and Steven D. ROSEN

Department of Psychiatry, University of California
San Diego, School of Medicine
La Jolla, California 92093 (USA)

Specific cellular recognition must be mediated by quantitative or qualitative properties of the surfaces of the interacting cells. Determination of the molecular basis of cellular recognition may prove difficult in complex metazoan organisms that contain many cell types and highly specific forms of cellular association. For this reason, a simple eukaryotic cell type, like the cellular slime mold, is a preferable starting point for an analysis of the molecular basis of cellular recognition.

This paper has two purposes: 1) to present the reasons for using cellular slime molds as a model system for studying the molecular basis of specific cellular associations; 2) to review the evidence from our laboratory that the species-specific cellular association observed in these organisms is due to interactions between developmentally regulated cell surface lectins and complementary oligosaccharides.

ADVANTAGES OF CELLULAR SLIME MOLDS FOR STUDIES OF CELL RECOGNITION

The life cycle of cellular slime molds such as Dictyostelium discoideum has been described in detail (6, 11, 14, 28). The organisms may be found as unicellular amoebae that consume bacteria and divide every three hours. In this condition the cells are called vegetative cells and this part of their life cycle is called the non-social phase, since there is no tendency for the amoebae to associate with one another. As long as ample food is available the amoebae remain in this state. However, when the food supply is exhausted the amoebae differentiate into an aggregation-competent state over the course of about 9-12 hr after food deprivation. During this phase of development, a chemotactic system draws the cells together. They are then capable of forming stable intercellular contacts. A multicellular structure (pseudoplasmodium) containing up to 10^5 cells is assembled. In the next 12 hr this aggregate undergoes further differentiation culminating in the formation of a fruiting body with a spore cap that contains spores. The spores are in a dormant state and are resistant to inclement environments. However, if spores are exposed to a favorable environment they germinate into amoebae, and a new life cycle is begun.

There are several advantages in using cellular slime molds for studies of specific cellular associations:

(1) A large number of identical cells can be raised in culture. Therefore, the problem of cellular heterogeneity which would complicate studies of cellular recognition in a complex tissue is eliminated;

(2) The culture conditions are simple and resemble the natural environment of these organisms. Therefore, slime mold cells retain cellular recognition properties under defined conditions where cells from higher organisms might lose them;

(3) The cells can be isolated in a non-cohesive form (as vegetative cells) and at various stages during development of cohesiveness;

(4) There are a number of species of cellular slime molds that have been shown to display species-specific cellular association (i.e., mixtures of cells from different species "sort out" into separate colonies each composed of slime mold cells of a single species) (7, 17, 24). Sorting out is presumably due, at least in part, to species-selective intercellular affinities. Thus, this system provides the opportunity for studying the molecular basis of selective cellular cohesion.

Beug, Gerisch and colleagues (2-4) have demonstrated the usefulness of cellular slime molds for studies of cellular association. Some of these studies are reviewed in this volume. They raised high affinity antibodies to cohesive *Dictyostelium discoideum* cells in rabbits and degraded the resultant immunoglobulin to univalent antibodies (Fab fragments). After adsorption with vegetative cells the unadsorbed Fab fragments were shown to block cohesion of *D. discoideum*. This result indicates that new antigens have appeared on the surface of cohesive slime mold cells and that these antigens termed contact sites presumably mediate cell cohesion. In recent studies it was shown that when particulate fractions of cohesive *D. discoideum* cells were solubilized with detergents a soluble component that adsorbs the active Fab fragments was fractionated by chromatographic procedures (12).

CARBOHYDRATE-BINDING PROTEINS MAY MEDIATE CELLULAR RECOGNITION IN SLIME MOLDS

In the past few years we have attempted to determine the precise mechanism for species-specific cellular association in cellular slime molds. This work is still in progress and will be summarized briefly here. Details will be found in published papers, although some preliminary studies will also be mentioned. These studies were guided by the general hypothesis derived from many lines of evidence (1) that cell recognition may be mediated by interaction between cell surface proteins and oligosaccharides. All our work supports the general hypothesis that cell surface carbohydrate-binding proteins and oligosaccharides mediate cellular recognition in slime molds.

Appearance of Carbohydrate-binding Proteins in Cellular Slime Molds as They Become Cohesive

Extracts of cohesive forms of the cellular slime molds *Dictyostelium discoidem* (20) and *Polysphondylium pallidum* (22) contain proteins that agglutinate erythrocytes. Formalinized erythrocytes of appropriate type are generally used for these agglutination assays since they are more stable than fresh erythrocytes. An example of the agglutination assay is shown in Fig. 1. The appearance of agglutinin in extracts of *D. discoideum* parallels the development of cohesiveness as the cells differentiate (Fig. 2). Similar results were found with *P. pallidum* (21).

The agglutinins are carbohydrate-binding proteins as indicated by the finding that agglutination can be blocked by specific sugars. For example, the agglutination of formalinized sheep erythrocytes by extracts from *D. discoideum* was completely inhibited by 0.3 M N-acetyl-D-galactosamine (Fig. 1). Appropriate simple sugars also blocked the agglutination of extracts of *P. pallidum* as considered further below. This suggested that the agglutinins were related to the carbohydrate-binding proteins called lectins (13) that have previously been observed primarily in plant seeds.

Different Properties of Purified Agglutinins from *D. discoideum* (discoidin) and *P. pallidum* (pallidin)

Agglutination activity from extracts of *D. discoideum* cells could be quantitatively adsorbed on columns of Sepharose 4B and eluted with D-galactose. The eluate from this column showed two

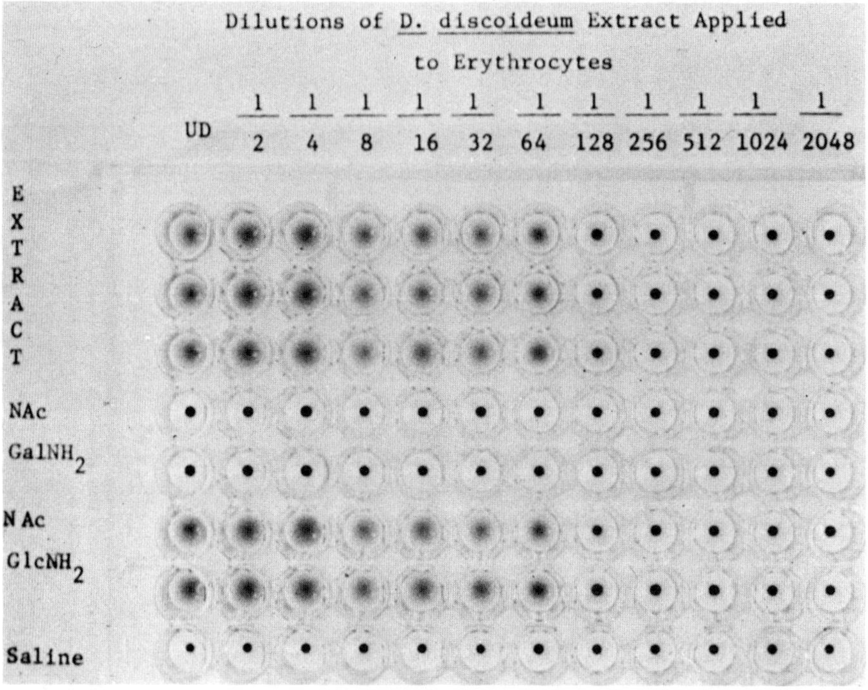

Fig. 1. Agglutination of formalinized sheep erythrocytes by extract of cohesive D. discoideum cells and effect of sugars. All wells contained dilutions of agglutinin in saline except the bottom row which contained only saline. Unagglutinated erythrocytes settle to form clear dots at the bottom of the plate; agglutinates settle as a heavy layer. The titer of the extract shown is taken as 1/64 (greatest dilution that produces agglutination). Addition of N-acetyl-D-galactosamine [final concentration 0.1 M] completely inhibits agglutination; but an identical concentration of N-acetyl-D-glucosamine has no effect. The assay may be performed in two ways as described (20, 21).

bands on polyacrylamide gel electrophoresis in a sodium dodecyl sulfate system, a major band containing more than 90% of the stained protein and a minor band of slightly lower molecular weight containing the remainder (25). The purified proteins, called discoidin I and II (10), will be discussed in more detail by Frazier et al. (elsewhere this volume). To purify agglutination activity from extracts of P. pallidum we made use of the finding that fixed erythrocytes that are agglutinated by a lectin can be used as an affinity adsorbent for its purification (18). We therefore adsorbed the agglutination activity of P. pallidum extracts with formalinized human Type O erythrocytes and eluted with D-galactose (26). The

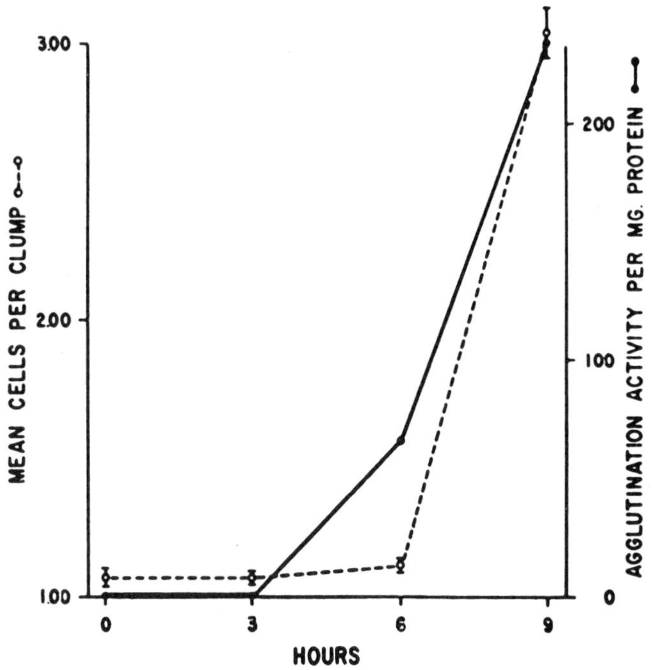

Fig. 2. Development of cohesiveness and agglutination activity with differentiation of *D. discoideum*. Vegetative *D. discoideum* cells were separated from bacteria at time 0 and differentiated on Millipore pads. At the indicated times cells were assayed for cohesiveness (mean cells per clump) with a quantitative cohesiveness assay; and aliquots were extracted and assayed for agglutination activity with formalinized sheep erythrocytes. For details see Rosen *et al.* (20).

resultant purified agglutinin, named pallidin, electrophoresced as a single band in a sodium dodecyl sulfate system (Fig. 3). Recent studies (see Frazier *et al.*, elsewhere in this volume) indicate that at least one other agglutinin is also present in extracts of *P. pallidum* that is not adsorbed by formalinized human Type O erythrocytes but is adsorbed by Sepharose. The major agglutinin from *D. discoideum* and that from *P. pallidum* differ in subunit molecular weights, isoelectric point and amino acid composition (26). The molecular weight of *pallidin* is about 250,000.

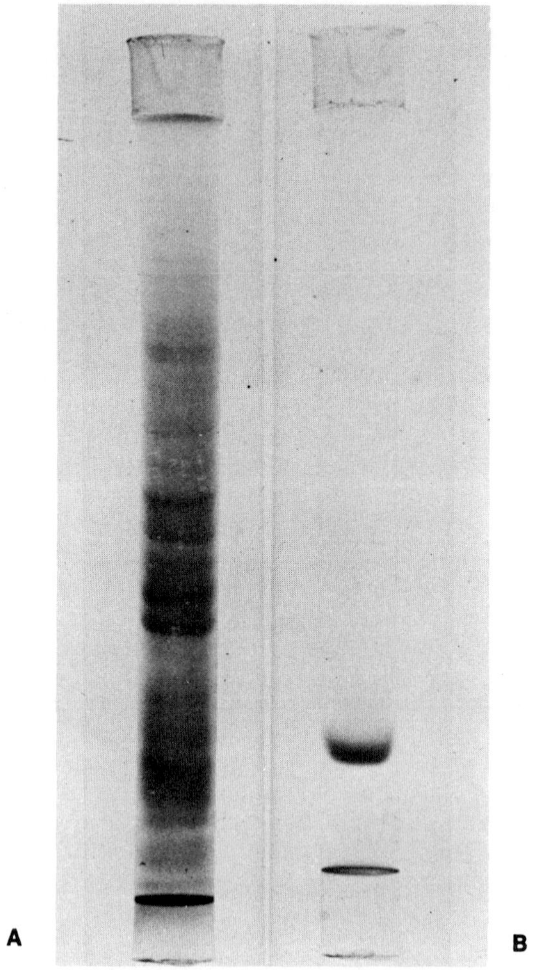

Fig. 3. Polyacrylamide gel electrophoresis in a sodium dodecyl sulfate system of: (A) a soluble extract of cohesive *Polysphondylium pallidum* cells; (B) pallidin that has been purified by adsorption of the soluble extract to formalinized erythrocytes followed by elution of the adsorbed protein with galactose. For details see Simpson *et al.* (26).

whereas the molecular weights of discoidins are about 100,000. The relative potency of a number of saccharides as inhibitors of agglutination of erythrocytes was different with these two lectins (Table I). This indicates that the active site of the major agglu-

TABLE I

Sugar effects on agglutination of erythrocytes by agglutinins from *P. pallidum* and *D. discoideum*.

Sugar	Sugar Concentration (mM) for 50% Inhibition of Agglutination*	
	P. pallidum agglutinin	*D. discoideum* agglutinin
Lactose	1.6	12.5
αmethyl-D-galactose	6.2	6.2
D-galactose	6.2	25
N-acetyl-D-galactosamine	25	1.6
D-fucose	25	3.1
L-fucose	50	12.5
3-O-methyl-D-glucose	100	1.6
αmethyl-D-glucose	≥100	12.5
βmethyl-D-glucose	≥100	100
D-glucose	≥100	≥100
D-glucosamine	≥100	≥100
N-acetyl-D-glucosamine	≥100	≥100
D-mannose	≥100	≥100
αmethyl-D-mannose	≥100	≥100

*Agglutination of formalinized human erythrocytes by serial dilutions of slime mold agglutinins and the sugar concentration that produced 50% inhibition of agglutination was determined as described in Rosen *et al*. (21).

tinin from *D. discoideum* and that from *P. pallidum* are different. Evidence for similar but discriminable agglutinins from four other species of cellular slime molds was recently presented (22; see also Frazier *et al*., elsewhere this volume).

The agglutinins are present on the cell surface.

If these lectins are to play a role in cell cohesion they must be present on the surface of differentiated (cohesive) slime mold cells. Discoidin's presence on the cell surface was first inferred from the finding that formalinized sheep erythrocytes formed rosettes around cohesive *D. discoideum* cells which could be blocked by N-acetyl-D-galactosamine but not by N-acetyl-D-glucosamine (20). Similar studies have been done with cohesive *P. pallidum* cells and again, rosette formation could be blocked by specific sugars but not by sugars that reacted poorly with pallidin. In subsequent studies antibodies were raised to discoidin (a mixture of discoidin I and discoidin II) and to pallidin (the major lectin whose purification was shown in Fig. 3). The antibodies were shown to bind to the surface of cohesive slime mold cells, using immunofluorescent and immunoferritin techniques. Examples, using an antibody to pallidin, are shown in Fig. 4 and 5. Similar results with an antibody to discoidin have been published (9). The antibodies did not bind to the surface of vegetative slime mold cells. Control immunoglobulin preparations obtained from the same rabbits before immunization did not bind to the surface of cohesive slime mold cells, indicating that the binding represented specific association of antibody to the appropriate lectin that was present on the surface of the cohesive slime mold cells. Using these immunological techniques, it was demonstrated that the lectin on the cell surface was diffusely distributed and not confined to sites of contact between slime mold cells (Fig. 4 and 5). The lectin on the cell surface can move laterally in the plane of the surface membrane (16) since capping occurs when specific antibody is mixed with living slime mold cells followed by a reaction with fluorescein isothiocyanate labelled goat anti-rabbit immunoglobulin (Fig. 4).

Evidence for developmentally-regulated species-specific high-affinity cell surface receptors for these lectins.

If these cell surface lectins mediate species-specific cell cohesion there must be specific complementary oligosaccharide receptors on the surface of differentiated (cohesive) slime mold cells. In an attempt to determine if there were cell surface receptors for these lectins in slime mold cells and if they were developmentally-regulated, the ability of purified discoidin I and II to agglutinate *D. discoideum* cells at progressive stages of differentiation were tested. The cells were first fixed with glutaraldehyde to destroy endogenous cell cohesiveness. It was found that fixed vegetative *D. discoideum* cells were not significantly agglutinated by discoidin I and II but that, as the cells differentiated, they became increasingly agglutinable by both discoidins (19) (Fig. 6). These results suggested that a developmentally-regulated cell surface receptors to discoidin, presumably an oligosaccharide, appeared with the development of cohesiveness.

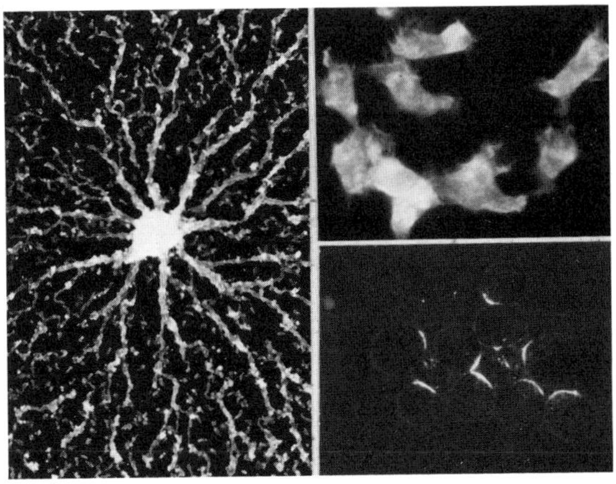

Fig. 4. Detection of pallidin on surface of cohesive P. pallidum cells with fluorescent antibody technique. Antibody to pallidin prepared as in Fig. 3 was raised in rabbits. P. pallidum cells were either differentiated on glass and fixed with brief exposure to glutaraldehyde (left and top right) or differentiated in suspension and studied without fixation (lower right). Antiserum was incubated with the cells and the excess was removed by careful washing. Goat anti-rabbit immunoglobulin labelled with fluorescein isothiocyanate was then added, the excess removed by washing and the cells were observed with a fluorescence microscope. Note the diffuse staining of the fixed cells. The living cells, when incubated with antibody, form "caps". (Unpublished experiments of C.-M. Chang and S. H. Barondes; for details of similar experiments with D. discoideum, see ref. 9).

The agglutinability of glutaraldehyde-fixed vegetative and cohesive D. discoideum cells with pallidin, Ricinis communis agglutinin I (RCA), concanavalin A, and wheat germ agglutinin were also determined. With pallidin and RCA I, it was also found that increased agglutinability with the fixed differentiated cells compared with the fixed vegetative cells (19). This suggests that the developmentally-regulated receptor that binds discoidin probably reacts with these lectins as well. This would not be surprising since these lectins, like the discoidins, bind simple sugars with a D-galactose configuration. With concanavalin A, which binds preferentially to mannose and glucose residues, there was only a slight change in agglutinability with differentiation; and with wheat germ agglutinin, which binds preferentially to N-acetyl-D-glucosamine residues, the fixed vegetative cells were found to be much more agglutinable than the fixed cohesive cells. Therefore, the developmentally-regulated cell surface receptor on D. discoideum cells has specificity for slime mold lectins and lectins with related carbohydrate binding specificities.

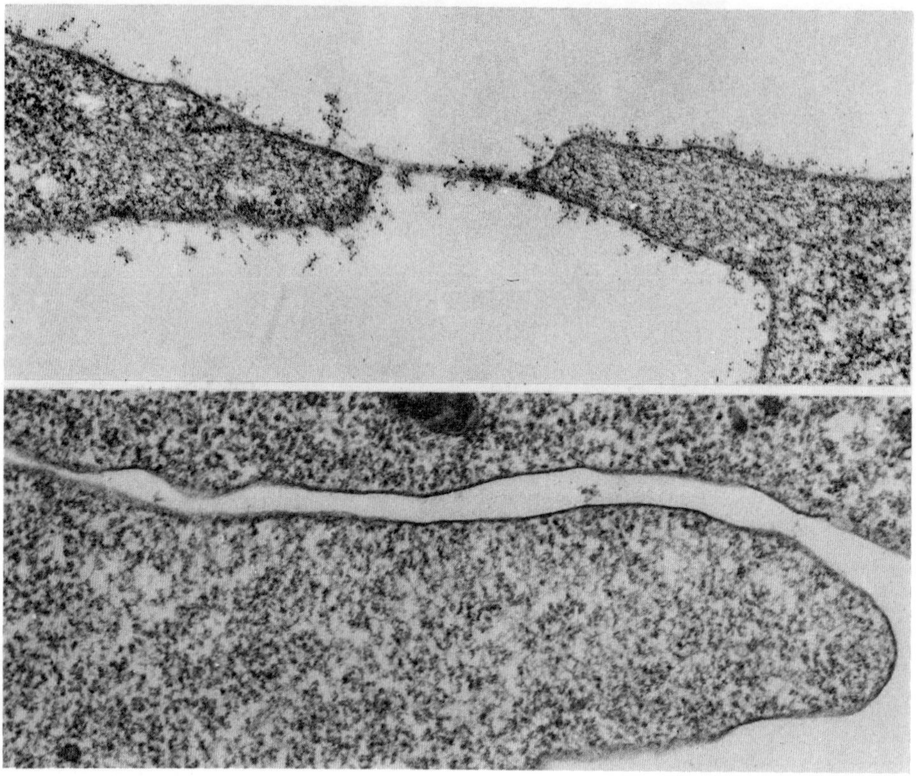

Fig. 5. Detection of pallidin on surface of cohesive P. pallidum cells with ferritin antibody technique. Fixation of cells was done as in Fig. 4. After incubation with antiserum to pallidin (top) or with control antiserum (bottom), both preparations were reacted with goat anti-rabbit immunoglobulin covalently bound to ferritin. The cells were then embedded for electron microscopy and stained. Only the cells treated with antibody to pallidin show surface labelling. (Unpublished experiments of C.-M. Chang and S. H. Barondes; for details of similar experiments with D. discoideum , see ref. 9).

The species specificity of discoidins I and II and pallidin was tested in direct binding experiments using these lectins and fixed vegetative and cohesive D. discoideum and P. pallidum cells. The amount of lectin bound was determined as the reduction in the concen-

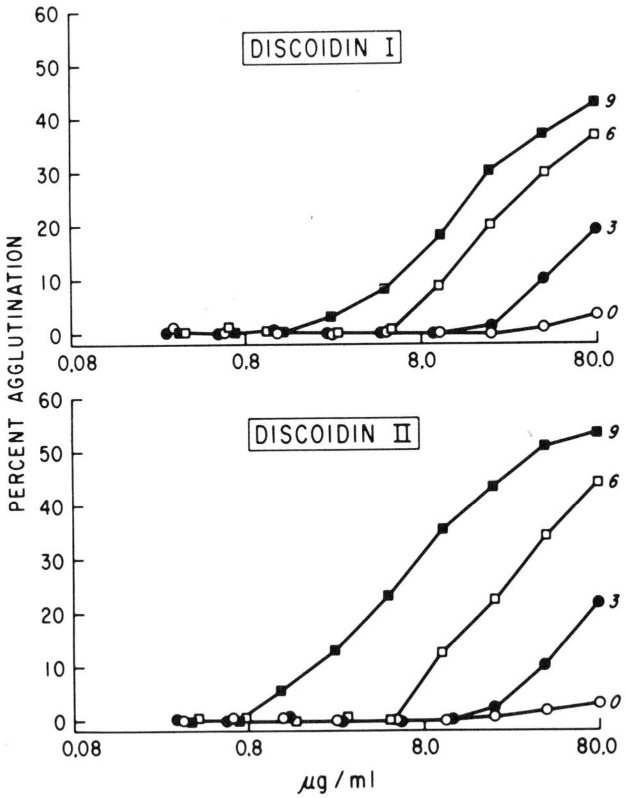

Fig. 6. Increased agglutinability of D. discoideum cells by discoidin I and II with differentiation. D. discoideum cells were separated from bacteria at time 0 and then differentiated for up to 9 hr in suspension. At the indicated times cells were fixed with glutaraldehyde and their agglutinability by discoidin I and discoidin II was determined using a quantitative agglutination assay. Effects of a series of concentrations of discoidin I and II on agglutination of the cells at different stages of differentiation are shown. For details see Reitherman et al. (19).

tration of lectin from a standard solution produced by adsorption with increasing numbers of fixed slime mold cells. Lectin concentration was determined from standard curves of red cell agglutination activity versus concentration by determining the residual (unadsorbed) cell agglutination activity in a quantitative assay

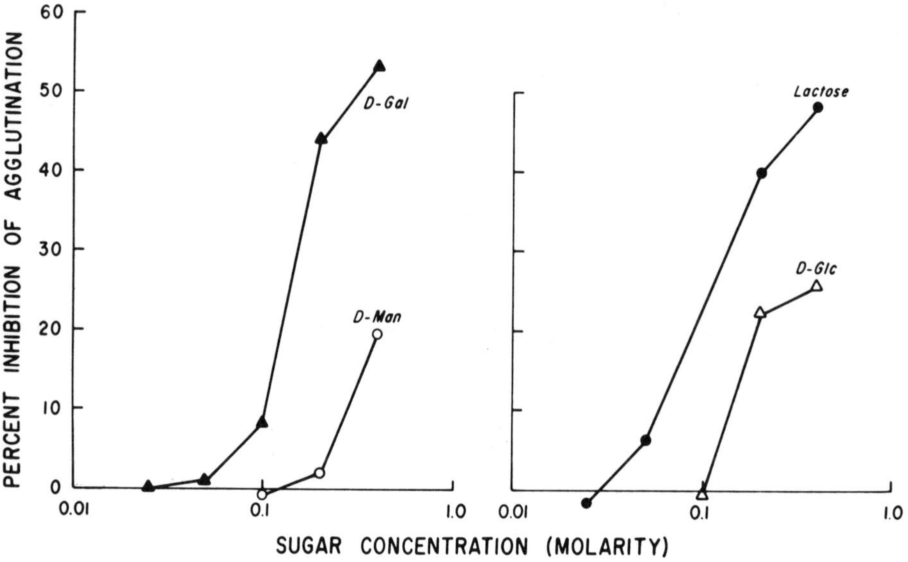

Fig. 7. Effect of sugars on aggregation of cohesive P. pallidum cells. Cohesive P. pallidum cells were gyrated in suspension and agglutination was monitored by determining disappearance of single cells with a quantitative assay using a Coulter counter in the presence and absence of the indicated concentration of sugars. Two separate experiments are shown. For details see ref. 21.

employing a Coulter counter. Association constants and numbers of binding sites were determined for the binding of all three lectins to both D. discoideum and P. pallidum cells using the graphic analysis of Steck and Wallach (27).

Fixed cohesive (9 hr) D. discoideum cells bound discoidin I and II with high affinity. In both cases, the association constant of these lectins for fixed cohesive P. pallidum cells was an order of magnitude lower. With pallidin, the association constant for fixed differentiated P. pallidum cells was in the range of 4 x 10^9 M^{-1} and an order of magnitude lower for fixed differentiated D. discoideum cells, again indicating clear species-specificity. Specificity was also shown by the finding that the association constant of discoidin I for fixed rabbit erythrocytes was about three orders of magnitude lower than that for fixed differentiated D. discoideum cells; and that the association constant of concanavalin A for fixed differentiated D. discoideum cells was about three orders of magnitude lower than the association constant of discoidin I for these cells. The asso-

TABLE II

Association Constant and Number of Receptors per Cell

Cell Type*	Discoidin I	Discoidin II	Pallidin	Concanavalin A
		K_a (M^{-1})		
D.d (0)	5×10^7	6.5×10^7	1.3×10^8	
D.d (dif.)	1.3×10^9	1.5×10^9	5.0×10^8	1.8×10^6
P.p (dif.)	1.6×10^8	1.9×10^8	3.8×10^9	
Rabbit RBC	1.3×10^6			
		Binding sites per cell		
D.d (0)	3×10^5	3×10^5	1.4×10^5	
D.d (dif.)	5×10^5	4×10^5	2.7×10^5	4×10^5
P.p (dif.)	3×10^5	3×10^5	3.2×10^5	
Rabbit RBC	9×10^5			

*All cells were fixed with glutaraldehyde. D.d (0): vegetative D. *discoideum* cells; D.d (dif.) and P.p (dif.): differentiated, cohesive D. *discoideum* and P. *pallidum* cells. For details see Reitherman *et al.* (19).

ciation constants of discoidin I or II for fixed vegetative D. discoideum cells was about twenty-fold lower than for fixed cohesive D. discoideum cells, confirming our conclusion (from the developmental study of D. discoideum agglutinability by discoidin) that the receptors are developmentally regulated. These studies indicate that as slime mold cells become cohesive high affinity receptors appear on their surface; and these receptors have a higher affinity for the carbohydrate binding protein from the same species as compared with that from another species of slime mold. The solubilization, purification and chemical characterization of these developmentally-regulated receptors is in progress.

Cohesion is blocked by specific sugars and glycoproteins

Given the presence of both carbohydrate binding proteins and receptors on the surface of cohesive cells, it still remains to be shown that the interaction of these substances mediates cell cohesion. If the interaction of the carbohydrate binding proteins and the receptors mediates cell cohesion then cohesion should be blocked by appropriate saccharides. This result would occur only if saccharides are added in concentrations sufficient to block the high affinity of carbohydrate-binding proteins for cell surface receptors, as will be discussed below. This condition has been successfully met in studies with cohesive Polysphondylium pallidum cells (21). When these cells were gyrated in a suspension, they aggregated. The formation of aggregates was determined quantitatively using a Coulter electronic particle counter by determining the disappearance of single cells from a suspension. Addition of lactose or D-galactose, sugars which have a relatively high affinity for pallidin when compared with other simple sugars, blocked the cohesion of heat-treated or living P. pallidum cells in this assay at substantially lower concentrations than D-mannose or D-glucose, which have relatively low affinity for pallidin (Fig. 7). These studies indicate that interaction between pallidin and oligosaccharide receptors on the surface of cohesive P. pallidum cells mediates the agglutination of these cells as determined in this assay. In studies with D. discoideum cells (20), very high concentrations of N-acetyl-D-galactosamine or D-galactose were required to block the agglutination of cohesive cells in suspension; and at these concentrations the osmotic effect of the sugars (which cause cell shrinkage) might have been responsible, since D-glucose was also effective. However, recent unpublished experiments with an axenic mutant of D. discoideum suggests that cohesion of these cells can be blocked with much lower concentrations of specific sugars, whereas equal concentrations of other sugars are ineffective.

In preliminary studies we have also tested the action of desialated fetuin and bovine $alpha_1$-acidic serum glycoprotein on the aggregation of P. pallidum cells on a surface as measured by direct visual observation of formation of aggregates. Both these modified glycoproteins, which contain terminal beta-linked galactose residues, interact with pallidin since they are potent inhibitors of its erythrocyte agglutination activity; and both glycoproteins inhibit the formation of aggregates in cultures of cohesive *P. pallidum* cells at concentrations as low as 10 μg/ml. The inhibitory activity in this cell aggregation assay and in the erythrocyte agglutination assay is markedly reduced by gentle periodate treatment of the glycoproteins which presumably modifies the critical saccharides. Therefore, it appears that *in situ* aggregation can be inhibited by competitive binding of cell surface pallidin by a glycoprotein whose oligosaccharide moieties are related to the native receptor.

SUMMARY OF EVIDENCE

The present results indicate that as slime mold cells differentiate from a vegetative nonsocial form to a cohesive form, species-specific lectins appear that are detectable both in their cell surface and as extractable soluble proteins. The appearance of the lectins both in extracts and on the cell surface is correlated with the development of cohesiveness, as is the appearance on the cell surface of species-specific high affinity receptors. On the basis of these findings, we infer that species-specific cell cohesion of differentiated slime mold cells is mediated by the association of cell surface lectins with complementary oligosaccharide receptors. The findings that in *P. pallidum* cohesiveness of cells can be blocked by appropriate saccharides and that *in situ* aggregation can be blocked by glycoproteins which bind to pallidin support these inferences.

The relationship of these findings to the work of Gerisch and colleagues is not presently clear. One possibility is that the Fab fragments used by Gerisch and colleagues to block cohesion of *D. discoideum* include antibodies to either the lectins or the oligosaccharide receptors or possibly also to complexes of these moieties. The recent purification of the antigens that can adsorb the blocking activity of the mixture of Fab fragments (12) should allow a direct test of this hypothesis.

CONCLUSION

The experiments with slime molds provide strong evidence that species-specific cellular recognition can be mediated by interactions between specific cell surface lectins and complementary oligosaccharides. These is evidence for a similar process in species-specific "self" recognition in sponges (15, and Jumblatt *et al.*, this volume elsewhere). There is also substantial evidence that complementary interactions involving specific protein-carbohydrate association may underlie certain viral-host cell interactions (23), nitrogen-fixing bacteria-plant cell associations (5), and Acanthamoebae-yeast cell interactions (8). The consequence of these interactions range from infection (23) to symbiosis (5) to phagocytosis (8). Evidence for interactions of this type in the mating of yeast has been presented by Ballou (elsewhere this volume). Since reactions of this type are so common, it seems plausible that they have also been adapted to mediate specific interactions in the development of complex metazoa.

ACKNOWLEDGEMENT

This work was supported by a grant from the National Institutes of Health (NIMH #18282).

REFERENCES

1. BARONDES, S. H., The Neurosciences, Second Study Program, (F. O. Schmitt, ed.), Rockefeller University Press, New York, (1970) 747.
2. BEUG, H., GERISCH, G., KAMPFF, S., RIEDEL, V., and CREMER, G., Exp. Cell Res. 63 (1970) 147.
3. BEUG, H., KATZ, F. E., STEIN, A., and GERISH, G., Proc. Nat. Acad. Sci. USA 70 (1973) 3150.
4. BEUG, H., KATZ, F. E., and GERISCH, G., J. Cell. Biol. 56 (1973) 647.
5. BOHLOOL, B. B., and SCHMIDT, E. L., Science 185 (1974) 269.
6. BONNER, J. T., The Cellular Slime Molds, 2d Ed. Princeton University Press, Princeton, New Jersey, (1967).
7. BONNER, J. T. and ADAMS, M. S., J. Embryol. Expt. Morphol. 6 (1958) 346.
8. BROWN, R. C., BASS, H., and COOMBS, J. P., Nature 254 (1975) 434.
9. CHANG, C.-M., REITHERMAN, R. W., ROSEN, S. D., and BARONDES, S. H., Exp. Cell Res. (1975) in press.
10. FRAZIER, W. A., ROSEN, S. D., REITHERMAN, R. W., and BARONDES, S. H., J. Biol. Chem. 250 (1975) 7714.
11. GERISCH, G., Current Topics in Developmental Biology, (A. Monroy and A. Moscona, eds.) Vol. 3, Academic Press, New York, (1958) 157.
12. HUESGEN, A., and GERISCH, G., FEBS Letters 56 (1975) 46.
13. LIS, H., and SHARON, N., Ann. Rev. Biochem. 42 (1973) 541.
14. LOOMIS, W. F., Jr., Dictyostelium discoideum, a Developmental System, Academic Press, New York, (1975).
15. MacLENNAN, A. P., Arch. Biol. (Bruxelles) 85 (1974) 53.
16. NICOLSON, G. L., Int. Rev. Cytol. 39 (1974) 89.
17. RAPER, K. B., and THOM, F., Am. J. Botany 28 (1941) 69.
18. REITHERMAN, R. W., ROSEN, S. D., and BARONDES, S. H., Nature 248 (1974) 599.

19. REITHERMAN, R. W., ROSEN, S. D., FRAZIER, W. A., and BARONDES, S. H., Proc. Nat. Acad. Sci. USA (1975) in press.
20. ROSEN, S. D., KAFKA, J., SIMPSON, D. L., and BARONDES, S. H., Proc. Nat. Acad. Sci. USA 70 (1973) 2554.
21. ROSEN, S. D., SIMPSON, D. L., ROSE, J. E., and BARONDES, S. H., Nature 252 (1974) 128, 149.
22. ROSEN, S. D., REITHERMAN, R. W., and BARONDES, S. H., Exp. Cell Res. (1975) in press.
23. SCHULZE, I. T., Adv. Virus Res. 18 (1973)1.
24. SHAFFER, B. M., Quart. J. Micros. Science 98 (1957) 377.
25. SIMPSON, D. L., ROSEN, S. D., and BARONDES, S. H., Biochemistry 13 (1974) 3487.
26. SIMPSON, D. L., ROSEN, S. D., and BARONDES, S. H., Biochim. Biophys. Acta (1975) in press.
27. STECK, T. L., and WALLACH, D. F. H., Biochim. Biophys. Acta 97 (1965) 510.
28. SUSSMAN, M., Progress in Molecular and Subcellular Biology, (F. Hahn, ed.), Vol. IV, Springer, Berlin, 1975.

MULTIPLE LECTINS IN TWO SPECIES OF CELLULAR SLIME MOLDS

William A. FRAZIER, Steven D. ROSEN,
Richard W. REITHERMAN and Samuel H. BARONDES

Department of Psychiatry, University of California,
San Diego, School of Medicine, La Jolla,
California 92093 (USA)

Evidence has been presented in the preceding communication (Barondes and Rosen, elsewhere this volume) that specific cell cohesion in the cellular slime molds is accomplished by the interaction of carbohydrate-binding lectins with complementary cell surface oligosaccharide-containing receptors. In the species *Dictyostelium discoideum* and *Polysphondylium pallidum* both the lectins (1, 3, 5) and their surface receptors (2) are developmentally regulated. That is, both are absent in feeding, vegetative amoebae and appear as the starved cells differentiate and become cohesive. The proteins responsible for the lectin (red cell agglutination) activity of crude extracts of five species of cellular slime mold have been purified and partially characterized. Several of these lectins exhibit heterogeneity. In particular, cohesive *D. discoideum* cells contain two structurally and functionally distinct lectins and cohesive *P. pallidum* cells contain three.

Figure 1 shows a typical affinity column purification of lectin from the soluble fraction of broken cohesive slime mold cells. The affinity adsorbent is underivatized Sepharose 4B, a galactose containing polymer, and lectin is specifically eluted with a D-galactose solution (6). D-Galactose also inhibits the red cell agglutination activity of the lectins (4). This purification method has been found to be effective for lectins from all species examined to date with the exception of those of *P. pallidum*. For lectins of this species, a purification method has been devised using fixed erythrocytes as the affinity adsorbent (7).

Lectins purified from the four species *D. discoideum*, *P. violaceum*, *D. purpureum* and *D. mucoroides* by affinity chromatography all

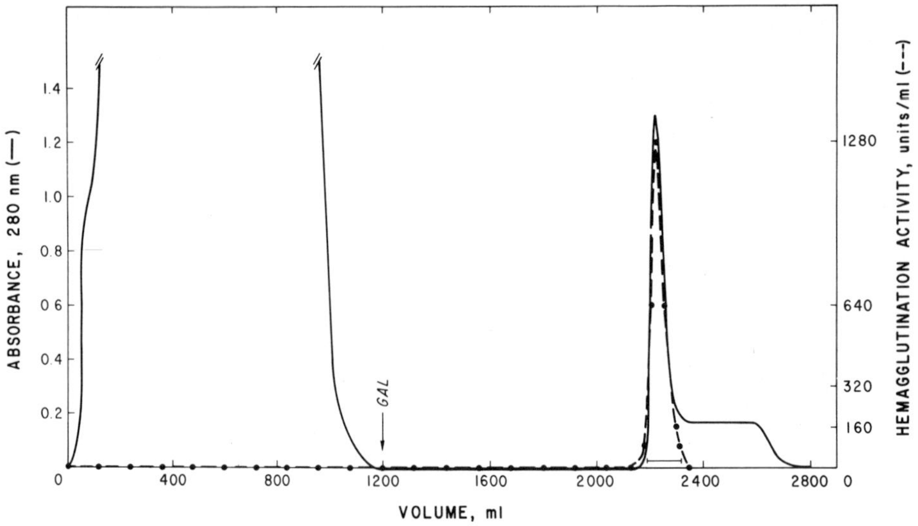

Fig. 1. Affinity chromatography of crude extracts (100,000 x g supernatant of cells broken by rapid freezing in liquid N_2 and thawing) of cohesive slime mold cells on a Sepharose 4B column (1,000 ml bed volume, flow rate 72 ml/hr) (1). The solid line is the recording of absorbance at 280 nm and the points and dashed line indicate the hemagglutination activity measured with rabbit erythrocytes in a V-plate assay (3). Fractions of 12 ml were collected starting when the sample and rinse (∿400 ml) had been loaded on the column. Active fractions were pooled as indicated by the horizontal bar. This material is referred to as the "Sepharose 4B pool."

have subunit molecular weights in the range of 23,000 to 26,000 (Fig. 2). The Sepharose 4B pool material from D. discoideum and D. mucoroides clearly shows a major band of higher molecular weight and a more rapidly moving minor band on SDS 10% polyacrylamide gels (Fig. 2) while that of P. violaceum and D. purpureum exhibits broad single bands.

Purification and Properties of Discoidin I and II

The two components of the Sepharose 4B pool of D. discoideum have been designated discoidin I (major band) and discoidin II (minor band) (1). These two components can be separated under non-denaturing conditions by ion exchange chromatography on DE-52 cellulose as shown in Fig. 3. The first peak to elute from the column (pool 1) consists of the lower molecular weight discoidin II and the second peak is the major component discoidin I. The profile of fixed sheep erythrocyte agglutination activity in Fig. 3 is coincident with discoidin I, while discoidin II does not agglutinate these cells.

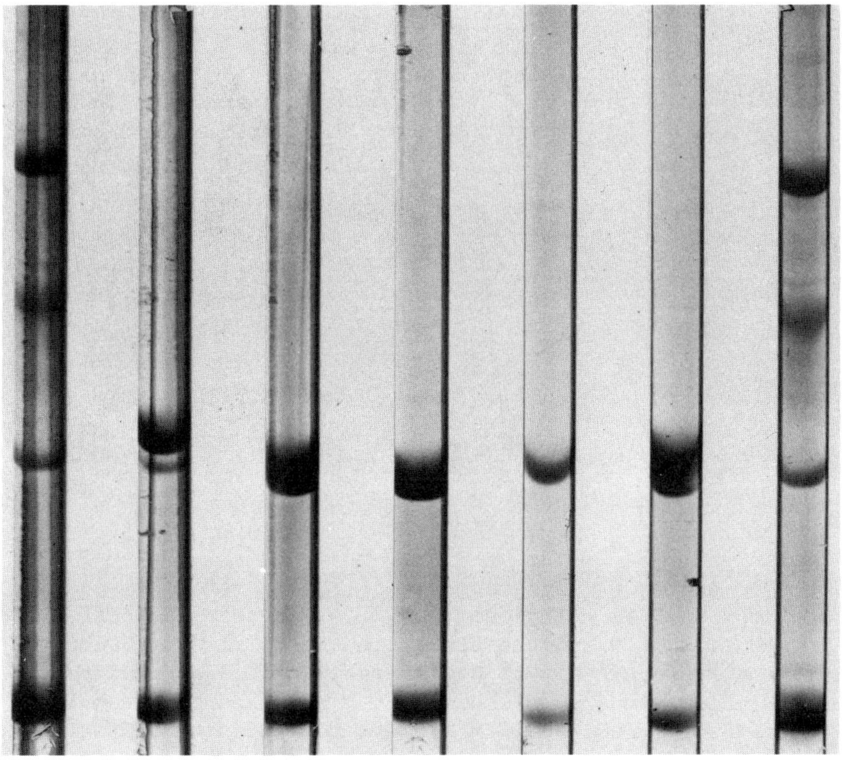

Fig. 2. Electrophoresis on 10% polyacrylamide gels containing SDS of a β-mercaptoethanol SDS, 8 M urea treated protein of the Sepharose 4B pools obtained from four species of cohesive slime mold. The left and right gels are standard proteins (bovine serum albumin, ovalbumin, α-chymotrypsinogen and cytochrome c, top to bottom) Gels of the Sepharose 4B pools are from the species (left to right) D. discoideum, P. violaceum, D. purpureum, P. pallidum (purified on formalinized human Type O erythrocytes, see text) and D. mucoroides. Cytochrome c has been included in all samples. Staining was with Coomassie blue.

It does, however, agglutinate fixed or fresh rabbit or human Type O erythrocytes more effectively than discoidin I (not shown). Discoidin II has an isoelectric point of 6.8-7.0 and that of discoidin I is 6.0 (1, 6). Discoidin II runs a single compact band on pH 8.9 non-SDS 7% polyacrylamide gels while discoidin I does not run in this system. Both proteins are tetramers in the native state. The subunit molecular weight of discoidin I is about 26,000 and that of discoidin II about 24,000.

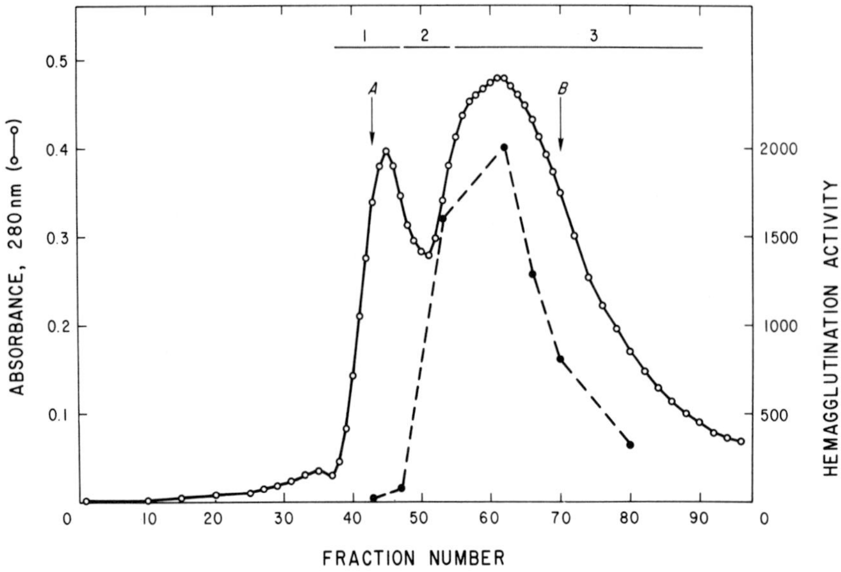

Fig. 3. Ion exchange chromatography of the Sepharose 4B pool of D. discoideum on DEAE cellulose (DE-52). The starting buffer was 5 mM Tris-HCl, pH 8.0 and the flow rate was 60 ml/hr. Under these conditions all protein of the Sepharose 4B pool was bound to the column. Elution was accomplished with a 2-chambered parabolic gradient (300 ml) from 0-0.25 M NaCl in the starting buffer applied at fraction 1 and extending to fraction 96. The open circles and solid line indicate the absorbance at 280 nm and the closed circle and dashed line indicate the hemagglutination activity measured with formalinized sheep erythrocytes in the V-plate assay. Fractions marked A and B were those assayed with three species of erythrocytes as indicated in the text. Fractions were pooled as indicated by the horizontal bars.

The amino acid compositions of discoidins I and II are shown in Table I. The data have been calculated to yield molecular weights in agreement with the peptide chain size determined by SDS polyacrylamide gel electrophoresis (Fig. 2). Although the amino acid compositions are similar in many respects, some major differences distinguish the two proteins. For example, discoidin I contains 24 threonine residues and 15 serine residues while this ratio is reversed in discoidin II, which has only 12 threonine residues and 23 serine residues. Other differences include the number of glutamic acid residues and the alanine, lysine and arginine content. These data indicate that discoidin II is not derived from discoidin by proteolysis. This conclusion was further substantiated by the results of

TABLE I. Amino Acid Compositions of Discoidins I and II[a]

	Discoidin I		Discoidin II	
	mean residue number	integral number	mean residue number	integral number
Aspartic acid	34.6	35	36.1	36
Threonine	24.0	24	12.4	12
Serine	15.0	15	23.3	23
Glutamic	20.1	20	13.0	13
Proline	9.9	10	7.8	8
Glycine	15.0	15	13.8	14
Alanine	17.6	18	12.2	12
Cysteine	(3.6)[b]	(4)	ND	ND
Valine	18.0	18	17.4	17
Methionine	1.0	1	0.7	1
Isoleucine	12.8	13	12.7	13
Leucine	13.0	13	13.6	14
Tyrosine	10.8	11	7.9	8
Phenylalanine	11.3	11	10.0	10
Histidine	4.9	5	6.1	6
Lysine	8.0	8	11.9	12
Arginine	13.1	13	10.4	10
Glucosamine	<0.2[c]	0	<0.2[c]	0
Galactosamine	<0.2[c]	0	<0.2[c]	0
TOTAL RESIDUES[d]		234		209
Calculated molecular weight[d]		26,301		23,059

[a]Based on duplicate analyses of 4 preparations of discoidin I and 3 preparations of discoidin II, 24 hr hydrolysis.

[b]From ref. 6.

[c]Estimated detection limit

[d]Does not include tryptophan.

comparative tryptic peptide maps of ^{14}C-S-carboxymethyl discoidins I and II which indicated only three potentially identical tryptic peptides out of the approximately 20 from each protein (1).

As noted above, the red cell species specificity of discoidins I and II is quite distinct. However, both lectins agglutinate fresh rabbit erythrocytes to a comparable extent at the same concentration (1). Using this cell type for assay, the sensitivity of the agglutination activity of the lectins to inhibition by a 5 mM concentration of ten simple sugars was tested. The pattern of hapten inhibition obtained for the two proteins was quite distinct, reflecting differences in their carbohydrate binding sites. For example, D-galactose, lactose and D-fucose were the best inhibitors of discoidin II, while N-acetyl-D-galactosamine was the most potent inhibitor of discoidin I (1).

In addition to these structural and functional differences between discoidin I and II, the two lectins appear to be under different regulation during the transition from the vegetative, non-cohesive amoebae to the multicellular stage. During this period (about 12 hr) discoidin I increased very slowly from 0 to 6 hr of differentiation and then rapidly between 6 and 12 hr. Discoidin II, on the other hand, reaches a higher proportion of its final value early in this 12 hr period. The result of these two patterns of appearance is that at early times (0 to 6 hr of development) discoidin II comprises nearly 35% of the total lectin of differentiating cells. At later times discoidin I is ten times more abundant than discoidin II (1).

Purification of Pallidin I, II and III

The human Type O erythrocyte agglutinating activity of the soluble fraction of cohesive *P. pallidum* cells can be quantitatively adsorbed by fixed (formalinized) Type O cells. However, rabbit erythrocytes are still agglutinated by the supernatant. After washing the red cells, a lectin can be eluted from them with a 0.6 M solution of D-galactose (7). This lectin is a better agglutinin of Type O cells than of rabbit erythrocytes and the agglutination is inhibited most strongly by lactose. It exhibits a single band on 7% polyacrylamide gels (pH 8.9) has an isoelectric point of about 7.0 and runs a single species of spproximate molecular weight 24,000 on 7, 10 and 12.5% polyacrylamide gels containing SDS. The protein exhibits a range of native molecular weights, the predominant species being about 250,000, a decamer (7). This lectin has been designated pallidin I.

If the crude extract is adsorbed with formalinized rabbit erythrocytes, all of the human Type O agglutination activity and more of the rabbit agglutination activity is bound. The material eluted

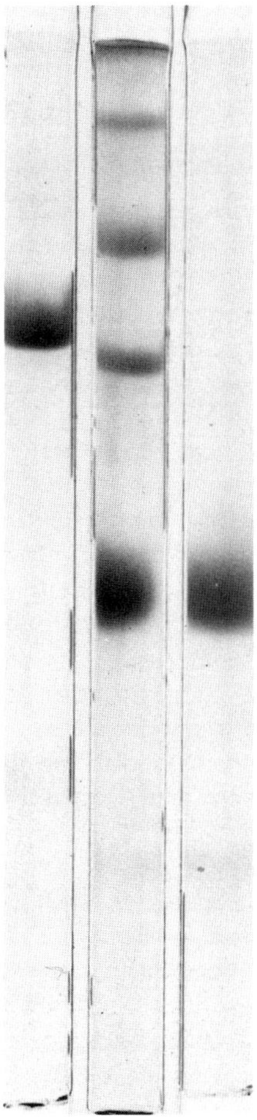

Fig. 4. Electrophoresis of pallidin I (left), standard proteins bovine serum albumin, ovalbumin, α-chymotrypsinogen and sperm whale myoglobin (center), and pallidin III (right) on 12.5% polyacrylamide gells containing SDS. Staining was with Coomassie blue.

from these cells with 0.6 M galactose is a better agglutinin of rabbit erythrocytes than of human Type O. It exhibits a single band on 10 and 12.5% polyacrylamide SDS gels which coelectrophoreses with pallidin I. However, on nonSDS 7% polyacrylamide gels (pH 8.9) multiple bands are seen, one of which coincides with the single band of pallidin I. This material prepared by affinity adsorption on rabbit erythrocytes is designated pallidin II.

When the rabbit erythrocyte agglutination activity remaining in the crude supernatant after adsorption with either Type O or rabbit cells or after sequential adsorption with both is tested for inhibition of agglutination activity by monosaccharides, it is found to be most sensitive to N-acetylglucosamine. This property immediately distinguishes this lectin activity from discoidins I and II, pallidins I and II and mucoroidin and purpurin (4)(see Fig. 2). This lectin, which does not appear to bind with high affinity to either human O rabbit erythrocytes, can be purified by affinity chromatography on Sepharose 4B, specific elution being accomplished with a solution of 0.1 M N-acetyl D-glucosamine. After elution from the affinity column, the purified material has extremely poor agglutination activity when assayed with rabbit erythrocytes. This material runs as a single band on 7% polyacrylamide gels (pH 8.9), has a discrete isoelectric point near neutrality and appears as a single species of molecular weight 18,000 on 12.5% polyacrylamide SDS gels. This carbohydrate-binding proteins has been designated pallidin III.

Fig. 4 shows the results of electrophoresis of pallidin I and III in 12.5% polyacrylamide SDS gels. Standard proteins are shown for comparison. Pallidin II comigrates with pallidin I in the presence of SDS. The amino acid compositions of pallidins I and III are seen in Table II. The compositions have been calculated to yield subunit moletular weights near the values determined for the peptide chains with SDS gel electrophoresis (Fig. 4). The two proteins appear to be distinct in the relative amounts of several amino acids and the greater number per peptide chain of aspartic acid and valine in pallidin III indicates that pallidin III is not a derivative of pallidin I. The amino acid composition of pallidin II (not shown) is very similar to that of pallidin I, perhaps indicating that it is composed of some subunits which are identical and some which are closely related to those of pallidin I. Further structural data is needed to determine the precise relationship of pallidin I and II.

Correlations and Conclusions

The lectins of the two species of cellular slime mold which have been most extensively studied, *D. discoideum* and *P. pallidum*, have been shown in each case to be at least two distinct gene products. Preliminary evidence for a multiplicity of lectins in other species has also been obtained. For example, the Sepharose 4B purified material of *D. mucoroides* exhibits a major and minor component on SDS gels in identical positions with discoidins I and II (Fig. 2). The mucoroidins are very similar in carbohydrate binding specificity to the discoidins (4). Furthermore, tryptic peptide maps reveal that about half of the peptides of the major mucoroidin are probably identical to peptides of discoidin I, indicating extensive sequence homology. Amino acid composition data for the Sepharose 4B pool from *P. violaceum* (not shown) indicates that it is more

TABLE II. Amino Acid Compositions of Pallidins I and III[a]

	Pallidin I		Pallidin III	
	mean residue number	integral number	mean residue number	integral number
Aspartic acid	25.0	25	29.2	29
Threonine	18.6	19	16.7	17
Serine	16.0	16	10.8	11
Glutamic acid	15.7	16	11.0	11
Proline	11.2	11	5.6	6
Glycine	15.9	16	11.3	11
Alanine	13.4	13	7.0	7
Cysteine[b]	3.9	4	ND	ND
Valine	14.7	15	16.5	17
Methionine	2.2	2	1.2	1
Isoleucine	9.7	10	6.6	7
Leucine	14.8	15	10.8	11
Tyrosine	6.0	6	7.4	7
Phenylalanine	9.0	9	7.9	8
Histidine	5.0	5	2.8	3
Lysine	17.7	18	4.0	4
Arginine	8.5	9	4.2	4
TOTAL RESIDUES[c]		209		154
Calculated molecular weight[c]		23,282		16,887

[a] Based on duplicate analyses of 2 preparations of pallidin I and III, 24 hr hydrolysis.

[b] From ref. 7.

[c] Does not include tryptophan.

closely related to discoidin I and the putative mucoroidin I than to discoidin II or pallidin I. Thus it appears likely that the lectins from these four species will fall into two or three classes on the basis of structure and carbohydrate binding specificity. Proteins within each class have probably evolved from a common ancestral lectin structural gene. The relationships between these classes and the relationship of slime mold lectins to plant lectins will provide interesting and useful insights on lectin evolution and function.

With regard to the apparent role of the slime mold lectins in specific cell cohesion, the reason for the presence of multiple lectins of different specificity in a single species remains unclear at present. The difference in the time course of appearance of discoidins I and II during differentiation may indicate a temporal segregation of function for these two lectins. They may also serve to specify spatial information as markers of the positions of cells within the multicellular slug. Alternatively, the combinations of specificities achieved by multiple lectins may serve to refine the recognition of homospecific versus heterospecific cells. This possibility is particularly attractive if the cellular slime molds are to serve as a useful model system for cell adhesion in higher, more complex systems.

REFERENCES

1. FRAZIER, W. A., ROSEN, S. D., REITHERMAN, R. W. and BARONDES, S. H., J. Biol. Chem. 250 (1975) 7714.
2. REITHERMAN, R. W., ROSEN, S. D., FRAZIER, W. A. and BARONDES, S. H., Proc. Nat. Acad. Sci. USA, (1975) in press.
3. ROSEN, S. D., KAFKA, J. A., SIMPSON, D. L. and BARONDES, S. H., Proc. Nat. Acad. Sci. USA 70 (1973) 2554.
4. ROSEN, S. D., REITHERMAN, R. W. and BARONDES, S. H., Exp. Cell Research (1975) in press.
5. ROSEN, S. D., SIMPSON, D. L., ROSE, J. E. and BARONDES, S. H., Nature 252 (1974) 128, 149.
6. SIMPSON, D. L., ROSEN, S. D. and BARONDES, S. H., Biochemistry 13 (1974) 3487.
7. SIMPSON, D. L., ROSEN, S. D. and BARONDES, S. H., Biochim. Biophys. Acta (1975) in press.

RECEPTORS MEDIATING CELL AGGREGATION IN *DICTYOSTELIUM DISCOIDEUM*

Gunther GERISCH

Biozentrum der Universität Basel
4056 Basel (Switzerland)

Cell-Surface Changes During Cell Development

Several hours after the end of the growth phase, single cells of *Dictyostelium discoideum* aggregate into a multicellular organism. Cell differentiation during the interphase between growth and aggregation is expressed in cell surface changes which include the following markers: 1) an increase in the number of cyclic-AMP binding sites which are believed to act as receptors mediating chemotaxis and other developmental functions of extracellular cyclic-AMP (13, 14, 18); 2) a fast increase in the activity of cyclic-AMP phosphodiesterase (19). This enzyme is involved in the control of the cyclic AMP released from, and diffusing between, the interacting cells in an aggregation field (17); 3) the appearance of antigenic sites ("contact sites A") related to cell adhesion (2).

Immunochemical Analysis of Cell-to-Cell Adhesion

The definition of "contact sites" is an operational one which is based on the following experimental results: first, cell-to-cell adhesions in aggregation-competent cells can be completely blocked by univalent antibody fragments (Fab) directed against the antigens present in crude membrane preparations (1). IgG cannot be used for blocking because it induces "cap" formation and cell agglutination (3).

Secondly, complete blockage of cell adhesion requires the binding of Fab to two different cell-surface sites, termed contact sites A and B (2, 6). These sites differ with respect to developmental control, pattern of cell assembly and EDTA-sensitivity of their action. Contact sites B are already active in growth-phase cells which are able to associate into loose irregular assemblies. The same occurs in a mutant (aggr 50-2) which is specifically defective in developmentally controlled markers including contact sites A. Immunological detectability of contact sites A is correlated with the ability of the cells to adhere tightly end-to-end. This leads to aggregation into cell chains, which are characteristic elements of the normal aggregation pattern.

The function of contact sites A can be specifically investigated either in the presence of EDTA which preferentially blocks the activity of contact sites B, or in the presence of anti-B Fab which has the same effect. The function of contact sites B can be investigated either in growth-phase cells or in aggregation-competent ones after blockage of their A sites. These can be blocked by Fab directed against membranes of aggregating cells, which, for removal of anti-B activity, has been absorbed with growth-phase cells (2). The specific blockage of either contact sites A or B indicates that these sites are different molecular entities which function independent of each other.

Thirdly, binding of ^{3}H-labeled Fab shows that contact sites A are completely blocked by 3 x 10^{5} Fab molecules per cell which cover less than 2% of the cell surface area (3). If ^{3}H-Fab from another anti-serum is used which mainly reacts with the carbohydrate moieties of glycosphingolipids (25), then binding reaches a plateau at 2 x 10^{6} Fab molecules per cell, without any effect of the bound Fab on the activity of contact sites A. These results on cell adhesion, and not simply the number of Fab molecules bound to the surface.

Fourthly, fluorescent Fab, that does not block contact sites A, nevertheless labels the whole surface of aggregating cells. It is of particular interest that this Fab also remains bound to their ends where the cells tightly adhere to each other (3). It can be concluded, therefore, that molecules of a size of 60 x 35Å can fit within the space between attached membranes without interfering with cell adhesion. This again shows that the crucial process in membrane-membrane interaction is restricted to specific loci of the cell surface.

Electron microscope localization of antigens has been performed with ferritin-labeled antibodies directed against membrane of aggregating cells, from which the antibodies directed against growth-phase cells had been removed (6, 24). This antibody preparation is not necessarily monospecific for contact sites A, since antibodies may also bind to any other cell-surface site specific for aggregation-competent cells. The following conclusions are therefore tentative. No association of any antigen into plaques has been detected, indicating that the loci of actual interaction between cells are of macromolecular size rather than of larger dimensions. No accumulation of antigens at the ends of single aggregation-competent cells has been found. This possibly means that contact sites A are present everywhere at the cell surface, and that a specific mechanism may exist which regulates the activity of these sites according to their location. The presence of contact sites A all over the cell surface would agree with the finding that by local applications of cyclic AMP, a new front of pseudopods can be induced within a period of a few seconds. This means that every part of the surface is a potential tip of the cell (9, 11).

The extension of different antigenic sites in the direction perpendicular to the cell surface has been compared using antibodies of different specificities as the first layer, together with one batch of ferritin-labeled anti-IgG. According to these measurements, the cell surface is covered by a homogeneous layer of short carbohydrate chains, apparently part of glycolipids. Only relatively few sites of different specificity extend up to 30 to 40 Å beyond the carbohydrate layer (6, 24). Although the identity of these sites with contact sites A is not proven, the results, nevertheless, favor the idea that the space between the membranes of adhering cells is locally bridged by interacting macromolecules.

Solubilization and Partial Purification of Contact Sites A

Contact sites A can be solubilized by sodium desoxycholate (DOC), and an Fab-binding assay (2) can be used for their purification (15). By conventional DEAE cellulose chromatography and gel filtration in the presence of 0.1% DOC, contact sites A can be separated both from the major membrane proteins and from membrane-bound cyclic-AMP phosphodiesterase. On Sephadex G-200, the DOC-solubilized A-sites elute in one symmetrical peak in the region of MW 130,000. The peak fractions completely absorb the Fab species that, in living cells, block the contact-site A system (15). Thus the total antigenic activity attributed to contact sites A appears to reside in this particular peak.

Both pronase and periodate destroy the antibody binding activity of solubilized contact sites A (10). This result might indicate that carbohydrate is required for the structural integrity of these sites. Since discoidin activity (see Barondes and Rosen, elsewhere in this volume) which is present in the unfractionated DOC-extract, is not recovered together with contact sites A, the relation of discoidin to the contact sites A remains unresolved (15).

Functions of Cyclic-AMP receptors in Cell Aggregation

Since the properties of cyclic-AMP receptors are discussed by Malchow et al. in this volume, this communication will be focussed on receptor-mediated cellular functions. These functions are: 1) chemotaxis, 2) relaying of cyclic-AMP signals and 3) phase-resetting of the cellular oscillator responsible for the periodic generation of cyclic AMP.

Chemotaxis. The first discovered function of cyclic AMP in cell aggregation of *D. discoideum* was its chemotactic action (4, 16). When exposed to a spatial concentration gradient of cyclic AMP, the cells orientate along the gradient and move towards the higher concentrations. Chemotactic responsiveness increases during the interphase from growth to aggregation-competence (4), in coincidence with the number of cyclic-AMP binding sites at the cell surface (18). Since local application of cyclic AMP from microcapillaries induces pseudopod formation from any part of the cell surface, the receptors must be distributed over the whole surface (9, 11). The mechanism used by the cells for measuring the gradient is still unknown; two different possibilities have been discussed recently (9). It is also unknown how the cell-surface receptors process the signal through the plasma membrane, so that it is eventually translated into a localized activation of the contractile system of the cell.

Signal relaying. Pulses of cyclic-AMP are transmitted in aggregation fields from cell to cell (20). The visible effect of transmission of pulsatile signals is the propagation of waves of chemotactic activity from the center to the periphery of an aggregation territory. Thus large aggregation territories are organized around the centers. Because of losses of the transmitter due to diffusion and enzymatic inactivation, any cell in such a system has to function as a signal amplifier (22). Recently it has been shown that in fact the cells release cyclic AMP on stimulation by cyclic AMP (23), and that small pulses (5×10^{-10} M) induce the release of a number of cyclic-AMP molecules two orders of magnitude higher than the stimulus (21). The induced release is preceded by a sharp increase of the intracellular cyclic-AMP concentration within the first 1 to 2 minutes after stimulation. The conclusion is that activation of cell surface receptors by cyclic-AMP not only regulates the transport of cyclic-AMP through the plasma membrane, but also the activity of adenylate cyclase.

Periodicity of cyclic AMP production. A prerequisite of signal transmission is the pulsatile release of cyclic AMP from aggregation centers which are the origins of the waves of chemotactic activity. Periodic activity of the centers is often seen in time-lapse movies, particularly in the early stages of aggregation. A convenient way to study periodic cyclic-AMP synthesis and release is to keep the cells in a stirred suspension, where they synchronize each other as they do in aggregation centers (12). Since pulses of nanomolar amplitude shift the phase of the oscillator, cell-surface receptors appear to process signals to the oscillating system responsible for periodic cyclic-AMP generation (8). The cyclic-AMP receptors are supposed, therefore, to participate in the synchronization of the periodic cellular activities in an aggregation center.

Cyclic-AMP Actions in Cell Differentiation

So far the function of cyclic-AMP receptors in actually aggregating cells has been discussed. The receptors, however, seem to interact also with the system that controls cell differentiation from the growth-phase state to aggregation-competence. This can be shown by using cells harvested from axenic medium during the late stationary phase (7), or by using a certain class of non-aggregating mutants (5). In such cells differentiation is blocked; but when stimulated by cyclic AMP, these cells become aggregation-competent, and hence, in the absence of further stimulation, they aggregate normally. The most effective stimuli are provided by cyclic-AMP pulses, e.g. by pulses of 5 nmolar amplitude given at intervals of 7 minutes. Continuous cyclic-AMP application at the same average rate has, if any, only a weak effect. At the molecular level the action of cyclic-AMP pulses on cell differentiation can be resolved into their effects on the regulation of single cell-surface constituents (10, 11). All the markers studied are under the control of extracellular cyclic-AMP pulses. These markers are contact sites A, cyclic-AMP receptors and cell-surface phosphodiesterase. The time courses of increase of the individual markers in response to cyclic-AMP pulses is different. This renders it unlikely that the different markers of "differentiated" cells are regulated *en bloc*.

Of particular interest is the regulation of cyclic-AMP receptors by cyclic AMP. It implies a positive regulatory interaction in the control pathway of cell differentiation. By increasing the receptor number, cyclic AMP increases the sensitivity of the cells for cyclic AMP. Since the receptors are functionally coupled to the cyclic-AMP generating system, the cyclic-AMP output should be increased when the number of receptors rises. Thus not only the sensitivity, but also the signal level itself is increased. Further analysis of this positive interaction between constituents of the cyclic-AMP signal system may reveal a basic feature of the control of cell differentiation.

REFERENCES

1. BEUG, H., GERISCH, G., KEMPFF, S., RIEDEL, V. and CREMER, G., Exptl. Cell Res. 63 (1970) 147.
2. BEUG, H., KATZ, F. E. and GERISCH, G., J. Cell. Biol. 56 (1973) 647.
3. BEUG, H., KATZ, F. E., STEIN, A. and GERISCH, G., Proc. Nat. Acad. Sci. USA 70 (1973) 3150.
4. BONNER, J. T., BARKLEY, D. S., HALL, E. M., KONIJN, T. M., MANSON, J. W., O'KEEFE, G. III and WOLFE, P. B., Devel. Biol. 20 (1969) 72.
5. DARMON, M., BRACHET, P. and PEREIRA DA SILVA, L. H., Proc. Nat. Acad. Sci. USA 72 (1975) 3163.
6. GERISCH, G., BEUG, H., MALCHOW, D., SCHWARZ, H. and STEIN, A., in: Biology and Chemistry of Eukaryotic Cell Surfaces, Miami Winter Symposia 7 (1974) 49.
7. GERISCH, G., FROMM, H., HUESGEN, A. and WICK, U., Nature 255 (1975) 547.
8. GERISCH, G. and HESS, B., Proc. Nat. Acad. Sci. USA 71 (1974) 2118.
9. GERISCH, G., HÜLSER, D., MALCHOW, D. and WICK, U., Phil. Trans. Roy. Soc. Lond. B 272 (1975) 181.
10. GERISCH, G., HUESGEN, A. and MALCHOW, D., Proc. of the Tenth FEBS Meeting (Y. Raoul ed.) Elsevier, Amsterdam. (1975) 257.
11. GERISCH, G., MALCHOW, D., HUESGEN, A., NANJUNDIAH, V., ROOS, W., WICK, U. and HÜLSER, D., Proc. ICN-UCLA Symposium on Devel. Biol. (D. McMahon and D. F. Fox eds.) Benjamin, Cal. (1975).
12. GERISCH, G. and WICK, U., Biochem. Biophys. Res. Comm. 65 (1975) 364.
13. GREEN, A. A. and NEWELL, P. C., Cell 6 (1975) 129.
14. HENDERSON, E. J., J. Biol. Chem. 250 (1975) 4730.
15. HUESGEN, A. and GERISCH, G., FEBS LETT. 56 (1975) 46.
16. KONIJN, T. M., Adv. Cyclic Nucleotide Res. 1 (1972) 17.
17. MALCHOW, D., FUCHILA, J. and NANJUNDIAH, V., Biochim. Biophys. Acta 385 (1975) 421.
18. MALCHOW, D. and GERISCH, G., Proc. Nat. Acad. Sci. USA. 71 (1974) 2423.
19. MALCHOW, D., NÄGELE, B., SCHWARZ, H. and GERISCH, G., Eur. J. Biochem. 28 (1972) 136.
20. ROBERTSON, A., DRAGE, D. J. and COHEN, M. H., Science 175 (1972) 333.
21. ROOS, W., NANJUNDIAH, V., MALCHOW, D. and GERISCH, G., FEBS LETT. 53 (1975) 139.
22. SHAFFER, B. M., Adv. Morphogenesis 2 (1962) 109.
23. SHAFFER, B. M., Nature 255 (1975) 549.
24. SCHWARZ, H., Thesis. Universität Tübingen. (1973).
25. WILHELMS, O.-H, LÜDERITZ, O., WESTPHAL, O. and GERISCH, G., Eur. J. Biochem. 48 (1974) 89.

CELL SURFACE COMPONENTS MEDIATING THE REAGGREGATION OF SPONGE CELLS

James E. JUMBLATT, George WEINBAUM*, Robert TURNER**, Kurt BALLMER and Max M. BURGER

Department of Biochemistry, Biocenter, University of Basel, CH 4056 (Switzerland)

Proposed mechanisms for the directed, morphogenetic movement of cells and establishment of defined cell-cell contacts during embryogenesis include molecular gradients and selective cell adhesion (8, 31, 37). Pioneered by the classic studies of H. V. Wilson on reaggregation of dissociated marine sponge cells (42-44), investigations of cell sorting *in vitro* have demonstrated cellular recognition and/or selective adhesion at the tissue, organ, and species levels in a variety of developing metazoan systems (14, 15, 20, 25, 34). Although there is increasing evidence that developmental regulation of cellular affinities is mediated by components of the cell surface (1, 2, 16, 18, 30, 35), the biochemistry of intercellular recognition sites remains largely obscure.

Sponge Reaggregation as a Biochemical Model for Cellular Recognition

Wilson first reported that suspensions of sponge cells obtained by passing pieces of sponge tissue through bolting cloth (*mechanical dissociation*) would reaggregate and, in some cases, reconstitute a complete, functional sponge (43). Reaggregation was subsequently found to be species-specific, and to require divalent cations (9, 12, 36, 52). Humphreys and Moscona extended these studies by devis-

* Present address: Albert Einstein Medical Center, Northern Division, York and Tabor Roads, Philadelphia, Pa. 19141

** Present address: Department of Biology, Wesleyan University, Middletown, Conn. 06457

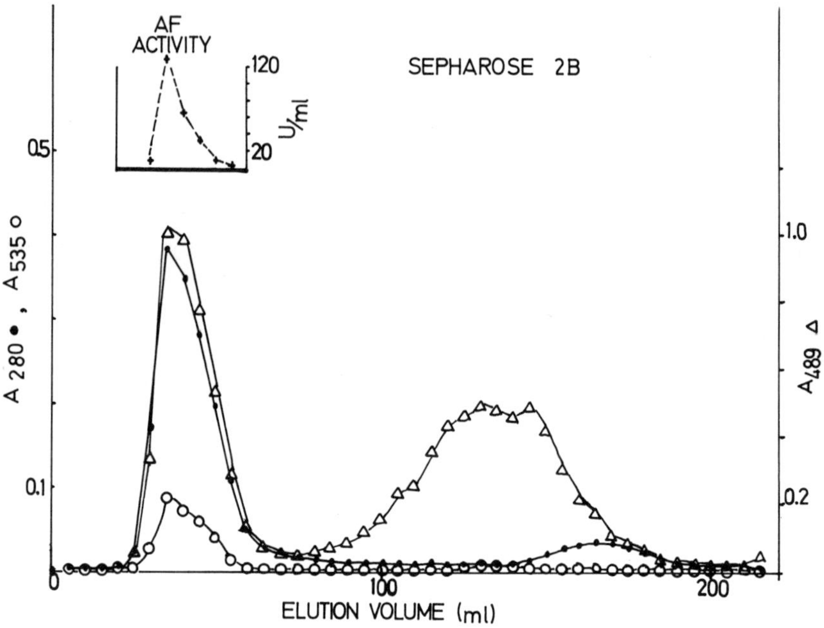

Fig. 1. Sepharose 2B chromatography of *M. prolifera* AF. *M. prolifera* AF (5 ml) concentrated by differential centrifugation as described by Henkart *et al.* (19) were applied to a 250 ml column of Sepharose 2B and eluted with CMF-SW (2 mM $CaCl_2$). Fractions were analyzed for protein (A_{280}), sugar (A_{489} ref. 11), uronic acids (A_{535}, ref. 10), and aggregation-promoting activity (19).

ing improved techniques for dissociating sponge cells in the absence of divalent cations (*chemical dissociation*), employing rotation culture methods to minimize the contribution of cell migration to aggregate formation, and developing a semiquantitative assay for cellular reaggregation (22, 32). It was shown that species-specific cellular reaggregation is promoted by macro-molecular factors (AF's) released from the homotypic cell surface during chemical dissociation. Suspensions of mechanically dissociated cells (still containing AF on their surface) would reaggregate spontaneously in seawater, whereas reaggregation of chemically dissociated cells (washed free of AF) required addition of both the homotypic AF and divalent cations (22).

In initial studies, AF's isolated from several sponge species were found to be glycoproteins with considerable interspecific variation in carbohydrate composition and sugar protein ratios (23, 28, 29). More recently a detailed study by Henkart, Cauldwell and Humphreys (6, 19) showed the AF from *Microciona parthena* to be a large, heterodisperse, protein-polysaccharide complex with an estimated molecular weight of 2 x 10^7 and a protein:hexose:uronic acid ratio of 49:46:6. Electron microscopy of the purified AF revealed fibrous complexes arranged in "sunburst" configurations, each consisting of an 800Å circle with 11-15 radiating "arms" (19). Treatment of the M. parthena AF with EDTA caused immediate, irreversible loss of activity and gradual dissociation of the complex into smaller units, suggesting that the long-recognized importance of Ca^{++} and Mg^{++} ions in sponge cell reaggregation might reflect cation requirements for conformational and chemical stability of the AF molecule (6).

To purify the AF isolated from Microciona prolifera, we have used the methods described by Henkart and Humphreys for purification of the M. parthena AF (19). Figure 1 shows a representative Sepharose 2B chromatographic separation of concentrated M. prolifera AF. All of the aggregation-promoting activity eluted as a single peak near the excluded volume. Based on colorimetric determinations, the active fractions contained protein and carbohydrate at a ratio of approximately 1:1, with 10-15% of total sugar present as uronic acid (data not shown). The purified AF banded sharply in $CsCl_2$ gradients at a buoyant density of 1.56.

The apparent physical and chemical similarities between the AF's from M. parthena and M. prolifera are not surprising, since species-specific recognition does not occur between these two sponges[1]. Although both AFs bear some structural and chemical resemblance to proteoglycan complexes isolated from vertebrate cartilage (6, 17), the detailed arrangements of protein and polysaccharide subunits within the AF has not been determined, and is of obvious importance in understanding the nature of AF-cell interactions.

Involvement of Carbohydrate in the Reaggregation of Microciona Prolifera Cells

We have examined the possible role of carbohydrate in AF-promoted cell reaggregation. A priori, such a possibility is reasonable in light of the high carbohydrate content and periodate sensitivity of the AF molecule (32, 38), and the wide variety of recognition phenomena (e.g., lymphocyte homing (13), platelet aggregation (3, 24), mating reactions in yeast and Chlamydomonas (7, 41), aggregation of slime-mold amoebae (36) and certain virus-host interactions (21)), in which carbohydrate specificity has now also been implicated.

A number of carbohydrates commonly found in glycoproteins were tested for their ability to inhibit the Af-promoted aggregation of

---

[1] Humphreys, T., personal communication.

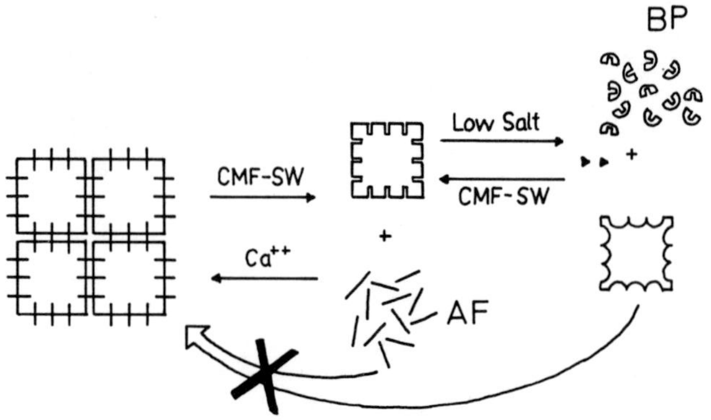

Fig. 2. Requirement of a membrane-bound baseplate for AF-promoted reaggregation of sponge cells. Sponge cells dissociated in Ca^{++}/ Mg^{++} free seawater (CMF-SW) release aggregation factor (AF). Both AF and Ca^{++} must be added for reaggregation. Treatment of AF-free cells with low salt releases a receptor or baseplate (BP) which interacts with the AF. Hypotonically treated cells lacking BP do not aggregate in response to AF, but can be restored to do so by preincubation of cells with BP. (From Burger, ref. 4).

M. prolifera cells. Only glucuronic acid proved to be an inhibitor of aggregation when incubated with chemically dissociated cells followed by addition of AF (5, 38). When the order of addition was reversed, i.e. if the carbohydrate hapten was added first to the M. prolifera factor, inhibition was generally weak or absent. Furthermore, stronger inhibition was observed using cellobiuronic acid, a disaccharide of glucuronic acid. Chelation of essential Ca^{++} ions by glucuronic acid seemed an unlikely explanation, since (a) galaturonic acid (an even stronger calcium chelator than glucuronic) had no inhibitory effect, and (b) glucuronic inhibition could not be reversed by addition of excess Ca^{++} (38). The specificity of inhibition of AF-cell interaction by glucuronic acid is consistent with the substantial amount of uronic acids present in the AF molecule (Figure 1), and with the inability of glucuronic acid to inhibit aggregation of a second species (Cliona celata) (38). Further evidence that carbohydrates present on the M. prolifera AF are involved in AF-cell interations is summarized below:

1. Studies by Kuhns indicate that treatment of M. prolifera AF with plant lectins specific for several different sugars causes inhibition of aggregation (26, 39). These results suggest that particular carbohydrates, to which the lectins bind (or others in their vicinity), are involved in the aggregation process. The possible role of glucuronic residues on the AF molecule could not be investigated directly by this technique, since no lectin specific for uronic acids has yet been found.

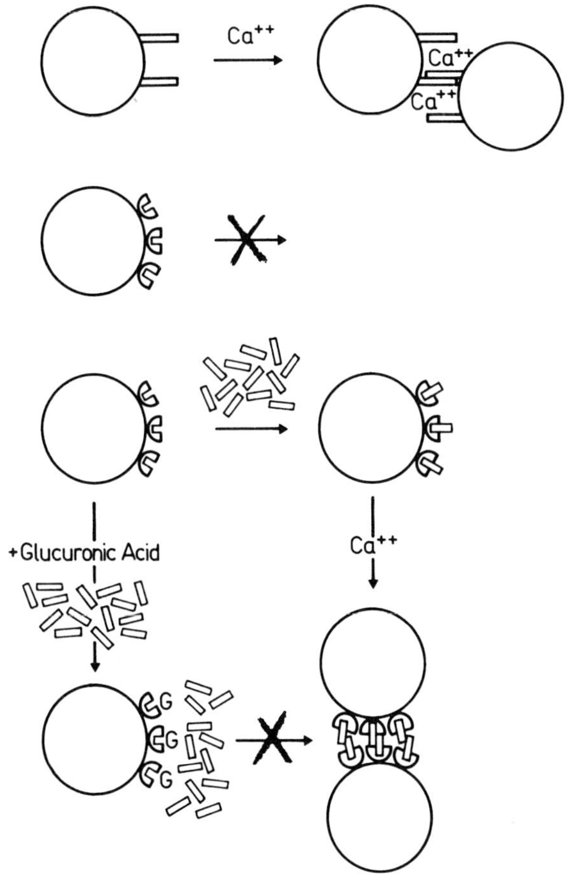

Fig. 3. Aggregation of agarose beads conjugated with baseplate or aggregation factor. On the first line, beads conjugated with AF are shown to aggregate in the presence of Ca^{++}. This aggregation could not be prevented by glucuronic acid. On the second line, beads conjugated with baseplate (shown as grooved, kidney-shaped receptors) do not aggregate in the presence of calcium. If, however, aggregation factor (bars) is presented and Ca^{++} is added, aggregation occurs (lines 3 and 4, right). Such aggregation could be inhibited by glucuronic acid (lines 3 and 4, left). (From Weinbaum and Burger, ref. 40).

2. Incubation of M. prolifera AF with a crude mixture of glycosidases from Helix pomatia completely destroyed factor activity (38). This could be prevented by including high concentrations of glucuronic acid (but not galacturonic acid or glucose) in the incubation mixture. Failure of purified β-glucuronidase to inactivate AF (6) suggests (among other possibilities) that

enzymatic degradation of active glucuronic sites may require elimination of other terminal or neighboring saccharides.

Identification of a Surface Component Involved in AF-cell Interaction

The evidence described above suggested a mechanism in which a cell surface component (not removed during chemical dissociation) is capable of specific recognition of uronic groups on the M. prolifera AF molecule. We subsequently found that hypotonic swelling of chemically dissociated cells caused the release of surface material having the predicted properties of such a "receptor" (40). Hypotonically treated cells would no longer aggregate in response to AF and divalent cations, but could be restored to do so by preincubation with the isolated surface material. Furthermore, the crude "receptor" preparation (termed baseplate) was found to inactivate AF and to inhibit the aggregation of mechanically dissociated cells. Addition of baseplate to chemically dissociated cells, however, had no inhibitory effect on subsequent aggregation in the presence of AF (40). A schematic summary of these findings is shown in Fig. 2.

To demonstrate direct interaction between the baseplate and AF, the proteins released by hypotonic treatment were covalently coupled to agarose beads (40). Such beads did not aggregate in the presence of divalent cations alone, but were rapidly aggregated by addition of AF, or by incubation with AF-coupled beads. As predicted from the sugar inhibition studies with intact cells, the AF-promoted aggregation of baseplate-beads could be inhibited by preincubation of the beads with glucuronic acid. In addition, the baseplate-beads were capable of binding to live, mechanically dissociated cells, but not to chemically dissociated cells lacking AF on their surfaces. These results (summarized diagrammatically in Fig. 3) clearly indicated that one or more receptor-like components present in the crude surface extract are capable of direct interaction with AF.

Partial Purification and Characterization of the Baseplate Component

To monitor the purification of baseplate from crude surface extracts, two kinds of assays have been utilized. The first is purely qualitative, and involves the ability of baseplate to reconstitute the aggregation of hypotonically treated cells in response to AF (40). We have developed a second, semiquantitative assay for the inhibition of AF-promoted cellular reaggregation by preincubation of AF with baseplate. The inhibition assay is based on the endpoint technique described by Humphreys for titration of AF activity (19, 22). Serial dilutions of baseplate are incubated with a known amount of AF for

TABLE I

PROPERTIES OF GLUTARALDEHYDE FIXED VS. LIVE CHEMICALLY DISSOCIATED CELLS

	Live CD-cells	Gluteraldehyde-fixed CD-cells
AF-aggregation (end-point titer)	+	+
Inhibition of aggregation by glucuronic acid	+	+
Inhibition of aggregation by pretreatment of AF with baseplate	+	+
Morphology of aggregates	compact round	loose irregular
Sensitivity to hypotonic treatment	+	-

Chemically dissociated *M. prolifera* cells (CD-cells) were washed for 6 hours in CMF-SW at 10^7 cells/ml, then fixed for 12 hrs at 4° in CMF-SW containing 1% glutaraldehyde (pH 7.2). Fixed cells were washed for 30 min in CMF-SW, 0.1 M glycine, then washed 5 times in 20 volumes of CMF-SW. Cells were stored at 4° in CMF-SW, 0.05% NaN_3, 0.001% glutaraldehyde.

10 minutes, after which cells are added and reaggregation allowed to proceed on a rotary shaker for 20 minutes at room temperature. The dilution of baseplate showing marginal inhibition of aggregation multiplied by the AF concentration (19) used in the assay is taken as the relative titer of inhibition.

Because *M. prolifera* sponges collected from the Woods Hole area have shown seasonal and regional variability in the yield of AF per cell and the responsiveness of cells to given concentrations of AF, we have employed glutaraldehyde-fixed, chemically dissociated cells

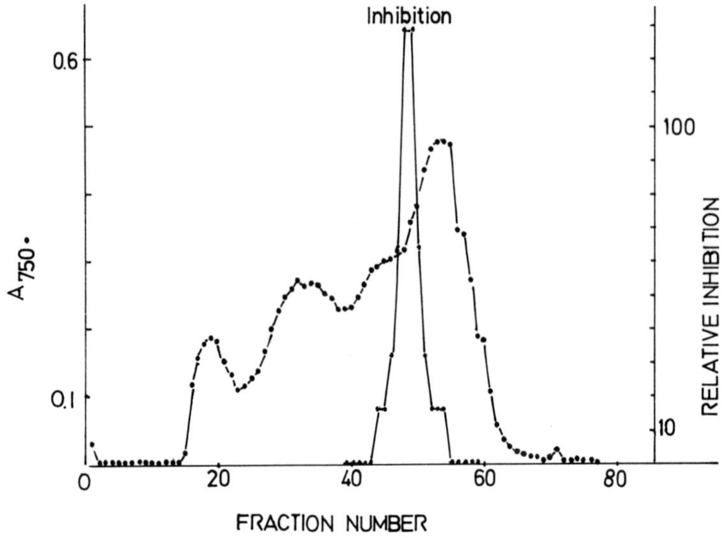

Fig. 4. Sephadex G-100 chromatography of baseplate-containing material released by hypotonic swelling. 5 ml of baseplate were clarified by centrifugation at 105,000 x g, dialyzed against CMF-SW and applied to a 275 ml column of Sephadex G-200. Fractions eluted with CMF-SW (2 mM Ca^{++}) were assayed for protein (A_{750}, ref. 27) and inhibitory activity.

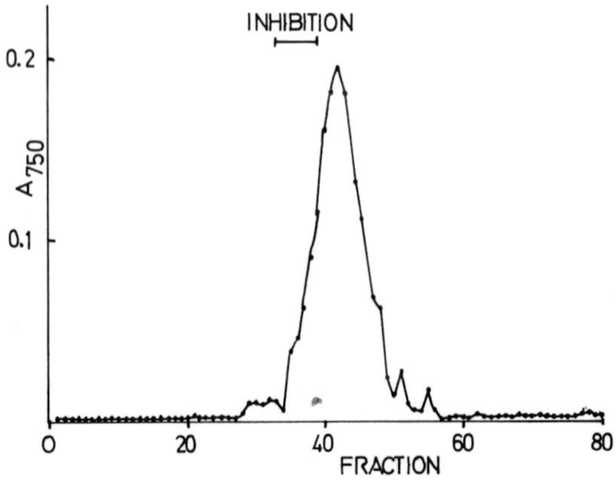

Fig. 5. Sephadex G-100 chromatography of inhibitory fractions from Sephadex G-200 column. Pooled inhibitory fractions from Sephadex G-200 chromatography (Fig. 4) were concentrated to 1 ml by lyophilization, applied to a 100 ml Sephadex G-100 column and eluted with CMF-SW (2 mM Ca^{++}). Column fractions were assayed for protein (A_{750}, ref. 27) and inhibitory activity.

for the inhibition assays (33). A range of glutaraldehyde concentrations was tested, and 1-2% was found to be optimal for preservation of chemically dissociated cells. A comparison of the relevant properties of fixed vs live cells is summarized in Table I. The glutaraldehyde-fixed cells were unaffected by 6 months storage at 4° in seawater containing 0.05% NaN_3 and 0.001% glutaraldehyde.

The optimal hypotonic treatment for release of baseplate has varied considerably (0.1 M - 0.3 M NaCl) from season to season. This result is not entirely unexpected in view of the fact that AF activity and responsiveness to AF also varied seasonally. Thus, it is important that the minimum, non-lytic NaCl concentration be determined for each preparation of sponge cells. The degree of cell lysis is monitored by (a) counting surviving cells after hypotonic shock, (b) observing the release of pigment into the extracellular medium, and (c) staining by Trypan Blue. Acceptable preparations of baseplate and hypotonically treated cells are those in which more than 90% of the original cells are left intact based on the above criteria.

Our initial studies indicated the baseplate component to be non-dialyzable and soluble after centrifugation at 105,000 x g for 90 minutes. As judged by the inhibition assay, the baseplate activity is sensitive to heat inactivation at 60° for 10 min. Unlike AF, baseplate appeared stable to EDTA (5 mM) repeated freezing and thawing, lyophilization, and a pH range of 3-12. Partial purification of the base plate was achieved by differential centrifugation and molecular sieve chromatography. The inhibitory activity appeared as a single, symmetrical peak during chromatographic separation on Sephadex G-200 (Figure 4). When the pooled inhibitory fractions from the G-200 column were compared with preceding fractions for the ability to restore aggregation to fresh, hypotonically treated cells, only the inhibitory fractions were active in reconstitution, suggesting that both activities reside in the same component.

As shown in Figure 5, further purification of baseplate was obtained by Sephadex G-100 chromatography on the active material from the G-200 column. The baseplate component eluted in a discreet zone at the leading edge of a broad protein peak. Based on calibration of the G-100 column with several protein standards, the molecular weight of the baseplate was estimated to be in the range of 45-60,000.

Table II summarizes our progress toward baseplate purification. In addition, preliminary experiments indicate that baseplate binds strongly to DEAE-Sephadex (pH 6.7) and to glutaraldehyde-fixed, mechanically dissociated cells (containing AF on their surfaces). We are therefore optimistic that considerably greater purification can be obtained by a combination of ion exchange and affinity-adsorption techniques. Purification of the baseplate component should

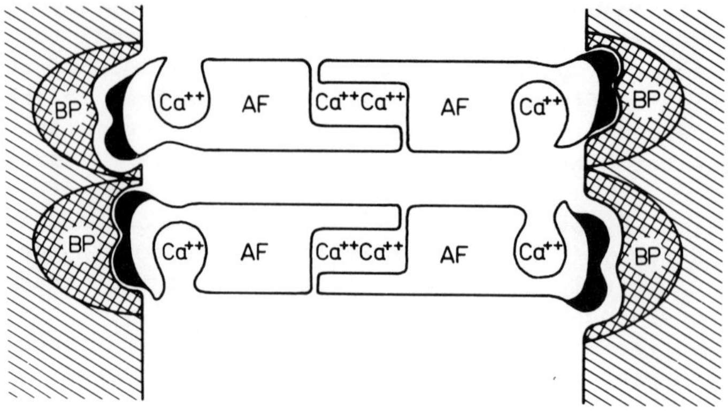

Fig. 6 A

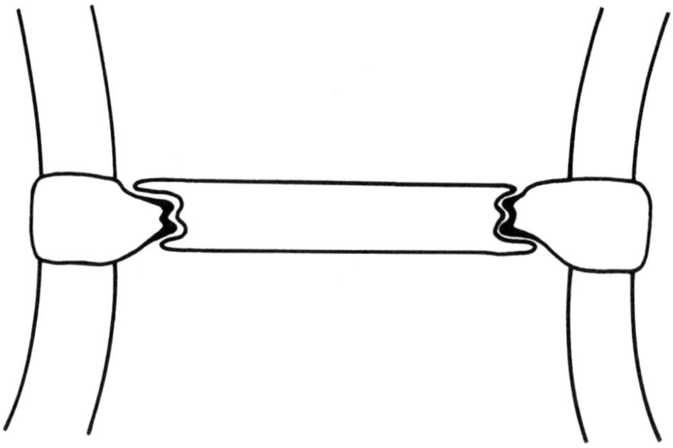

Fig. 6 B

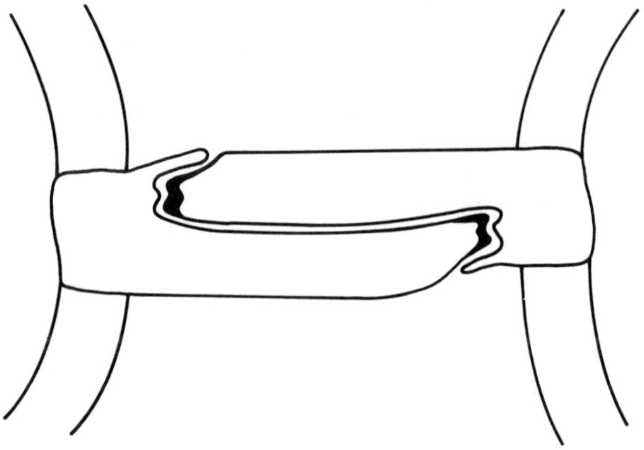

Fig. 6 C

Fig. 6 A, B, C. Three possible molecular models for surface specific cell-cell recognition.

(A) Tentative model for cell-cell recognition in Microciona prolifera. Two macromolecular aggregation factors (AF) are illustrated each consisting of at least two subunits. The black termini at each pole carry the carbohydrates that are recognized by the baseplate (BP) anchored in adjoining cell surfaces. The calcium which keeps the subunits together in this model is required for function of the aggregation factor and is removed by EDTA or EGTA. The calcium at the periphery of the aggregation factor is removed by calcium-magnesium free seawater and is meant to help in stabilizing the aggregation factor at the cell surface. It does not have to be bound to the aggregation factor though. It can fulfill such a stabilization function also by being bound to other molecules in the neighborhood of the aggregation factor (from ref. 40).

(B) Inversion of the polarity: the recognizing component lies between the cells and the antigenic or recognized component in the surface.

(C) Recognizing and recognized component are in the same unit. They are both anchored in close contact or possibly the same cell membrane and not easily extractable.

---

allow (A) a detailed study of AF-baseplate interaction, (B) the development of specific probes (e.g. antibodies) to localize baseplate on the cell types, and (C) the possibility of detecting additional surface components involved in sponge cell adhesion.

Models for Cell-Cell Recognition via Surface Components

Our present working model for the reaggregation of M. prolifera cells is depicted in Figure 6. The adhesion mechanism contains a minimum of three components (AF, baseplate, and divalent cations), and involves AF-AF as well as AF-receptor interactions. Clearly, this is only one of several possible mechanisms for cell-cell recognition, and two alternative mechanisms are suggested in Figure 6B and 6C. It is uncertain whether the model we have proposed for cell recognition in Microciona prolifera is applicable to other sponge systems or to higher metazoans. Due to their relative simplicity, sponge systems are well suited for biochemical studies, and will hopefully yield additional insights and experimental approaches for the study of cellular recognition in higher organisms.

TABLE II

PARTIAL PURIFICATION OF BASEPLATE FROM MICROCIONA PROLIFERA

Treatment	Specific Activity (Relative Inhibitory units per µg protein)	Ability to restore aggregation to Hy-CD cells (Reconstitution)
CD-cells		
↓ 0.15 M NaCl, 15 min. 22°C, 600 x G, 10 min.		
Supernatant	-	-
↓ 12,000 x G, 15 min.		
Crude baseplate	1.5	+
↓ 105,000 x G, 90 min.		
Supernatant	2.2	+
↓ G 200 Sephadex		
Pooled inhibitory fractions (G 200)	6.4	+
↓ G 100 Sephadex		
Pooled inhibitory fractions	16.0	N.D.

Samples taken at each stage of purification of a crude baseplate preparation were assayed for protein (27), inhibitory activity, and ability to reconstitute AF-promoted aggregation of hypotonically treated cells (see text for details). N.D. = not determined. Hy-CD = hypotonically treated, chemically dissociated cells.

ACKNOWLEDGEMENT

This work was supported by the Swiss National Foundation, Grant No. 3.1330.73.

REFERENCES

1. BALSAMO, J. and LILIEN, J., Biochemistry 14 (1975) 167.
2. BEUG, H., GERISCH, G., KEMPFF, S., RIEDEL, V. and CREMER, G., Expl. Cell Res. 63 (1970) 147.
3. BOSMANN, H. B., Biochem. Biophys. Res. Comm. 43 (1971) 1118.
4. BURGER, M. M., The Neurosciences Third Study Program, (F. O. Schmitt, ed.) MIT Press, Cambridge, (1973) 773.
5. BURGER, M. M., LEMON, S. M., and RADIUS, R., Biol. Bull. 141 (1971) 380.
6. CAULDWELL, C. B., HENKART, P. and HUMPHREYS, T., Biochemistry 12 (1973) 3051.
7. CRANDALL, M. A. and BROCK, T. D., Science 161 (1968) 473.
8. CRICK, F. H. C., Nature 224 (1970) 420.
9. DE LAUBENFELS, M. V., J. Elisha Mitchell Sci. Soc. 44 (1928) 82.
10. DISCHE, Z., J. Biol. Chem. 167 (1947) 189.
11. DUBOIS, M., GILLES, K. A., HAMILTON, J. K., REBERS, P. A. and SMITH, F., Anal. Chem. 28 (1956) 350.
12. GALTSOFF, P. S., J. Exp. Zool. 42 (1925) 223.
13. GESNER, B. M. and GINSBURG, V., Proc. Nat. Acad. Sci. USA 52 (1964) 750.
14. GIERE, A., BERKING, S., BODE, J., DAVID, C. N., FLICK, K., HANSMANN, G., SCHALLER, H., and TRENKNER, E., Nature New Biol. 239 (1972) 98.
15. GIUDICE, G., Devel. Biol. 5 (1962) 402.
16. GOTTLIEB, D. I., MERRELL, R. and GLASER, L., Proc. Nat. Acad. Sci. USA 71 (1974) 1800.
17. HASCALL, V. C. and SAJADERA, S. W., J. Biol. Chem. 244 (1969) 2384.
18. HAUSMAN, R. E. and MOSCONA, A. A., Proc. Nat. Acad. Sci USA 72 (1975) 916.
19. HENKART, P., HUMPHREYS, S. and HUMPHREYS, T., Biochemistry 12 (1973) 3045.
20. HORIKAWA, M. and FOX, A. S., Science 145 (1964) 1437.
21. HUGHES, R. C., Progr. Biophys. Molec. Biol. 26 (1973) 191.
22. HUMPHREYS, T., Dev. Biol. 8 (1963) 27.
23. HUMPHREYS, T., Symp. Zool. Soc. Found. 25 (1969) 325.
24. JAMIESON, G. A., URBAN, C. L., and BARBER, A. J., Nature New Biol. 234 (1971) 5.
25. KONDO, K. and SAKAI, H., Devel. Growth Diff. 13 (1971) 1.
26. KUHNS, W. J. and BURGER, M. M., Biol. Bull. 141 (1971) 393.
27. LOWRY, D. H., ROSEBROUGH, N. J., FARR, A. L., and RANDALL, R. J., J. Biol. Chem. 193 (1951) 265.

28. MAC LENNAN, A. P., Symp. Zool. Soc. London 25 (1969) 279.
29. MARGOLIASH, E., SCHENCK, J. R., HARGIE, M. P., BUROKAS, S., RICHTER, W. R., BARLOW, G. H. and MOSCONA, A. A., Biochem. Biophys. Res. Commun. 20 (1965) 383.
30. MERRELL, R. and GLASER, L., Proc. Nat. Acad. Sci. USA 70 (1973) 2794.
31. MOSCONA, A. A., J. Cell. Comp. Physiol. Suppl. 1 60 (1962) 65.
32. MOSCONA, A. A., Proc. Nat. Acad. Sci. 49 (1963) 742.
33. MOSCONA, A. A., Dev. Biol. 18 (1968) 250.
34. MOSCONA, A. and MOSCONA, H., J. Anat. 86 (1952) 287.
35. ROSEN, S. D., KAFKA, J. A., SIMPSON, D. L. and BARONDES, S. H., Proc. Nat. Acad. Sci. USA 70 (1973) 2554.
36. SPIEGEL, M., Biol. Bull. 107 (1954) 130.
37. STEINBERG, M. S., J. Exp. Zool. 173 (1970) 395.
38. TURNER, R. and BURGER, M. M., Nature, 244 (1973) 509.
39. TURNER, R. S., WEINBAUM, G., KUHNS, W. J. and BURGER, M. M., Arch. Biol. 85 (1974) 35.
40. WEINBAUM, G. and BURGER, M. M., Nature 244 (1973) 510.
41. WIESE, L. and SHOEMAKER, D. W., Biol. Bull. 138 (1970) 88.
42. WILSON, H. V., J. Exp. Zool. 5 (1907) 245.
43. WILSON, H. V., Bull. Bur. Fisheries 30 (1910) 1.
44. WILSON, H. V., J. Exp. Zool. 11 (1911) 281.

ISOLATION OF A *HANSENULA WINGEI* MUTANT WITH AN ALTERED SEXUAL AGGLUTININ

Venancio SING, Yar-Fen YEH and Clinton E. BALLOU

Department of Biochemistry, University of California
Berkeley, California 94720 (USA)

Haploid cells of opposite mating type of the yeast *Hansenula wingei* possess complementary cell-surface factors that mediate a specific sexual agglutination reaction (4), the function of which appears to be that of enhancing the efficiency of mating. Regulation of the synthesis of these two factors is tightly linked to the mating type locus.

The factor on the mating type 5-cell can be released by subtilisin digestion, and it is a multivalent glycoprotein (more accurately a manno-protein) with the ability to agglutinate 21-cells. It is called 5-agglutinin. The 21-factor, which is released from 21-cells by trypsin digestion, is a univalent manno-protein that inhibits the agglutination of 21-cells by 5-agglutinin.

The 5-agglutinin prepared by subtilisin digestion is a heat-stable molecule of about 10^6 molecular weight that possesses 6 binding sites, of 12,000 molecular weight each, attached by disulfide linkage to the central core of about 9×10^5 molecular weight (11, 13). 5-Agglutinin is 85% mannose, 5% phosphate and 10% protein. The protein is of unusual composition, containing 55% serine and 9% threonine. Most of the mannose is released from 5-agglutinin by β-elimination (0.1 N NaOH at 40° for 24 hrs), and can be recovered as an assortment of oligosaccharides from mono- to pentadecasaccharide. The 5-agglutinin is reversibly inactivated by dithiothreitol, but the activity is irreversibly destroyed by digestion with a bacterial exo-α-mannanase (13) and pronase (4). The inactivation by these enzymes is prevented by adding an appropriate competing substrate (15).

The central core recovered after reduction of 5-agglutinin is enriched in serine, whereas the binding fragments have a reduced serine content and are enriched in threonine. The binding fragment possesses about 28 moles of 11 different amino acids, as well as a large amount of mannose. The model for 5-agglutinin, first presented by Taylor and Orton (11), consists of a large central core with 6 small binding sites attached in disulfide linkage. 21-Factor is only poorly characterized, but it is heat labile, it has a molecular weight of about 30,000, and it appears to contain 35% carbohydrate and 65% protein (4).

The sexual agglutination factors are only a minor component of the bulk cell wall manno-protein of H. wingei, and the latter material differs fundamentally in structure from the sex factors. In contrast to 5-agglutinin in which all of the mannose is linked to hydroxyamino acids, most of the mannose in the bulk manno-protein is attached to protein as alkali-stable polymannose chains, presumably linked to asparagine by way of a chitobiose unit as in Saccharomyces cerevisiae mannan (14). The structural relatedness between the carbohydrate components of the agglutinin and the bulk cell-wall manno-protein is demonstrated by the identity of the carbohydrate fragments produced by partial acetolysis and by the observed immunochemical cross-reactivity (15).

Yeast manno-proteins are known to be the principal immunogens on the cell surface (6), and in several species the carbohydrate fragments produced by partial acetolysis of the mannan give complete inhibition of the precipitin reaction between isolated manno-protein and the homologous antiserum (1). Hansenula wingei mannan is different in that the acetolysis fragments give only about 30% inhibition, whereas some unidentified acid-labile determinant makes up the remainder of the immunochemical reaction (14). On the assumption that this acid-labile determinant is related to the phosphate in the manno-protein, perhaps involving a phosphodiester bond, mutants of H. wingei 5-cells have been selected that lack the acid-labile determinant with the aim of obtaining a mutant that lacks phosphate in its manno-protein.[1] Such a mutant has been obtained, and as we demonstrate here it makes phosphate-deficient sexual agglutinin as well as phosphate-deficient bulk cell wall mannan.

EXPERIMENTAL PROCEDURES

Materials--Hansenula wingei Y-2340 diploid and the haploid mating types 5 and 21 are the same strains used previously (4). The yeasts were grown on a medium containing 0.7% yeast extract, 3% D-glucose and 0.5% KH_2PO_4, with 2% agar added for plates and slants. Whatman #1 powdered cellulose, Bio-Gel A5m and A50m, DEAE-cellulose, Dextran

[1] Yeh, Y.-F., Master's thesis (unpublished), Univ. of California, Berkeley, CA (1974).

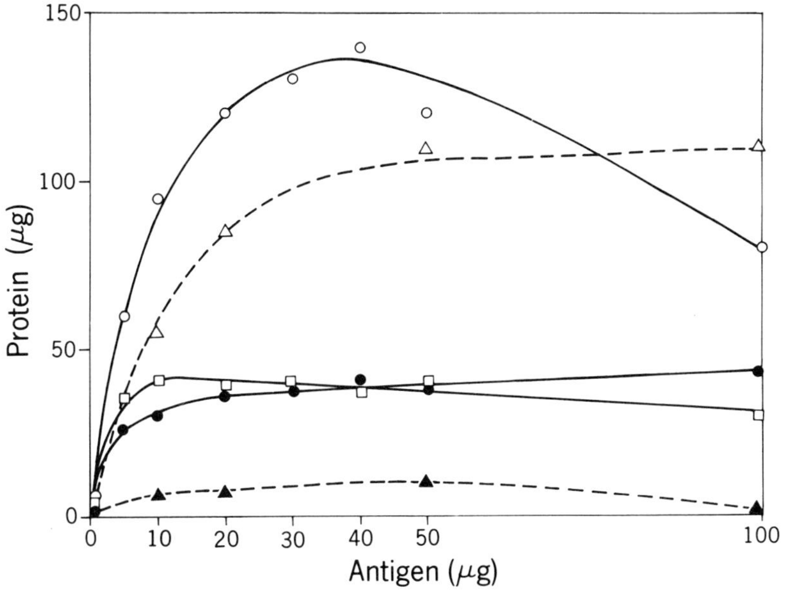

Fig. 1. Precipitin curves of wild-type and mutant manno-proteins with antiserum against 5-cells (solid lines) and with antiserum specific for the acid-labile determinant (dashed lines). The curves are for whole wild-type 5-cell mannan (O—O), whole 5-cell mannan heated in 0.1 N HCl at 100° for 30 min (●—●), whole 5-40-cell mannan (□—□), wild type 5-agglutinin (Δ—Δ), and mutant 5-40-agglutinin (▲—▲).

T2000, and other reagents were from commercial sources.

Mannan characterization. Yeast mannan was isolated by extraction of cells with hot citrate buffer followed by ethanol precipitation, and it was purified by DEAE-cellulose chromatography (14). Digestions with bacterial exo-α-mannanase (7), β-mannosidase (12) and with pronase, and the determination of mannose-phosphate ratios, products of β-elimination, acetolysis patterns and immuno-chemical properties followed published procedures (14). The binding of alcian blue dye by intact cells was used as a qualitative test for the presence of phosphate in the cell wall (5).

Mutant isolation. The procedure of Raschke et al. (9) was used. A suspension of H. wingei 5-cells was mutagenized with ethyl methane sulfonate, the mutant cells were allowed to multiply in liquid medium, and the unchanged cells in the culture were agglutinated by antiserum specific for the acid-labile determinant of the wild-type strain (14) prepared by adsorption of H. wingei 21-cell antiserum with S. cerevisiae X2180 cells. The remaining suspension of single cells was plated out and the isolated clones that grew up were tested for agglutinability with the same antiserum. Cells that failed to agglutinate with antiserum, but that retained the sexual agglutination property, were then cultured in quantity sufficient for isolation of the cell wall mannan and determination of its phosphate content and acetolysis pattern.

Isolation of 5-agglutinin. An affinity column was prepared by packing a dry mixture of 25 g of heat-treated (100°, 30 min) 21-cells and 150 g of Whatman #1 cellulose powder in a 4.5 x 35 cm glass column. The column was washed thoroughly with water and buffer, after which the subtilisin digest of 100 g of wet yeast cells (15) was applied in a buffer at pH 4. The column was washed extensively with this buffer, and the 5-agglutinin was then eluted by shifting to a pH 1.8 buffer. Assay of the 21-cell agglutination activity was performed in a Microtiter plate (Cooke Engineering Co.) on serial dilutions of the agglutinin solution in 10 mM KH_2PO_4, pH 5.5, containing 20 mM $MgSO_4$ (15).

RESULTS AND DISCUSSION

Following mutagenesis with ethyl methane sulfonate and removal of unchanged wild-type cells with specific rabbit antiserum directed against an acid-labile surface determinants, we obtained a mutant of Hansenula wingei type 5-cell (designated strain 5-40) that has an altered manno-protein structure but that retains full sexual agglutination activity. The bulk cell wall mannan of the mutant gives an acetolysis pattern indistinguishable from the wild-type strain, but it contains no detectable phosphate and it lacks a major immunochemical determinant that in the wild-type mannan is characteristically acid-labile (Fig. 1). We conclude that this immunochemical determinant must be associated with a mannosylphosphate linkage, since that is the most probable phosphate structure with the observed acid lability. Saccharomyces cerevisiae X2180, a strain similar to baker's yeast, synthesizes mannan with mannosylphosphate and mannobiosylphosphate groups attached to the mannan side chains (2), and the mannosylphosphate groups are important immunochemical determinants in this mannan as demonstrated by the potent hapten activity of α-D-mannose 1-phosphate. However, the acid-labile phosphate-containing determinant in H. wingei Y-2340 is clearly different from that in S. cerevisiae and is yet to be defined (14). In agreement with the absence of the acid-labile determinant, the precipitin reaction between the isolated 5-40 mannan and the antiserum against wild-type cells was completely inhibited by the acetolysis oligosaccharides.

H. wingei wild-type cells of both mating types bind the basic dye alcian blue owing to the acidic charge resulting from phosphate

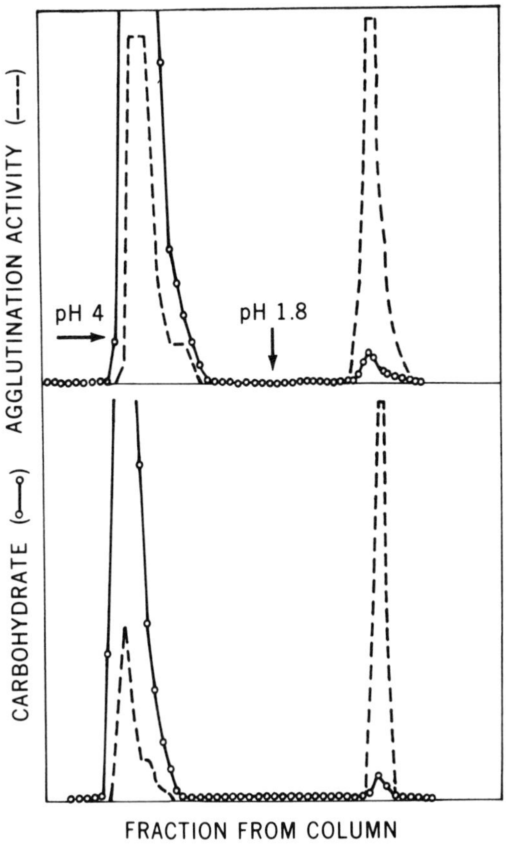

Fig. 2. Isolation of mutant 5-40-agglutinin on an affinity column of 21-cells. The subtilisin digest of 5-40-cells was passed through the column at pH 4 and eluted at pH 1.8. Excess agglutinin that failed to bind to the column in the first passage, did bind when the material was passed a second time through the regenerated column.

in the mannan, whereas the 5-40 mutant strain binds not at all. Because the diploid obtained by mating 21-cells with 5-40-cells also binds dye, the mutation in strain 5-40 is recessive. Interestingly, a dominant mutant of S. cerevisiae has been isolated with the same nonbinding phenotype (3). Thus, the latter mutation is probably regulatory in nature whereas the former one may involve the structural gene for a presumed mannosylphosphate transferase (8).

Subtilisin digestion of 5-40 cells releases a soluble agglutinin with most of the properties of the wild-type 5-agglutinin. The 5-40-agglutinin was isolated by adsorption to an affinity column of 21-cells at pH 4 followed by elution at pH 1.8 (Fig. 2). This material was then purified by passage through a Bio-Gel A5m column

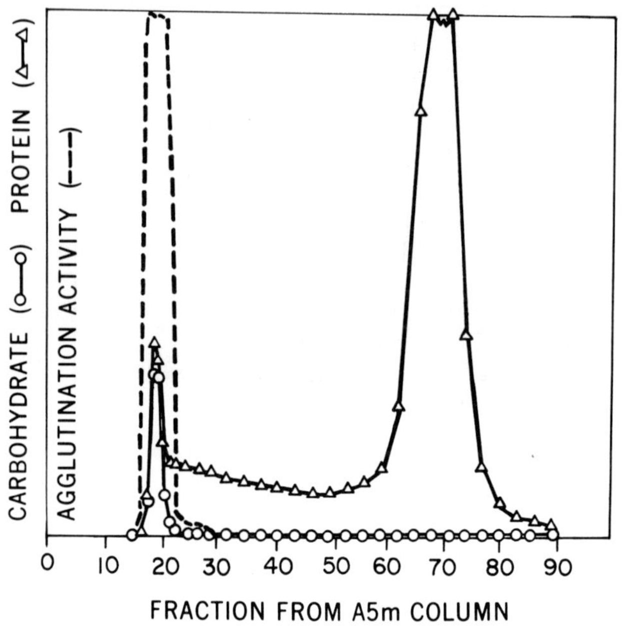

Fig. 3. Purification of mutant 5-40 agglutinin on a Bio-Gel A5m column. The agglutinin is eluted near the void volume and is separated from a protein contaminant.

(Fig. 3), a step in which it appears to be excluded. The purification is summarized in Table I.

The mutant 5-40-agglutinin was isolated in a yield of about 6 mg per 100 g of wet cells. Because of the nature of the isolation procedure its homogeneity is not certain, but the biological purity is high as indicated by the fact that the material in the preparation is quantitatively adsorbed by 21-cells. Like the bulk cell wall mannan, the 5-40-agglutinin lacks bound phosphate (Table II), but its carbohydrate and protein compositions are similar to the wild-type 5-agglutinin. Its precipitin reaction with wild-type H. wingei 5-cell antiserum is completely inhibited by the acetolysis oligosaccharides, and it lacks the acid-labile determinant (Fig. 1).

Although wild-type 5-agglutinin is excluded on Bio-Gel A5m, it was shown by sedimentation studies to have a molecular weight of about 10^6 (11, 15). Thus, this manno-protein has anomalous gel filtration properties that suggest it may possess a highly extended conformation. Both wild-type and mutant 5-agglutinins are included on Bio-Gel A50m, and they are eluted ahead of Dextran T2000, a polysaccharide with a light scattering molecular weight of 2 x 10^6. Thus, elimination of the phosphate in the 5-40-agglutinin does not

TABLE I
PURIFICATION OF WILD-TYPE AND MUTANT 5-AGGLUTININS

Treatment	Volume (ml)	Carbohydrate (mg)	Protein (mg)	Phosphate (mg)	Total activity (units x 10^{-4})	Specific activity (units/mg CHO)	Purification (fold)
WILD-TYPE 5-AGGLUTININ							
Subtilisin digestion	230	1,840	2,011		236	1,280	1
21-Cell column	280	7.2			143	200,000	169
A-50m column	122	5.0	0.48	0.34	123	244,000	195
MUTANT-TYPE 5-AGGLUTININ							
Subtilisin digestion	200	1,420	2,340		102	720	1
21-Cell column	213	5.5			109	198,000	284
A-50m column	122	3.2	0.32	0	62.5	194,000	273

TABLE II

COMPARISON OF WILD-TYPE AND MUTANT 5-AGGLUTININS

Property	Wild-type	Mutant
Molecular weight	10^6	a
Protein	10%	10%
Serine (mole %)	55%	
Threonine (mole %)	9%	
Carbohydrate (mannose)	85%	90%
Phosphate	5%	0
Mannose released by mannanase digestion	38%	65%
Mannose released by β-elimination	90%	

[a]Similar to wild-type agglutinin by gel filtration.

alter its molecular size appreciably, an indication that the phosphate is not involved in cross-linking large subunits of the manno-protein.

Wild-type 5-agglutinin is inactivated by digestion with pronase and with a bacterial exo-α-mannanase (Table III), the latter enzyme removing 38% of the carbohydrate from the manno-protein (15). The mutant 5-40-agglutinin has similar properties, although the exo-α-mannanase releases about 65% of the mannose, an understandable result if the bound phosphate in the wild-type agglutinin is attached to mannose and inhibits action of the enzyme. The carbohydrate still attached to protein in the mannanase-digested 5-40-agglutinin is included on a Bio-Gel A5m column and is eluted in a continuous broad band from the void volume to the position at which the mannose appears. Thus, the residue of manno-protein is very heterogeneous.

Wild-type 5-agglutinin undergoes a β-elimination reaction in mild alkali with the release of an assortment of carbohydrate fragments that range in size from mono- to pentadecasaccharides, a fraction of which is acidic owing to the presence of esterified phosphate. Under the same conditions, the 5-40-agglutinin yields an assortment of neutral mannooligosaccharides of similar size (Fig. 4). After exo-α-mannanase digestion, however, the sole product of β-elimination is mannose. Thus, all but the last mannose attached to serine and threonine can be removed by the action of this enzyme. Whether the

TABLE III

STABILITY OF DIFFERENT 5-AGGLUTININ PREPARATIONS

Treatment	Wild-type agglutinin[a]	Mutant agglutinin	Residual activity of mannanase-digested mutant agglutinin
	% of original activity		
100°, 30 min, water	98	100	100
100°, 30 min, 0.1 N HCl	50	50	50
25°, 30 min, 0.1 N NaOH	40	6	3
Pronase digestion	30	0	0
Pronase + serum albumin	95	0	0
Exo-α-mannanase digestion	20	3	
Exo-α-mannanase + mannan	90	100	

[a]Taken from reference 15.

failure to remove the last mannose is a result of steric hinderance or a difference in anomeric configuration, is still to be determined.

Sexual agglutination is a general phenomenon in heterothallic yeasts, and these systems provide convenient sources of material for investigation of the biochemical basis of the intercellular recognition and agglutination process. Moreover, formation of the agglutinins is under genetic control and their structures can be modified through selective mutation. Although *Hansenula wingei* agglutinin is produced constitutively in haploid cells, its formation is shut down in the heterozygous diploid (4). In *Saccharomyces cerevisiae*, the agglutination factors are induced as a consequence of the reciprocal action of diffusible sex pheromones (10), thus making their study somewhat more difficult. Further application of the combined genetic and biochemical approaches to analysis of the *H. wingei* sexual agglutination factors should lead to a complete chemical elucidation of the structure of the binding sites, from which we hope to be able to divine something of the fundamental nature of this specific interaction.

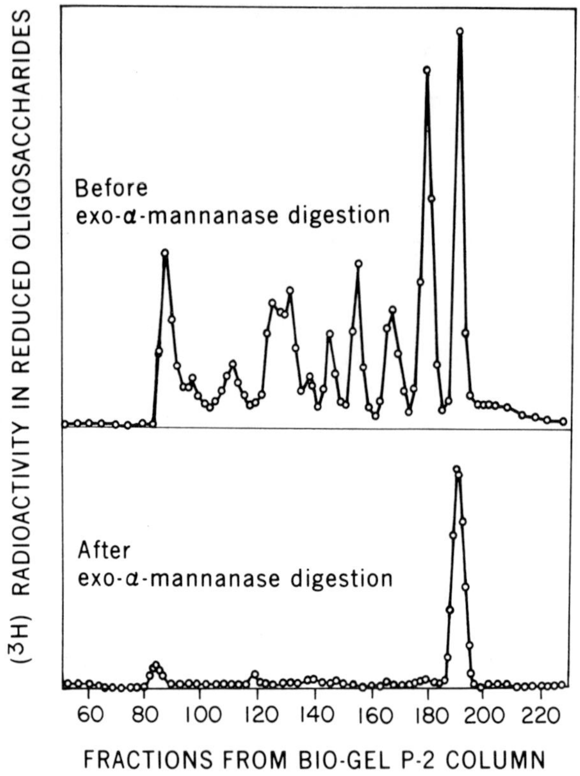

Fig. 4. Separation of the β-elimination products from 5-40-agglutinin. The $NaBT_4$-reduced oligosaccharides were fractionated on a Bio-Gel P-2 column, the top figure showing those from intact agglutinin and the bottom figure those from the agglutinin after exhaustive digestion with a bacterial exo-α-mannanase during which over two-thirds of the carbohydrate was released as mannose.

SUMMARY

A mutant of Hansenula wingei type 5-cell, designated strain 5-40, has been isolated that possesses altered surface antigenic properties but that retains the normal wild-type sexual agglutination phenotype. The change in antigenic structure is associated with a loss of the phosphate that is esterified in the wild-type cell wall manno-protein. The 5-agglutinin isolated from this mutant also lacks phosphate; and, because it retains full agglutination activity toward 21-cells, we conclude that the phosphate is not required for the agglutination reaction. Treatment of the 5-40-agglutinin with 0.01 N NaOH causes β-elimination of a series of manno-oligosaccharides from attachment to serine and threonine residues in the protein, a property also shown by the wild-type agglutinin.

Whereas a bacterial exo-α-mannanase will remove only about 40% of the mannose from wild-type 5-agglutinin, apparently owing to inhibition by the phosphate, it will remove 65% of the mannose from the mutant 5-agglutinin. When this digested 5-agglutinin is treated with 0.01 N NaOH mannose is the only sugar released, indicating that the enzyme was able to hydrolyze all but the last mannose link to the hydroxyamino acids.

ACKNOWLEDGEMENTS

We thank Dr. Y.-T. Li for a sample of purified β-mannosidase. This work was supported by NSF Grant BMS72-02208 and NIH Grant AM 884.

REFERENCES

1. BALLOU, C. E., J. Biol. Chem. 245 (1970) 1197.
2. BALLOU, C. E., Adv. Enzymol. 40 (1974) 239.
3. BALLOU, D. L., J. Bacteriol. (1975) in press.
4. CRANDALL, M. A. and BROCK, T. D., Bacteriol. Rev. 32 (1968) 139.
5. FRIIS, J. and OTTOLENGHI, P., C. R. Trav. Lab. Carlsberg 37 (1970) 327.
6. HASENCLEVER, H. F. and MITCHELL, W. O., Sabouradia 3 (1964) 288.
7. JONES, G. H. and BALLOU, C. E., J. Biol. Chem. 243 (1968) 2442.
8. KOZAK, L. P. and BRETTHAUER, R. K., Biochemistry 9 (1970) 1115.
9. RASCHKE, W. C., KERN, K. A., ANTALIS, A. and BALLOU, C. E., J. Biol. Chem. 248 (1973) 4660.
10. SENA, E. P., RADIN, D. N. and FOGEL, S., Proc. Nat. Acad. Sci. USA 70 (1973) 1373.
11. TAYLOR, N. W. and ORTON, W. L., Biochemistry 10 (1971) 2043.
12. WAN, C. C., 'URSO, M. D., MULDREY, J. E., LI, S. C., LI, Y-T., Fed. Proc. 34 (1975) 678.
13. YEN, P. H. and BALLOU, C. E., J. Biol. Chem. 248, (1973) 8316.
14. YEN, P. H. and BALLOU, C. E., Biochemistry 13 (1974) 2420.
15. YEN, P. H. and BALLOU, C. E., Biochemistry 13 (1974) 2428.

THE MATING PHENOMENON IN *ESCHERICHIA COLI*

Mark ACHTMAN

Max-Planck-Institut für molekulare Genetik
Berlin (West Germany)

Mating between *Escherichia coli* was first demonstrated in the 1940's and somewhat defined during the 1950's and early 1960's (for review see 7). These results were obtained using primarily donor cells carrying the sex factor F (see below) and recipient cells not carrying it. Under laboratory conditions matings are performed by mixing actively growing donor and recipient cultures at 37° and allowing DNA transfer to occur for one to two hours. Early observations on mating mixtures led to the general concept, still accepted today, that mating consists of pairing between donor and recipient cells in a form sensitive to mild shear forces. The results presented here indicate that this concept is inadequate and a modified definition of mating will be presented below.

The ability to act as a donor is coded by genes on a sex factor. Genetic analyses of the F sex factor (for review see 1) have resulted in the map presented in Fig. 1. One-third of the F DNA (one-third of 65 x 10^6 molecular weight) carries cistrons coding for mating-related phenomena. Donor cells produce sex pili (for review see 5). Those coded by the F sex factor are protein organelles of 8 nm diameter and up to several μ long. All the *tra* cistrons between *traA* and *traH* on Fig. 1 (8 cistrons) are necessary for F pili production. The removal of F pili from donor cells by mechanical or mutational means prevents the donor cells from binding to recipient cells. Thus F pili may be considered as donor mating receptors, which require at least 8 proteins for their synthesis. Their possible further role in DNA transfer is still ambiguous.

The normal sequel to binding of donor and recipient cells is DNA transfer from the donor to the recipient. The transfer is of single-stranded DNA starting at a specific nucleotide sequence on

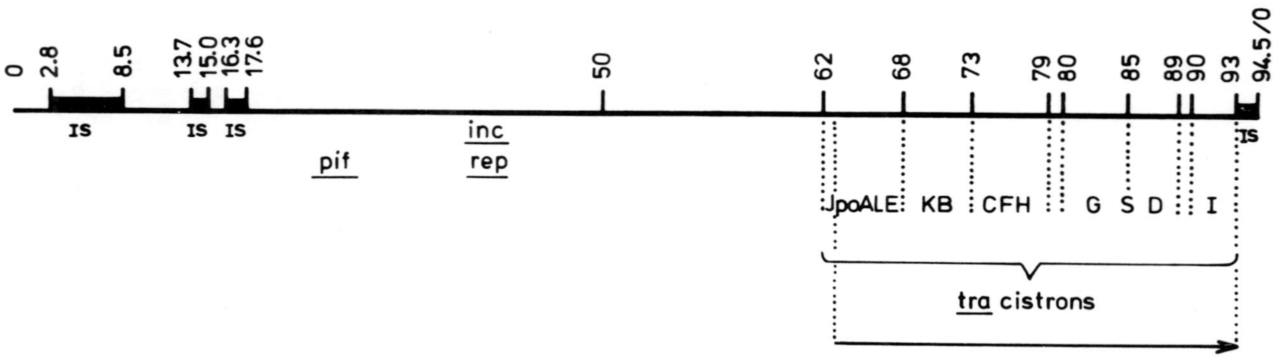

Fig. 1. The F genetic map. Distances are in kilobases (1000's of base pairs). The references for the various assignments are given in references 8 and 9. IS represents insertion sequences, pif represents the cistrons coding for T7 resistance, inc and rep represent the cistrons involved in incompatibility and replication.

the sex factor called the transfer origin. The complementary single-strand remains in the donor cell and the resulting single DNA strands in donor and recipient are converted to double-stranded DNA. At least 3 proteins on F (coded by cistrons traG, traD, and traI, Fig.1) are involved in the DNA transfer since mutants in these cistrons make F pili and bind to recipient cells but do not promote any DNA transfer (1). Events subsequent to DNA transfer are poorly defined; some results are presented here on disaggregation of the donor-recipient aggregates.

The F sex factor is an example of a group of DNA molecules called sex factors which can exist semi-independently of the bacterial chromosome. All carry genes coding for DNA transfer, primarily that of the sex factor itself. As many of these sex factors also carry genes coding for antibiotic resistance, or production of bacteriocins (similar to antibiotics but much more specific), catabolic enzymes, resistance to heavy metal ions and ionizing radiation, etc. (6) these properties are also transferred together with the sex factor. Sex factors have now been isolated from numerous gram-negative bacterial species. Their responsibility for the interspecific and intergenic spread of antibiotic resistance by mating is now generally accepted and is of great importance to medicine. It is also clear that there are numerous distinct sex factors with only minimal DNA homology to each other and presumptively of distinct evolutionary origin.

The analysis of sex factors other than F is still rudimentary. However, several have been demonstrated to promote the production of sex pili by their host cells (5). These sex pili are inferred to be involved in mating although the proof is still lacking for other than F pili. Immunological analysis indicates that the different sex pili produced by different sex factors are usually distinct with no immunological cross-reactions (5).

The matings discussed above are heterosexual crosses between sex factor-carrying donor cells and sex factor-less recipient cells. Most sex factors, including F, also promote a phenomenon called surface exclusion which inhibits homosexual mating between cells carrying the same sex factor. The existence of the surface exclusion phenomenon may be rationalized on the basis that DNA transfer of a sex factor is superfluous to cells already carrying that sex factor and surface exclusion reduces this superfluous DNA transfer. The surface exclusion phenomenon is highly specific and only DNA transfer promoted by closely related sex factors is affected. To date, four specificities have been identified among plasmids coding for sex pili similar to those produced by F (13) and it is to be expected that many more will be distinguished. These results are most easily explained on the basis of a receptor on female cells necessary for DNA transfer which is specifically altered when cells carry a sex factor. Genetic analyses of F have indicated that traS (3, 12)

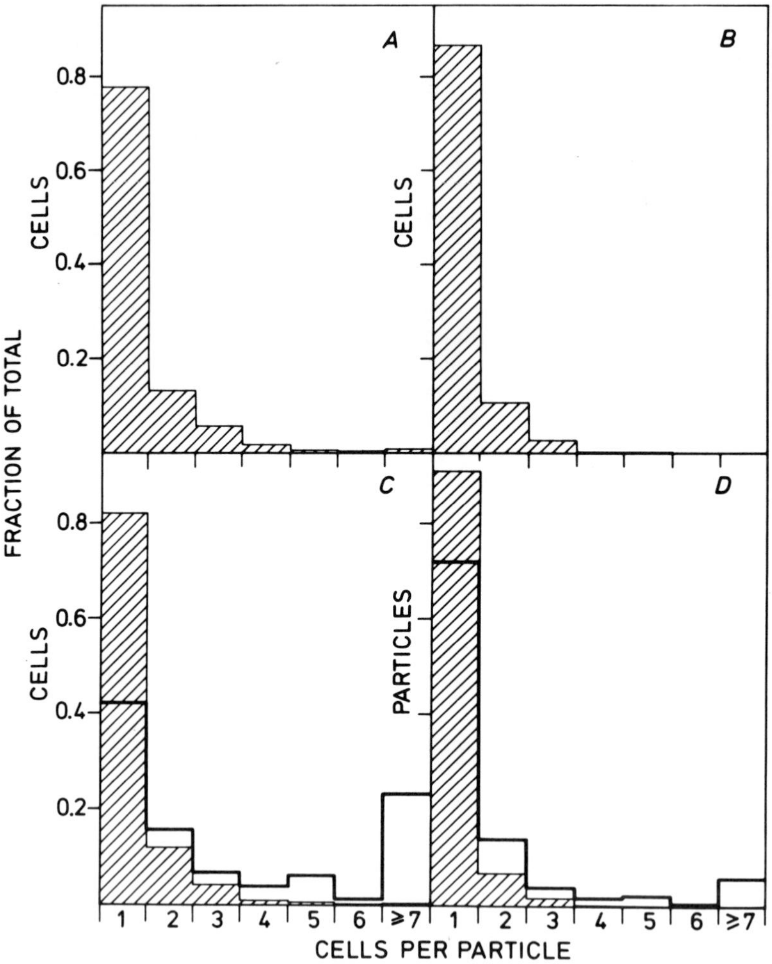

Fig. 2. Size distribution of mating aggregates. A (Flac)$^+$ donor was mated with an F$^-$ recipient for 30 min at 37° (1 : 1 donor : recipient; both at approx. 3.5 x 10^8 cell/ml). Samples were fixed with formaldehyde and examined by light microscopy for the number of particles in each size class. A) (Flac)$^+$ culture alone. B) F$^-$ culture alone. C) Average of A) and B) is hatched. (Flac)$^+$ x F$^-$ mating is unhatched. D) The data from C) is expressed in terms of the fraction of particles scored rather than in terms of the fraction of cells scored.

(see Fig. 1) and at least one more cistron (unpublished results) are directly involved in the surface exclusion phenomenon.

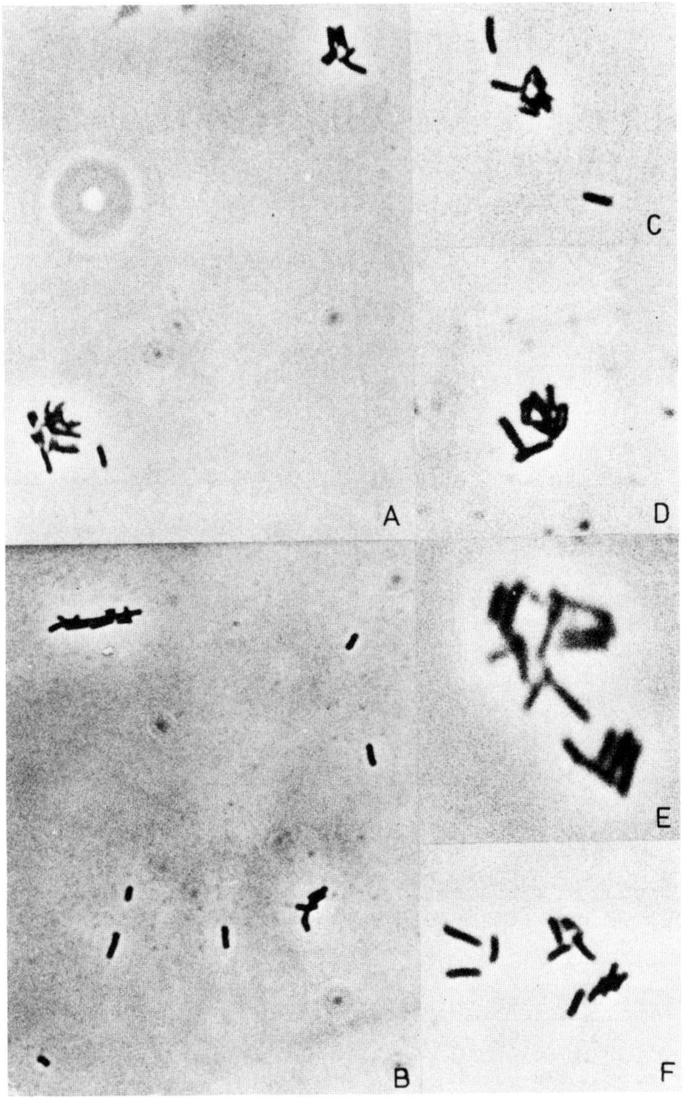

Fig. 3. Mating aggregates. Some typical larger mating aggregates from a mating similar to that analyzed in Fig. 2 are shown.

A New Definition of Heterosexual Matings

As a preliminary to a more detailed analysis of mating, a quantitative physical analysis of the nature of mating in E. coli was performed. In contrast to the generally accepted definition of mating as donor-recipient pairs, it was found that mating populations normally contain larger aggregates as well as pairs. A similar but quanti-

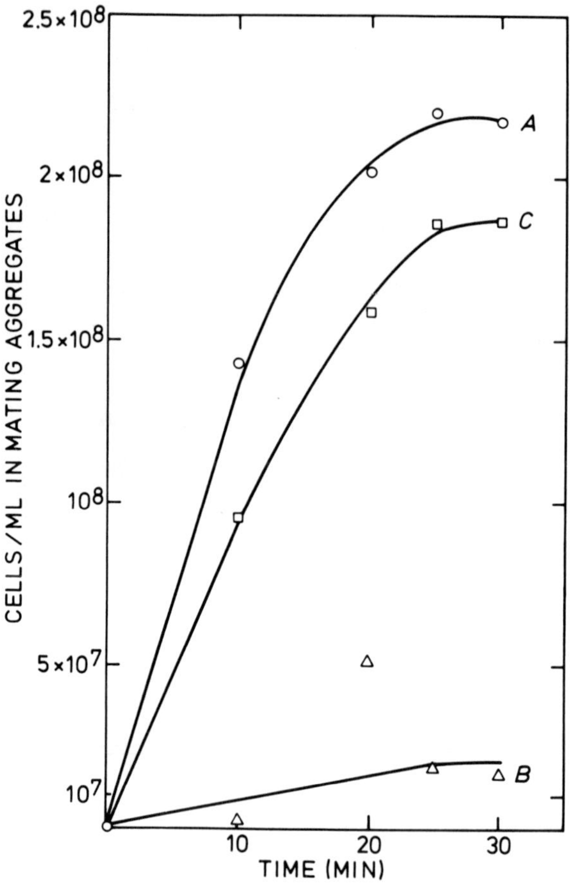

Fig. 4. Effects of surface exclusion on mating aggregate formation. Three matings were performed as described in Fig. 2 between an F^- (Flac)$^+$ donor culture and three different recipient cultures: A) F^- B) Flac traAl which does not produce F pili but is surface-exclusion positive C) Flac traJ90 which does not produce F pili and is surface-exclusion deficient.

tative observation has previously been published (4). Fig. 2 presents the size distribution of cell aggregates in a mating mixture as determined by light microscopy (2). Under conditions where most of the cells in a population are mating, most of the mating cells are in aggregates containing 5 or more cells each. Examples of such aggregates are presented Fig. 3. The largest aggregates observed contained twenty cells (2) but still larger aggregates have been reported (4). The kinetics of aggregate formation is shown by curve A in Fig. 4.

The generally accepted shear-sensitivity of mating aggregates also proved false. Mating aggregates of all sizes are remarkably stable structures. They are not appreciably disrupted by pipetting, centrifugation through a sucrose gradient, or analysis by Coulter counter. The size distribution presented in Fig. 2 has been qualitatively confirmed by sucrose gradient and Coulter counter analyses (2). Coulter counter analysis proved to be the simplest and most accurate of the three methods and was used to obtain most of the results presented below.

The formation of mating aggregates can be prevented by the addition of Zn^{2+} to 10^{-3} M, CN^{-} to 5×10^{-3} M, or sodium dodecyl sulfate to 0.1 % w/v (2). Mating aggregate formation also does not occur at 0° (unpublished observation). Disaggregation of mating aggregates occurs quite quickly after DNA transfer (unpublished observations). However, disaggregation can largely be prevented by treatment with CN^{-} or storage at 0° (unpublished observations). Disaggregation can be provoked at any time by very vigorous shaking or sodium dodecylsulfate treatment without affecting cell viability. These observations are incorporated into a model further in this presentation.

Aggregate Size and DNA Transfer Efficiency

Although both large (5 to 20 cells) and small (2 to 4 cells) aggregates are formed, their proportion depends upon the ratio of donor and recipient cells available. Using excess recipients and mating times of only 5 to 10 min at 37°, it is sometimes possible to obtain mating populations in which most of the aggregates are small. At later times (20 to 40 min), or at all times under 1 : 1 donor : recipient input ratios, larger aggregates are also present. The ratio of donors to recipients within the aggregates is also dependent on the input ratios. At equal input, the ratio of donor : recipients within the aggregates is 1 : 1 whereas under excess donor or recipient input it is possible to shift the ratio to 2 : 1 or 1 : 3 respectively (2).

The small and large aggregates have identical biological properties. The entire genome of <u>Flac</u>, or F sex factor with integrated chromosomal DNA of total molecular weight of 100×10^6 is transferred with an efficiency of 35% per aggregated cell (donor and recipients both) regardless whether the population contains predominantly small or large aggregates. Furthermore, 2% of aggregated Hfr (in which F is integrated into the bacterial chromosome) and F^{-} cells succeeded in transferring single-stranded chromosomal DNA of molecular weight as large as 5×10^8 (2). Many more cells were successful in transferring lesser amounts of choromosmal DNA (2). Again the same kinetics and efficiency of DNA transfer are observed regardless of which aggregate type predominates.

Homosexual Matings

The availability of the techniques described above has enabled me to analyze the surface exclusion phenomenon. Figure 3 presents a typical experiment on mating aggregate formation. Curve A presents the mating aggregate formation in a heterosexual cross while curves B and C present two homosexual crosses. To avoid possible complications due to non-specific stickiness of F pili, the F carrying cells used as recipients for curves B and C both contained mutated F factors incapable of producing F pili. The mutated F factor used for curve C also does not produce surface exclusion. It may be seen from Fig. 2 that surface exclusion reduces the number of mating aggregates in a homosexual cross.

F factor carrying cells are generally 5-to 20-fold worse at forming mating aggregates when used as recipients in homosexual crosses with other F factor carrying donor cells than are F^- cells. However, when the DNA transfer efficiency per aggregated cell is calculated, it becomes clear that the few homosexual aggregates formed are 5 to 20 fold worse than heterosexual aggregates in allowing DNA transfer. Thus surface exclusion coded by the F factor acts to prevent mating at the aggregate formation level and again at the DNA transfer level.

Analysis of a number of mutant F factors affecting surface exclusion has permitted the following conclusions. Mutants which totally remove surface exclusion remove both the barrier to aggregate formation and the barrier to DNA transfer within the aggregates. The production of F pili is irrelevant to surface exclusion. Mutants which do not produce F pili may still be surface exclusion-positive while surface-exclusion deficient mutants which still produce F pili have been isolated. Mutants have also been isolated which still form as few aggregates when used as recipients as the parental F sex factor-carrying cells but do allow DNA transfer within the aggregates at the F^- efficiency. Thus it is possible to separate the two processes.

These results necessarily lead to a subdivision of mating into two sequential stages: 1) aggregate formation and 2) DNA transfer. Aggregate formation is prerequisite but not sufficient to ensure DNA transfer. This subdivision is also necessitated by the following observations. A) F^- chromosomal mutants have been isolated (10, 11) which are incapable of acting as recipients in heterosexual matings with F sex factor-carrying cells. An analysis of three such mutants with the Coulter counter method demonstrated that one mutant did not allow aggregate formation whereas the other two did allow aggregate formation but prevented DNA transfer within the aggregates (2). B) Various DNA transfer-deficient mutants of the F sex factor are still capable of promoting F pilus synthesis. When tested as donors in heterosexual matings to normal F^- cells with the Coulter counter method, mating aggregates were formed but no DNA transfer ensued. Thus F pili on the donor cell are sufficient to allow aggregate formation whereas they play no role on the recipient cell. D) Although surface exclusions coded by F factor-carrying cells reduces both aggregate formation and DNA transfer, surface exclusion of a different specificity by a related sex factor, R100-1, only affects DNA transfer. R100-1 carrying cells are strongly aggregated with each other during culture growth but demonstrate a 20-fold reduction in DNA transfer when used as recipients in mating with a very closely related sex factor. Thus R100-1 could be considered a spontaneous mutant affecting the barrier to aggregate reduction. (Alternatively F has an extra facet to surface exclusion than that normally manifested by sex factors).

A Model for the Mating Phenomenon

The observations described above can be incorporated in the following model. F pili (or sex pili of other sex factors) are recognition organs synthesized by the sex-factor carrying cells which bind to a specific receptor on appropriate recipient cells. The receptors recognized by different types of sex pili may well be different. Binding leads to a series of enzymatic steps resulting in stable aggregation. Aggregates containing many cells result because the recipient cells have more than one receptor on the cell surface, and because many donor cells have more than one sex pilus. After stable aggregation has occurred, a signal triggers a series of enzymatic processes of unknown nature in both donor and recipients which result in DNA transfer. After DNA transfer has occurred further enzymatic steps are necessary to separate the mating cells.

The surface exclusion phenomenon implies that both mating aggregation and the DNA transfer are active and specific processes. Surface exclusion by F results in a specific alteration of the recipient receptors and the DNA transfer processes in the recipient such that F-carrying cells are incapable of transferring DNA to other F-carrying cells.

REFERENCES

1. ACHTMAN, M., Current Topics in Microbiol. and Immun. 60 (1973) 79.
2. ACHTMAN, M., J. Bacteriol 123 (1975) 505.
3. ACHTMAN, M., and HELMUTH, R., Microbiology - 1974 (ASM Publications, Washington, D. C. (1975) 95.
4. ANDERSON, T. F., Cold Spring Harbor Symp. Quant. Biol. 23 (1958) 47.
5. BRINTON, C. C., Jr., Critical Microbiol. 1 (1971) 105.

6. CLOWES, R. C., Bacteriol. Rev. 36 (1972) 361.
7. CURTISS, R. III, Ann. Rev. Microbiol. 23 (1969) 69.
8. DAVIDSON, N., Microbiology - 1974 (ASM Publications, Washington, D. C.) (1975) 56.
9. HELMUTH, R., and ACHTMAN, M., Nature (1975) in press.
10. REINER, A. M., J. Bacteriol. 119 (1974) 183.
11. SKURRAY, R. A., HANCOCK, R. E. W., and REEVES, P., J. Bacteriol. 119 (1974) 726.
12. WILLETTS, N., J. Bacteriol. 118 (1974) 778.
13. WILLETTS, N., and MAULE, J., Genet. Res. (Cambr.) 24 (1974) 81.

NEURONAL CELL ADHESION

Luis GLASER, Ronald MERRELL, David I. GOTTLIEB,
Dan LITTMAN, Morris W. PULLIAM and Ralph A. BRADSHAW

Department of Biological Chemistry
Division of Biology and Biomedical Sciences
Washington University, St. Louis, Missouri 63110 (USA)

Investigations in a variety of systems have shown that dissociated single cells will, under appropriate conditions, aggregate and, if maintained in culture, will morphologically differentiate to resemble the original organ or tissue (23, 33, 34). It can further be shown that in mixed aggregates of cells prepared from different organs (for example, neural retina and liver) there occurs cell segregation such that cells originally derived from the same organ attached to each other and separate from the cells derived from a different organ. The simplest explanation for this observation is that cells have a preferential or higher affinity for homologous cells, and if free to move within the aggregate, will rearrange so as to be adjacent to homologous rather than heterologous cells. This simple view is probably an oversimplification, because not only do cells migrate to be adjacent to homologous cells, but they also take up characteristic positions within the aggregate which cannot be explained simply by differential affinities (33).

If cells have specific recognition sites on their surface for homologous cells, then one would like to know what components of the cell surface are involved in cell recognition, in order to be able to study this phenomenon at the molecular level.

There are three biological systems which have received attention in recent years for the study of cell adhesion. These are aggregation of slime molds (see Barondes *et al.*, Malchow *et al.* and Gerisch, elsewhere in this volume); the aggregation of sponge cells (Jumblatt *et al.* elsewhere in this volume), and the aggregation of dissociated embryonal cells, which is the subject of this presentation. In recent years, the aggregation of embryonic cells has been the subject of intensive investigation in a number of laboratories, notably Roseman

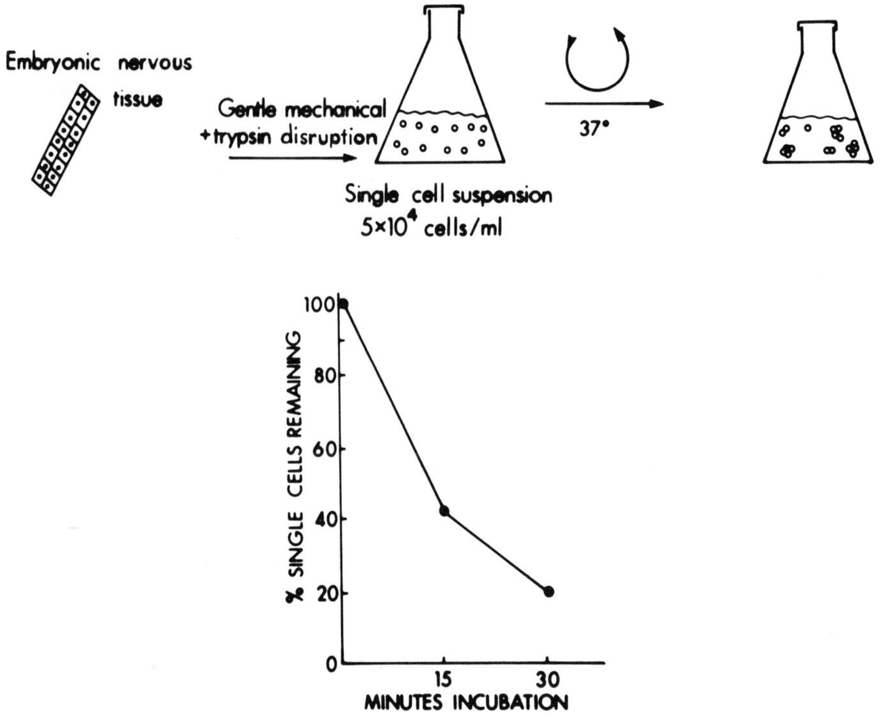

Fig. 1. Rotation mediated cell aggregation. Single cell suspensions are prepared by gentle dissociation of a given organ; the cells are then incubated at 37° in a rotating flask in suitable medium and the disappearance of single cells is determined either with a Coulter counter or a hemocytometer (24). The data shown are for neural retinal cells obtained from 8-day old chick embryos, incubated in calcium and magnesium-free Hank's solution.

(8, 9, 24, 25), Roth (4, 29, 30) and Moscona (13, 12, 18), as well as our own (14, 15, 20, 21 and Gottlieb et al., succeeding chapter) Most of this work has been concerned with neuronal cells from young chick embryos. The choice of neuronal cells for this purpose is partially a technical one, in that single neuronal cells can readily be obtained from young chick embryos by gentle methods, and these cells will reaggregate rapidly under suitable conditions.

As in many young fields of investigation, the precise methods used to measure cell aggregation are an important consideration. All the variables that effect cell recognition are not yet known and it is quite possible that different methods are in fact measuring different parameters involved in cell aggregation. It therefore seems worthwhile to review briefly the basic methods which have been

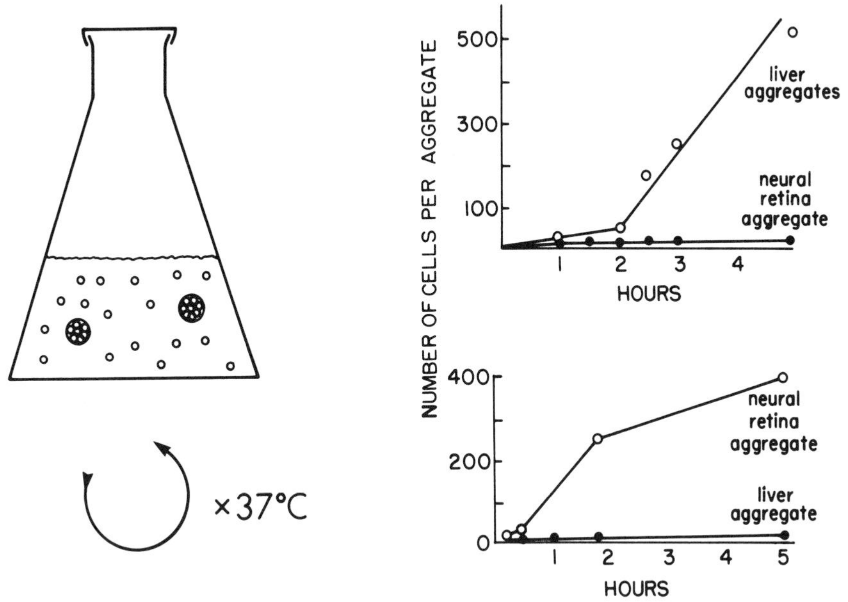

Fig. 2. Adhesion of cells to large aggregates. The method illustrates the rate of attachment of radioactive single cells to preformed large aggregates. O - liver cells, ● - neural retinal cells, the data have been redrawn from reference (30).

used to study cell aggregation and their limitations.

The simplest method to measure cell aggregation is to measure the disappearance of single cells as they form aggregates. This is a method which can be made quantitative (24) (Fig. 1), but in its simplest form is not suitable for measurement of specificity, since if one mixes different cell populations it provides no information on which cell types are adhering to each other. This assay can provide information regarding the specificity of agents which block cell aggregation, for example this assay could be used to show that certain carbohydrates will block the aggregation of liver cells but not retina, etc.

Two basic methods have been used in which the adhesive specificity of single cells can be determined. In the first, large aggregates of cells are prepared and these are then incubated in a suspension of radioactive single cells (Fig. 2). The aggregates by virtue of their size can readily be separated from single cells or small aggregates and the number of radioactive cells adhering to the aggregate can be determined (Fig. 2). This method for the first time showed

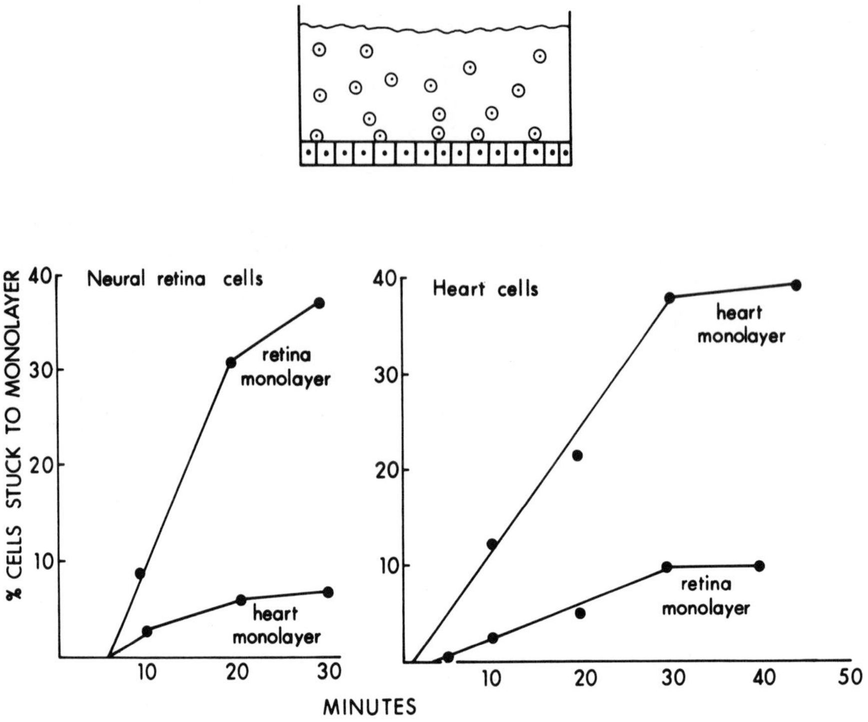

Fig. 3. Binding of cells to monolayers. Dense monolayers of cells are prepared by 48 hour incubation of embryonal cells in tissue culture medium, and then radioactive probe cells are added, and the rate of attachment of these cells to the monolayer is measured. The data are redrawn from reference (35).

that homotypic cells would adhere preferentially to aggregates, and is again a quantitative method (28-30). The possible limitations of this method are a) that the formation of large aggregates requires about 24 hours incubation in culture, which may change surface specificity; b) that the efficiency of pick up of single cells by aggregates is relatively poor; and c) that this method assumes that the surface specificity of cells exposed on the surface of the aggregate is the same as the surface specificity of isolated cells and, therefore, ignores the possibility that if recognition sites are mobile on the cell surface, then most of the sites on the cells in the aggregate may be occupied by binding to adjacent cells, and would not be available to bind to free cells.

The second method for measuring adhesive specificity uses cells bound to an inert support such as glass (Fig. 3) and measures the rate with which radioactive single cells bind to these immobilized cells. Some of the same objections apply to this method as to the aggregate-pick up method (35). A variant of the method in which stable monolayers of cells can be formed extremely rapidly, and used to measure adhesive specificity is discussed by Gottlieb et al. (15, and elsewhere in this volume). The possible effects of interaction between the cells in the monolayers on this system is not known.

Finally, considerable work has been done measuring factors which effect the size of aggregates obtained after 24 hour incubation of cells in rotating culture. Factors specific for neural retina and telencephalon have been obtained in this way (3, 17, 18), one of these as a pure protein (18). It is a limitation of this assay that one cannot distinguish an aggregation promoting factor, which works like a polyvalent lectin or an antibody from a trophic factor which enhances cell viability, or any aspect of cellular metabolism. The organ specificity of the factors which have been obtained is intriguing, and an elucidation of their mode of action will be a very important event in the understanding of cell interactions and developmental embryology.

The simplest assumption with which one can proceed in this field is that cell recognition involves the direct attachment of complementary sites on the surface of two adhering cells, without the necessity for intermediate multivalent ligands to crosslink the cells. Although such lectin like molecules have been isolated and characterized both for sponges (7, 36) and for slime molds (26, 27 and Barondes et al., elsewhere in this volume), it is quite possible that these crosslinking molecules are in fact originally multivalent membrane components, as shown diagramatically in Fig. 4. When these components become dissociated from the membrane, they will act as agglutinating agents. Although this may appear to be a semantic argument, biologically it is very different to assume that a secreted molecule crosslinks cells and has organ specificity, or to assume that certain multivalent molecules are firmly attached to the plasma membrane and can recognize complementary structures on the other cell surfaces.

A useful simple model for cell recognition is that the complementary structures on the cell surface have the characteristics of an enzyme and its specific substrate. The useful feature of this model is that it points out quite clearly that only one of the complementary components need be a protein with a specific binding site, but that the other component could be a protein, could be the carbohydrate moiety of a glycoprotein, or could be a glycolipid, or a combination of these structures.

A specific version of this model has been presented by Roseman (25) and elaborated by Roth (31). This version assumes that the enzymes involved are glycosyl transferases binding to their substrates. On this specific assumption, Roseman and coworkers have linked

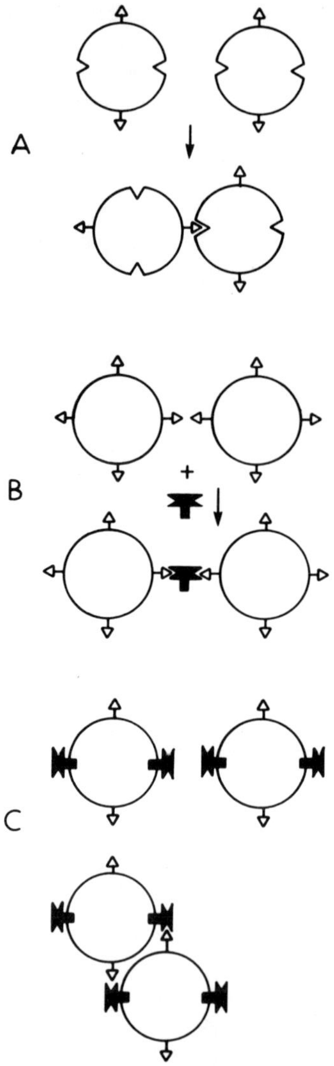

Fig. 4. Diagramatic representation of cell adhesion models. Model A assumes direct attachment of complementary structures on the cell surface. Model B assumes that attachment occurs with the aid of a multivalent soluble ligand. Model C shows that if this ligand is originally attached to the membrane model B becomes experimentally identical to model A.

different monosaccharides to Sepharose beads and measured cell adhesion to such beads. Specific adhesion of SV40 transformed 3T3 cells to galactose containing beads has been seen (8, 9) but the relation of this interesting phenomenon to cell-cell adhesion is not yet known.

The biological significance of specific cell adhesion is not clear, except in the case of slime molds (5, Barondes *et al.*, Malchow *et al.* and Gerisch, elsewhere in this volume). It seems highly unlikely that some of the choices that we ask cells to make in the laboratory, for example to distinguish between a retinal cell and a liver cell, has any biological meaning. It seems more likely that at least in the early stages of development it is important for cells to recognize boundaries between different organs (10), or within the same organ, but positive proof that specific cell adhesion is required for normal development is not yet available. Some suggestive evidence in this direction should be mentioned.

DeLong and Sidman (11) have examined the pattern of reaggregation of cerebellar cells from normal mice and reeler mutants. This mutant lacks proper cellular organization in the cerebellum. They found that at day 17 of gestation cerebellar cells from normal litter mates and reeler mice formed aggregates but the mutants failed to segregate different cell types within the aggregate. These data imply, but certainly do not prove, that selective adhesion is required for proper organization in the cerebellum.

A different and non-neuronal example comes from a study of t mutants in mice (1). Eleven mutations are known in the t locus which in homozygotes interfere with normal embryonal development and are fatal. The morphological changes observed have been interpreted as factors in cell interaction. The t^{12} antigens have been detected on sperm surface (6) and are also detected in early mouse embryos at the 8 cell stage and disappear thereafter. Homozygous t^{12}/t^{12} embryos fail to develop past the 8 cell stage (2). The clear implication of these observations is that a surface molecule specified by the t^{12} locus appears transiently on the embryonal cell surface and failure of this antigen to appear is fatal. The extrapolation that this is a failure in cell-cell interaction is tempting but not conclusive. Attention should be drawn to the very transient appearance of this antigen (2, 19) as well as other early embryonal antigens.

In the case of neuronal cells with which we have been primarily concerned, there is a continuous temptation to relate cell surface adhesive specificities to synaptic specificities. It is well to remember in this connection that all the adhesion assays which we have described measure adhesion between cell bodies, while synaptic connections involve specialized structures such as axon tips. Furthermore even the smallest neuronal area that we or anybody else has so far been able to assay for cell adhesion specificity contains a variety of cell types or precursors of various cell types, some of which are destined to make highly specific synaptic connections well below the resolution power of our crude methodology.

The simple model of cell adhesion on which we have proceeded is to assume that cell adhesion involves direct contact between two

TABLE I

Analysis of Retinal Membrane Fractions

Membranes were prepared from 8 day neural retinal cells which had been labeled with [^{3}H]glucosamine. The data represent the relative specific activity of the enzyme markers relative to the initial homogenate taken as unity. Radioactivity is expressed as dpm/mg protein. Membrane fraction 1 is enriched in plasma membrane markers and is used in the experiments described. For details see reference (20).

	Membrane Fractions		
	1	2	3
	Relative Specific Activity		
Mg^{++} ATPase	13.4	0.47	0.7
Alkaline phosphatase	9.2	0.26	1.6
Phosphodiesterase	3.2	0.05	0.11
NADH diaphorase	0.4	0.04	0.27
	dpm/mg protein		
[^{3}H]Glucosamine	2×10^6	4×10^5	4.0×10^5

surfaces, that the initial adhesion can be considered formally as the formation of an enzyme-substrate complex, and that it should be possible to isolate plasma membranes which contain at least one of the complementary components required for cell adhesion. We realize that initial events in cell adhesion are sometimes followed by complex morphological changes on the cell surface, which we cannot quantitate. This simple model also ignores the fact that cell surfaces are complex active structures, and do not look like a smooth sphere. The fact that cell adhesion is extremely temperature sensitive and does not occur at 0° already suggests a more complicated process than the formation of a simple enzyme-substrate complex (15, 24). Nevertheless these assumptions allow the design of some simple experiments which are summarized below. Experimental details for the most part have been published (14, 20, 21).

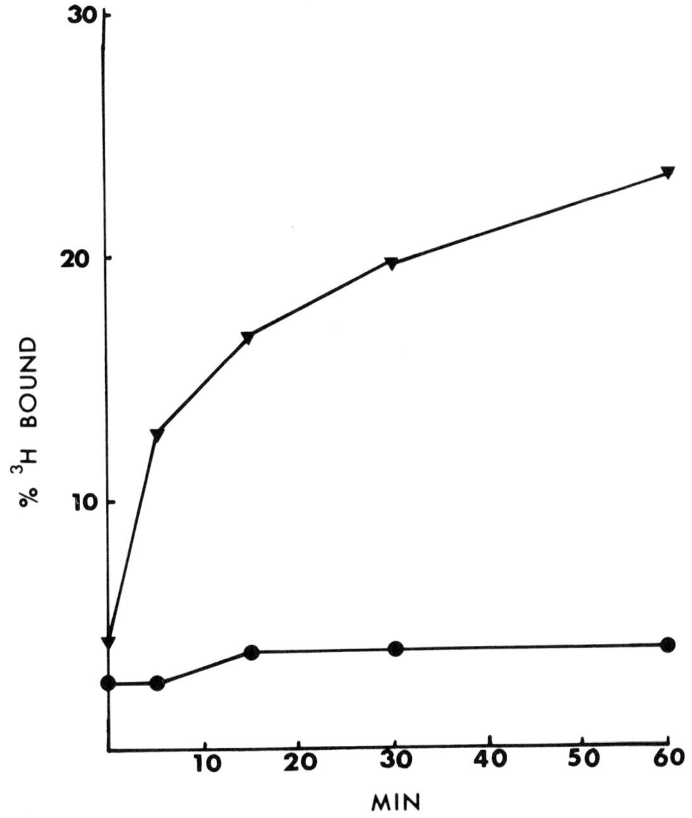

Fig. 5. Binding of radioactive membranes to cells. 6 x 10^5 cells were incubated in 3 ml of medium HH with [^{3}H]glucosamine-labeled plasma membrane (20 μg of protein, 15,000 dpm). At each time point the aggregation was stopped by 3-fold dilution in HH medium; cells with attached membranes were collected by centrifugation at 300 x g for 5 min. The pellet was washed with 3 ml of 5% trichloroacetic acid, dissolved in 1 ml of Protosol (New England Nuclear Corp.), and counted. ●, cerebellar cells; ▼, retinal cells (20).

<u>Binding of retinal plasma membranes to retinal cells.</u> Based on these considerations plasma membrane fractions were prepared from 8-day chick neural retinal cells. These membrane fractions were enriched in some of the usual cell surface markers (Table I). When these membranes were prepared from cells incubated with [^{3}H]glucosamine, it was shown that these membranes bound to retinal cells (Fig. 5) but not to cerebellar cells or embryonic liver cells.

The effect of these membrane fragments on cell aggregation (Fig. 6) was also tested and it was found that they appeared to

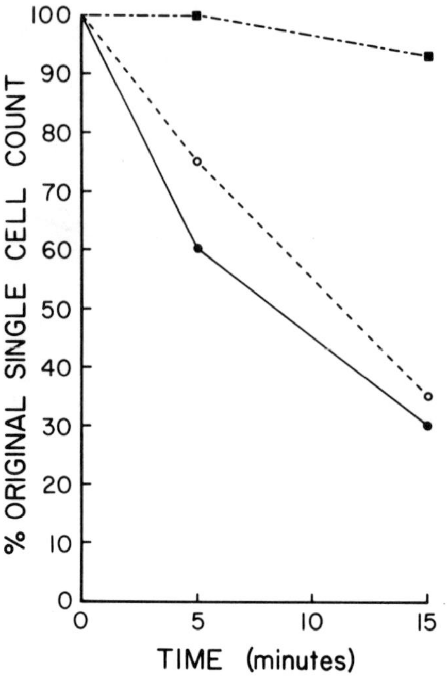

Fig. 6. Inhibition of retinal cell aggregation by membranes. 1.5 x 10^5 neural retinal cells were incubated in 3 ml of medium HH with no additions, ●; with cerebellar plasma membrane (0.1 mg of protein), O; with retinal plasma membrane (0.1 mg of protein), ■. The percent remaining single cells is plotted as a function of time. Data are the average of five separate experiments with different cell suspensions and different membrane preparations; different experiments agreed within 5% (20).

specifically inhibit homotypic but not heterotypic cell aggregation. The interpretation of these observations is schematically illustrated in Fig. 7. It assumed that the membrane fragments functionally contain only one of the two complementary recognition elements, and that by coating the cells, they effectively prevent adhesion. It should also be clear that recognition sites on the surface of the retinal cells may exist for ligands which occur only rarely. Heterologous membrane fragments may contain such ligands, but their binding would not interfere with cell adhesion.

The inhibition of cell aggregation has been used as an assay for membrane and cell specificity. It should be clear that the assay selects for abundant recognition sites to which membranes can bind rapidly, since in the assay, once cell to cell binding occurred, the cells were no longer counted as single cells.

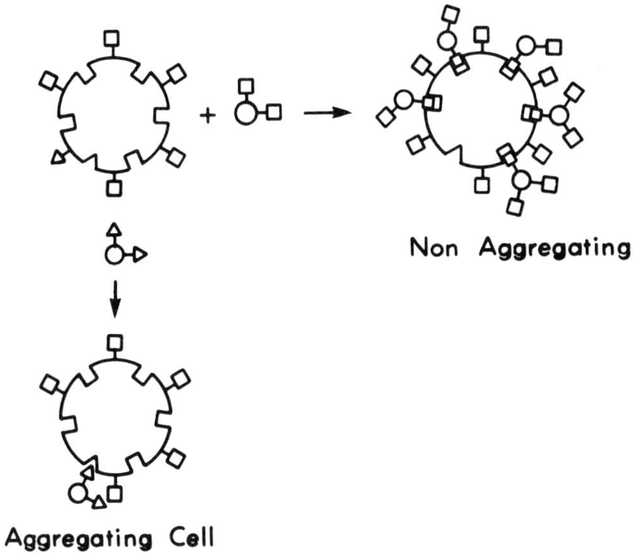

Fig. 7. Role of cell ligand density in aggregation assay. Membrane fragments which contain adhesive elements which occur on the cell surface with high frequency will coat the cell surface and prevent aggregation. Membrane fragments which contain adhesive elements which occur infrequently on the cell surface will bind to the cell but will not prevent aggregation.

The membranes are extremely trypsin sensitive and pretreatment of the membranes with trypsin abolishes activity (Fig. 8). This suggests that one of the components present in the membrane is a protein, but more complex interpretations such as changes in membrane structure due to proteolytic cleavage are not ruled out.

Temporal changes in cell surface specificity. Two examples of apparent temporal changes in cell surface markers during development have been described already, one in the cerebellum of the reeler mutant (11) and one in the cleavage stages of the mouse embryo (2). Thus the question of whether, as detected by the inhibition of aggregation assay, cell surface specificity changes during development can be asked. The data in Fig. 9 show that membranes obtained from 8-day retinal cells preferentially inhibit the aggregation of 8-day retinal cells, as compared to 7- or 9-day retinal cells. These same data can be obtained whether the membranes are prepared from the whole organ or from single isolated cells, and suggest that there are substantial changes in some surface recognition during retinal development. The assay however does not say that 8-day markers are not present on 9-day cells, but only that their frequency has become low enough so that 8-day membranes do not interfere with cell aggregation.

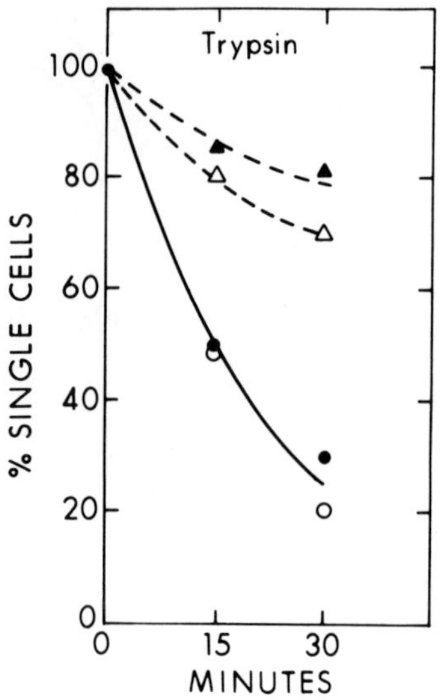

Fig. 8. Effect of trypsin on membrane activity. A suspension of 8-day retinal plasma membrane (0.1 ml) containing 0.1 mg of protein was incubated for 10 min at 37° with 0.1 mg of crystalline trypsin. Soybean trypsin inhibitor was added and the membranes assayed for their ability to inhibit aggregation of 8-day retinal cells. O - no addition, ▲ - intact membranes, ● - trypsin located membranes, Δ - membranes treated with trypsin + trypsin inhibitor.

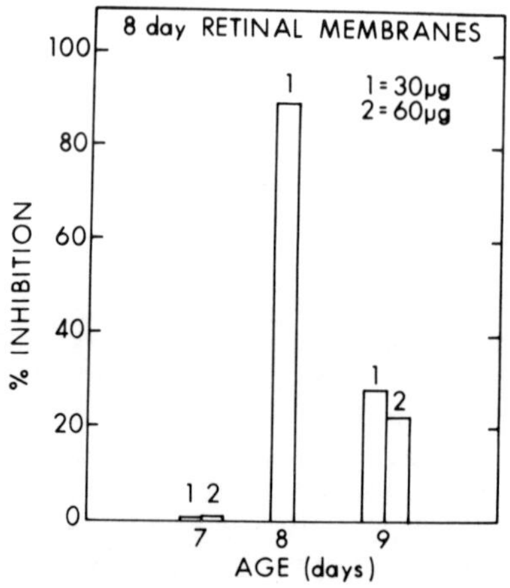

Fig. 9. Effect of retinal cell age on membrane inhibition. Neural retinal cells were obtained from 7, 8 or 9-day old embryos and allowed to aggregate either in the presence or absence of the indicated quantities (μg of protein) of 8-day neural retinal plasma membranes for 30 min. The percent inhibition of aggregation by the membrane is shown (64).

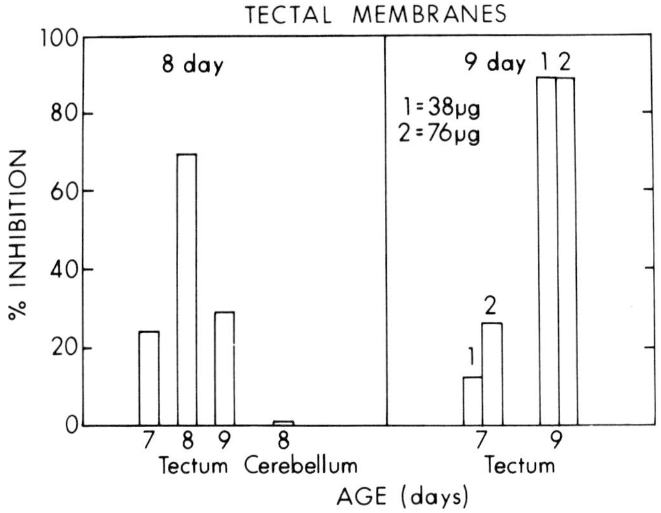

Fig. 10. Age specificity of tectal cells. In the left panel is shown the inhibition of aggregation obtained with 8-day tectal membranes (90 μg of protein) using 7, 8 and 9-day old tectal cells and 8-day cerebellar cells. In the right panel is shown the inhibition obtained with two different concentrations of 9-day tectal membrane concentrations on the aggregation of 7 and 9-day tectal cells (14).

In similar experiments plasma membranes prepared from 7 or 9-day retinal cells preferentially recognize cells obtained from embryos at the same stage of development.[1] Very similar data can be obtained with the optic tectum where cell surface specificity also changes daily from day 7 to day 9 (Fig. 10).

Retinal tectal cross-reactivity. The ganglion cells in the neural retina send projections to the optic tectum. The connections that are made are highly specific and involve an inversion of both the dorsal ventral and nasal-temporal axis of the retina on the tectum. Work in S. Roth's laboratory (4) has shown that cells obtained from the dorsal half of the retina, when incubated with the tectal half preferentially adhere to the ventral half of the tectum and, conversely, that ventral retinal cells preferentially adhere to the dorsal half of the tectum. This adhesion is apparently not to cell bodies of the tectum, but to the surface layer of the tectum. Neural retina and pigmented retina show the same difference in specificity, suggesting that at early developmental stages there are common

---

[1] Embryos on day 7 are at stages 28-29 of the Hamburger-Hamilton development series (16); embryos at day 8 are at stages 30-31 and at day 9 at stages 33-34. More details are given in reference (21).

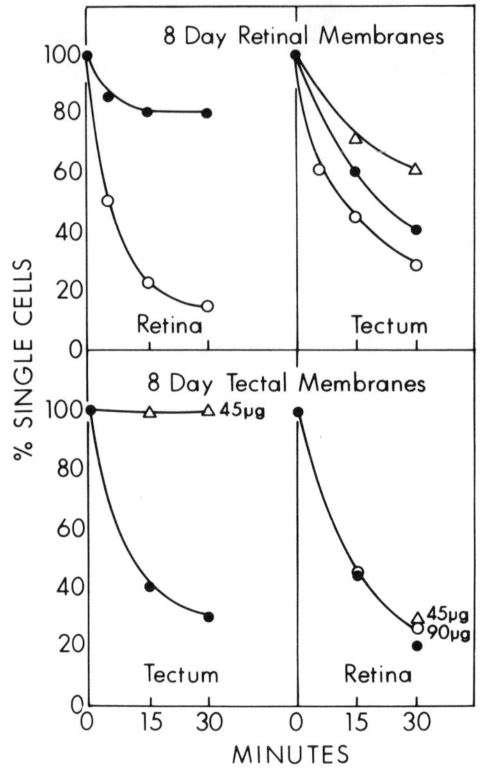

Fig. 11. Retina-tectum cross reactivity. The top portion of the figure shows the time course of retinal and tectal cell aggregation in the presence of retinal membranes. O, no added membranes; O, 31 μg of membrane protein; Δ , 62 μg of membrane protein. The bottom half shows similar data with tectal membranes, with the quantity of membrane protein indicated (14).

recognition sites on the surface of these two very distinct components of the retina.

Without wishing to make any implications regarding synaptogenesis, we have tested the reactivity of plasma membranes from retina and tectum with cells obtained from each of these organs. It is unclear whether adhesion of plasma membrane to cell bodies, in any way mimics the specificity of synaptogenesis which involves the adhesion of axons to specific receptors. As shown in Fig. 11, retinal membranes inhibit both retinal and tectal cell aggregation while tectal cell members inhibit only tectal cell aggregation. This cross inhibition occurs in apparent temporal synchrony in both organs (Fig. 11).

These data suggest the presence of a component on the retinal cell surface which is recognized by tectal cells; they do not prove that the same component is recognized by retinal cells and tectal cells on retinal membranes.

Soluble factors can be extracted from an acetone powder of the membranes by low concentrations of lithium diiodosalicylate. These trypsin sensitive factors specifically inhibit aggregation of homoologous cells and show the same temporal and spatial specificity as the membranes from which they were obtained (21). Adsorption studies indicate that either tectal or retinal cells can adsorb out the factor from retinal extract which inhibits the aggregation of retinal cells or tectal cells, suggesting that the same molecule in retinal extracts is responsible for the inhibition of aggregation of both cell types.

Changes in temporal specificity in tissue culture. A few attempts at purification of the lithium diiodosalicylate extract of membranes made it clear that the factors we are dealing with are not the major components of the cell surface, and that it would be extremely difficult to obtain adequate quantities of material for purification by conventional techniques. A simple calculation will show 10^8 cells containing as much as 10^5 adhesive molecules of molecular weight 100,000 will only contain 10 μg of such a protein.

Two other approaches to the partial purification of such molecules therefore have been considered. The first was to obtain temporal changes in tissue culture, on the assumption that this would allow the preparation of radioactive membranes of high specific activity, and allow attempts to purify by specific adsorption. The second approach was to consider the possibility that neuronal tumorous cells either already in existence or new ones that might be generated would share surface specificity with the embryonal cells, and would provide a source of recognition factors in unlimited supply.

In attempting to obtain a temporal change in tissue culture, attention has been focussed initially on the transition between 7 and 8 days in the optic tectum because it is a sharp transition and the tectum is easily dissected. After a number of partially successful trials in which 7 day tectal cells were incubated in tissue culture medium with embryo extracts and serum, by serendipity, the embryo extracts were replaced with large concentrations of nerve growth factor (NGF) obtained from mouse submaxillary gland, which proved to be a specific requirement for this temporal transition (Fig. 12). The transition is time dependent, and the effect is specific for NGF; a variety of other hormones, or protein with high isoelectric point have no effect in this system (22).

The requirement for NGF in this system differs significantly from the well studied requirement for NGF for survival and neurite

TABLE II

Comparison of the Effects of Mouse Submaxillary Gland Nerve Growth Factor (NGF) on Dorsal Root Ganglia and Optic Tectum

The data are a summary of observations reported in detail elsewhere (22). Cell maintenance is defined by the cell survival and by the rate of incorporation of radioactive leucine and uridine into cells. NGF (B_2) is a form of NGF which contains only B chains and is prepared by a different procedure* than the usual NGF preparation which contains equal quantities of A and B chains (5). 2-ox and 3-ox NGF are derivatives in which 2 or 3 of the tryptophan residues have been oxidized with N-bromosuccinimide (12).

	7-8 day tectal transition	Neurite outgrowth in dorsal root ganglia
Concentration of NGF for maximal effect	10^{-7} M	10^{-9} M
Cell maintenance by NGF	-	+
NGF derivatives:		
NGF(B_2)	+	+
Naja naja NGF	-	+
2-ox and 3-ox	+	-
Performic acid oxidized NGF	-	-
Other Proteins:		
Proinsulin	-	-
Cytochrome c	-	-
Insulin	-	-
Lysozyme	-	-
Glucagon	-	-

*Jeng, I. M. and Bradshaw, R. A., in preparation

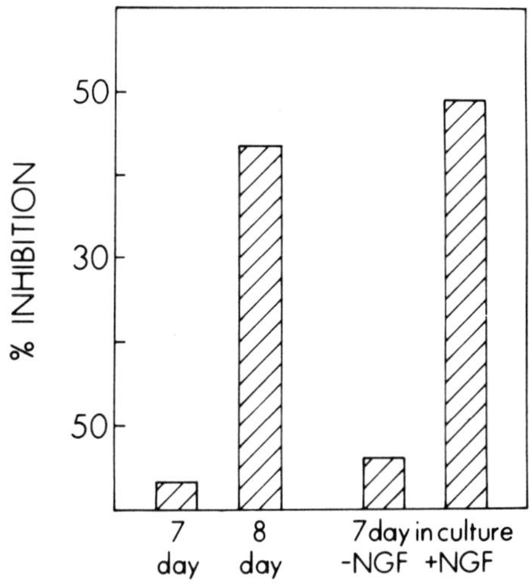

Fig. 12. Temporal change in surface specificity during cell culture. Seven day tectal cells were cultured under standard conditions with and without NGF. The bars represent the average of 48 separate experiments in which the degree of inhibition of cell aggregation by 8-day tectal membranes was measured in control tectal cells obtained from 7 and 8-day old chick embryos and in 7-day tectal cells cultured for 24 hours in the presence and absence of NGF. For convenience in cell counting, submaximal concentrations of membranes are used in these experiments (22).

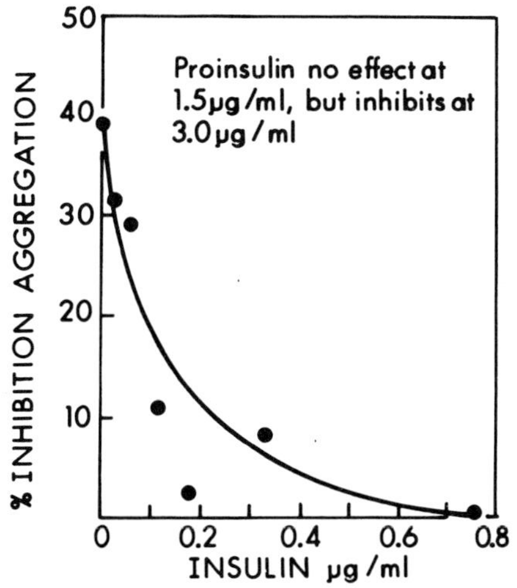

Fig. 13. Effect of insulin concentration on the change in tectal surface specificity. The data show the effect of added insulin on the change of surface specificity induced by NGF, determined by inhibition of cell aggregation by 8-day tectal membranes.

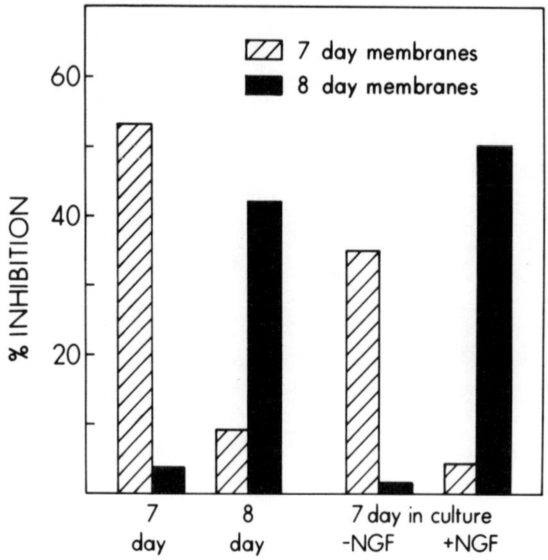

Fig. 14. Disappearance of 7-day specificity in the presence of NGF. Standard conditions were used, but cells were tested for inhibition of aggregation by 7-day tectal membranes, cross-hatched bars; and 8-day tectal membranes, solid bars (22).

outgrowth in the sympathetic ganglia. These differences are summarized in Table II. The tectal transitions require 100-fold higher concentrations of NGF than the neurite outgrowth. NGF maintains viability of sympathetic neurons, but not of tectal cells. NGF in which 2 or 3 tryptophans have been oxidized to the oxindole derivative are inactive in the sympathetic system but active with the tectum, and the cobra venom NGF which is active in the sympathetic system is inactive in the tectal system.

Insulin, but not proinsulin, at levels of 10-fold lower than NGF abolishes the NGF effect (Fig. 13). Concomitant with the appearance of 8-day specificity, the tectal cells lose specificity as determined by a membrane inhibition assay (Fig. 14).

The effect of NGF is totally resistant to cycloheximide addition, and thus the limited transition that has so far been observed in culture is not suitable as a method of radioactively labelling recognition factors. The appearance of 8-day specificity in culture induced by NGF represents the induction of a post-transcriptional event in membrane assembly or maturation. Attempts to obtain conditions were new protein synthesis is required for a temporal transition to take place, primarily by starting with earlier embryonal cells, are underway.

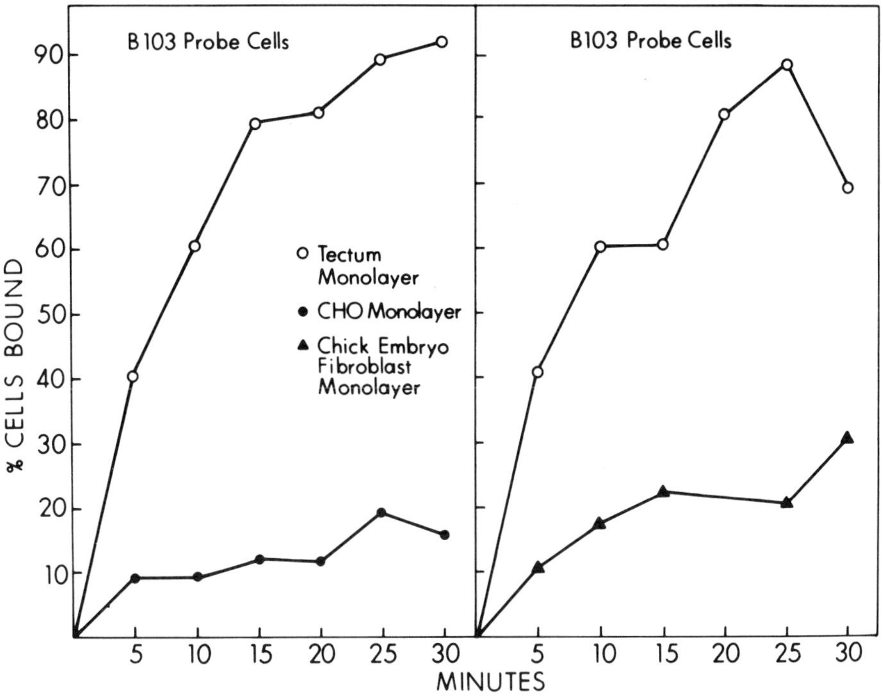

Fig. 15. Binding of B103 cells to monolayers of tectal cells or chick embryo fibroblasts. Binding of radioactive B103 cells to monolayers prepared from 9-day tectal cells and chick embryo fibroblasts was carried out by standard procedure. The same results were obtained whether the fibroblast monolayers were freshly prepared after trypsinization of the cells or the monolayers were allowed to recover from trypsinization.

The high concentrations of NGF which are required, as well as the differences in specificity between tectal cells and sympathetic neurons for NGF derivatives, taken together with the fact that it has been impossible to demonstrate the existence of NGF in the central nervous system,[2] all would seem to argue that NGF is acting in this system as an analog of a normal trophic agent which is required for neuronal cell differentiation. The fact that the observed transition is cycloheximide resistant suggests that what is observed is either the insertion into the membrane of a pre-existing component or the modification of a pre-existing membrane component. The data suggest that a search for trophic factors which influence neuronal development would be a fruitful endeavor.

---

[2] Jeng, I. M., and Bradshaw, R. A., in preparation.

Tumor cells as a source of cell recognition factors. Since some tumor cells are known to express embryonal antigens (19), it seemed worthwhile to check existing tumor lines to ascertain whether they would specifically adhere to embryonal neural cells. In the absence of chicken neural cell tumors, a number of neuronal tumor cell lines from rats, obtained through the kindness of Dr. David Shubert at the Salk Institute (32), were examined. This inter-species approach is not totally unreasonable since a large amount of literature on cell recognition suggests that during early embryonal development organ specificity overrides species specificity (29).

Of a number of cell lines tested, one designated B103 showed striking adhesion to all chicken neuronal cells, but not to non-neuronal cells of various types. The data in Fig. 15 were obtained by a monolayer adhesion assay described in detail (15), and show that B103 adhers to optic tectum cells but not to fibroblasts. It also does not adhere to chick embryo liver cells or Chinese hamster ovary cells. A plasma membrane enriched fraction can be obtained from these cells which binds specifically to neuronal cells but not to non-neuronal cells and this is illustrated for optic tectum and liver in Fig. 16.

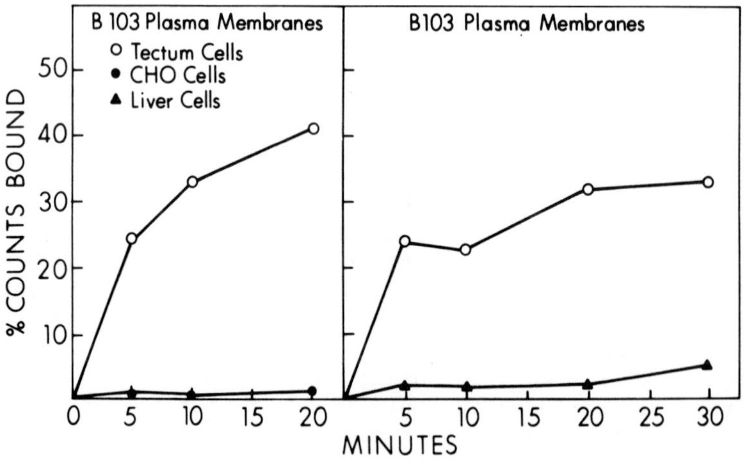

Fig. 16. Binding of B103 plasma membranes to tectal cells or liver cells. Binding assays were carried out by a modification of the methods described previously (20). Both the tectal cells and the liver cells were obtained from 9-day old chick embryos.

This cell line cannot differentiate between different neuronal regions, but seems to show a general neuronal specificity. Although this is by no means an ideal cell line, it appears that it will be a suitable model for the purification of recognition factors, encouraging further attempts to devise methods to generate cell lines which are more specific.

CONCLUSIONS

It has been shown that cell recognition can be studied at the subcellular level and that plasma membranes retain some of the components of cell surface specificity. Cell surface adhesive specificity is not a constant parameter but varies rapidly during development, and these changes appear to be under the control of trophic factors. The natural trophic factors are not known but in the case of the transition between 7 and 8-day specificity in the optic tectum this transition can be mimicked by mouse submaxillary gland NGF.

A rat tumor cell line has been shown to bind specifically to neuronal cells. This opens the possibility to specifically extract recognition components from the membrane of a cell available in large quantities. Even if this cell line does not show the precise temporal and spatial specificities of the embryonal cell lines, it would be an enormous advance in understanding the adhesion process at the molecular level.

As is shown in the ensuing article, it appears likely that operationally defined neuronal cell adhesive specificities can be further subdivided. This is biologically interesting but it also indicates that as the number of different factors within any organ keeps growing, the work involved in purification of these factors in order to identify them chemically becomes more difficult. This difficulty is compounded by the inadequacy of present methodology, in which a complex rate determined both by affinity of the complementary sites as well as by the number of available sites is being measured. The complications that arise because of the three dimensional complexity of the plasma membrane are recognized but their contribution to cell adhesion and to the measurement of cell adhesion remain to be explored.

ACKNOWLEDGEMENTS

This work has been supported by research grants GM 18405 (L.G.) and NS 10229 (R.A.B.) from N.I.H., and 74-22638 from N.S.F. (L.G.). Training Grant support was received from grant GM-60371 (R.M.), F22-NS 166 (D.I.G.), F22-NS 00757 (M.W.P.), and TO-5-GM 02016 (D.L.).

REFERENCES

1. ARTZT, K. and BENNETT, D., Nature 256 (1975) 545.
2. ARTZT, K., BENNETT, D. and JACOB, F., Proc. Nat. Acad. Sci.USA 71 (1974) 811.
3. BALSANO, J. and LILIEN, J., Proc. Nat. Acad. Sci. USA 71 (1974) 727.
4. BARBERA, A. J., MARCHASI, R. B. and ROTH, S., Proc. Nat. Acad. Sci. USA 70 (1973) 2482.
5. BOCCHINI, V. and ANGELETTI, P. U., Proc. Nat. Acad. Sci. USA 64 (1969) 787.
6. BUC-CARON, M., GACHELIN, G., HOFNUNG, M., and JACOB, F., Proc. Nat. Acad. Sci. USA 71 (1974) 1730.
7. CAULDWELL, C. G., HENKART, P. and HUMPHREYS, T., Biochemistry 12 (1973) 3051.
8. CHIPOWSKY, S., LEE, Y. C. and ROSEMAN, S., Proc. Nat. Acad. Sci. USA 70 (1973) 2309.
9. CHIPOWSKY, S., SCHNAAR, R. and ROSEMAN, S., Fed. Proc. 34 (1975) 614.
10. CRICK, F. H. C. and LAWRENCE, P. A., Science 184 (1975) 340.
11. DE LONG, G. R. and SIDMAN, R. L., Developmental Biol. 22 (1970) 584.
12. FRAZIER, W. A., HOGUE-ANGELETTI, R. A., SHERMAN, R. and BRADSHAW, R. A., Biochemistry 12 (1973) 3281.
13. GARBER, B. B. and MOSCONA, A. A., Developmental Biol. 27 (1972) 235.
14. GOTTLIEB, D. I., MERRELL, R. and GLASER, L., Proc. Nat. Acad. Sci. USA 71 (1974) 1800.
15. GOTTLIEB, D. I. and GLASER, L., Biochem. Biophys. Res. Comm. 63 (1975) 815.
16. HAMBURGER, V. and HAMILTON, H. L., J. Morphology 88 (1951) 49.
17. HAUSMAN, R. E. and MOSCONA, A. A., Proc. Nat. Acad. Sci. USA 70 (1973) 3111.
18. HAUSMAN, R. E. and MOSCONA, A. A., Proc. Nat. Acad. Sci. USA 72 (1975) 916.
19. MARTIN, G. R., Cell 5 (1975) 229.
20. MERRELL, R. and GLASER, L., Proc. Nat. Acad. Sci. USA 70 (1973) 2794.
21. MERRELL, R., GOTTLIEB, D. I. and GLASER, L., Topics in Neurobiology, Vol. 2 (S. Barondes ed.) Plenum Press, New York (1976) in press.
22. MERRELL, R., PULLIAM, M. W., RANDONO, L., BOYD, L. F., BRADSHAW, R. A. and GLASER, L., Proc. Nat. Acad. Sci. USA 72 (1975) 4270.
23. MOSCONA, A. A., Cells and Tissues in Culture (F. N. Willmer ed.) Academic Press, New York, Vol. 1 (1965) 197.
24. ORR, C. W. and ROSEMAN, S., J. Membrane Biol. 1 (1968) 110.
25. ROSEMAN, S., Chem. Phys. Lipids 5 (1970) 270.
26. ROSEN, S. D., KAFKA, J. A., SIMPSON, D. L. and BARONDES, S. H., Proc. Nat. Acad. Sci. USA 70 (1973) 2554.

27. ROSEN, S. D., SIMPSON, D. L., ROSE, J. E. and BARONDES, S. H. Nature 252 (1974) 128, 149.
28. ROTH, S. and WESTON, J. A., Proc. Nat. Acad. Sci. USA 58 (1967) 974.
29. ROTH, S., Developmental Biol. 18 (1968) 602.
30. ROTH, S., McGUIRE, E. J. and ROSEMAN, S., J. Cell Biol. 51 (1971) 525.
31. ROTH, S. and WHITE, D., Proc. Nat. Acad. Sci. USA 69 (1972) 485.
32. SCHUBERT, D., HEINEMANN, S., CARLISLE, W., TARIKAS, H., KIMES, B., PATRICK, J., STEINBACH, J. H., CULP, W. and BRANDT, B. L., Nature 249 (1974) 224.
33. STEINBERG, M. S., Science 141 (1963) 401.
34. TOWNES, P. S. and HOLTFRETER, J., J. Expt. Zool. 128 (1955) 53.
35. WALTHER, B. T., OHNAN, R. and ROSEMAN, S., Proc. Nat. Acad. Sci. USA 70 (1973) 1569.
36. WEINBAUM, C. and BURGER, M. M., Nature 244 (1973) 510.

EVIDENCE FOR A GRADIENT OF ADHESIVE SPECIFICITY IN THE DEVELOPING CHICK RETINA

David I. GOTTLIEB, Kenneth ROCK and Luis GLASER

Departments of Biological Chemistry and Neurobiology
Division of Biology and Biomedical Sciences
Washington University, St. Louis, Missouri 63110 (USA)

The evidence summarized in the preceding chapter by Glaser *et al.* shows that cell-cell receptors occur on developing brain cells. As might be expected from such a complex tissue, there appears to be a multitude of receptor specificities and these have both spatially and temporally different distributions. Recent success in extracting elements of these recognition systems has raised the hope of answering one of the basic questions about cell recognition by characterizing a receptor at the molecular level. Another major question, perhaps even more difficult to answer, concerns the functional role of cell-cell receptors during embryonic development. Work in this field proceeds on the assumption that such receptors play an essential role in the formation of tissues from cells, and while this is a reasonable guess, direct evidence showing cell-cell receptor involvement in tissue morphogenesis is lacking. Consequently experiments have been initiated to examine the relationship between patterns of cell-cell adhesion and normal cellular patterns which occur during the development of the brain.

The first step in this research was to develop an assay with which the adhesive specificities of relatively small numbers of developing brain cells could be measured. Such an assay was essential because the presumed "units of specificity" of the developing brain, such as a cellular laminae and cytoarchitectonic areas, subdivide each major region of the brain into a formidable number of compartments. The monolayer adhesion assay developed in the laboratory of Saul Roseman (21) seemed well suited in that the adhesive specificity of small numbers of cells (a few thousand) could be measured. This assay suffered from one disadvantage if it was to be applied to developing brain cells, namely that cells had to be cultured for 48 hours in order to form mechanically stable monolayers, and as we have shown, some surface specificities change dramatically during even this brief period. To overcome this drawback, the monolayer adhesion assay was modified by forming the monolayers on glass surfaces derivatized with a protein binding reagent, glutaraldehyde activated γ-amino-propyl triethoxysilane. Using the procedure illustrated in Figure 1, stable monolayers of brain cells could be formed within an hour after the tissue had been dissected (11). These monolayers can be formed in small glass vials which only require about 2 x 10^6 cells to completely cover the bottom and consequently the adhesive specificity of small defined regions can easily be measured (11). For instance, in recent studies we have been able to divide the 12 day embryonic chick retina into six equal parts and measure the adhesive specificity of each.

A number of control experiments show that the cells adhering to derivatized glass have not been damaged by the procedure involved

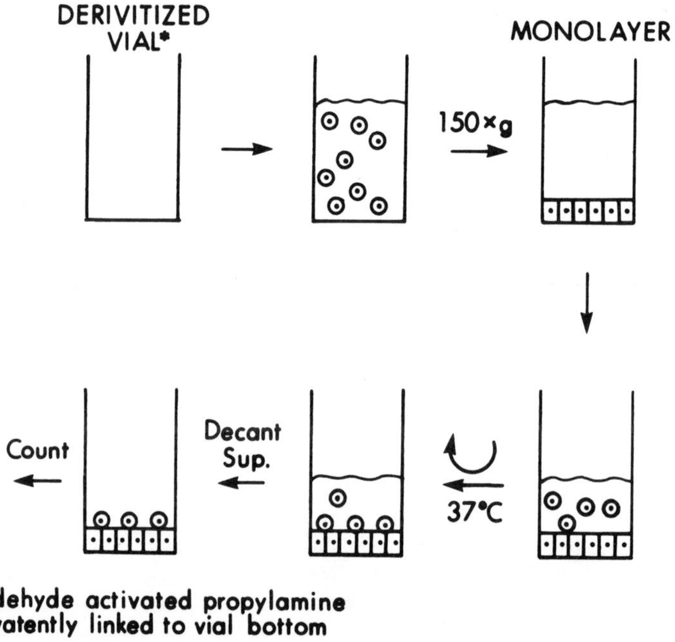

Fig. 1. Schematic representation of monolayer adhesive assay. Propylamine groups are covalently linked to the bottom of a glass vial. The amino groups are reacted with glutaraldehyde and a cell suspension is gently centrifuged onto the derivatized surface to yield a monolayer of cells. Radioactive single cells are then incubated with the monolayer in a rotary shaker. At the end of the experiment the supernatant fluid is removed, the monolayer is gently washed and the number of cells adhering to the monolayer is determined by measuring the bound radioactivity in a liquid scintillation counter. For details see (11).

in forming the monolayer. Figure 2 demonstrates that cells adhering to glass incorporate $^{32}P_i$ into TCA insoluble material at the same rate as cells in suspension culture for up to 2 hours. They also continue to exclude trypan blue for that period of time. Figure 3 shows that preincubation of either monolayer or radioactive probe cells for one hour does not diminish their binding capacity. It has been shown that C6 cells (glial tumor line) (11) grow well on derivatized glass, demonstrating the lack of a general toxic effort resulting from the derivatization procedure. Other controls (11) demonstrate that binding of radioactive probe cells to glass doesn't occur during the assay and that leakage of radioactivity from probe cells does not distort the results.

In all of the experiments probe cells labelled by short incubation (3 hours) in medium containing $^{32}P_i$ have been used. Two types of control experiments have been carried out to show that binding

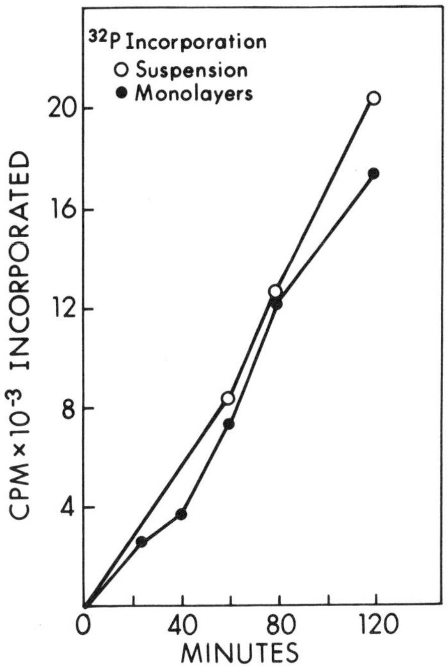

Fig. 2. Incorporation of $^{32}P_i$ into tectal cells in suspension and in monolayers. Each vial contained 2 x 10^6 cells and was incubated for the identical times in Dulbecco's modified Eagle's medium (10) + 10% heat inactivated chicken serum and 200 µCi of $^{32}P_i$. At the indicated time intervals the quantity of trichloroacetic acid precipitable counts was determined in the vials, as well as in an equivalent quantity of cells which had been kept in suspension in the same medium.

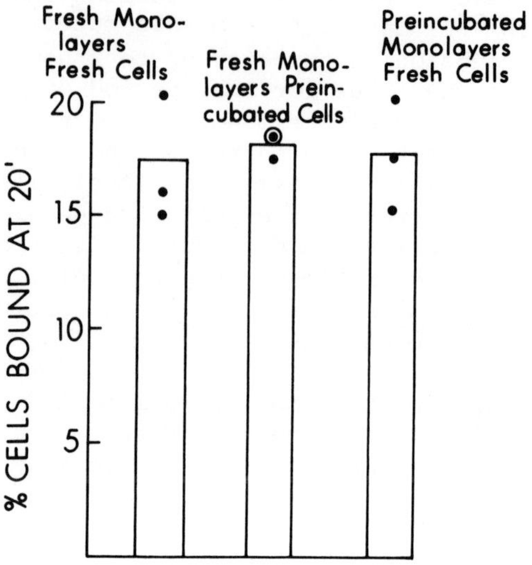

Fig. 3. Effect of preincubation on cell binding. The monolayer was prepared from telencephalic cells from 9 day old embryos and the probe cells were tectal cells from 9 day old embryos. Samples were preincubated for 45 min in calcium and magnesium-free Hanks solution containing 0.5% bovine serum albumin. The percent of cells bound at twenty minutes is shown. The closed circles represent individual samples and the bars the average value.

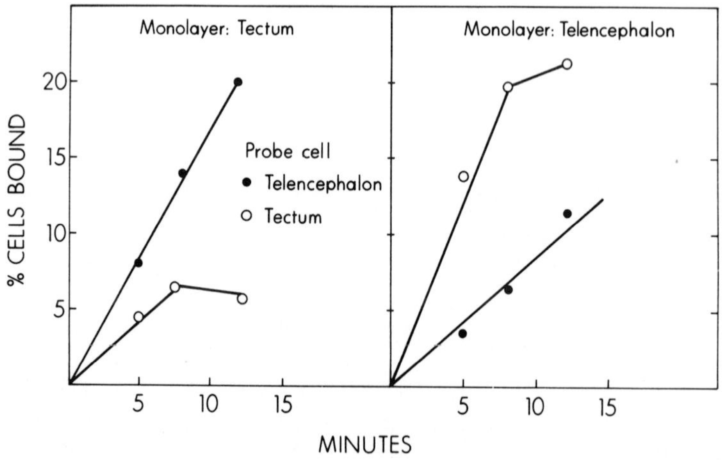

Fig. 4. Binding of tectal and telencephalic cells to monolayers. All neuronal cells were obtained from 9 day chicken embryos and incubated with the appropriate monolayers as indicated.

of ^{32}P to the monolayer represents binding of cells rather than transfer to the monolayer of a ^{32}P labelled molecule. In the first control, it has been shown that by gentle trypsinization the cells in the monolayer remained attached to the glass, but ^{32}P labelled cells were removed and showed the sedimentation characteristics of intact cells. In the second control, cells were labelled *in vivo* with [^{3}H]thymidine and used as probe cells, which adhered to the monolayer as effectively as ^{32}P labelled cells.

With this methodology in hand, attempts to demonstrate adhesive specificity between cells from the developing chick brain were made. The easiest pair to work with, namely cells from the optic tectum and the telencephalon, both of which are large regions and easy to dissect, were chosen. These cells were readily distinguishable in the adhesion assay (Fig. 4), but rather than the expected result, namely self-recognition, the cells appear to bind preferentially to cells from the other region.

At this point several aspects of the result should be discussed. First, the assay measures rates of cell-cell binding but does not show whether these rates reflect numbers of cell-cell binding sites, their relative affinities, or a combination of both parameters, so that the term "recognition" really refers to a complex rate measurement. While measurements of the relative rate of cell adhesion are very reproducible the data on the *extent* of binding is less so. This is not surprising for two reasons: 1) the extent of binding may be the same to two different monolayers that differ in the number of adhesion sites because these are present in excess, and 2) the extent of cell binding also reflects the fraction of cells that may have been damaged during isolation. Secondly, it is important to note that while homotypic rates (tectum-tectum, telencephalon-telencephalon) are lower than heterotypic (tectum-telencephalon) rates, the homotypic rates are still much higher than the binding of these cells to Chinese hamster ovary cells, a cultured fibroblastic line (11). The results therefore only tell us that heterotypic recognition is faster than homotypic, not that the first is present and the second absent.

Finally it is important to emphasize that the probe cells are binding to a confluent monolayer. It is conceivable that single cells from either tectum or telencephalon bind best to single homotypic cells and that when a confluent monolayer is created, it allows the homotypic cells to saturate all of their cell-cell receptors. This possibility would require that the cell-cell receptors and their ligands were free to move in the plane of the membrane and that they be present in roughly equal concentration. If this were the case, then in the monolayer adhesion assay one would be measuring residual sites not already occupied by the interaction of adjacent cells in the monolayer. Attempts to determine if this was

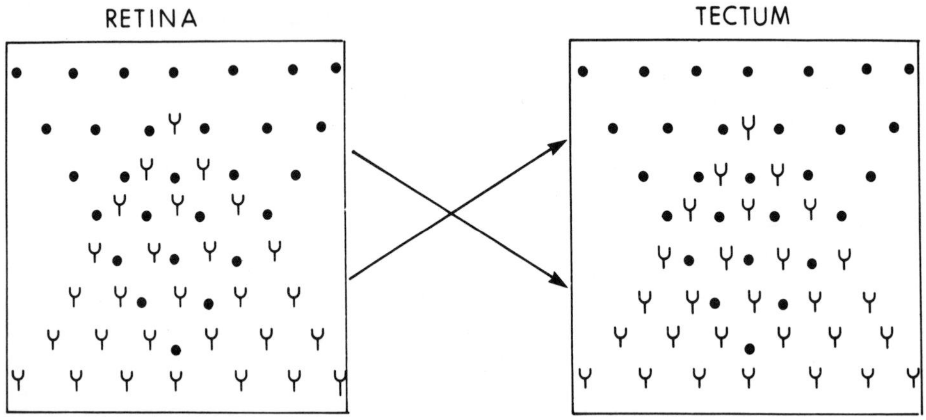

Fig. 5. Model of a gradient of cell surface markers in retina and tectum. The model assumes that the neural retina has on its surface two complementary ligands, unevenly distributed so that one ligand is present predominately on dorsal cells and other ligand on ventral cells. A similar distribution of ligands is assumed to occur on the surface of tectal cells. Thus when axons from the retina reach the tectum, axons originally from dorsal cells will connect to ventral cells in the tectum and axons from ventral cells will connect to dorsal cells. A second gradient in the medial-lateral direction (not shown) is required to account for the pattern of retina-tectal connectivity. The particular representation of the model is taken from Barbera (1), but is based on the original work of Sperry (20). A similar model has been suggested previously by Barondes (3).

so have been made by constructing subconfluent monolayers, but the results have been difficult to interpret. Although the potential difficulties are evident, we continue to use the monolayer method because it does demonstrate regional differences in the brain and because a confluent monolayer is no more an "unnatural" test object than isolated single cells.

It was of particular interest to study the specificities within a single region of neural tissue and attention was first focussed on the developing retina for several reasons. The developing retina is large and relatively homogeneous in composition when compared to other brain regions. The development of the retina has been carefully studied from an anatomical viewpoint (for example see 6, 16, 17), but perhaps most importantly, the retina has been extensively used as an experimental model in studies on the basis of synaptic specificity (for example see 1-5, 13-15, 20). As a result several models have been proposed to account for the mechanism by which the retina establishes specific synaptic connections with the optic tectum.

The connections between retina and optic tectum are such that dorsal retinal cells connect to the ventral tectum and ventral cells to the dorsal area of the tectum. A similar inversion occurs along the medial lateral axis. The first model to explain this inversion was that of R. Sperry (20) who proposed matching gradients of recognition sites on retina and tectum. A model first proposed by Barondes (3) and later by Barbera et al. (1, 2) was more explicit and is illustrated in Figure 5. This model suggests a mechanism for determining dorsal-ventral specificity by suggesting that both retina and tectum have two opposed gradients of complementary molecules. This system would allow for retina-tectal (dorsal retina to ventral tectum, ventral retina to dorsal tectum) **recognition** but also makes a prediction about intra-retinal recognition, namely that the cells from the dorsal retina should have a higher affinity for cells of the ventral retina than for themselves. A second independent medial-lateral gradient is postulated to account for the total retinal-tectal specificity.

The model is clearly an oversimplification in that if taken literally it would suggest that dorsal retina cells would connect to ventral retinal cells, which is clearly not the case. However the model makes defined predictions about retinal surface specificity which are testable on the assumption that in early development all retinal cells in a given area share the same surface markers, and that these markers are not only restricted to the ganglion cells. Indeed, evidence that this is not an unreasonable assumption has been presented by a number of laboratories (for example see 2, 18), and we have therefore proceeded to test this model in the following way.

The first set of experiments on intra-retinal adhesive specificity was done on the 9 day retina. The retina was divided into dorsal and ventral halves. Monolayers were prepared from both dorsal and ventral cells and radioactive probe cells were also prepared from both halves. The results from these experiments were clean-cut and confirmed the prediction in the Barondes-Barbera model, namely that dorsal probe cells adhered preferentially to ventral monolayers and that ventral probe cells adhered preferentially to dorsal monolayers (Fig. 6). This result was highly reproducible and when more rapid kinetics were examined it can be seen to be due in part to the initial rates of adhesion.

Analogous experiments were performed on the 12 day retina and they too showed cross-recognition between dorsal and ventral halves. The cross-reaction did not change between days 9 and 12 of development (Figure 7). Since retinal connections with the tectum must be specified in two dimensions, we were naturally curious to see if there was recognition between anterior and posterior retina. Experiments in which anterior and posterior monolayers were challenged with anterior and posterior probes failed to show any recognition.

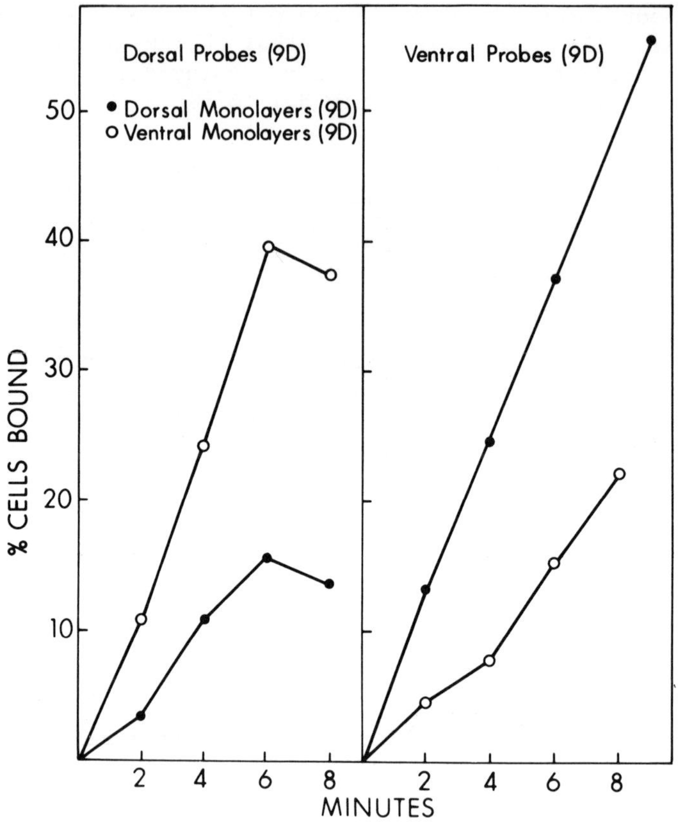

Fig. 6. Rate of adhesion of dorsal and ventral retinal cells from 9 day chick embryos. Single cell suspensions were prepared from the dorsal and ventral halves of the 9 day neural retina, and the rate of adhesions was determined using monolayer prepared from the dorsal and ventral half of the neural retina of 9 and 12 day old chicken embryos (12).

This result was true with both the 9 and 12 day retina and it must be concluded that any anterior-posterior gradient of adhesive specificity in the retina that might be present must be considerably weaker than the dorsal-ventral gradient.

The results on dorsal and ventral retinal halves are compatible with the gradient models or with a binary division of the retina into dorsal and ventral specific fields. Experiments on the 12 day retina indicate that the adhesive specificities are indeed arranged in a dorsal-ventral gradient. For these experiments, the retina was dissected into six horizontal strips as indicated in Figure 8.

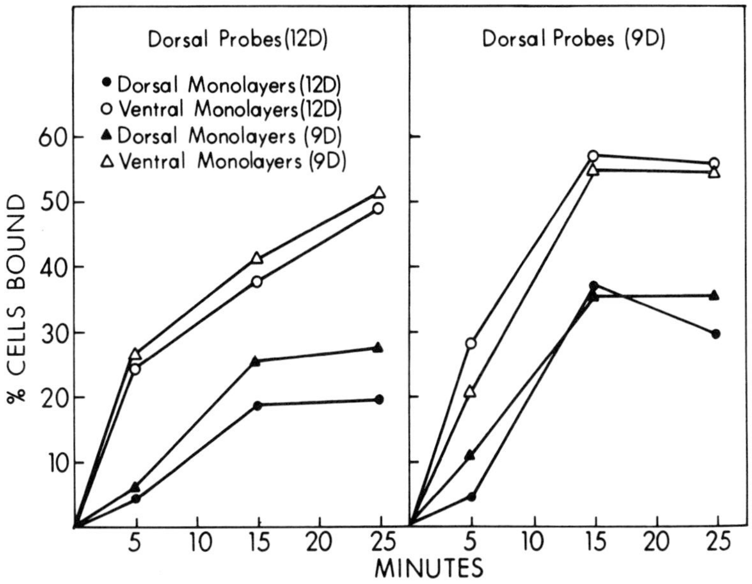

Fig. 7. Rate of adhesion of dorsal and ventral retina cells from 9 and 12 day chick embryos. Radioactively labeled probe cells were prepared from the dorsal half of 9 and 12 day old chicken embryos and the rate of adhesion was determined using monolayers prepared from the dorsal and ventral half of the neural retina of 9 and 12 day old chicken embryos (12).

Monolayers were made from each of the six strips and probe cells from the most dorsal and most ventral (i.e., strips 1 and 6). The affinity of the two types of probe cells for the series of monolayers was determined. As Figure 8 shows, dorsal cells had a relativly low affinity for all three strips from the dorsal retina but an increasing affinity for strips 4, 5 and 6. Ventral probe cells showed the reciprocal behavior having a uniformly low affinity for monolayers from all three ventral strips and an increasing affinity for more dorsal monolayers.

Because individual experiments show considerable scatter, additional experiments were done to show that the major features of the data in Figure 8 were statistically significant. Table I summarizes all of the experiments on intra-retinal recognition done to date. The data show that dorsal cells have a statistically significant preference for cells from region 6 over those from region 5 but cannot distinguish between cells from regions 1 and 2. Reciprocal experiments show that probe cells from region 6 have higher affinity for cells from segment 1 than from 2 but cannot distinguish between cell populations from regions 5 and 6.

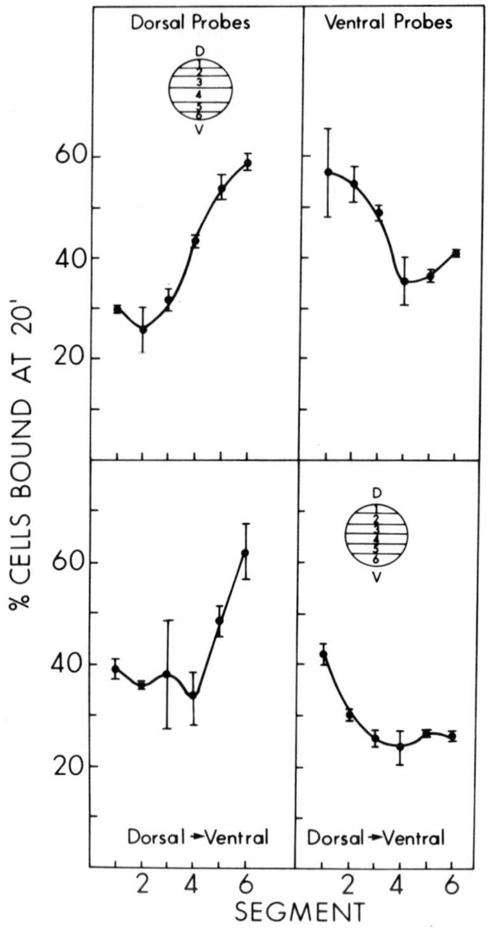

Fig. 8. Adhesive gradient across the retina. The 12 day neural retina was cut into six strips as shown and monolayers were prepared from each strip. The adhesion of ^{32}P labelled neural-retinal cells prepared from the extreme dorsal and extreme ventral strips to these monolayers are shown. The bars connect the values obtained in replicate samples. Two separate experiments are shown. A more extensive series of experiments is listed in Table I.

The original purpose of this work was to determine if there is a relationship between cell-cell adhesion and morphogenetic patterns which are important during the development of the brain. We have been able to show that there is a dorso-ventral gradient of cell-cell affinity in the developing retina which bears a resemblance to the type of gradient proposed to account for retino-tectal specificity. The critcal question is "Does the gradient of adhesive recognition have anything to do with the specificity of synapse formation?"

TABLE I

A series of experiments was carried out in which the rate of adhesion of radioactively labelled extreme dorsal cells (area 1, Fig. 8) was measured against monolayers prepared from areas 1, 2, 5 and 6 of the neural retina. Similar experiments were carried out using extreme ventral cells (area 6, Fig. 8) as probe cells. Each experiment was carried out in triplicate. The data are presented as the ratio of cells adhering to monolayers from area 5/6 (12).

Probe Cell	No. of Experiments	Ratio of Adhesion to Monolayer 1/2	5/6
Extreme Dorsal (Area 1)	6	--	0.77 ± .07[1]
Extreme Ventral (Area 6)	5	-	1.006 ± .06
Extreme Dorsal (Area 1)	4	1.01 ± .10	-
Extreme Ventral (Area 6)	4	1.35 ± .22[1]	-

[1]Significantly different from unity at the 0.01 level of confidence.

There are several points which argue against equating the two gradients. The first is simply that it has been found that there is a functional gradient by measuring the affinity of retinal cell bodies for one another. In real life, a cell body from the dorsal retina would never encounter one from the ventral retina, so the assay is based on an "unnatural" pairing. Furthermore, the retinal ganglion cells, the only retinal cells to project to the optic tectum, comprise only a minor fraction of the retina, but the differential between dorsal and ventral retina seems to involve a substantial fraction of cells. To argue that the adhesive gradient is related to synaptic specificity, we would have to assume that most all cells in the retina become marked with a common positional indicator. This is not unreasonable and is in fact suggested by the columnar organization of many multilayered neural structures. The chick retina itself receives an ordered projection from the isthmo-optic nucleus which seem to terminate on amacrine cells lying superficial to the ganglion cell layer, but is nevertheless in register with the retinal map (7, 9, 19). The failure to find an anterior-posterior gradient offers serious difficulty to the hypothesis that the dorso-ventral gradient is involved in synaptogenesis since retino-tectal connections are specified in two dimensions. It is possible to argue that this gradient is present but too weak to be detected by our assay, but at this point the discrepancy between theory and experimental finding remains.

One alternative interpretation is that the dorsal-ventral gradient represents a developmental gradient. In fact, recent observations in frog retina by M. Jacobson (personal communication) suggest the presence of a dorso-ventral developmental gradient. If such a gradient exists in the chicken retina, then our surface adhesive gradient may reflect this developmental gradient. It should be noted, however, that the gradient of surface specificity which has been observed appears not to change between days 9 and 12 of development, while one would expect to see changes in a developmental gradient during this time period.

Orderly embryonal development has been repeatedly postulated to require either gradients of diffusable effectors and/or gradients of surface markers (8, 20). The precise function of the gradient of adhesive specificity clearly remains to be elucidated, but the finding of this gradient and the techniques used are clearly of considerable relevance in furthering understanding of the development of the retina. The techniques used to find this gradient should be directly applicable to other systems.

ACKNOWLEDGEMENTS

This work has been supported by Grant BMS-74-22638 from the National Science Foundation and Grant GM 18405 from the National Institutes of Health. During part of this investigation, D. I. G. was supported by Grant NS 10943 from the National Institutes of Health.

REFERENCES

1. BARBERA, A. J., Developmental Biology 46 (1975) 167.
2. BARBERA, A. J., MARCHASE, R. B. and ROTH, S., Proc. Nat. Acad. Sci. USA 70 (1973) 2482.
3. BARONDES, S. H. in The Neurosciences Second Study Program (F. O. Schmitt ed.) Rockefeller University Press (1970) 747.
4. CHUNG, S. A., Cell 9 (1974) 201.
5. CROSSLAND, W. J., COWAN, W. M., ROGERS, L. A. and KELLY, J. P., J. Comp. Neurology 155 (1974) 127.

6. COULOMBRE, A. J., Amer. J. Anat. 96 (1955) 153.
7. COWAN, W. H., ADAMSON, L. and POWELL, T. P. S., J. of Anatomy 95 (1961) 545.
8. CRICK, F. H. C. and LAWRENCE, P. H., Science 189 (1975) 340.
9. DOWLING, J. E. and COWAN, W. M., Z. Zellfosch. Mikrok. Anat. 71 (1966) 14.
10. DULBECCO, R. and FREEMAN, G., Virology 12 (1959) 185.
11. GOTTLIEB, D. I. and GLASER, L., Biochem. Biophys. Res. Comm. 63 (1975) 815.
12. GOTTLIEB, D. I., ROCK, K. and GLASER, L., Proc. Nat. Acad. Sci. USA 73 (1976) *in press*.
13. HUNT, R. K. and JACOBSON, M., Proc. Nat. Acad. Sci. USA 69 (1972) 780.
14. HUNT, P. R. and JACOBSON, M., Proc. Nat. Acad. Sci. USA 70 (1973) 507.
15. JACOBSON, M., Developmental Neurobiology, Rinehart and Winston, New York (1970).
16. KAHN, A. J., Brain Res. 63 (1973) 285.
17. KAHN, A. J., Develop. Biol. 38 (1974) 30.
18. MARCHASE, R. B., BARBERA, A. J. and ROTH, S., Ciba Foundation Symposium No. 29, Cell Patterning, Elsevier, Amsterdam. (1975) 315.
19. MC GILL, J. I., POWELL, T. P. S. and COWAN, W. M., J. of Anatomy 100 (1966) 35.
20. SPERRY, R. W., Proc. Nat. Acad. Sci. USA 50 (1963) 703.
21. WALTHER, B. T., OHNAN, R. and ROSEMAN, S., Proc. Nat. Acad. Sci. USA 70 (1973) 1569.

Section III

Cell-Macromolecule Interactions

MEMBRANE RECEPTORS FOR BACTERIAL TOXINS

W. E. VAN HEYNINGEN

Sir William Dunn School of Pathology
University of Oxford (United Kingdom)

Paul Ehrlich may well have been right when he said "Corpora non agunt nisi fixata - Things won't work unless they're fixed." But there are various ways of looking at fixing, and the receptors that do the fixing: I used to think that receptors were the final targets that active substances latched on to and altered in some way. I remember seventeen years ago when I investigated Wasserman-Takaki phenomenon, that is the apparently specific fixation of tetanus toxin by nervous tissue, and discovered that it was ganglioside that fixed tetanus toxin. I proclaimed that ganglioside was the receptor for tetanus toxin in nervous tissue, and confidently expected that tetanus toxin would change it in some way - that is, tetanus toxin had to be an enzyme and ganglioside its substrate. But the ganglioside appeared to be quite unaltered by its avid and specific encounter with tetanus toxin. The biochemist in me was greatly disappointed - I had reached a so what? situation. But my pharmacologist friends took quite a different view. "What are you grumbling about?" they asked - "you've identified the tetanus toxin receptor!" To them, I had reached the top of the ladder; to me, I was only at the bottom. I couldn't understand them at all; but I have since come to realize that some receptors are targets on the cell surface while others are no more than receptionists that recognize and receive the active substance and facilitate its access to a target beyond the cell surface. Unless active proteins act on substances in solution in plasma, the way for example that coagulases and fibrinolysins do, they have to act on cells, and in acting on cells they first have to cope with their membranes. They work either by altering or completely destroying the membrane, or by getting through the membrane to act on a target beyond. At least this seems to be the case with protein toxins - small molecular substances probably act differently, and the way they work is pharmacology.

MEMBRANE RECEPTORS THAT ARE TARGETS

There are a number of cytolytic bacterial toxins that directly attack the membranes of cells, generally including red blood cells, and so bring about their destruction in one way or another. These cytolytic toxins can be classified according to whether they 1) cause cell membranes to shrink, 2) cause membranes to become permeable to cations or 3) cause membranes to become permeable to macromolecules.

Membranes destroyed by shrinkage

The first bacterial toxin whose mode of action was defined in chemical terms was the alpha-toxin of *Clostridium perfringens*, which was shown in 1941 to be a lecithinase C (41, 42). Since then many other bacterial lecithinases C have been recognized, the most interesting of which is the staphylococcus beta-toxin which was shown to be a sphingomyelinase (20). The *Cl. perfringens* alpha-toxin splits phosphorylcholine from phosphatidyl choline (and also attacks other phospholipids) and the staphylococcus beta toxin splits phosphorylcholine from sphingomyelin (and also attacks other phospholipids). Both these toxins are "hot-cold" hemolysins - that is, when incubated with susceptible red cells no hemolysis takes place at 37° (provided there is not an excess of toxin), but as soon as the temperature is lowered, even to room temperature, hemolysis sets in.

The attack by *Cl. perfringens* or *Bacillus cereus* phospholipases C, or by staphylococcus sphingomyelinase, on susceptible red cells brings about a release of water-soluble phosphorylated amines, and leaves shrunken ghosts that are in a fragile state. The cell membrane shrinks, with a 45% reduction in its area, but remains intact and the average protein conformation is unchanged (13, 14, 16, 39). However, the shrunken membrane is very fragile, possibly because the stripping of the outer layer of choline phospholipids leaves regions that are essentially monolayers of phosphatidyl serine and phosphatidyl ethanolamine. These regions are stable enough to persist at 37°, but as soon as the temperature is lowered they collapse and hemolysis takes place (see 9). This is the "hot-cold" hemolysis that characterizes these hemolysins. Cells that have been made fragile like this by the staphylococcal beta-hemolysin can also be made to hemolyze, at 37°, by a number of other agents, such as lipases, proteases, glycerin (see 62).

Membranes made permeable to cations

The hemolytic toxins that cause cell membranes to become permeable to cations act in two phases, prelytic and lytic. During the prelytic phase the toxin is bound to the cell membranes and cations are released from the cell. During this phase the action of the toxin can be stopped with antitoxin if it is added soon enough. During the lytic phase, when hemoglobin is released from red cells, the action can no longer be stopped with anti-toxin, but it can be stopped by immersing the cells in an osmotic stabilizer, such as a sucrose solution. Two of the cytolysins to be discussed are unlikely to be enzymes, and there is no evidence that the third is an enzyme.

Staphylococcal alpha-toxin, Streptolysin S and Cl. septicum hemolysin. The staphylococcal alpha-toxin, (see 1, 9) is hemolytic to red cells of a number of species, but those of the rabbit are by far the most susceptible. The sensitivity of the red cell to the toxin can be correlated with its ability to bind the toxin.

There is no evidence that hemolysis is due to any enzymic action of the toxin, and indeed the alpha-toxin is used up during hemolysis, which suggests a stoichiometric rather than an enzymic reaction. Recent evidence supports the suggestion that the action of the toxin is non-enzymic. Staphylococcal alpha-toxin has a sedimentation coefficient of about 3 S, and a molecular weight of about 28,000. The 3 S form polymerizes spontaneously to a 12 S form with a molecular weight of about 300,000, and this polymerization can be induced and hastened by a number of lipids. The 12 S form appears on electron microscopy to consist of rings with an external diameter of about 10 nm, made up, on the average, of six subunits per ring, each 2-2.5 nm in diameter, each subunit apparently being a dimer of the 3 S form. It is possible that the ability of the alpha-toxin to form these rings is responsible for its hemolytic activity.

When red cells are lysed by the 3 S toxin, ring-like structures similar to those of the 12 S form appear on, or in, the membranes (Fig. 1). Freeze-etching of toxin-treated membranes suggests that the rings may penetrate to the central hydrophobic plane of the membrane lipid bilayer. It is not yet known whether these rings are formed before lysis of the cells, or only on membranes resulting from lysis. If it were the former, and the rings were to penetrate the membranes, the resulting holes (about 5 nm internal diameter?) in the membranes could explain the altered permeability of the membranes to ions, which would lead to osmotic lysis. Such a mechanism (the doughnut model, 8.8 nm holes) has also been suggested for the lysis by complement (21).

Hemolysis of red cells by streptolysin S (see 2, 7, 27) and Clostridium septicum hemolysin (see 7, 9) is similar to that caused by staphylococcal alpha-toxin in that the cells are made permeable to cations during a prelytic phase when antitoxin is still effective, and lysis can be stopped by sucrose in the second phase. There has

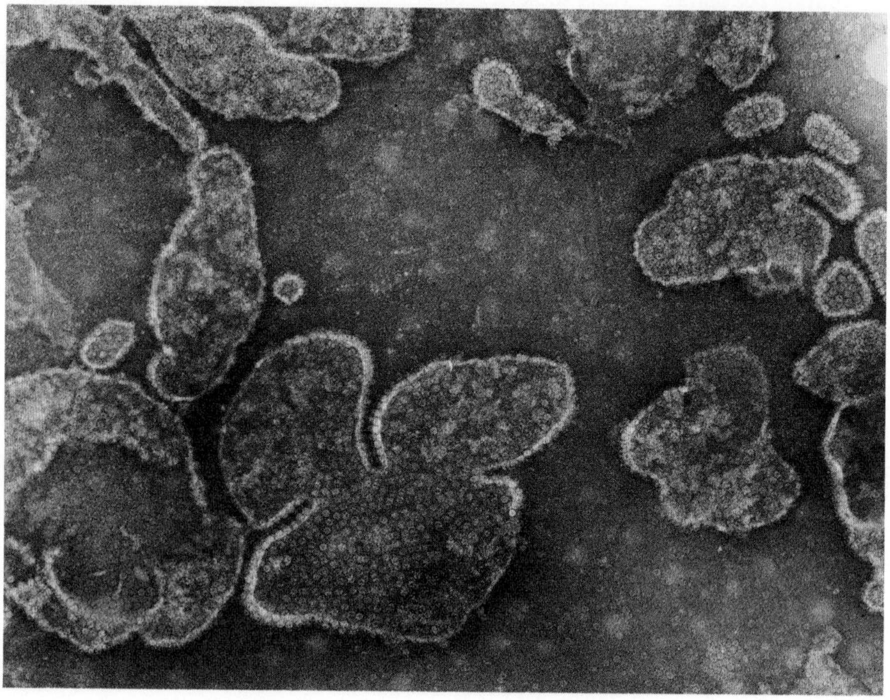

Fig. 1. 12 S aggregates of staphylococcal alpha-toxin on membranes of hemolysed human red cells. Figure kindly provided by Dr. J. H. Freer of the University of Glasgow and reproduced with his permission.

been no evidence of any holes appearing on the cell membrane. Streptolysin S, which is a complex of oligonucleotide and polypeptide components, does not act catalytically (vide infra).

Cl. septicum hemolysin appears to be a simple protein, and there is no evidence whether or not it is an enzyme.

Staphylococcal leukocidin

The staphylococcal leukocidin (see 72) is toxic to polymorphonuclear leukocytes and macrophages in man and the rabbit, but no other cell type has been found to be susceptible. The toxin causes an increase in the permeability of the cell membranes to potassium ions, which results in the secretion of protein, the accumulation of calcium, stimulated ATPase activity and stimulated nucleotide-orthophosphate exchange. The leukocidin consists of two proteins, the F (32,000 M. W.) and S (38,000 M. W.) components, which are inactive singly, but act together synergistically. At low ionic strength both F and S components undergo partial polymerization. The site of action of the toxin is the outside of the cell membrane, and during the action of the toxin on the membrane it is only briefly bound and is returned to solution in an inactivated condition. The interaction of the leukocidin with the cell membrane appears to involve phospholipids, but only one phospholipid, triphosphoinositide which occupies 1% of the membrane surface, is involved in the inactivation of the toxin. During this process it undergoes a small conformational change.

The mode of action of the staphylococcal leukocidin is very complicated, and not yet completely understood. It seems to involve first the interaction of F with the cell membrane, then the interaction of S with F, then the interaction of S with the membrane. The following mechanism has been proposed: 1) Part of the F component molecule enters the membrane in a region of low dielectric constant. 2) The F component is transformed into an expanded condition, analogous to that found at low ionic strength. 3) Esterified fatty acid enters the hydrophobic regions of the F molecule, producing a change analogous to that produced by phospholipid micelles. 4) The S component is adsorbed on to the altered surface of the F component. 5) The S component alters the conformation of triphosphoinositide molecule by interacting with its inositolphosphate region. 6) Electrolyte-permeable channels are created, leading to loss of the regions of low dielectric constant in the membrane and subsequent return of the F component to the compact state. 7) The inactivated F component is deadsorbed from the esterified fatty acid side chains of triphosphoinositide, and the inactivated S component is deadsorbed from the compact F component. 8) The conformation of triphosphoinositide in the membrane is restored with expenditure of cellular ATP. 9) The cycle is repeated with fresh leukocidin molecules.

Membranes Made Permeable to Macromolecules

There are a number of toxins, produced by widely different bacteria (Table I), that have many similarities - similarities so marked

TABLE I

THE OXYGEN-LABILE HEMOLYSINS

Name	Producing organism
Streptolysin O	Streptococci of groups A, B, C and G
Pneumolysin	*Streptococcus pneumoniae*
Tetanolysin	*Clostridium tetani*
Theta-Toxin	*Clostridium perfringens*
Delta-Toxin	*Clostridium septicum*
Epsilon-Toxin	*Clostridium histolyticum*
Delta-Toxin	*Clostridium oedematiens* Type A
Hemolysin	*Clostridium bifermentans*
Hemolysin	*Clostridium botulinum*
Cereolysin	*Bacillus cereus*
Thuringiolysin	*Bacillus thuringiensis*
Hemolysin	*Bacillus alvei*
Hemolysin	*Bacillus laterosporus*
Listeriolysin	*Listeria monocytogenes*

as to have prompted the statement (9): "the detailed mode of action of all the SH-activated lysins will be understood at the moment any one of them is." These common properties are: 1) they are proteins; 2) they are serologically cross-reactive; 3) they are reversibly inactivated by mild oxidation and reversibly reactivated by reduction (their inability to act in the oxidized state seems to be due to their inability to be bound to the cell membrane in this state); 4) they are hemolytic (and some, at least, are lytic to other cells); 5) they are cardiotoxic; 6) they are bound and inactivated by cholesterol and certain other digitonin-precipitable steroid derivatives.

An entirely different class of compounds - also steroid derivatives, viz. saponin and certain bile acids, share the last three properties with these protein toxins. We shall consider hemolytic and antihemolytic properties of steroid derivatives first.

It has been known for a long time that antihemolytic steroid derivatives (including cholesterol) have a hydroxyl group at position 3 cis to the methyl group at position 10, and either a trans fusion of rings A and B, or a double bond between positions 5 and 6 (as in cholesterol); bile acids (but not sterols) having arrangements of the substituents at position 3 and 5 opposite of those of the protective compounds are hemolytic (6) (Fig. 2).

The oxygen-labile hemolysins are bound and blocked, on the whole, by the same steroid derivatives that block the hemolytic steroid derivatives (see 8, 9); this, and the fact that the hemolytic steroid derivatives seem to have similar effects to those of the hemolytic proteins on the red cell membrane, and on the heart, and the further fact that the oxygen-labile hemolysins seem to have a common antigenic determinant, suggest that these protein hemolysins have a common structure that behaves biologically very much like a hemolytic steroid derivative. Since steroid derivatives are not enzymes, this suggests that the biological action of the oxygen-labile enzymes is not enzymic. Many years ago an experiment was done (30) that suggests that streptolysins O and S are not enzymes. From dose-response curves the concentrations of hemolysins were determined that produced 80% hemolysis of a red cell suspension after incubation for 2 hours at 37°; then the time course of hemolysis by these concentrations of hemolysins were plotted. Eighty per cent hemolysis was attained in about 20 min, and further incubation up to 60 min produced no furthur hemolysis. In other words, a given amount of hemolysin produced only a given degree of hemolysis, no matter how long the incubation period. This is a stoichiometric reaction, not an enzymic reaction.

That cholesterol is the receptor for these oxygen-labile hemolysins (and the non-protein hemolysins) is suggested by the finding that all cells known to be susceptible contain cholesterol in their membranes, and cells not containing membrane cholesterol (e. g. bacterial protoplasts and spheroplasts) are insensitive. Furthermore, the various hemolytic substances known to bind to cholesterol, viz. oxygen-labile hemolysins, saponin and the polyene antibiotic filipin (which has no structural relation to steroid compounds), reciprocally interfere with each other's binding to natural and artificial membranes containing cholesterol (9).

It seems that cholesterol is not only the membrane receptor for the stoichiometrically acting oxygen-labile hemolytic proteins, but also the target. Oxygen-labile hemolysins produce holes in susceptible cell membranes, 50-70 nm in diameter, lined with material of different electron density (after negative staining) from that of the surrounding matrix. On the basis of this and other information the following mechanism has been proposed for hemolysis induced by these hemolysins (9): firstly, the hemolysin is bound to the cholesterol to form complexes; then the complexes are translocated laterally

Bile acids and sterols

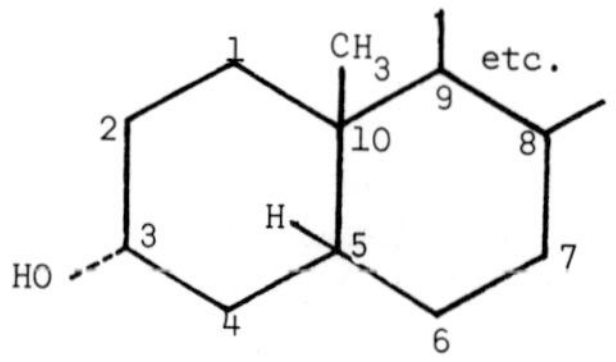

lithocholic acid
hemolytic

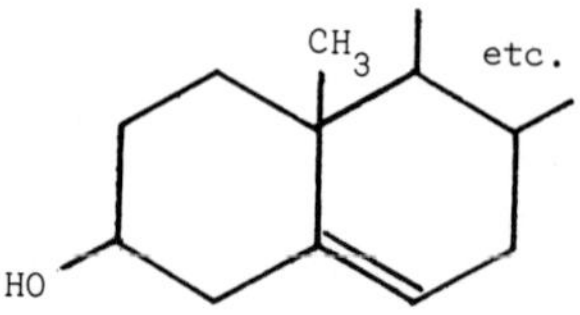

cholesterol
anti-hemolytic

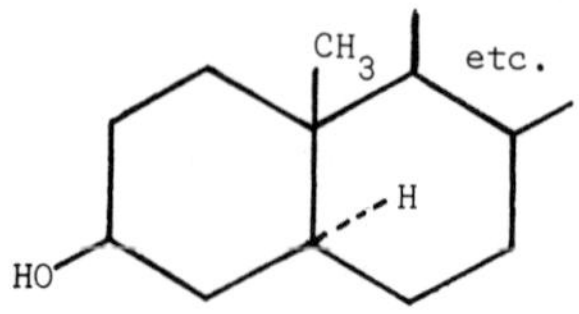

dihydrocholesterol
anti-hemolytic

Saponin aglycones

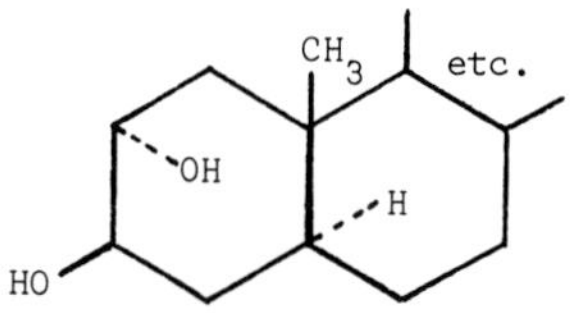

digitogenin, gitogenin
? hemolytic

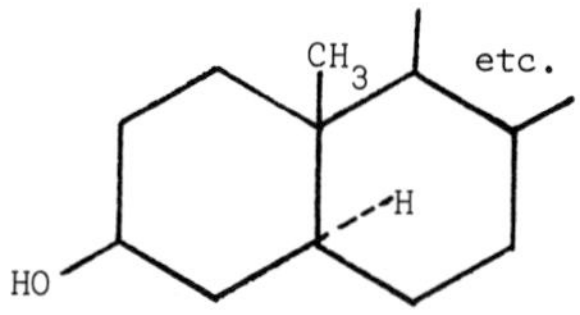

tigogenin
? anti- or non-hemolytic

Cardiac aglycones

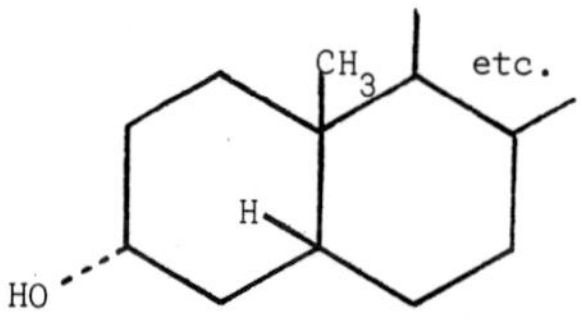

digoxigenin
? hemolytic

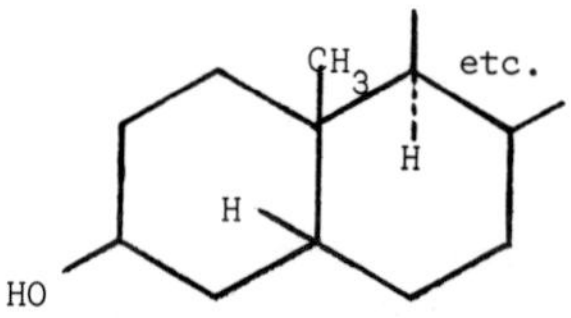

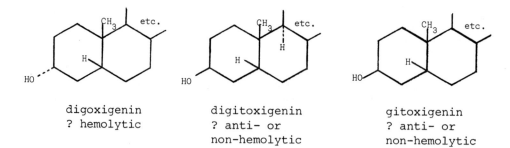

digitoxigenin
? anti- or
non-hemolytic

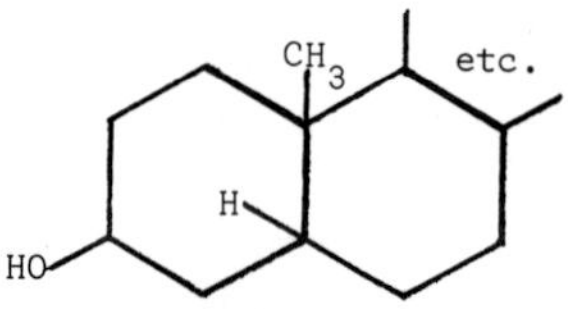

gitoxigenin
? anti- or
non-hemolytic

Fig 2. A and B Rings of Hemolytic and Anti- or Non-Hemolytic Steroid Derivatives.

to form holes or channels lined by these complexes. Finally the hemoglobin escapes either through the holes or channels, or through the other areas in which the lipids may have become disorganized by the loss of cholesterol.

MEMBRANE RECEPTORS THAT ARE NOT TARGETS

Diphtheria toxin

Diphtheria toxin (see 49) is produced only by lysogenic strains of *Corynebacterium diphtheriae* infected with a bacteriophage carrying the *tox* gene, which carries the structural information for synthesis of the toxin. The expression of this information is regulated, at least in part, by the bacterial host. The toxin acts by blocking protein synthesis in susceptible cells (e.g. Hela cells) by inhibiting polypeptide chain synthesis, provided NAD^+ is present, and it does this by catalyzing the following reaction:

$$\underset{\text{active}}{NAD^+ + EF\text{-}2} \rightleftharpoons \underset{\text{inactive}}{ADPR - EF\text{-}2} + \text{nicotinamide} + H^+$$

The elongation factor EF-2, which is required for translocating polypeptidyl-transfer RNA from the acceptor site to the donor site on the eukaryotic ribosome, is inactivated by being coupled to the adenosine diphosphate ribose (ADPR) resulting from the cleavage of NAD^+.

Diphtheria toxin is a protein with a molecular weight of about 62,000 consisting of a single polypeptide chain with two non-overlapping cystine bridges. One of these bridges is about 40% along the chain from the amino-terminal glycine, and spans a loop of about 10 amino acids, three of which are arginine residues. This loop is sensitive to trypsin at the peptide links after any of the three arginine residues. The resulting product is still fully toxic, and is called the "nicked" toxin. If the disulfide bridge is broken by reduction, two large peptides are formed: the amino-terminal fragment A, and the carboxyl fragment B. The molecular weights of the A and B fragments are about 24,000 and 38,000 respectively, varying a little according to which of the arginine peptide links was "nicked" (Fig. 3). After reduction and alkylation of the -SH groups, the A and B chains are firmly held together by weak interactions, and the toxicity of the complex is unimpaired.

Fragments A and B can be separated by electrophoresis or gel-filtration in denaturing solvents. The B fragment, especially the region near the carboxyl terminal, appears to contain many hydrophobic residues, and therefore is unstable and denatures and aggregates easily and cannot be recovered in the native state. The

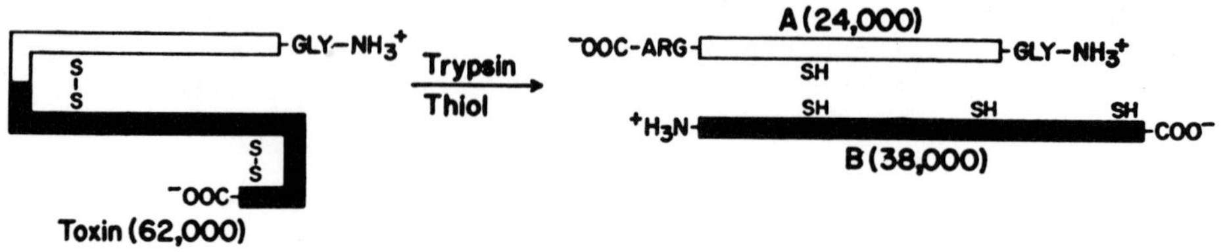

Fig. 3. The diphtheria toxin molecule before and after reduction followed by mild hydrolysis with trypsin. From A. M. Pappenheimer, Jr. and D. M. Gill (49). Reproduced with permission of the authors and the editorial office of Science. Copyright 1973 by the American Association for the Advancement of Science.

A fragment, however, is easily renatured. It is stable to heat, severe extremes of pH and proteolytic enzymes.

The enzyme activity of diphtheria toxin resides in the A fragment. Intact toxin will not catalyze the interaction of NAD^+ and EF-2 in cell-free solution of the reactants, but the A fragment will do so; on the other hand, the A fragment is not toxic to whole cells, whereas the intact (or the nicked) toxin is; the complex of A and B fragments held together by non-covalent bonds is rather less toxic than intact toxin to whole cells, and is also enzymically active (Table II).

Covalent attachment of the B fragment to the A fragment thus appears to block the enzymic activity of the latter; on the other hand the A fragment can only exert its effect on the intact cell when it is bound, either covalently or non-covalently, to the B fragment. The function of the B fragment seems to be to recognize the toxin receptor in the sensitive cell membrane, to bind to it, and thus to facilitate the passage of the A fragment into the cell. The cell membrane receptor for diphtheria toxin is the receptor for the B component; the target of the active component of the toxin is the elongating factor, together with NAD^+.

Although it has not been possible to make any soluble preparations of the B fragment, which would have been useful in studying the interaction of diphtheria toxin and the sensitive cell membrane, and which would be of more interest in this conference, it has been possible to prepare certain diphtherial proteins that have been useful for this purpose. When replicating phage is treated with a mutagen, e. g. nitrosoguanidine, some survivors carry modifications of the tox gene, which, when expressed by the host bacterium, result in the production of proteins of partial or totally reduced activity that cross-react with diphtheria antitoxin. In these proteins, known as CRM's (cross-reactive material), the A and B components

TABLE II

PROPERTIES OF DIPHTHERIA TOXIN AND RELATED PROTEINS*

	Toxicity	(1) Enzymic activity	(2) Blocking activity
AB	full	full	nt
A/B	full	full	nt
(A+B)	half?	full	nt
A	none	full	none
B	na	na	na
CRM_{45}	none	full	none
CRM_{197}	none	none	full
CRM_{228}	none	none	some
$A_{45}B_{197}$	full	full	nt
$A_{45}B_{228}$	some	full	some

*AB = intact toxin; A/B = nicked toxin; (A+B) = nicked and reduced toxin; CRM = cross-reactive material. 1) enzymic activity after nicking and reduction; 2) ability to inhibit competitively the action of intact toxin on Hela cells; na = not available; nt = not tested. (Adapted from 49).

fall into three groups (i) A good in enzymic activity, B poor in binding (CRM_{45}); (ii) A poor in enzymic activity, B good in binding (CRM_{197}); (iii) both A and B poor (CRM_{228}) (Table II). It is possible to prepare hybrids of the A and B fragments of these CRM's by nicking them with trypsin in the presence of a thiol and then allowing them to reoxidize together by dialysis. Table II shows that hybrids of "good" A and "good" B are indistinguishable from intact toxin ($A_{45}B_{197}$), and a hybrid of "good" A and "poor" B has potential enzymatic activity but little toxicity ($A_{45}B_{228}$). This table also shows that a CRM containing a "good" B is capable of blocking the action of toxin on Hela cells (CRM_{197}), whereas CRM's 45 and 228 have little or no blocking capacity. In the case of diphtheria toxin we know what the target is within the cell, but we do not know what the receptor is on the cell membrane, or how the binding of the B component to this receptor facilitates the passage of the A component across the membrane. Nor do we know what happens to the B component during this process, nor whether the nicking and reduction of intact toxin takes place at, or in, the membrane, nor whether the entire toxin molecule enters the cell before the liberation of the A component.

No apparent damage is done to the membrane during this process; there is no leakage of potassium or phosphate ions, and the cells can still take up fluorescein acetate and accumulate fluorescein.

Cholera toxin

Cholera toxin acts on all animal cells so far tested by stimulating adenylate cyclase and so bringing about a rise - after a characteristic lag of 20-30 minutes - in the cellular level of cyclic AMP. In the brush border cells of the epithelium of the small intestine this results in increased secretion of chloride ions into the lumen, followed by water and various ions, leading to diarrhea (55). When introduced into the skin it causes an increase in capillary permeability (see 22). It is not yet known whether the toxin acts directly on the adenylate cyclase, or whether an intermediate is involved. The presence of NAD^+ and an unidentified macromolecular substance is apparently necessary for the action of toxin (25, 26). It is thought that adenylate cyclase is located on or in the cytoplasmic side of the cell membrane (35).

Cholera toxin is a protein with a molecular weight of about 84,000. During the isolation of the toxin, a biologically inert material with a molecular weight of about 54,000, called choleragenoid, may also be isolated. Choleragenoid is nearly, but not quite immunologically identical with cholera toxin, and apparently is derived from it. Recent work from several laboratories suggests that cholera toxin, like diphtheria toxin, consists of a binding component B and an active component A. B corresponds to choleragenoid and is composed, according to different interpretations, of 4 to 6 subunits bound together by non-covalent bonds. B is bound to A by non-covalent bonds. A_2 consists of an inactive A portion (molecular weight about 5,000) attached to an active A_1 portion (molecular weight about 25,000) (19, 23, 24, 26, 40, 60).

The cell membrane receptor for cholera toxin is now known: it appears to be a complex containing the sialidase-stable monosialosyl ganglioside, galactosyl-N-acetylgalactosaminyl-sialosyllactosylcermide, generally known as G_{M1}, but referred to here as GGnSLC (36, 65, see also 18, 33) (Fig. 4). The rather easy identification of ganglioside as the receptor for cholera toxin (67) owed its origin to the rather less easy identification of ganglioside as the receptor for tetanus toxin many years previously (63), when ganglioside was hardly known and the multiplicity of gangliosides not recognized, much less their structures worked out (66).

When loops of dog small intestine are exposed to choleragenoid, and then to cholera toxin, the action of the toxin in promoting secretion of fluid into the intestinal loop is blocked (54), one molecule of choleragenoid blocking the activity of one molecule of

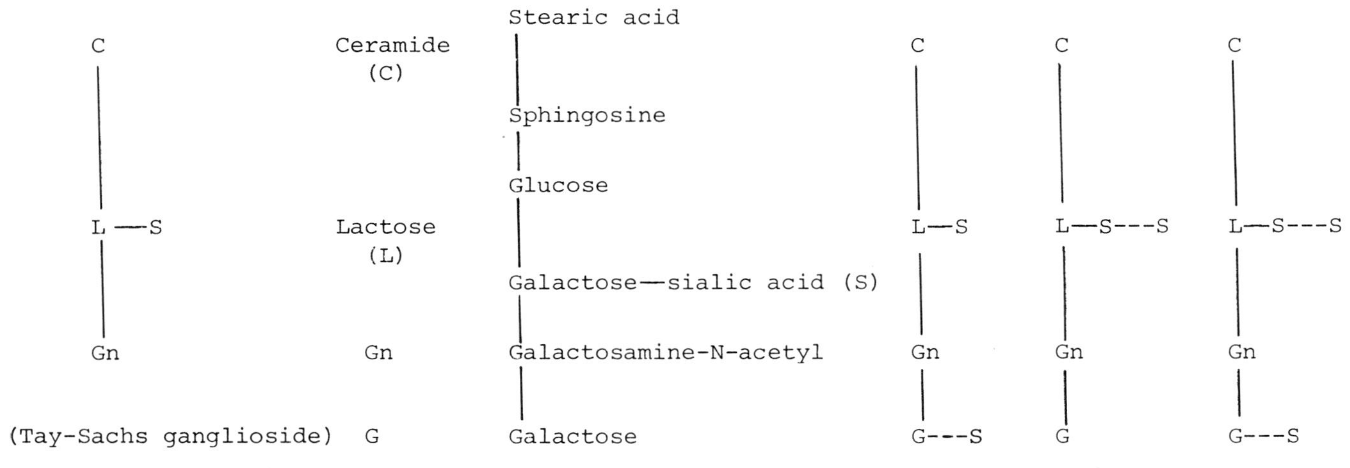

Notation:

Svennerholm (59)	G_{M2}	G_{M1}	G_{D1a}	G_{D1b}	G_{T1}
McCluer (44)	GnSLC	GGnSLC	SGGnSLC	GGnSSLC	SGGnSSLC

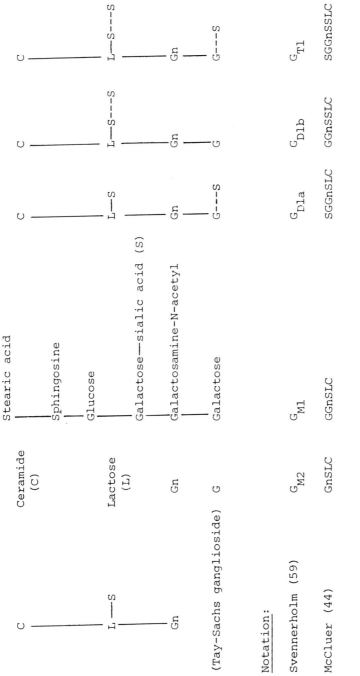

Fig. 4. Trivial structures of the best-known gangliosides, in descending order of R_f, on thin-layer chromatography. The abbreviated trivial names of McCluer are self-explanatory. —S and ---S signify sialidase-resistant and -sensitive bonds respectively.

toxin. This suggests that choleragenoid occupies the toxin receptors on the epithelial cell membranes (65). Similarly, if homogenates of intestinal epithelial cells are pretreated with choleragenoid their capacity to bind cholera toxin is diminished, one molecule of choleragenoid blocking the binding of 0.69 molecules of toxin. If ganglioside GGnSLC in free solution is mixed with choleragenoid, the binding of only 0.31 molecules of toxin is blocked by one molecule of choleragenoid; but if the ganglioside is complexed with cerebroside (which has no binding capacity for cholera toxin), the binding of 0.62 molecules of toxin is blocked (37). This suggests that the cholera toxin receptor is a complex containing GGnSLC, but not necessarily that cerebroside is involved. Possibly phospholipids may be involved. It is likely that the ceramidyl hydrophobic regions of gangliosides would be buried among the ceramidyl regions of the ceramidyl phospholipids in the outer layer of the membrane lipid bilayer, leaving the oligosaccharide moieties exposed on the outside of the membranes (14, 66). The specificity of ganglioside for cholera toxin and tetanus toxin resides in the oligosaccharide moiety, although the ceramidyl moiety (or at least part of it) is also necessary for binding (34, 57).

It may be of significance in the development of the disease of cholera that *Vibrio cholerae* produces a sialidase, and that some sialidase-sensitive gangliosides, which do not bind cholera toxin, are converted to the toxin-binding GGnSLC; that toxin binding by intestinal epithelial homogenates is increased after treatment with sialidase (65); that pretreatment of dog small intestinal loops with sialidase increases their output of fluid in response to cholera toxin (56); and that treatment of adrenal cells with sialidase increases their response (formation of corticosterone, cyclic AMP) to cholera toxin (30). It is also very interesting that when exogenous ganglioside is added to cells it is incorporated in them and increases their binding of cholera toxin, and their response to it (18). In the case of the pigeon red cell, not all the exogenous ganglioside that can be incorporated binds the toxin, and not all the toxin that is bound by incorporated ganglioside stimulates adenyl cyclase, although about half the toxin bound by endogenous ganglioside is effective in stimulating adenyl cyclase (35).

The A component of cholera toxin can be detached from the B component in a number of ways, including acid treatment, and binding of toxin to ganglioside-cerebroside followed by elution of the A component with 8M urea (19, 23, 24, 26, 40, 60). The A component has a slight activity in increasing capillary permeability in the rabbit skin, and in stimulating adenylate cyclase in the pigeon red blood cell, but this activity, unlike that of the intact toxin, cannot be blocked with ganglioside.

When the pigeon red cell is perforated with the theta-hemolysin of *Clostridium welchii* (26) both the A component and the intact toxin act vigorously on the ghosts without the characteristic delay seen with whole cells, and neither A nor the intact toxin are blocked by ganglioside (26, 61).

All this work suggests that the ganglioside GGnSLC is an essential component of the cell membrane receptor which recognizes the B component of cholera toxin and attaches the toxin to the surface. The B component binds very rapidly and, once bound, appears to remain attached to the cell surface. This is suggested by the demonstration by immunochemical staining that choleragenoid remains at the cell surface (52); by high binding constants of cholera toxin for cells (13, 26) and for gangliosides (Table III); and by the finding that intact cells become saturated with toxin and cannot be further stimulated by the addition of more toxin, whereas perforated red cell ghosts can be further stimulated by concentrations of toxin that would saturate intact cells (26, 29).

TABLE III

INTERACTIONS OF GANGLIOSIDES WITH CHOLERA AND TETANUS TOXINS

	No. of Molecules of ganglioside binding 1 molecule toxin		
	Cholera	tetanus	
conc. toxin	1) 1.2×10^{-9} M	2) 3.6×10^{-5} M	3) 1×10^{-11} M
Ganglioside			
GGnSLC	3.3	35	75000 (GGnSLC + SGGnSLC)
SGGnSLC	> 2000	21	
GGnSSLC	> 2000	4.0	5000 (GGnSSLC + SGGnSSLC)
SGGnSSLC	> 2000	3.5	

1) Ganglioside in free solution, deactivation of toxin measured by injecting toxin-ganglioside into ligated ileal loops of rabbits (36, 67).
2) Ganglioside in free solution, binding measured in analytical ultracentrifuge (64)
3) Toxin bound to insoluble 25% ganglioside cerebroside complex, spun down, supernates tested for toxicity in mice (68).

While the B component of cholera toxin may remain at the cell surface it is evident that the A component rapidly passes into the membrane, then, more slowly, further into it and perhaps right through it. After addition to cells, toxin remains neutralizable by antitoxin for only a few minutes, after which it is "eclipsed" for most of the lag period until the stimulation of adenylate cyclase (27). It has been suggested that toxin combines with receptor to form an inactive complex which diffuses laterally through the cell membrane, leading ultimately to the formation, within the phospholipid bilayer, of a complex of the A component with adenylate cyclase, which is then stimulated (5). According to this suggestion the adenylate cyclase is presumably approached by the A component from within the cell membrane. That an approach from this direction is not essential for the stimulation adenylate cyclase is suggested by the finding that adenylate cyclase can rapidly be stimulated by A component added to red cell ghosts, when the adenylate cyclase is approached from the cytoplasmic side. In the case of the epithelial cells of the small intestine, it is thought that the adenylate cyclase is not located on the inner side of the brush borders, to which the toxin is bound (71), but in basal and lateral membranes of the plasma membrane (51). This would suggest that in these cells the A component must pass into the cytoplasm in order to reach adenylate cyclase.

Some strains of *Escherichia coli* produce a toxin similar in its action to cholera toxin in many respects (3). This toxin is now being studied vigorously in several laboratories.

Abrin and Ricin

It is instructive enough that diphtheria and cholera toxins, although derived from widely different species of bacteria, should be so similar in being, in each case, composed of an active component attached to a cell binding component; it is even more instructive that the toxic proteins, abrin and ricin, derived from the seeds of two taxonomically unrelated plant families (*Leguminosae* and *Euphorbeaceae*) should be similarly composed (see 46-48, 50). These highly interesting proteins are only mentioned in the context of diphtheria and cholera toxins. The possibility that certain protein hormones (4, 15, 53) may be similarly constituted should also be noted.

MEMBRANE RECEPTORS NOT YET UNDERSTOOD

Tetanus and Botulinum Toxins

Tetanus and botulinum toxins are the two most potent poisons known, being about a million times as toxic as strychnine. They are both produced by anaerobic clostridia, and they are both neurotoxins without any apparent effect on any tissue other than nervous tissue. Both inhibit neuromuscular transmission, apparently by blocking the output of acetylcholine from the motor nerve endings at the neuromuscular synapse. In addition, tetanus toxin also blocks the autonomic system, and especially the central nervous system. It exerts its central action by blocking synaptic inhibition at the inhibitory synapses of the inhibitory interneurons with the motor nerves in the spinal cord and the cerebellum (12, 38, 69). Again it is thought that the output of transmitter (in this case inhibitory transmitter) from the ends of the inhibitory interneurones is blocked. It is not known at the molecular level how these two extremely powerful toxins block the output of transmitter substances at synapses. It is possible that what we learn about one of these two neurotoxins (and what we learn about cholera toxin, see below) will be helpful in understanding the other.

In the case of tetanus toxin, we do at least know what its receptor may be. It binds to the disialosyllactosyl gangliosides GGnSSLC and SGGnSSLC (64) and it is possible that they are the cell membrane receptors for tetanus toxin. In the case of cholera toxin, the binding of the toxin to its ganglioside GGnSLC leads to deactivation of the toxin; but in the case of tetanus toxin this is not so. It is possible that this may be due to the cleavage of the disialosyl bond, which is essential to optimal binding of the toxin, by endogenous sialidases in the host tissue. If enough ganglioside is injected into mice they can be protected against small doses of tetanus toxin (45). Binding of tetanus toxin to ganglioside in solution can be demonstrated in the analytical ultracentrifuge (70). In this case it is necessary to use high concentrations of reactants (5 mg toxin/ml, 0.5 mg ganglioside/ml). Binding can also be shown by making water-insoluble complexes of ganglioside and cerebroside (which does not react with tetanus toxin), mixing decreasing amounts of the complex with constant very small amounts of toxin (20 LD_{50}, 0.3 ng), spinning down and testing the supernates for toxicity in mice. This will give the least amount of ganglioside binding 10 LD_{50} of toxin (68).

The capacity of ganglioside-cerebroside complexes to bind tetanus toxin in this way would account quite well for the toxin-fixing capacity of nervous tissue (68).

Table III compares the binding of tetanus toxin at high concentrations of toxin with its particular gangliosides in free solution with the binding of toxin in low concentration by ganglioside-cerebroside complexes; and it also compares the binding of cholera toxin at low concentrations by gangliosides. In free solution 3 to 4 molecules of GGnSSLC (column 2) and GGnSLC (column 1) respectively are required to bind one molecule tetanus and cholera toxins respectively.

However, far more molecules of GGnSSLC are required to bind a molecule of tetanus toxin when the ganglioside is in the form of a complex with cerebroside (column 3). This difference in binding tetanus toxin could be due to the 6 orders of magnitude difference in toxin concentration in the two tests, or it could be due to the ganglioside being less efficient at binding tetanus toxin in the form of a complex with cerebroside (column 3). If this were so, and since the toxin-fixing capacity of nervous tissue, per unit ganglioside, is about the same as that of ganglioside-cerebroside (68), it would suggest that not all the ganglioside in nervous tissue takes part in the binding of tetanus toxin. The comparison of the figures of GGnSSLC in columns 2 and 3 would suggest that only 4 out of 5000, or about 1 in 1250, molecules of GGnSSLC in nervous tissue actively bind tetanus toxin.

Tetanus toxin, like diphtheria toxin, can be cleaved by nicking of peptide links and reduction of disulfide bonds into two non-toxic peptide chains, but so far no particular properties have been found to be associated with either of these chains (10, 11, 17, 43). A papain-stable fragment of tetanus toxin has been isolated that accounts for most of the antibody-combining power of the intact toxin, but is neither toxic nor capable of binding to ganglioside(31). This confirms earlier work showing that the receptor and antibody-binding sites of tetanus toxin are different (38, 63). Botulinum toxin can also be cleaved into two peptide chains by nicking and reduction (58).

Little is known about the binding of botulinum toxin. It used to be thought that botulinum toxin was not bound by nervous tissue or by gangliosides (see 28, 29), but now it appears that binding can be shown under certain circumstances. The binding capacity of nervous tissue can be reduced by treating the tissue with sialidase (28, 29). Since exogenous sialidase has little or no effect on gangliosides in nervous tissue, it might be advisable to consider other compounds containing sialic acid residues, such as glycoproteins.

What is the significance of the binding of tetanus toxin to the ganglioside GGnSSLC? And what is the significance of the binding of botulinum toxin to nervous tissue? Is ganglioside a receptor for tetanus toxin in the same sense that it is for cholera toxin? That is, does it recognize the toxin and bind it to the membrane of a susceptible cell and thereby facilitate the passage beyond the surface of the cell of the toxin, or an active component of it? Or are the receptors of these toxins also their targets? We have no idea. In the case of tetanus toxin there is no change in the constitution of the ganglioside as a result of its binding the toxin, but on the cell membrane there could conceivably be a change in its position or its orientation in the membrane; or no change at all. Perhaps the toxin acts simply by sitting tight on the membrane of the nerve terminal, and in so doing, either directly or indirectly blocks the passage of the transmitter substance through it to the synapse. If this were so, there would be implications to consider. For example, reference has already been made to the binding of antitoxin to toxin already bound to receptor (38, 63) - this could mean that once toxin has been bound to the susceptible membrane the addition of antitoxin, which is already known to be of little or no therapeutic (as distinct from prophylactic) value, might actually aggravate the action of the toxin.

REFERENCES

1. ARBUTHNOT, J. P., Microbial Toxins, Vol. 3. Academic Press, New York and London (1970) 189.
2. BANGHAM, A. D., STANDISH, M. M. and WEISSMANN, G., J. Pol. Biol. 13 (1965) 253.
3. BANWELL, J. G. and SHERR, H., Gastroenterology 65 (1973) 467.
4. BELLISARIO, R., CARLSEN, R. B. and BAHL, O. P., J. Biol. Chem. 248 (1973) 6796.
5. BENNETT, V., O'KEEFE, E. and CUATRECASAS, P., Proc. Nat. Acad. Sci. USA 72 (1975) 33.
6. BERLINER, F. and SCHOENHEIMER, R., J. Biol. Chem. 124 (1938) 525.
7. BERNHEIMER, A. W., J. Gen. Physiol. 30 (1947) 337.
8. BERNHEIMER, A. W., Microbial Toxins, Vol. 1, Academic Press, New York and London (1970) 183.
9. BERNHEIMER, A. W., Biochim. Biophys. Acta. 344 (1974) 27.
10. BIZZINI, B. and RAYNAUD, M., Ann. Immunol. 126c (1975) 159.
11. BIZZINI, B., TURPIN, A. and RAYNAUD, M., Naunyn-Schmiedeberg's Arch. Pharmakol. 276 (1973) 271.
12. BOROFF, D. A., and DAS GUPTA, B. R., Microbial Toxins, Vol. 2A, Academic Press, New York and London (1971) 1.
13. BOWMAN, M. H., OTTOLENGHI, A. C. and MENGAL, C. E., J. Membrane Biol. 4 (1971) 156.
14. BRETSCHER, M. S., Science 181 (1973) 622.
15. CARLSEN, R. B., BAHL, O. P. and SWAMINATHAN, N., J. Biol. Chem. 248 (1973) 6810.
16. COLEMAN, R., FINEAN, J. B., KNUTTON, S. and LIMBRICK, A. R., Biochim. Biophys. Acta. 219 (1970) 81.
17. CRAVEN, C. J. and DAWSON, D. J., Biochim. Biophys. Acta 317 (1973) 277.
18. CUATRECASAS, P., Biochemistry 12 (1973) 3547.
19. CAUTRECASAS, P., PARIKH, I. and HOLLENBERG, M., Biochemistry 12 (1973) 4253.
20. DOERY, H. M., MAGNUSSON, B. J., CHEYNE, I. M. and GULASEKHORAM, J. Nature 198 (1963) 1091.
21. DOURMASHKIN, R. R. and ROSE, W. F., Am. J. Med. 41 (1966) 699.
22. FINKELSTEIN, R. A., C. R. C. Critical Reviews in Microbiology. 2 (1973) 553.

23. FINKELSTEIN, R. A., BOESMAN, M., NEOTT, S. H., LA RUE, M. K. and DELANEY, R., J. Immunol. 113 (1974) 145.
24. FINKELSTEIN, R. A., LA RUE, M. K. and LOSPALLUTO, J. J., Infect. Immun. 6 (1972) 934.
25. GILL, D. M., Proc. Nat. Acad. Sci. USA 72 (1975) 2064.
26. GILL, D. M. and KING, C. A., J. Biol. Chem. (1975) in press.
27. GINSBURG, I., Microbial Toxins, Vol. 3, Academic Press, New York and London (1970) 99.
28. HABERMANN, E., Naunyn-Schmiedeberg's Arch. Pharmakol. 281 (1974) 47.
29. HABERMANN, E. and HELLER, I., Naunyn-Schmiedeberg's Arch. Pharmakol. 287 (1975) 97.
30. HAKSAR, A., MAUDSLEY, D. V., PERON, F. G., Nature. 251 (1974) 514.
31. HELTING, T. V. and ZWISLER, O., Proc. Fourth Internat. Conference on Tetanus, Dakar, (1974) in press.
32. HERBERT, D. and TODD, W. E., Brit. J. Exp. Path. 25 (1944) 242.
33. HOLMGREN, J., LÖNNROTH, I. and SVENNERHOLM, L., Scand. J. Infect. Dis. 5 (1973) 77.
34. HOLMGREN, J., LÖNNROTH and SVENNERHOLM, L., Infection and Immunity 8 (1973) 208.
35. KING, C. A., VAN HEYNINGEN, W. E. and GASCOYNE, N., J. Infect. Dis. in press.
36. KING, C. A. and VAN HEYNINGEN, W. E., J. Infect. Dis. 127 (1973) 639.
37. KING, C. A. and VAN HEYNINGEN, W. E., J. Infect. Dis. 131 (1975) 643.
38. KRYZHANOVSKY, G. N., Naunyn-Schmiedeberg's Arch. Pharmakol. 276 (1973) 247.
39. LENARD, J. N. and SINGER, S. J., Science 159 (1968) 738.
40. LÖNNROTH, I. and HOLMGREN, J., J. Gen. Microbiol. 76 (1973) 417.
41. MACFARLANE, M. G., and KNIGHT, B. C. J. G., Biochem. J. 35 (1941) 884.
42. MACFARLANE, R. G., OAKLEY, C. L., and ANDERSON, C. G., J. Path. Bact. 52 (1941) 99.
43. MATSUDA, M. and YONEDA, M., Biochem. Biophys. Res. Comm. 57 (1974) 1257.
44. MC CLUER, R. H., Cham. Phys. Lipids 5 (1970) 220.
45. MELLANBY, J., MELLANBY, H., POPE, D. and VAN HEYNINGEN, W. E., J. Gen. Microbiol. 54 (1968) 161.
46. NICOLSON, G. L., Nature 251 (1974) 628.
47. OLSNES, S., PAPPENHEIMER, A. M., JR., and MEREN, R., J. Immunol. 113 (1974) 842.
48. OLSNES, S., REFSNES, K., and PIHL, A., Nature 249 (1974) 627.
49. PAPPENHEIMER, A. M., JR. and GILL, D. M., Science 182 (1973) 353.
50. PAPPENHEIMER, A. M., JR., OLSNES, S. and HARPER, A. V., J. Immunol. 113 (1974) 835.
51. PARKINSON, D. , EBEL, H., DIBONA, D. R. and SHARP, G. W. G., J. Clin. Invest. 51 (1972) 2292.

52. PETERSON, J. W., LOSPALLUTO, J. J. and FINKELSTEIN, R. A., J. Infect. Dis. 126 (1972) 617.
53. PIERCE, F. G., Endocrinology 89 (1971) 1331.
54. PIERCE, N. F., J. Exp. Med. 137 (1973) 1009.
55. SHARP, G. W. G. and HYNIE, S., Nature 229 (1971) 266.
56. STAERK, J., RONNEBERGER, H. J. and WIEGANDT, H., Behring Inst. Mitt. 55 (1974) 145.
57. STAERK, J., RONNEBERGER, H. J., WIEGANDT, H. and ZIEGLER, W., Europ. J. Biochem. 48 (1974) 103.
58. SUGIYAMA, H., DAS GUPTA, B. R. and YANG, K. H., Proc. Soc. Exp. Biol. Med. 143 (1973) 589.
59. SVENNERHOLM, L., J. Neurochem. 10 (1963) 613.
60. VAN HEYNINGEN, S., Science 183 (1974) 656.
61. VAN HEYNINGEN, S. and KING, C. A., Biochem. J. 146 (1975) 269.
62. VAN HEYNINGEN, W. E., Bacterial Toxins, Oxford (Blackwell's Scientific Publications) (1950) 53.
63. VAN HEYNINGEN, W. E., J. Gen. Microbiol. 20 (1959) 310.
64. VAN HEYNINGEN, W. E., J. Gen. Microbiol. 31 (1963) 375.
65. VAN HEYNINGEN, W. E., Naunyn-Schmiedebergs Arch. Exp. Path. Pharmak. 276 (1973) 289.
66. VAN HEYNINGEN, W. E., Nature 249 (1974) 5456.
67. VAN HEYNINGEN, W. E., CARPENTER, C. C. J., PIERCE, N. F. and GREENOUGH, W. B., J. Infect. Dis 124 (1971) 415.
68. VAN HEYNINGEN, W. E. and MELLANBY, J., J. Gen. Microbiol. 52 (1968) 447.
69. VAN HEYNINGEN, W. E. and MELLANBY, J., Microbial Toxins, Vol. 2A Academic Press, New York and London (1971) 69.
70. VAN HEYNINGEN, W. E. and MILLER, P., J. Gen. Microbiol. 24 (1961) 107.
71. WALKER, W. A., FIELD, M., and ISSELBACHER, K. J., Proc. Nat. Acad. Sci. USA 71 (1974) 320.
72. WOODIN, A. M., Microbial Toxins, Vol. 3 Academic Press, New York and London (1970) 327.

GANGLIOSIDES AS POSSIBLE MEMBRANE RECEPTORS FOR CHOLERA TOXIN

C. A. KING

Sir William Dunn School of Pathology
University of Oxford (United Kingdom)

The disease of cholera is caused by the action of an exotoxin released by the Vibrio cholerae in the lumen of the small intestine. The toxin causes an increase in the secretion of chloride into the gut lumen by the mucosal cells (5) - a process which is accompanied by the excessive outpouring of fluid which is characteristic of the disease. However, cholera toxin can also cause a variety of other biological reactions under laboratory conditions, the nature of the response depending on the cell type which is being challenged. Examples are increased lipolysis in fat cells (13), glycogenolysis in liver (12), steroidogenesis in adrenal (4) and increased capillary permeability in the skin (1). The latter response is the basis of a widely used assay for cholera toxin (1). All of these effects have been attributed to an elevation in the concentration of cyclic AMP in the cell, due to the activation of adenylate cyclase on the inner surface of the cell membrane (6, 7).

MOLECULAR STRUCTURE

Cholera toxin (also known as cholera exotoxin, enteroexotoxin, choleragen (8)) is a protein of molecular weight of 84000 with a subunit structure which enables a division of the process of intoxication of the cell into two stages - binding of toxin and activation of adenylate cyclase. The binding of the toxin to the cell membrane is accomplished by the portion of the molecule known as choleragenoid (8), which is an aggregate (molecular weight 56000) of 4 to 6 B subunits, each of molecular weight about 10000 to 14000. Choleragenoid can be isolated during the purification of intact toxin from V. cholerae culture filtrates, and has no biological action other than that of binding to the cell membrane. The toxin also has one copy of the A subunit (molecular weight about 28000) which has low activity towards intact cells but which can activate adenylate cyclase if added to lysed cells under certain conditions (29). The activation of adenylate cyclase is the second stage of the intoxication process and it occurs after a lag period of about 30 mins, the significance of which will be discussed later.

Subunit A is associated with weak noncovalent bonds to the choleragenoid portion of the toxin molecule, and can be dissociated from it by acid and by urea (27). Subunit A itself can be dissociated by disulfide reducing agents into two peptide chains: A (molecular weight about 22000) thought to be responsible for the activation of adenylate cyclase, and A_2 (molecular weight about 5000) which probably forms the link between A_1 and the B subunits (20, 28).

BINDING OF CHOLERA TOXIN

Identification of the Receptor

In 1971, it was found by van Heyningen and coworkers (31) that if cholera toxin was treated with homogenized preparations of various tissues, the toxin in the resulting mixture had lost its ability to promote capillary permeability in the rabbit skin and diarrhea in the isolated intestinal loop i.e., the toxin had been deactivated. It was also observed that brain homogenates were the most effective in bringing about the deactivation of toxin - an observation which was rather unexpected, bearing in mind that the small intestine is the natural site of action of cholera toxin. In view of the already known ability of another toxin - tetanus - to react with nervous tissue, and particularly with gangliosides (32), the ability of various lipid extracts and of ganglioside to deactivate cholera toxin was tested. It was confirmed that gangliosides were very effective in the deactivation of cholera toxin, and the much greater content of ganglioside in brain compared with intestinal mucosa was found to account for the apparently much greater affinity of the toxin for brain tissue (see Table I).

Gangliosides are a class of amphipathic glyco-sphingolipids with the general structure shown in Fig. 1 (19). They consist of a ceramide portion (containing fatty acid - mainly stearic - and sphingosine), and an oligosaccharide portion which always contains at least one sialic acid residue, with glucose, galactose and galactosamine. The parent ganglioside is the monosialosyl ganglioside GGnSLC (McCluer notation (21); also known as GM1, GGtet1 or GI). More complex gangliosides have additional sialic acid residues attached to the terminal galactose and to the existing sialic acid

TABLE I

Deactivation of cholera toxin by various tissues with decreasing ganglioside contents

Tissue	Total ganglioside ng/ μg wet weight	ng toxin deactivated by 1 μg tissue
Rabbit brain	2.60	4.00
Guinea pig brain	2.38	6.70
Guinia pig colon	0.22	0.83
Guinea pig cecum	0.09	0.01
Rabbit colon	0.07	0.03
Rabbit ileum	0.07	0.05
Guinea pig ileum	0.06	0.01
Rabbit liver	0.04	0.002
Pigeon red cells	0.02	0.0005

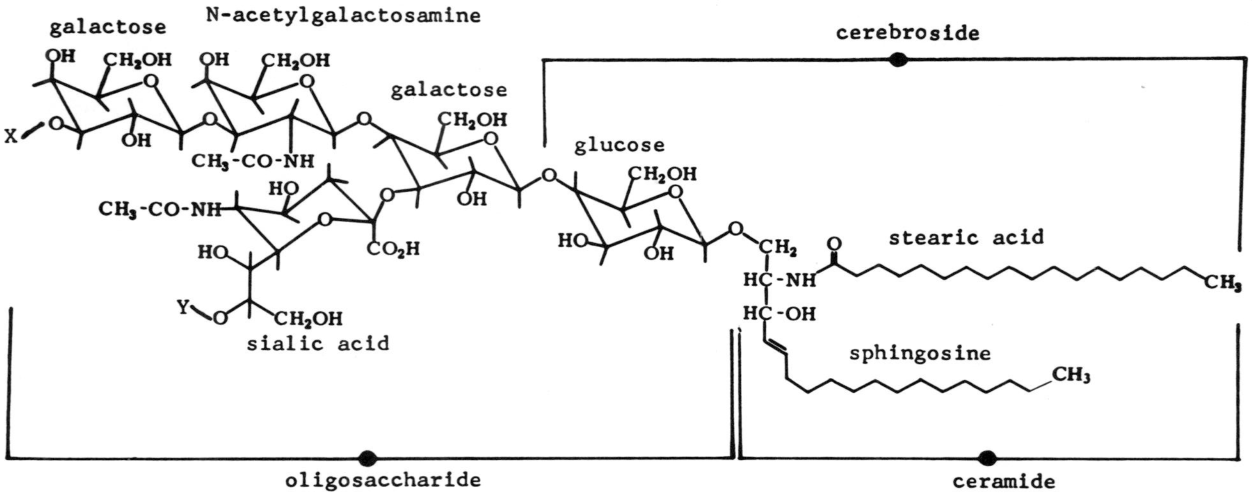

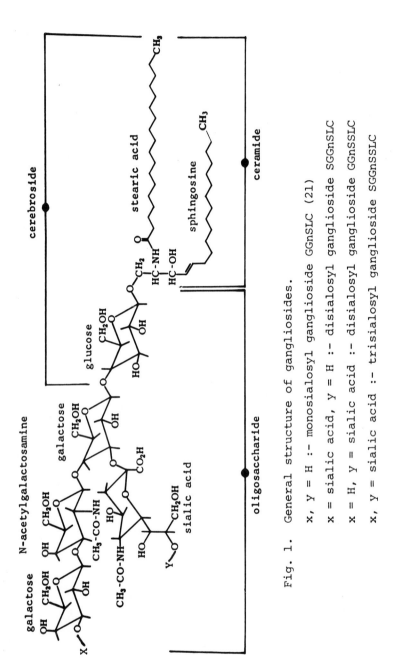

Fig. 1. General structure of gangliosides.

x, y = H :- monosialosyl ganglioside GGnSLC (21)

x = sialic acid, y = H :- disialosyl ganglioside SGGnSLC

x = H, y = sialic acid :- disialosyl ganglioside GGnSSLC

x, y = sialic acid :- trisialosyl ganglioside SGGnSSLC

to give the increasingly polar species SGGnSLC, GGnSSLC, SGGnSSLC and SSGGnSSLC.

In 1973, it was found that the ganglioside which was specifically responsible for the deactivation and binding of cholera toxin by membrane preparation and by mixed gangliosides was the sialidase resistant monosialosyl ganglioside GGnSLC (2, 15, 16, 30). No other ganglioside had any significant interaction with cholera toxin, and it was therefore postulated that this ganglioside was part of, or resembled very closely, the receptor for cholera toxin on the cell surface. The specificity of the recognition of cholera toxin by the ganglioside appears to reside in the oligosaccharide portion, but it seems that the intact ceramide portion is necessary for the reaction with the toxin which leads to its deactivation (in vitro) or to biological activity (in vivo) (26).

A certain amount of indirect evidence also suggests that GGnSLC is the receptor for cholera toxin. Firstly, it has been found that the treatment of intestinal mucosal homogenates with sialidase increases the ability of this tissue to deactivate cholera toxin (9). Secondly, treatment of adrenal cells with sialidase increases the degree of stimulation of steroidogenesis by cholera toxin (14). Finally, it has been shown that low concentrations of GGnSLC can be incorporated into the membranes of fat cells (3) or pigeon red cells (11), and thereby causes an increased capacity for binding of cholera toxin, and an enhanced response of the adenylate cyclase. This will be discussed in a later section.

Role of choleragenoid

If intestinal loops are preincubated for 4 hours with choleragenoid, cholera toxin can no longer elicit a diarrheagenic response (25). Similarly, choleragenoid can prevent the effect of cholera toxin on pigeon red cell adenylate cyclase, and can inhibit the toxin-deactivating ability of intestinal mucosal homogenates. It can also prevent the reaction of toxin with ganglioside GGnSLC - although this is most effective when the GGnSLC is in the form of a complex with cerebroside (17). This observation may indicate the possibly differing disposal of the B subunits in the isolated choleragenoid preparation compared with that in the native toxin. All of this evidence suggests that the role of the choleragenoid portion of the toxin molecule is to bind to the receptor area. Choleragenoid itself is not able to bring about any further effects and this fact immediately suggests that the ultimate biological response, i.e., the activation of adenylate cyclase, is the responsibility of the A subunit. Subunit A does not appear to have any reaction with the receptor ganglioside GGnSLC (27), but has been found to activate adenylate cyclase under certain conditions (see next section).

The entire toxin does not penetrate inside the cell during the intoxication process. It has been shown by fluorescent antibody techniques (24) that the choleragenoid portion at least remains firmly bound to the outside of the cell, thereby blocking the receptors which cannot then be reused. This is presumably the reason for the observed saturation of the biological response of intact cells at lower concentrations of cholera toxin than those which continue to activate adenylate cyclase to much greater levels in lysed cells, where the receptor mechanism is bypassed (11).

Treatments which interfere with the receptor mechanism, such as the incubation of cells with choleragenoid, or of toxin with high concentrations of ganglioside GGnSLC, do not reduce the ability of the toxin to activate adenylate cyclase when the toxin is presented with a lysed cell. Similarly, in the presence of low concentrations of GGnSLC, the reponse of intact cells to cholera toxin is enhanced but that of lysed cells is not (Table II).

TABLE II

Incubation of pigeon red cells with low and high concentrations of ganglioside GGnSLC. Effect on subsequent activation of adenylate cyclase by cholera toxin in intact and lysed cells.

		Adenylate cyclase % of activity with toxin alone	
GGnSLC µg/ml	Toxin µg/ml	Intact cells	Lysed cells
0	5	100	100
2.5	5	183	103
0	10	100	100
100.0	10	33	89

ACTIVATION OF ADENYLATE CYCLASE

Although the binding of the choleragenoid portion of cholera toxin to cells is very rapid, and the addition of antitoxin after about 3 min cannot reverse the development of the toxic reaction, there is always a lag of at least 30 min before the activation of adenylate cyclase and related effects can be measured. However, if the cell is ruptured under conditions which do not cause excessive dilution of the cell contents, cholera toxin (11) or the isolated subunit A (29) can activate adenylate cyclase without a lag to a much greater level of activity than is possible in the intact cell, where the biological response is limited by the number of toxin receptors. The activation process appears to require NAD (10, 16) and other unidentified components present in the cell sap. The requirement for NAD may explain why earlier attempts to make ruptured cells respond to cholera toxin were unsuccessful. Most cells contain the membrane-bound NAD glycohydrolase which is activated upon cell lysis and breaks down NAD very rapidly. However, the pigeon red cell which was used for the above experiments, lacks this enzyme and as a result the NAD content is maintained at a higher level in the lysed cell preparation. The role of the NAD is not yet known. It does not appear to be involved in the generation of ADP-ribose which is the case in the action of diphtheria toxin (22).

It has been suggested (11) that the lag period represents the time taken for the active subunit A to cross the cell membrane to the cell interior in which it is able to diffuse. It can then activate adenylate cyclase in the membrane at sites which are not necessarily in direct association with the original receptor site. This would explain the ability of cholera toxin to activate the adenylate cyclase which is located in the basal and lateral membrane of brush border cells, although the toxin is bound on to the mucosal border (23).

RESPONSE TO TOXIN OF CELLS TREATED WITH EXOGENOUS GGnSLC

It was first observed by Cuatrecasas (3) that if fat cells were soaked in ganglioside GGnSLC, they were then able to bind more cholera toxin, and were more sensitive to its lipolysis-stimulating action. This observation indicated that exogenous GGnSLC was capable of being incorporated into the cell membrane and could thereby provide additional receptors for cholera toxin. However, it would seem to be unlikely that all of the incorporated GGnSLC molecules would be able to fulfill the receptor function, and therefore it became of interest to determine that proportion of the incorporated GGnSLC which could bind extra toxin, and to determine whether all of the extra toxin was then capable of bringing about the extra activation of the biological response.

Using suspensions of pigeon red cells it was found that ^{3}H-GGnSLC was readily bound to, or incorporated in, the cell membrane after only 30 min of incubation at 37°. The process was complete after about 2 hours. The degree of incorporation was linearly dependent on the concentration of the soaking GGnSLC at low concentrations of GGnSLC (up to 10 μg/ml), when about 15% of the added GGnSLC was incorporated after 30 mins and 25% at completion. However, as the concentration of the soaking ganglioside increased towards and above the critical micellar concentration (around 20 μg/ml), the degree of incorporation began to diminish slightly.

Cells with increasing concentrations of incorporated GGnSLC were tested for their ability to bind and deactivate cholera toxin, and the degree of activation of adenylate cyclase by saturating concentrations (> 10 μg/ml of toxin) was measured. It was found that in the absence of GGnSLC treatment, about 0.2 to 0.4 μgs of toxin was bound per ml (4×10^9 cells). Cells with the lowest concentration of incorporated GGnSLC which was measured (0.2 μg GGnSLC/ml), were found to bind an extra 7 μgs of toxin/ml. This 30 fold increase was considerably more than had been expected on the basis of experiments with GGnSLC and toxin in solution (5 μgs toxin bound to 1 μg GGnSLC), and may reflect a greater affinity of toxin for GGnSLC which is incorporated in the more complex environment afforded by the cell membrane. The amount of extra toxin which was bound did not increase consistently further, even with much higher levels of incorporated GGnSLC. For example, cells with 9.4 μgs of incorporated GGnSLC/ml which would be expected to bind at least 47 μgs of extra toxin, in fact were found to bind only 5 μgs of extra toxin. It therefore appears that cells which have been treated with micellar solutions of GGnSLC cannot bind any more extra toxin, possibly because the individual binding sites are masked in the GGnSLC micelles associated with the membrane.

The activation of adenylate cyclase in GGnSLC treated pigeon red cells did not increase to more than twice the level of that in the untreated cells. A separate experiment in which increasing concentrations of toxin were added to untreated and GGnSLC-treated cells indicated that the extra response of the adenylate cyclase could be attributed to the action of an extra 0.5 μgs of bound toxin - approximately 10% of the amount of extra toxin which it was known could be bound by such cells. (Fig. 2).

It is concluded from these results that only a small proportion of the GGnSLC which can be incorporated into pigeon red cell ghosts is able to bind toxin in a manner that allows the active subunit A to pass into the cell where it can activate adenylate cyclase. This might be explained if the binding of the first few toxin molecules to added receptors caused some kind of irreversible change in the membrane which prevented the transfer of the A subunit of any toxin which was subsequently bound. Alternately, there may be only limited regions in the cell membrane which can accommodate extra GGnSLC molecules in the specialised environment which is needed for the fulfillment of the receptor function. The latter explanation would imply that the receptor for cholera toxin consists of GGnSLC plus a specific type of environment in the cell membrane. There is some tentative evidence (18) that also in the untreated pigeon red cell not all of the toxin which can be bound is involved in the activation of adenylate cyclase, and it could be that the same conditions

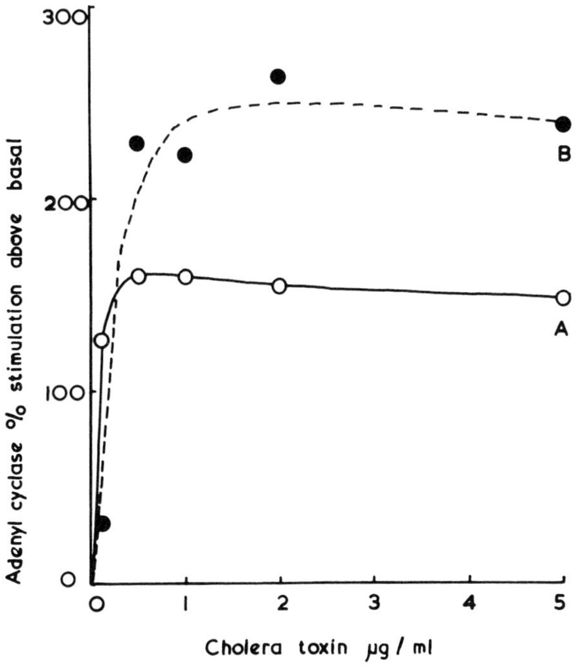

Fig. 2. Effect of cholera toxin on adenylate cyclase of intact pigeon red cells. A. untreated cells. B. cells with 6.3 µg extra GGnSLC/ml.

of specific environment for the GGnSLC receptor molecules are required.

ACKNOWLEDGEMENT

This work was supported by a grant from the Medical Research Council of Great Britain.

REFERENCES

1. CRAIG, J. P., J. Bacteriol. 92 (1966) 793.
2. CUATRECASAS, P., Biochemistry 12 (1973) 3547.
3. CUATRECASAS, P., Biochemistry 12 (1973) 3558.
4. DONTA, S. T., KING, M. and SLOPER, K., Nature New Biology (lond.) 243 (1973) 246.

5. FIELD, M., FROMM, D., WALLACE, C. K. and GREENOUGH, W. B. III, J. Clin. Invest. 48 (1969) 24a (Abstr.).
6. FIELD, M., PLOTKIN, G. R. and SILEN, W., Nature (Lond.) 217 (1968) 469.
7. FINKELSTEIN, R. A., CRC Critical Reviews in Microbology 2 (1973) 553.
8. FINKELSTEIN, R. A. and LO SPALLUTO, J. J., J. Infect. Dis. 121 (Suppl.) (1970) S 63.
9. GASCOYNE, N. and VAN HEYNINGEN, W. E., Infect. Immun. (1975) in press.
10. GILL, D. M., Proc. Nat. Acad. Sci. USA 72 (1975) 2064.
11. GILL, D. M. and KING, C. A., J. Biol. Chem. 250 (1975) in press.
12. GRAYBILL, J. R., KAPLAN, M. M. and PIERCE, N. F., Clin. Res. 18 (1970) 454.
13. GREENOUGH, W. B. III, PIERCE, N. F. and VAUGHAN, M., J. Infect. Dis. 121 (Suppl.) (1970) S 111.
14. HAKSAR, A., MAUDSLEY, D. V. and PERON, F., Nature (Lond.) 251 (1974) 514.
15. HOLMGREN, J., LÖNNROTH, I. and SVENNERHOLM, L., Infec. Immun. 8 (1973) 208.
16. KING, C. A. and VAN HEYNINGEN, W. E., J. Infect. Dis. 121 (1973) 639.
17. KING, C. A. and VAN HEYNINGEN, W. E., J. Infect. Dis. 131 (1975) 643.
18. KING, C. A., VAN HEYNINGEN, W. E. and GASCOYNE, N., J. Infect. Dis. (1975) (suppl.) in press.
19. KUHN, R. and WIEGANDT, H., Chem. Ber. 96 (1963) 866.
20. LO SPALLUTO, J. and FINKELSTEIN, R. A., Biochim. Biophys. Acta 257 (1972) 158.
21. MC CLUER, R. H., Chem. Phys. Lipids 5 (1970) 220.
22. PAPPENHEIMER, A. M., JR. and GILL, D. M., Science 182 (1973) 353.
23. PARKINSON, D. K., EBEL, H., DIBONA, D. R. and SHARP, G. W. G., J. Clin. Invest. 51 (1972) 2292.
24. PETERSON, J. W., LO SPALLUTO, J. L. and FINKELSTEIN, R. A., J. Infect. Dis. 126 (1972) 617.
25. PIERCE, N. F., J. Exp. Med. 137 (1973) 1009.
26. STAERK, J., RONNEBERGER, H. J., WIEGANDT, H. and ZIEGLER, W. Eur. J. Biochem. 48 (1974) 103.
27. VAN HEYNINGEN, S., Science 183 (1974) 656.
28. VAN HEYNINGEN, S., J. Infect. Dis. (Suppl.) (1975) in press.
29. VAN HEYNINGEN, S. and KING, C. A., Biochem. J. 146 (1975) 269.
30. VAN HEYNINGEN, W. E., Naunyn-Schmiedeberg's Arch. Pharmacol. 276 (1973) 289.
31. VAN HEYNINGEN, W. E., CARPENTER, C. C. J., PIERCE, N. F. and GREENOUGH, W. B., III, J. Infect. Dis. 124 (1971) 415.
32. VAN HEYNINGEN, W. E. and MELLANBY, J., Microbial Toxins Vol. 2A, Academic Press, London and New York (1971) 69.

BINDING AND UPTAKE OF THE TOXIC LECTINS ABRIN AND RICIN BY MAMMALIAN CELLS

Sjur OLSNES, Kirsten SANDVIG, Karin REFSNES,*
Øystein FODSTAD and Alexander PIHL

Norsk Hydro's Institute for Cancer Research
Montebello, Oslo (Norway)

Abrin and ricin are two toxic glycoproteins present in the seeds of *Abrus precatorius* and *Ricinus communis*. Both toxins are extremely potent, and upon intravenous injection into mice doses of about 13 ng of abrin or 65 ng of ricin are lethal in the course of a few days. With increasing amounts of toxin the survival time is reduced, but even very high doses fail to reduce the survival time to less than about 10 hours (Fig. 1). After intravenous injection of a certain amount of toxin the survival time can be predicted with surprising accuracy. Abrin and ricin are also very toxic to eukaryotic cells in culture, and a few picograms/ml will kill most cells within 24 hours. The toxins have no effect on the growth of bacteria.

The toxins, which have a molecular weight of 65,000 (Fig. 2), both consist of two polypeptide chains, A and B, which are linked by a disulfide bond (3, 12-15, 24). This linkage is necessary for toxic activity in living animals and cells in culture. Thus, mixtures of the two reduced chains are virtually nontoxic *in vivo* (11), but the toxicity is restored upon reformation of the disulfide bridge (16). Hybrid toxins consisting of abrin A chain and ricin B chain and *vice versa* can be formed (16). In all cases the reconstituted and hybrid toxins have about the same toxicity as the native toxins in living animals and cells in culture.

The A chains are enzymes. Experiments with cells in culture have shown that the toxic effect of abrin and ricin is due to their ability to inhibit protein synthesis (5, 14, 20). The toxins also strongly inhibit protein synthesis in cell-free system (9, 10). Whereas only the intact toxins inhibit protein synthesis in whole

*Fellow of the Norwegian Cancer Society.

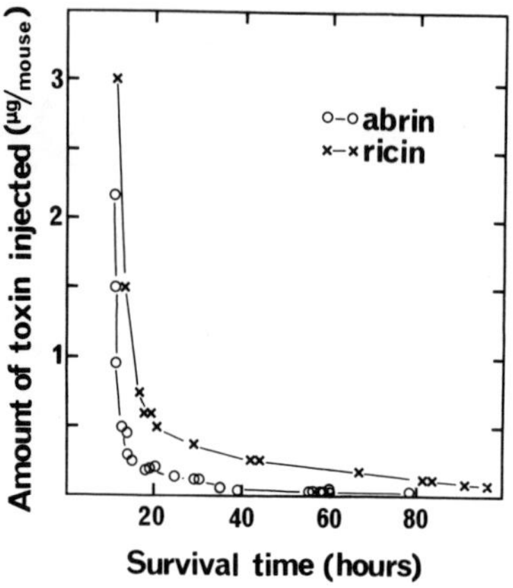

Fig. 1. Survival time after intravenous injections of abrin and ricin into mice.

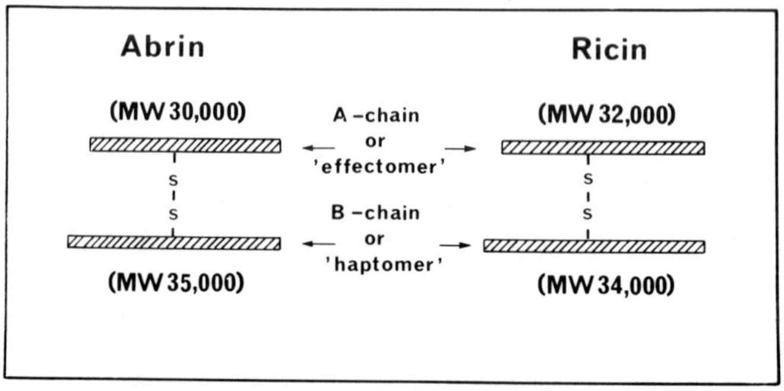

Fig. 2. Scheme of the structure of abrin and ricin (14).

cells, the reduced toxins also are able to inhibit in cell-free systems (11). Indeed in cell-free systems the reduced toxins have much higher inhibitory activity than the intact toxins. After separation of the reduced chains it was found for both toxins that the ability to inhibit cell-free protein synthesis resides exclusively in the shorter chain, the A chain or "effectomer" (12-14). The increased effect after reduction of the toxins indicate that the A chain is activated upon reduction of the interchain disulfide bridge. Indeed, immunological data indicate that ricin A chain changes conformation extensively after its liberation from the B chain (17). The target for the toxin A chains is the 60S ribosomal subunit which becomes irreversibly inactivated (22).[1] Actually the A chains are enzymes (9)[2] capable of inactivating 1500 purified ribosomes per minute[2] and 150 ribosomes per minute in a reticulocyte lysate.[3] The lesser molecular activity with ribosomes present in the lysate is probably due to protection by elongation factor 2 which, in the presence of GTP, binds to a site on the ribosomes which is close to (or overlapping) the target for abrin and ricin A chains.[4] The toxin-inactivated ribosomes have reduced ability to hydrolyze GTP (22)[1] and to bind enzymatically aminoacyl-tRNA (2). There is evidence that the target for the toxins is an 8S complex which can be released from the 60S ribosomal subunit by EDTA treatment and which consists of 5S RNA and one single ribosomal protein.[1]

The B chains are lectins. Abrin and ricin have the ability to bind strongly to receptor sites on cell surfaces. These sites appear to be glycoproteins with carbohydrate chains containing terminal non-reducing galactose residues. The ability of abrin and ricin to bind to such carbohydrates is used in the purification of the toxins. The toxins are allowed to bind to Sepharose beads (which contain galactose residues) and, after washing, the toxins are eluted with galactose (6, 12, 13). Experiments with the isolated peptide chains of abrin and ricin have shown that the binding ability (to cells and to Sepharose beads) resides exclusively in the B chains which have one binding site per molecule (18).

---

[1] Benson, S., Olsnes, S., Pihl, A., Skorve, J., Abraham, K. A., in press.

[2] Olsnes, S., Fernandez-Puentes, C., Carrasco, L. and Vasques, D., submitted for publication.

[3] Olsnes, S., Sandvig, K., Kefsnes, K. and Pihl, A., submitted for publication.

[4] Fernandez-Puentes, C., Olsnes, S., Benson, S., and Pihl, A., in preparation.

Kinetics of toxin binding to cells. Experiments with abrin and ricin labelled with ^{125}I in the presence of lactoperoxidase and H_2O_2[5] showed that the toxins bind rapidly to the cells with k_{+1} of $0.9\text{-}3.0 \times 10^5 M^{-1}s^{-1}$. The dissociation from cells is strongly temperature dependent, the k_{-1} being $3.4\text{-}45 \times 10^{-4}s^{-1}$ at 0° and $3.9\text{-}18 \times 10^{-3}s^{-1}$ at 37°. The presence of unlabelled toxin and particularly the presence of lactose strongly increases the rate of dissociation (Fig. 3). Binding studies carried out according to Steck and Wallach (23) indicated that each HeLa cell possesses about 3×10^7 receptors having uniform affinity to the toxins and a K_a of about $3 \times 10^8 M^{-1}$ at 0° for both toxins.[5] At 37° the K_a were $1.2 \times 10^8 M^{-1}$ and $2.6 \times 10^7 M^{-1}$ for abrin and ricin, respectively. Erythrocytes, which are smaller than HeLa cells, possess only about 10^6 receptors per cell and in this case two kinds of affinities were found at 20°, whereas at 0° the erythrocyte receptors exhibited uniform affinity for the toxins (Fig. 4).

Inhibiting effect of lactose. The binding of abrin and ricin to the cell surface receptors is necessary for the toxic effect on intact cells. Thus, if galactose or lactose which inhibit the binding to the surface receptors, is added to the cell culture medium before or together with abrin and ricin the toxic effect is strongly reduced (14, 20). In fact, the amount of toxin necessary to reduce protein synthesis to 50% of the control value after 3 hours incubation was found to be linearly related to the lactose concentration (Fig. 5). The amount of toxin actually bound to the cells at the various lactose concentrations was calculated using measured or calculated apparent association constants.[3,5] It was found (Fig. 5) that in all cases about the same amount of toxin was bound to the cells when protein synthesis was reduced to 50% after 3 hours, even when the concentration of toxin in solution varied by a factor of 300.

Inhibiting effect of ricin B chain. Ricin B chain binds to cells in a similar way as intact ricin. Probably this is also true for abrin B chain, but abrin B chain is more unstable, and we have not been able to label it with ^{125}I and carry out exact binding studies. Since only a small fraction of the bound toxin is actually taken up by the cells (see below) the possibility had to be considered that abrin and ricin bind with very high affinity to a low number of special receptor sites which are the only ones responsible for toxin uptake. In this case treatment of HeLa cells with isolated ricin B chains before adding the toxins should result in a strong reduction in toxicity. It was found, however, that the toxicity was reduced to the extent expected if the B chains compete equivalently with the toxins for the total number of binding sites.[3] It therefore appears that every toxin molecule bound to the cell surface has the

[5] Sandvig, K., Olsnes, S. and Pihl, A., submitted for publication.

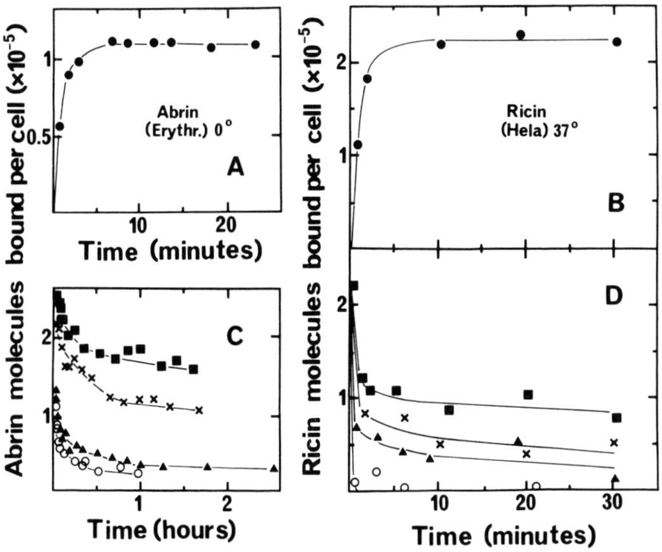

Fig. 3. Rate of abrin and ricin binding to (A,B) and dissociation from (C,D) human cells. ^{125}I-Abrin was added to human erythrocytes at 0° (A), ^{125}I-ricin was added to HeLa cells at 37° (B) and then aliquots were removed after various periods of time and the cells were rapidly sedimented to measure the amount of bound toxin. Other samples (C,D) were incubated for 15 minutes to allow maximal binding to take place and then the samples were diluted 100 fold with buffer alone (■); with buffer containing 75 µg/ml unlabelled toxins (χ), 10 mM lactose (O), or 0.1 mM lactose (▲). After various periods of time samples were removed and filtered through cellulose acetate filters to measure the amount of bound toxin.

same probability of entering the cell. Altogether the binding studies indicate that the toxic effect is closely related to the total amount of cell-bound toxin molecules.

Kinetics of protein synthesis inhibition in a cell-free system. To elucidate the relationship between the number of bound toxin mole-

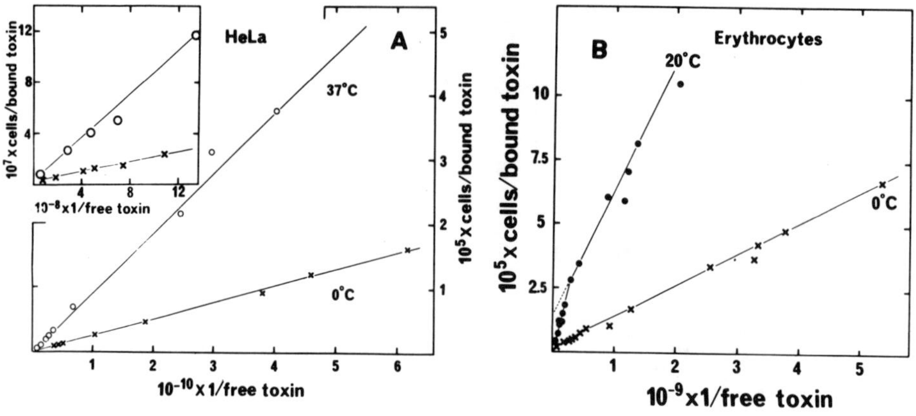

Fig. 4. Steck-Wallach plots of the binding of ricin to HeLa cells (A) and human erythrocytes (B). Various amounts of ricin were mixed with a constant amount of cells and after one hour incubation at 0°, 20° or 37° the cells were sedimented to measure the extent of binding. The ricin concentrations are given in M.

cules and the effect on protein synthesis, the relationship between the concentration of toxin and the concomitant protein synthesis inhibition in a cell-free system was studied first and then correlated with that seen in intact cells. For this purpose rabbit reticulocyte lysate which is reticulocyte cytoplasm diluted with 1 volume of H_2O and containing 10^{-5}M hemin was used. Due to its very high activity, it was felt that this system better reflects the situation in the cytoplasm than a cell-free system prepared from HeLa cells.

Abrin and ricin A chains were added to such a lysate and aliquots were transferred after various periods of time into tubes containing the specific anti A chain antibodies which immediately prevent further enzymatic activity of the A chains. The necessary components were then added and the lysate was tested for protein synthesizing ability. It was found[3] that the lysate loses its activity exponentially with time (Fig. 6A and B). The time required for 50% inhibition of protein synthesis was linearly related to the inverse of the A chain concentration (Fig. 7A). When on the other hand, intact toxins were added to the lysate, the rate of inactivation was first low and then increased (Fig. 6C and D). In this case the time required for 50% inhibition was linearly related to $1/\sqrt{\text{A chain}}$ (Fig. 7B). This is the kinetics expected for an enzyme which is activated with time after its addition to the system[3]. As pointed out above this activation is due to reduction of the interchain disulfide bond. It should be noted that the line in Fig. 7

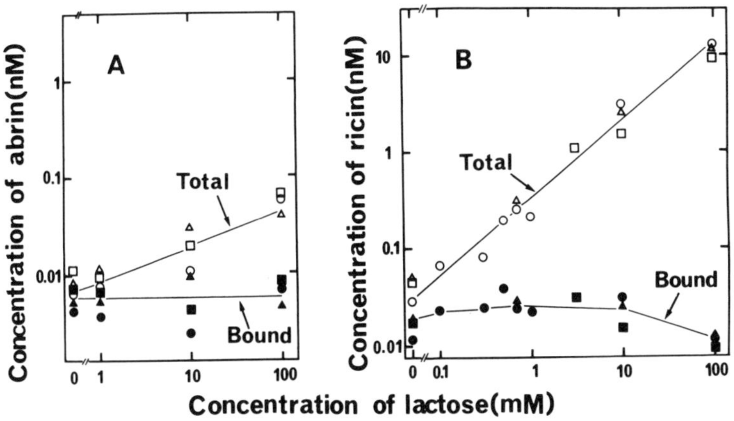

Fig. 5. Relationship between lactose concentration and toxicity. Increasing amounts of toxins were added to tubes containing 2 ml cultures of HeLa cells in the presence of various concentrations of lactose and the incorporation of [^{14}C] leucine was measured after 3 hours. The total concentration of toxin in the cell suspension required to reduce protein synthesis to half the control value was measured at each lactose concentration and plotted against the concentration of toxins (open symbols). The apparent association constants between cell surface receptors and toxins in the presence of various lactose concentrations were measured or calculated.[3,5] These constants were used to calculate the amount of toxins actually bound to the cells at the various lactose concentrations (closed symbols).

goes through the origin indicating that the inactivation occurred without any measurable lag time both when A chains and when intact toxins were added to the reticulocyte lysate.

Kinetics of protein synthesis inhibition in intact cells. When abrin and ricin are added to intact cells, there appears to be a lag time which increases with decreasing toxin concentration (insertions in Fig. 8). To determine if this lag time is also an apparent lag time similarly to those in Fig. 6C and D, the data were plotted in a similar way as in Fig. 7B as the time required for 50% inhibition of protein synthesis (t_{50}) against the square root of the inverse value of the number of toxin molecules bound per cell. The results (Fig.8) showed a close to linear relationship, but in this case the line did

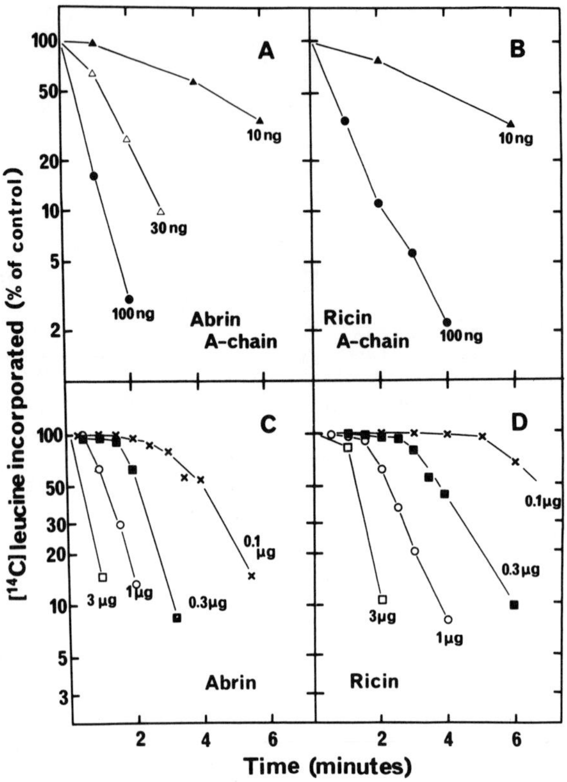

Fig. 6. Rate of inactivation of a rabbit reticulocyte lysate after addition of abrin A chain (A), ricin A chain (B), intact abrin (C) and intact ricin (D). The indicated amounts of A chains or intact toxins were added to rabbit reticulocyte lysates (1.15 ml) and incubated at 30°. After various periods of time (as indicated) aliquots were removed, the necessary components for cell-free protein synthesis were added and the ability to incorporate $[^{14}C]$ leucine was measured.

not go through the origin, but intercepted the time axis at about 30 minutes.[3] The most likely interpretation is that this period of

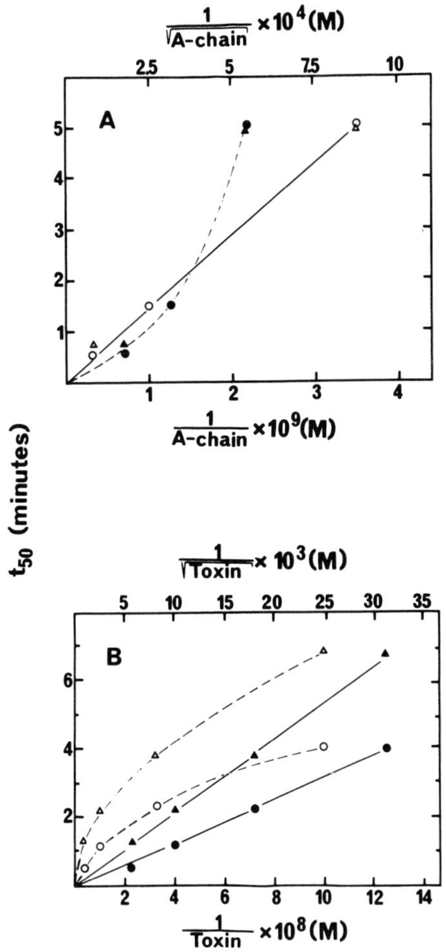

Fig. 7. Relationship between toxin concentration and the time required for 50% reduction in protein synthesizing activity. The time (t_{50}) required in the presence of various amounts of A chains (A) or intact toxins (B) was measured and plotted either against the inverse value (open symbols) or the square root of the inverse value (closed symbols) of the concentration of A chain or intact toxins. (O, ●), abrin A chain or abrin; (Δ , ▲), ricin A chain or ricin.

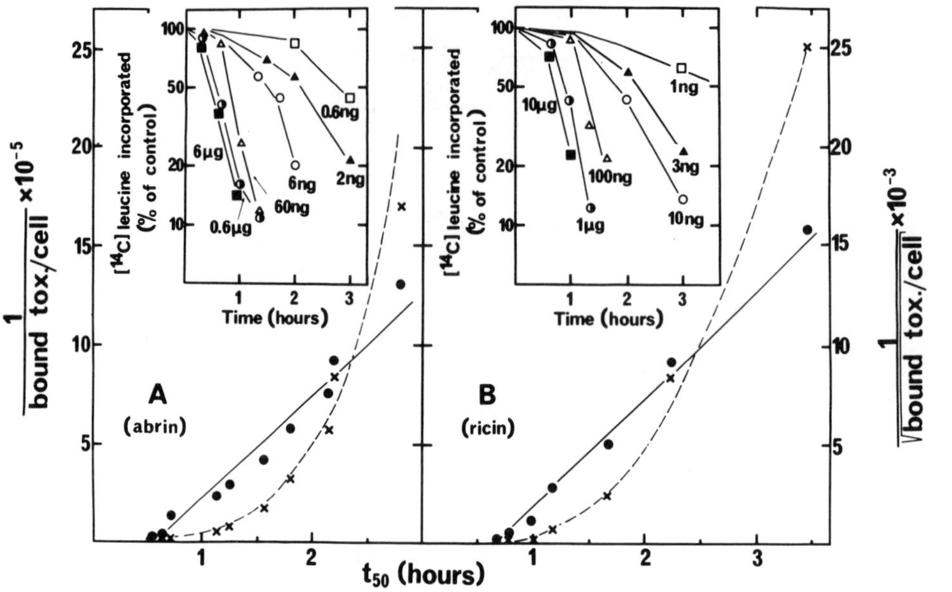

Fig. 8. Relationship between the amount of cell-bound toxins and inhibition of protein synthesis. The indicated amounts of toxins were added to 2.2 ml cultures of HeLa cells and, after various periods of time $[^{14}C]$ leucine was added and the radioactivity incorporated during 20 minutes was measured (inserted figures). The amount of cell-bound toxins was calculated using measured K_a values.[5] The time required for 50% reduction in protein synthesis (t_{50}) was plotted against the inverse (X) and the square root of the inverse (●) of the number of toxin molecules bound per cell.

time transpires from when the toxins are bound to the cell surface until protein synthesis inhibition can start. Experiments with antibodies specifically directed against the A and B chain (21) demonstrated that anti B chain antibody can only prevent the toxic effect when added almost immediately after the toxins, in accordance with our findings that the binding of abrin and ricin to cell surfaces occurs very rapidly. Then a certain period of time follows where the toxic effect can be prevented by the addition of anti A chain antibodies (Fig. 9). During the main part of the lag time, however, the toxic effect cannot be prevented by antibodies to either of the two constituent peptide chains of the toxins. Apparently during this time the toxins are neither on the cell surface nor have they penetrated completely into the cytoplasm where inactivation of ribosomes can take place.

How do the toxin A chains enter the cytoplasm? Experiments with ^{125}I-toxins showed that irreversible binding, i. e., binding which could not be released with lactose, occurs at a linear rate for a few hours (Fig. 10). Only a small fraction of the total number of bound toxin molecules becomes irreversibly bound per minute[5]. It is not yet known whether this reflects penetration of the toxins through the cell membrane or merely fixation to the membrane. Similar binding was found in the presence of various metabolic inhibitors. This irreversible binding is seen neither in HeLa cells at 0° nor in erythrocytes even at 37° (Fig. 10). We believe that the major part of this irreversible binding reflects uptake by pinocytosis. Whether it is relevant to the intoxication or merely a concomitant phenomenon is still obscure.

Several facts seem to favor the hypothesis that micropinocytosis or related processes are involved in the uptake. A lag time is always found even with high toxin concentrations. Obviously this may reflect the time required for the transport and release of the toxin (or the free A chain) into the cytoplasm (21). In fact, Nicolson (7) showed that ricin-ferritin complexes are present in pinocytotic vesicles and appear to be released into the cytoplasm after a certain lag time. Also the strong temperature dependence of the lag time may favour this hypothesis as pinocytosis is known

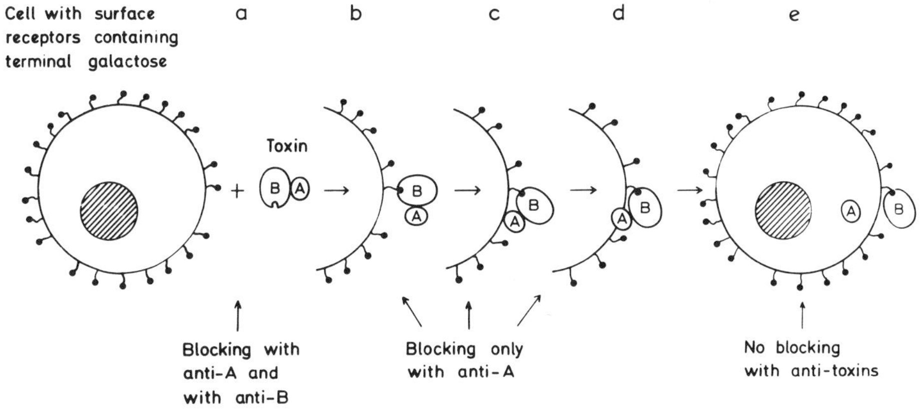

Fig. 9. Hypothetic scheme of abrin and ricin uptake (21).

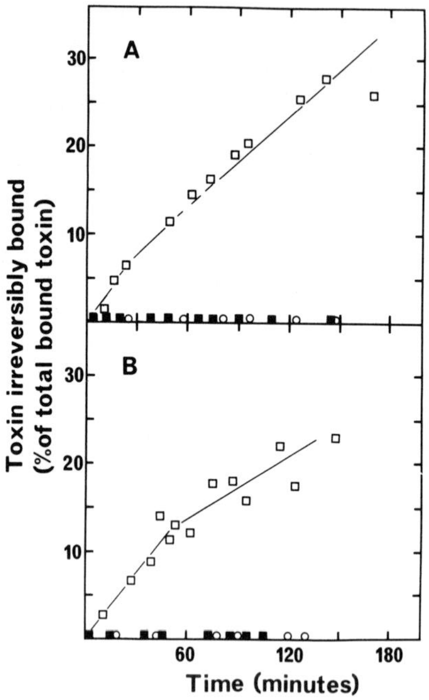

Fig. 10. Irreversible binding of abrin and ricin. HeLa cells (■,□) or human erythrocytes (O) were incubated at 0° (filled symbols) or at 37° (open symbols) with abrin (A) or ricin (B) and after the indicated periods of time samples were diluted into buffer containing 0.1 M lactose to liberate reversibly bound toxins. Then the cells were sedimented and washed and the bound radioactivity was measured.

to be strongly affected by the temperature. Furthermore, in intact reticulocytes, which have little pinocytotic activity, the toxins have no ability to inhibit protein synthesis although they bind rapidly to the cell surface (21). However, the reticulocyte cell membrane is in several respects different from that in most cells, a fact rendering it difficult to generalize from studies on reticulocytes.

On the other hand, several observations can be interpreted as evidence against the pinocytosis hypothesis. Thus, several components which are supposed to stimulate or inhibit pinocytosis do not change the rate of cell intoxication. Moreover, in the case of diphtheria toxin which in several respects is similar to abrin and ricin (19) the toxic effect is completely prevented in the presence of low concentrations of NH_4^+ ions (4) although the binding to cell surfaces and uptake by pinocytosis is not reduced (1). Whatever the mechanism we feel that it was hardly developed exclusively for the uptake of toxic components. It is conceivable that certain biologically important molecules, like hormones, enter the cytoplasm by these routes, and that in the course of development the toxins have adapted to the same uptake mechanism.

REFERENCES

1. BONVENTRE, P. F., SAELINGER, C. B., IVINS, B., WOSCINSKI, C. and AMORINI, M., Infect. Immunol. 11 (1975) 675.
2. CARRASCO, L., FERNANDEZ-PUENTES, C. and VAZQUES, D., Eur. J. Biochem. 54 (1975) 499.
3. FUNATSU, M., in Proteins, Structure and Functions (Funatsu, M., Hiromi, K., Imahori, K., Murachi, T. and Narita, K. eds.) (1972) John Wiley & Sons, New York.
4. KIM, K. and GROMAN, N. B., J. Bacteriol. 90 (1965) 1552.
5. LIN, J.-Y., KAO, W.-Y., TSERNG, K.-Y., CHEN, C.-C. and TUNG, T.-CH., Cancer Res. 30 (1970) 2431.
6. NICOLSON, G. L., BLAUSTEIN, J. and ETZLER, M. E., Biochemistry 13 (1974) 196.
7. NICOLSON, G. L., Nature 251 (1974) 628.
8. OLSNES, S., HEIBERG, R. and PIHL, A. Mol. Biol. Rep. 1. (1973) 15.
9. OLSNES, S. and PIHL, A., FEBS Lett. 20 (1972) 327.
10. OLSNES, S. and PIHL, A., Nature 238 (1972) 459.
11. OLSNES, S. and PIHL, A., FEBS Lett. 28 (1972) 48.
12. OLSNES, S. and PIHL, A., Biochemistry 12 (1973) 3121.
13. OLSNES, S. and PIHL, A., Eur. J. Biochem. 35 (1973) 179.
14. OLSNES, S., REFSNES, K. and PIHL, A., Nature 249 (1974) 627.
15. OLSNES, S., REFSNES, K., CHRISTENSEN, T. and PIHL, A., Biochim. Biophys. Acta 405 (1975) 1.
16. OLSNES, S., PAPPENHEIMER, A. M., JR. and MEREN, R., J. Immunol. 113 (1974) 842.

17. OLSNES, S. and SALTVEDT, E., J. Immunol. 114 (1975) 1743.
18. OLSNES, S., SALTVEDT, E. and PIHL, A., J. Biol. Chem. 249 (1974) 803.
19. PAPPENHEIMER, A. M., JR. and GILL, D. M., Science 182 (1973) 353.
20. PAPPENHEIMER, A. M., JR., OLSNES, S. and HARPER, A. A., J. Immunol. 113 (1974) 835.
21. REFSNES, K., OLSNES, S. and PIHL, A., J. Biol. Chem. 249 (1974) 3557.
22. SPERTI, S., MONTANARO, L., MATTIOLI, A. and STIRPE, F., Biochem. J. 136 (1973) 813.
23. STECK, T. L. and WALLACH, D. F. H., Biochim. Biophys. Acta 97 (1965) 510.
24. WEI, C. H., HARTMAN, F. C., PFUDERER, P. and YANG, W.-K., J. Biol. Chem. 249 (1974) 3061.

STUDIES ON AN HEPATIC MEMBRANE RECEPTOR SPECIFIC FOR THE BINDING AND CATABOLISM OF SERUM GLYCOPROTEINS

Gilbert ASHWELL and Anatol G. MORELL

National Institute of Arthritis, Metabolism, and Digestive Diseases, National Institutes of Health, Bethesda, Maryland 20014 (USA), and
Division of Genetic Medicine, Department of Medicine, Albert Einstein College of Medicine, Bronx, New York, N. Y. 10461 (USA)

The development of a new concept for the role of the carbohydrate moiety in regulating the serum survival time of plasma glycoproteins has been described in a recent review (3). Briefly, treatment with neuraminidase to remove sialic acid from circulating glycoproteins, followed by oxidation with galactose oxidase and reduction with KB^3H_4, introduced a tritum atom into carbon-6 of the terminal galactose residues (11). Upon injection of the resulting radioactive asialoglycoprotein into rabbits, the serum was promptly cleared of radioactivity and the intact macromolecule was recovered from the liver within 15 minutes after injection (10). The rapidity of removal was dependent upon the integrity of the exposed galactose; enzymatic oxidation by galactose oxidase or removal by hydrolysis with β-galactosidase diminished the rate of hepatic uptake and resulted in an increased serum survival time approaching that of the fully sialylated control (10). More specifically, replacement of the deficient sialic acid residues, by incubation of the asialoglycoprotein with sialyltransferase in the presence of CMP-sialic acid, restored the original serum survival time (6). From the foregoing study, it became clear that hydrolysis of only a small fraction of the total sialic acid residues was sufficient for the liver to recognize and remove the entire protein (17) which was subsequently shown to be catabolized in the hepatic lysosomes (4). The generality of this phenomenon was demonstrated by the injection into rats of the asialo derivatives of a number of plasma proteins such as orosomucoid, fetuin, haptoglobin, ceruloplasmin and α_2-macroglobulin. In each case, the desialylated material disappeared rapidly from the circulation and accumulated in the liver (9). Competitive inhibition of hepatic uptake was shown by the simultaneous injection of tracer amounts of [^{64}Cu] asialoceruloplasmin and substantive amounts of the above listed desialylated proteins or the glycopeptides derived from them. Uptake was not blocked by either the fully sialylated proteins or their corresponding glycopeptides (9).

With the realization that the viability of many of the serum glyco proteins depended on the presence of their normal complement of sialic acid, consideration was given to another category of biologically potent glycoproteins, i.e., the gonadotropic hormones. The biological activity of both human chorionic gonadotropin and follicle-stimulating hormone, is known to be destroyed by exposure to neuraminidase (12). Since the standard hormone assay involves injection of the active principle into an appropriately sensitized animal, it was thought possible that the potency loss might be correlated with their rapid removal by the liver. To test this hypothesis, both hormones were desialylated enzymatically, labeled with ^{125}I or tritium and injected into rats. Both preparations disappeared rapidly from circulation and were recovered in the liver (2, 9, 18, 19). The close parallel between these results and those with the serum glycoproteins was underlined by the marked increase in survival time in the presence of asialo-orosomucoid; intact orosomucoid was ineffective in prolonging serum viability. These results, which led to a better appreciation of the conditions essential for survival in the serum, prompted the development of a new and general method for the direct tritiation of terminal sialic acid residues without prior hydrolysis (16). The resulting radioactive glycoproteins were shown to possess a normal serum survival time (20).

Extension of these studies to an <u>in vitro</u> system revealed the hepatic plasma membranes to be a major locus of binding for circulating glycoproteins. Examination of the properties of this system revealed the binding process to involve a dual role for sialic acid in that its presence on the membrane was essential whereas its presence on the glycoprotein was incompatible with binding (13). Enzymatic restoration of the sialic acid residues onto partially desialylated membranes was accompanied by increased binding activity. In addition to the requirement for sialic acid, the binding of proteins by plasma membranes exhibited an absolute dependence upon the presence of calcium. Alternate removal or replacement of this ion, within the pH range of 6.5 to 7.5 resulted in a rapid reversal of the binding process (13). In contrast, binding was essentially irreversible in the presence of calcium at pH 7.5. The latter property provided the basis for the development of a quantitative assay to estimate the relative binding affinity of a variety of serum glycoproteins as well as their asialo- and agalacto-derivatives (21).

More recently, the successful solubilization and purification of an hepatic protein, which exhibited all of the characteristic binding properties associated with the membrane, was reported (8). The isolated material was water soluble and free from lipids. It was identified as a glycoprotein in which 10% of the dry weight consisted of sialic acid, galactose, mannose, and glucosamine in a molar ratio of 1:1:2:2. The integrity of the terminal sialic residues and the presence of calcium were shown to be absolute requirements for binding. Physical chemical studies indicated a high degree of aggregation in the detergent-free, water-soluble preparation. Subsequent studies[1] have shown the state of aggregation to result from the self-associating properties of a single oligomeric protein. The smallest functional unit identifiable in aqueous solution possessed an estimated molecular weight of 500,000 with each of the successive components increasing in size by an equal amount to form an oligomeric series bearing an integral ratio of 1:2:3:4:5. The tendence toward self-association was promptly and completely reversed by the addition of Triton X-100 with the concomitant appearance of a single component. Extensive treatment with sodium dodecyl sulfate permitted the identification and isolation of two subunits with estimated molecular weights of 48,000 and 40,000, respectively. Amino acid and carbohydrate analyses revealed both subunits to be glycoproteins with closely similar, but not identical, properties.[1]

In view of the presumed role of glycosyl transferases in the recognition and transient association of cells (14), the purified binding protein was examined for the presence of glycosyl transferase activity (7). Under optimally determined conditions, no transferase activity for sialic acid, galactose, N-acetylglucosamine, or fucose was detectable in the purified preparation. The inhibition of glycoprotein binding by α-lactalbumin, reported originally with hepatic plasma membranes by Aronson _et al_. (1) was confirmed and shown to result from competitive binding to the hepatic protein. In contrast to asialoglycoproteins, the binding of α-lactabumin was independent of calcium and unaffected by EDTA (7).

An unsuspected property of the solubilized binding protein proved to be its ability to agglutinate human and rabbit erythrocytes (15). The requirements for aggutination conformed, in every respect, to the above described conditions for successful binding to asialoglycoproteins. In addition, the observation that blood group A cells were more susceptible to agglutination than were blood group B cells, led to the realization that terminal N-acetylgalactosamine, as well as terminal galactose, residues were recognized by the binding protein. The physiological significance, if any, of the lectin-like properties of the hepatic binding protein is not known.

Current studies, describing work in progress, included a discussion of the effect of neoplastic transformation of liver cells whereby

---

[1]Kawasaki, T. and Ashwell, G., submitted for publication.

the binding activity of both primary and secondary (Morris) hepatomas was qualitatively identical to that seen in normal livers but quantitatively diminished (5). Although mammalian cell cultures have provided well defined conditions for metabolic studies, we have not been successful in obtaining a cell culture of normal adult hepatic cells which retained their specific binding capacity for asialoglycoproteins. However, freshly prepared hepatocytes, isolated from a well-perfused rat liver, have been shown to bind and to transport asialoglycoproteins across the cell membrane (5). Under the conditions used, the incorporation of labeled glycoprotein was linear with respect to the number of cells added and was both time- and temperature-dependent. At equilibrium, approximately 1 pmole of ^{125}I-asialo-orosomucoid was bound per 10^6 cells; of this amount of at least 50% was transported into intracellular loci. The latter process was shown to be energy-dependent in that it was differentially inhibited by a variety of metabolic antagonists such as cyanide, dinitrophenol, etc. Evidence for the catabolic destruction of ^{125}I-asialo-orosomucoid was obtained by monitoring the decrease in immune-precipitable total radioactivity released into the incubation medium over a period of several hours at 30° (5).

In addition to the above studies, evidence has recently become available to indicate that the hepatic binding protein is not uniquely located in the plasma membrane but is present in significant amounts in the microsomes, lysosomes, and the Golgi complex of mammalian liver cells.[2] In patients with liver diseases, such as cirrhosis or infectious hepatitis, a striking correlation has been observed between the onset and duration of the pathological condition and the appearance of measurable levels of desialylated glycoproteins in the serum.[3] The significance of this finding is interpreted as providing supportive evidence for the role of the hepatic binding protein in the physiological regulation of glycoprotein metabolism.

REFERENCES

1. ARONSON, N. N., TAN, L. Y., and PETERS, B. P., Biochem. Biophys. Res. Commun. 53 (1973) 112.
2. ASHWELL, G., Endocrinology: Proceedings of the Fourth International Congress on Endocrinology, Washington, (1972).
3. ASHWELL, G. and MORELL, A. G., Advances in Enzymology 41 (1974) 99.
4. GREGORIADIS, G., MORELL, A. G., STERNLIEB, I., and SCHEINBERG, I. H., J. Biol. Chem. 245 (1970) 5833.

[2] Pricer, W. E. and Ashwell, G., manuscript in preparation.
[3] Lunney, J. and Ashwell, G., manuscript in preparation.

5. HICKMAN, J. and ASHWELL, G., in Enzyme Therapy in Lysosomal Storage Diseases. (J. H. Tager, G. J. H. Hooghwinkel and W. Th. Daems eds.) North Holland Publishing Co., Amsterdam (1974) 169.
6. HICKMAN, J., ASHWELL, G., MORELL, A. G., VAN DEN HAMER, C.J.A. and SCHEINBERG, I. H., J. Biol. Chem. 245 (1970) 759.
7. HUDGIN, R. L. and ASHWELL, G., J. Biol. Chem. 249 (1974). 7369.
8. HUDGIN, R. L., PRICER, W. E., ASHWELL, G., STOCKERT, R. J. and MORELL, A. G., J. Biol. Chem. 249 (1974) 5536.
9. MORELL, A. G., GREGORIADIS, G., SCHEINBERG, I. H., HICKMAN, J. and ASHWELL, G., J. Biol. Chem. 246 (1971) 1461.
10. MORELL, A. G., IRVINE, R. A., STERNLIEB, I., SCHEINBERG, I. H. and ASHWELL, G., J. Biol. Chem. 243 (1968) 155.
11. MORELL, A. G., VAN DEN HAMER, C.J.A., SCHEINBERG, I. H. and ASHWELL, G., J. Biol. Chem. 241 (1966) 3745.
12. MORI, K. F., Endocrinology 85 (1969) 330.
13. PRICER, W. E. and ASHWELL, G., J. Biol. Chem. 246 (1971) 4285.
14. ROSEMAN, S., Chem. Phys. Lipids 5 (1970) 270.
15. STOCKERT, R. J., MORELL, A. G. and SCHEINBERG, I. H., Science 186 (1974) 365.
16. VAITUKAITIS, J., HAMMOND, J., ROSS, G. T., HICKMAN, J. and ASHWELL, G., J. Clin. Endocrinol. Metabolism 32 (1971) 209.
17. VAN DEN HAMER, C.J.A., MORELL, A. G., SCHEINBERG, I. H., HICKMAN, J. and ASHWELL, G., J. Biol. Chem. 245 (1970) 4397.
18. VAN HALL, E. V., VAITUKAITIS, J. L., ROSS, G. T., HICKMAN, J. and ASHWELL, G., Endocrinology 88 (1971) 456.
19. VAN HALL, E. V., VAITUKAITIS, J. L., ROSS, G. T., HICKMAN, J. and ASHWELL, G., Endocrinology 89 (1971) 11.
20. VAN LENTEN, L. and ASHWELL, G., J. Biol. Chem. 246 (1971) 1889.
21. VAN LENTEN, L. and ASHWELL, G., J. Biol. Chem. 247 (1972) 4633.

CONCANAVALIN A INDUCED CHANGES OF MEMBRANE-BOUND LYSOLECITHIN ACYLTRANSFERASE OF THYMOCYTES

Clay E. REILLY* and Ernst FERBER

Max-Planck-Institut für Immunbiologie
D-78 Freiburg (GFR)

One of the major problems of immunobiology today is the elucidation of the mechanism by which an antigen is able to specifically trigger lymphocytes into mitosis, thereby inducing the clonal expansion necessary for a competent immune response. However, the number of thymus dependent T-cells and bone-marrow derived B-cells which exhibit this sensitivity are so few in number that adequate biological studies are limited (see ref. 17). With the discovery of Nowell (20) that phytohemagglutinin (PHA), a protein extract of the red kidney bean, non-specifically induces blastogenesis in human peripheral blood lymphocytes, the door was opened for model studies of the antigen-lymphocyte interaction. The validity of such model studies is based on the similarity of the response of lymphocytes to PHA and the jack-bean mitogen concanavalin A (Con A), and that of antigen mediated blastogenesis. Some of these analogous reactions are the increase in RNA (16) and DNA synthesis (24), the induction of lymphocyte cytotoxicity in T-cells (21), and immunoglobulin synthesis in B-cells, when the latter are stimulated by Con A cross-linked to a solid surface (1). In addition, the excretion of such biologically active substances as macrophage inhibiting factor (MIF), lymphotoxin, chemotactic and mitogenic factors have been observed for both the antigen and lectin stimulation of lymphocytes (see ref. 15).

Greaves and Bauminger (11), by stimulating lymphocytes with Con A covalently bound to Sepharose, were also able to demonstrate that the triggering of lymphocytes to undergo mitosis occurs at the cell surface and that incorporation of mitogen is not a requirement

---

*Present address: Lehrsthul für Immunbiologie, D-78 Freiburg, Stefan-Meier-Str. 8, GFR.

for this process. One of the early biochemical changes in lymphocytes occurring immediately upon stimulation with a mitogen has been reported by Quastel and Kaplan (22), who have shown that the Na^+,K^+ pump of lymphocytes is activated by PHA. The activation of the surface membrane-bound Na^+-K^+-ATPase by Con A and PHA has also been established (14). In addition to changes in cationic pumps, Fisher and Mueller (10) have also observed that there is an increased incorporation of phosphate into phosphatidic acid and inositol phosphatides upon stimulation of lymphocytes with PHA. Resch and Ferber (23) were also able to show an increased inclusion of long chain fatty acids into phospholipids and that this increase is not based on the de novo synthesis of fatty acids, but upon the acylation of endogenous lysolecithin (see Fig. 1). Ferber and Resch (9) also developed a test to measure acyl-CoA:lysolecithin acyltransferase by employing acyl-CoA and [^{14}C]-lysolecithin as substrates and then separating the reaction product, lecithin, by thin layer chromatography. They found this enzyme to be activated only in the isolated plasma membranes of stimulated lymphocytes and not in other cell subparticles, such as mitochondria.

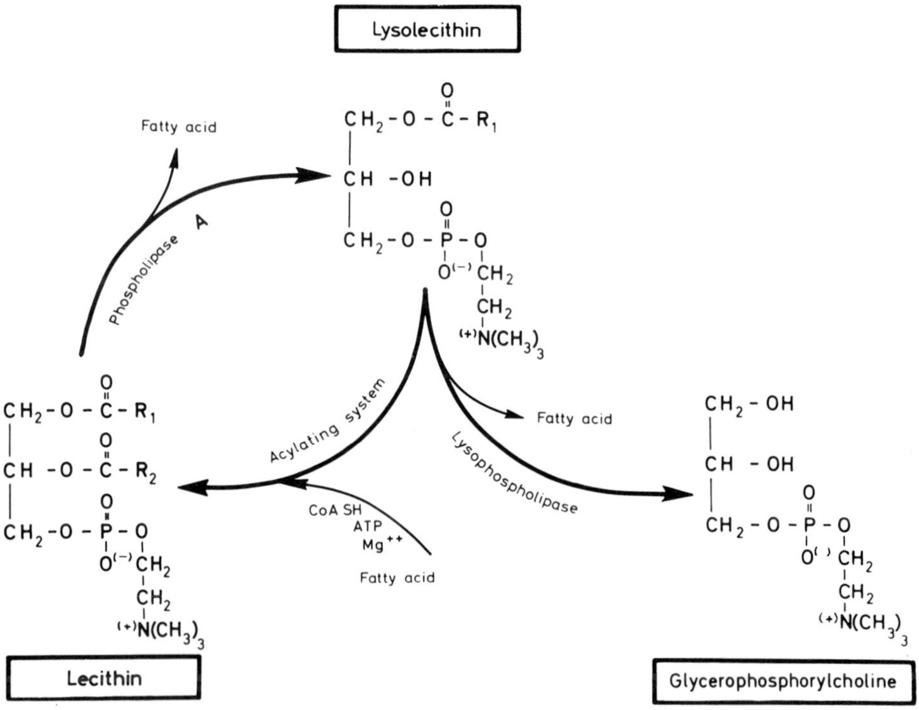

Fig. 1: Metabolic pathways of lysolecithin. The acylating system is catalyzed by acyl-CoA:lysolecithin acyltransferase, which is bound to the plasma membrane in lymphocytes and can be activated by Con A.

Cell Disruption and Membrane Preparation

Calf thymus was employed in these experiments because of the relatively homogeneous population of cells obtainable from this lymphoid organ, and because it can supply the large number of cells (5×10^9) necessary in acquiring sufficient membrane material for biochemical analysis.

The cell suspensions were incubated with Con A in Eagle's medium (Dulbecco modification) for no longer than 45 min at 0° or 37° incubation temperatures. The cells were subsequently washed by centrifugation and then disrupted by gas-bubble nucleation in a nitrogen pressure homogenizer at 30 atm, 4°, with an equilibration period of 20 min. The isolation of thymocyte microsomes, which contained about 80% plasma membranes, was achieved through differential centrifugation and it was these microsomes which were used in most enzymatic studies. If experiments demanded purified plasma membranes, they were isolated in a continuous dextran gradient (see Fig. 2).

Kinetics of Binding of Con A to Thymocytes

A. Dose-response behavior

In experiments studying the binding of ^{125}I-Con A, it was observed that plots of the dosage of the lectin to the amount of lectin bound to the thymocytes describe a hyperbola. This type of saturation kinetics is Michaelian and suggests an independence of binding sites, that excludes cooperativity as a function of binding to all sites present. Incubation of the cells with ^{125}I-Con A was at 37° for 45 min in Eagles medium buffered with 4-(hydroxyethyl)-1-piperazinyl-ethane-2-sulfonic acid (Hepes) at pH 7.2, with 5×10^7 cells/ml/assay. Parallel experiments were run in which 50 mM α-methyl-D-mannoside was present before Con A was administered. The values of these controls were used to correct for the aggregate binding of the lectin, as well as adsorption on the cell surface.

The reciprocal plots of these same data describe a straight line, and from the intercept of the ordinate it is possible to calculate the maximum amount of Con A bound to the cells. In the notation of the Sips equation, N, the maximum amount of Con A bound under the conditions of these experiments, as calculated from the reciprocal plot and also the Scatchard plot, is 10.7 μg/ml which is equal to 0.0991 nmol/ml, and thus 1.2×10^6 molecules/cell.

Applying the Sips equation, it is possible to calculate the degree of heterogeneity of binding. Thus if a(slope) = 1.0, then all sites are identical with respect to K, and the affinity of the ligand is the same for all receptors (see ref. 6). The Sips equation is

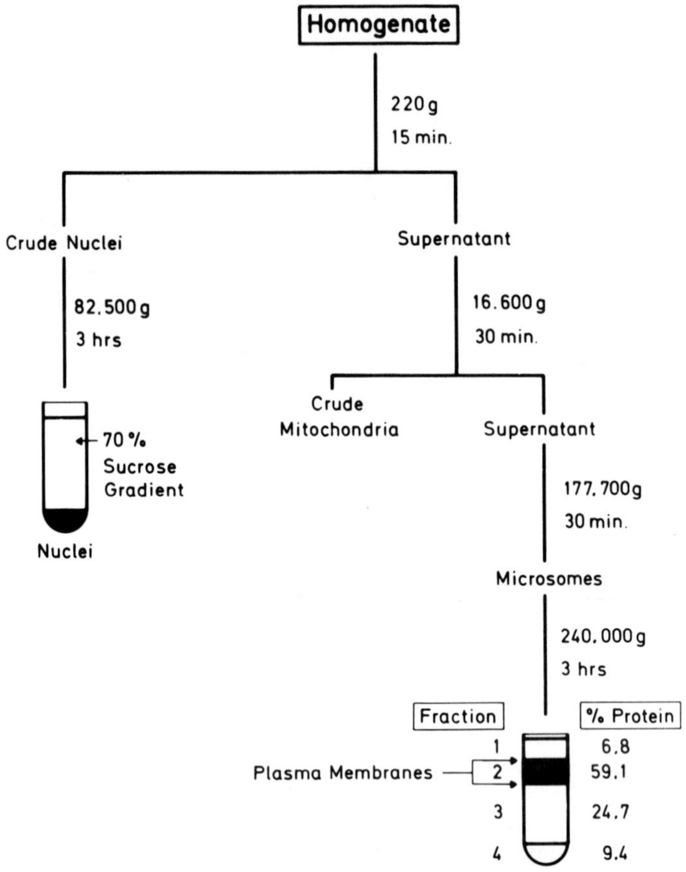

Fig. 2. Diagram of the subcellular fractionation of thymocytes. A dextran gradient (d = 1.088 - 1.092) was used.

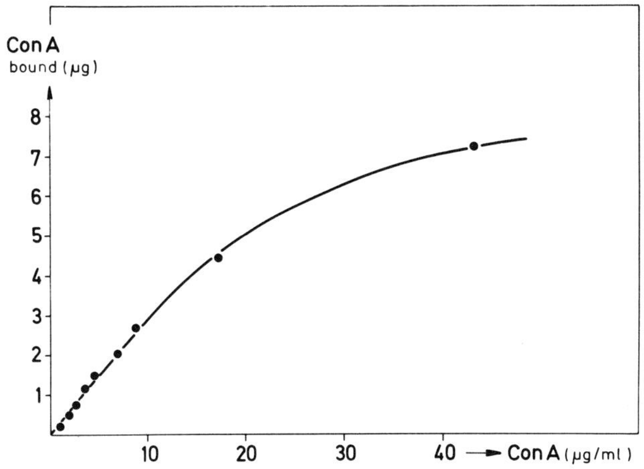

Fig. 3. Dose-response behavior of the binding of ^{125}I-Con A to calf thymocytes.

expressed as:

$$\frac{r}{N} = \frac{(Kc)^a}{1+(Kc)^a}$$

or

$$\log \frac{r}{N-r} = a \log K + a \log c$$

where N = maximum amount of Con A bound

r = Con A bound for a given concentration

K = equilibrium association constant

c = free Con A

This equation is equivalent to the Hill function of enzyme kinetics, and both are statistical distribution functions of the Mass Law (see ref. 27). The equilibrium association constant K is thus 6.2×10^6 l/mol and agrees with the data of Betal and van den Berg (3) for

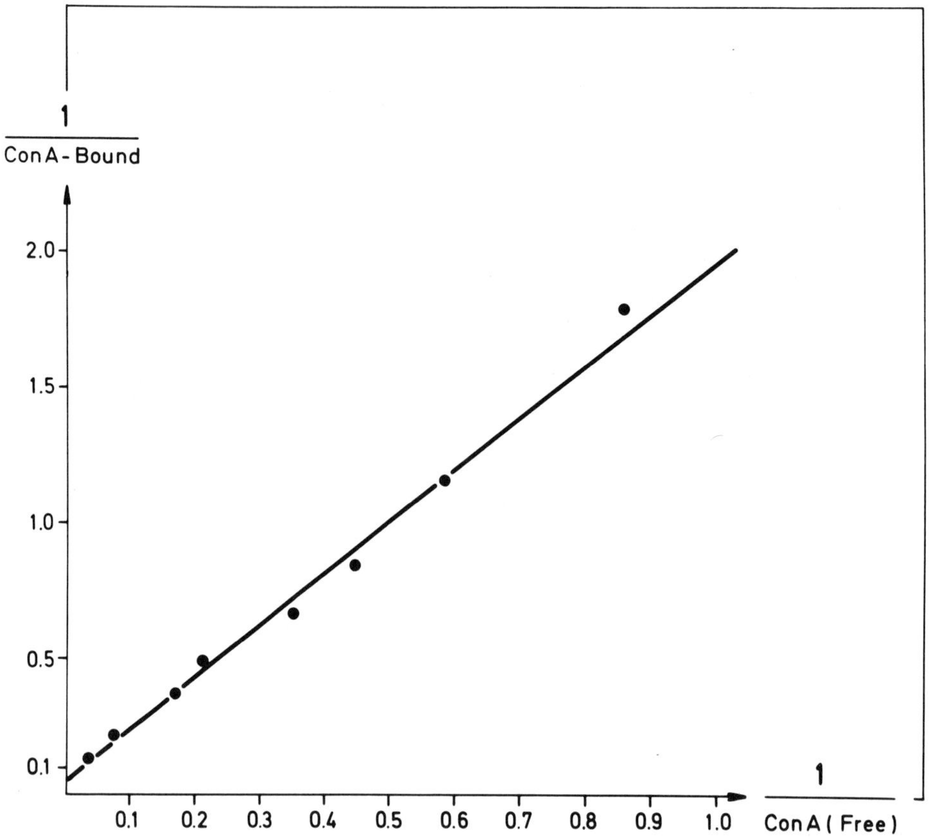

Fig. 4. Reciprocal plot of free ^{125}I-Con A to ^{125}I-Con A bound to the surface of thymocytes.

the binding of Con A to rat lymphocytes. The slope (a = 1.09) expresses a value of unity for the exponents of the Sips equation, thus demonstrating a lack of cooperativity of the Con A binding sites on the thymocyte cell surface.

B. Time-course studies

In the laboratory of the authors, Huber (13) studied the binding of ^{125}I-Con A to rabbit thymocytes as a function of time and concentration. He demonstrated that the maximum amount of Con A binding to the cells is reached at about 45 min after administering the lectin, and that only at very low dosage is the binding immediate. Cuatrecasas (5) found similar results for the binding of Con A to fat cells. Thus, in our enzyme studies, an incubation time of 45 min was employed in all experiments.

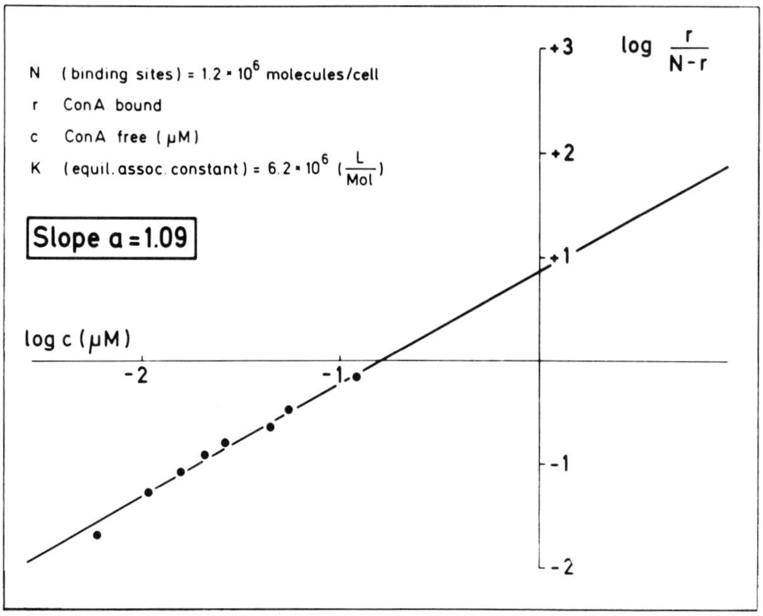

Fig. 5. Sips plot of the binding of ^{125}I-Con A to calf thymocytes.

Kinetics of the activation of acyl-CoA:lysolecithin acyltransferase by Con A

A. Temperature independence of response to Con A

If Con A is given to a suspension of thymocytes, there is a 3- to 4-fold increase in the acyltransferase activity of the isolated microsomes, and this stimulation of the enzyme occurs at both 37° and 0°. The levels of activation are the same for both temperatures, going from about 10 nmol/mg protein/min to 35 nmol/mg protein/min in the microsomes of cells incubated with Con A for 45 min at 10 µg/ml. This independence of incubation temperature suggests that metabolic factors are not involved in the activation of the enzyme and that the process is a direct one, i.e. surface receptor-Con A interaction triggers the activation.

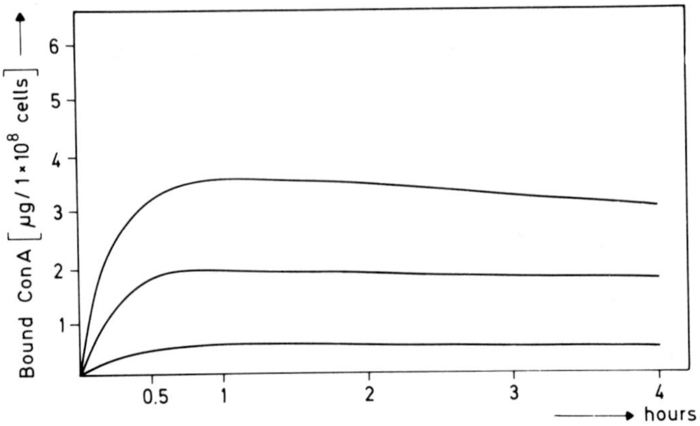

Fig. 6. Binding of Con A to rabbit thymocytes as a function of time and at various concentrations (13).

B. α-Methyl-D-mannoside inhibition of activation

The saccharide, α-methyl-D-mannoside (α-MM) is known to bind specifically to Con A (see ref. 19) and thus represents a competitive inhibitor of the binding of Con A to the glycoprotein and glycolipid receptors of the cell surface. In Fig. 7, 4 mg/ml of α-MM is sufficient to block the activation of a high dosage of Con A (20 µg/ml), if this saccharide is preincubated for 10 min before incubating with Con A for 60 min. If Con A is then incubated with the thymocytes for 15 min, then 45 min with α-MM, an increase in the acyltransferase of about 66% of that seen for the controls is observed. There is also a progressive increase in the acyltransferase in direct proportion to the time of incubation with Con A. In addition to this, the time necessary to activate the enzyme is the same for the progressive binding of the lectin to the thymocyte cell surface (see Fig. 6). Maximal lectin binding, as well as maximal enzyme activation, requires 45 min of incubation at the conditions of these experiments.

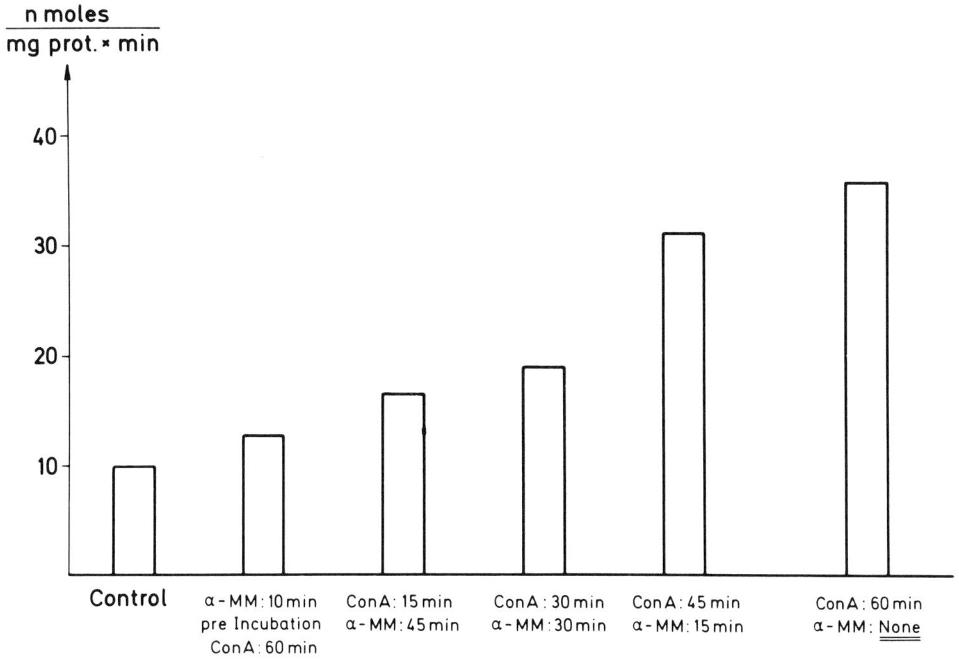

Fig. 7. Inhibition of the activation of lysolecithin acyltransferase by α-methyl-D-Mannoside. Incubation mixture contained 20 μg/ml ConA, 4 mg/ml α-methyl-D-mannoside, 1 x 10^8 cells/ml EM-H at 37°.

C. Dose-response behavior

When Con A is administered to several suspensions of thymocytes at concentrations of 0.8 μg/ml to 10.0 ug/ml, the microsomes isolated and tested for acyltransferase activity, there is a direct dependency of the dosage to the activation of the enzyme. Fig. 8 shows that this activation does not follow normal hyperbolic saturation kinetics, but is rather sigmoidal in form.

Moreover, if the inverse of the increase in the dosage of Con A is plotted against the inverse of the increase in enzyme activity, a parabola is described.

Because these data suggest positive cooperativity between Con A and lysolecithin acyltransferase, i.e., between Con A:surface recep-

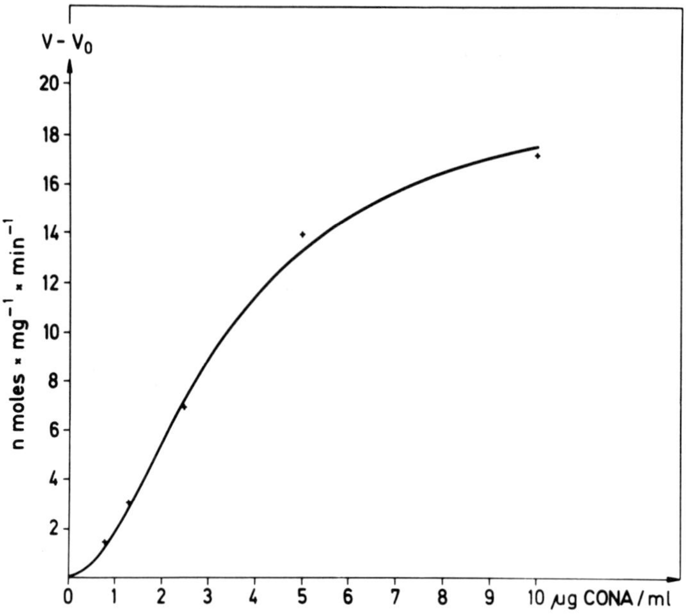

Fig. 8. Sigmoid relationship of the dose-response behavior of acyltransferase to Con A. V_o = unstimulated enzyme activity.

tors:enzyme, Hill plots were drawn in order to determine the degree of cooperativity involved (for the derivation of the Hill equation, see ref. 2). As seen in Fig. 10, the slope of this plot gives a value of n = 1.8 (Hill plot notation slope = n). Changeux (4) has referred to n as the coefficient of interaction between the binding sites. The stronger this interaction, the greater the approach of the coefficient to a whole number, so that in perfectly cooperative systems, Hill plots will have slopes of n = 2.0, 3.0, etc., depending upon the number of ligands bound and thus affecting 1 macromolecule; e.g. acetyl-CoA to pyruvate carboxylase (7) or oxygen molecules to hemoglobin (27). Thus it appears in the Con A:acyltransferase system, that at least two lectin molecules activate one enzyme molecule.

D. Substrate kinetics

Studies on the substrate kinetics of the acyltransferase showed that both lysolecithin and acyl-CoA are Michaelian in behavior. There is also no change in the affinity (Km) for the two acyl-CoA

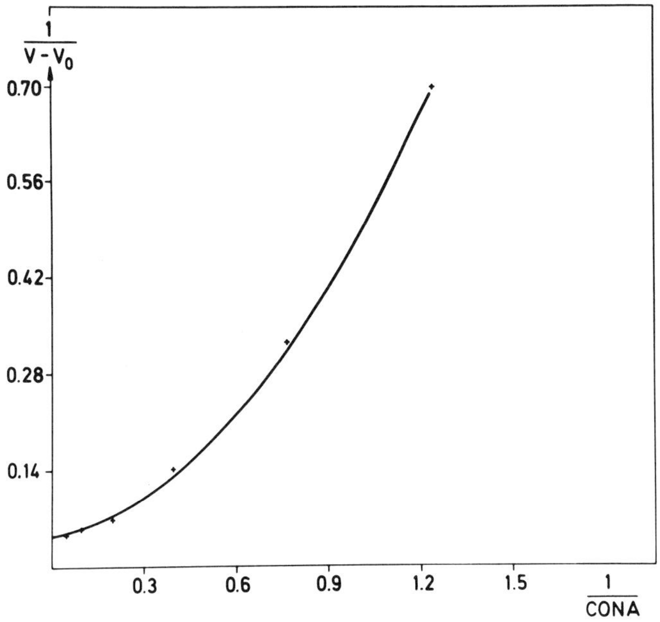

Fig. 9. Reciprocal plot of the increase in Con A dosage to the increase in acyltransferase activity.

substrates in the microsomes of either the Con A treated cells or in controls. Of the acyl-CoA compounds, arachidonoyl-CoA (20:4) exhibited a 4-fold increase in Vmax, whereas the oleoyl-CoA (18:1) only 2.0. The long chain highly unsaturated arachidonoyl-CoA exhibited moreover, a 15-fold higher affinity to the acyltransferase than the oleoyl-CoA.

Influence of Con A on other membrane-bound enzymes

A. Adenyl cyclase

Smith *et al.* (36) have reported that PHA is able to stimulate the increase in intracellular concentrations of cAMP in lymphocytes immediately upon binding the lectin. In order to determine whether adenyl cyclase plays a role in Con A stimulation of thymocytes, this enzyme was measured in the microsome preparations. It was found that both the microsomes of Con A treated thymocytes as well as controls exhibited specific activities of an average of 7.0 picomoles/mg protein/min. This lack of response of the adenyl cyclase to Con A agrees with the findings of Hadden *et al.* (12) that neither Con A nor PHA stimulate intracellular increases in cAMP in lymphocytes. This

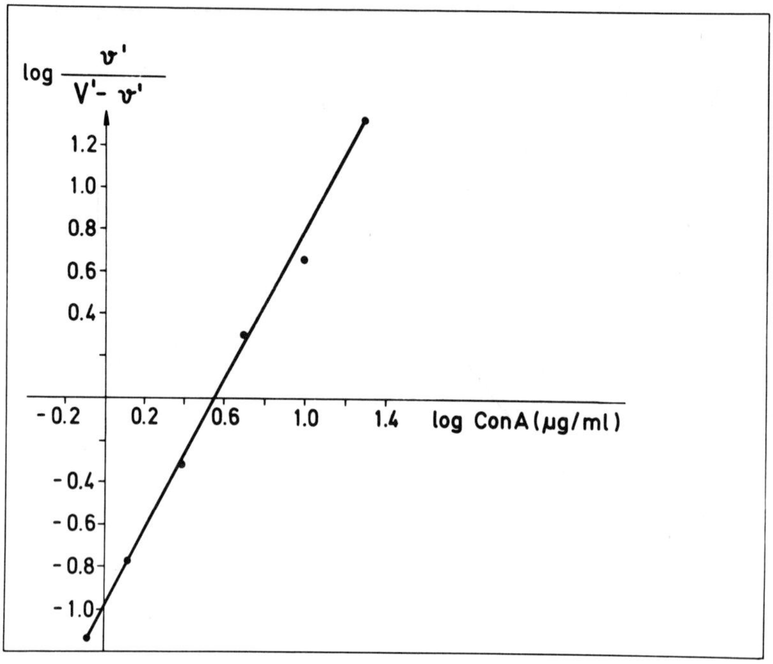

Fig. 10. Hill plot of the degree of cooperativity of Con A and lysolecithin acyltransferase. Hill equation:

$$\log \frac{v'}{V'-v'} = n \log S - n \log K,$$

where: $n = 1.8$, $v' = v - v_o$, and $V' = Vmax$.

group, however, was able to measure increases in intracellular concentrations of cGMP in cells stimulated by Con A and PHA, and have proposed that it is this cyclic nucleotide which is the carrier of the mitogenic signal within the cell.

B. Na^+-K^+-ATPase and Mg^{++}-ATPase

Novogrodsky (18) has reported that Con A is only capable of stimulating the Na^+-K^+-independent-Mg^{++}-ATPase, and not the Na^+-K^+-ATPase, i.e., the ouabain-sensitive ATPase. Lauf (14) on the other hand has reported that Con A, PHA and anti-lymphocyte sera bring about a 3- to 4-fold activation of the ouabain-sensitive Na^+-K^+-ATPase and a 2-fold increase in the Mg^{++}-ATPase in mature human lymphocyte

SUBSTRATE:	OLEOYL-CoA (18:1)		ARACHIDONOYL-CoA (20:4)	
	Vmax	Km (M)	Vmax	Km (M)
control	3.9	1.0×10^{-5}	7.6	6.4×10^{-7}
CONA	7.9	1.2×10^{-5}	31.7	8.5×10^{-7}

Fig. 11. Substrate specificity of microsomal acyl CoA: lysolecithin acyltransferase of Con A stimulated thymocytes.

V_{max} is given in nmols x mg protein^{-1} x min^{-1}

microsomes. Our studies show that in thymocytes, which are immature lymphocytes, that only the ouabain-sensitive Na^+-K^+-ATPase is activated 3-4 fold by Con A, but that the Mg^{++}-ATPase remains unaffected. It is interesting, however, to note that the Na^+-K^+-ATPase can be activated by Con A, as well as the acyltransferase, at an incubation temperature of 0°.

Conclusion

The data presented on the activation of lysolecithin acyltransferase by Con A describe this enzyme as a positive V_{max} system, i.e. there is an increase in V_{max} with a constant K_m, and cooperativity is exhibited by the effector Con A. It also appears that the process of activation is a direct one, and low amounts of energy are sufficient to trigger the response, e. g. activation can occur at 0°. Also, a form of desensitization can be observed, in that only the plasma membrane as it exists in the intact cell is capable of activating the enzyme; apparently the regulatory and catalytic subunits are disconnected upon cell disruption and vesicle formation. On the other hand, the higher affinity for the arachidonoyl-CoA (20:4), with its 4 double bonds in a chain of 20 carbon atoms, suggests the possibility that this enzyme is capable of regulating the number of double bonds in the lipid plane of the cell surface membrane, (see ref. 25 for the fluid mosaic model of cell membranes) and thereby

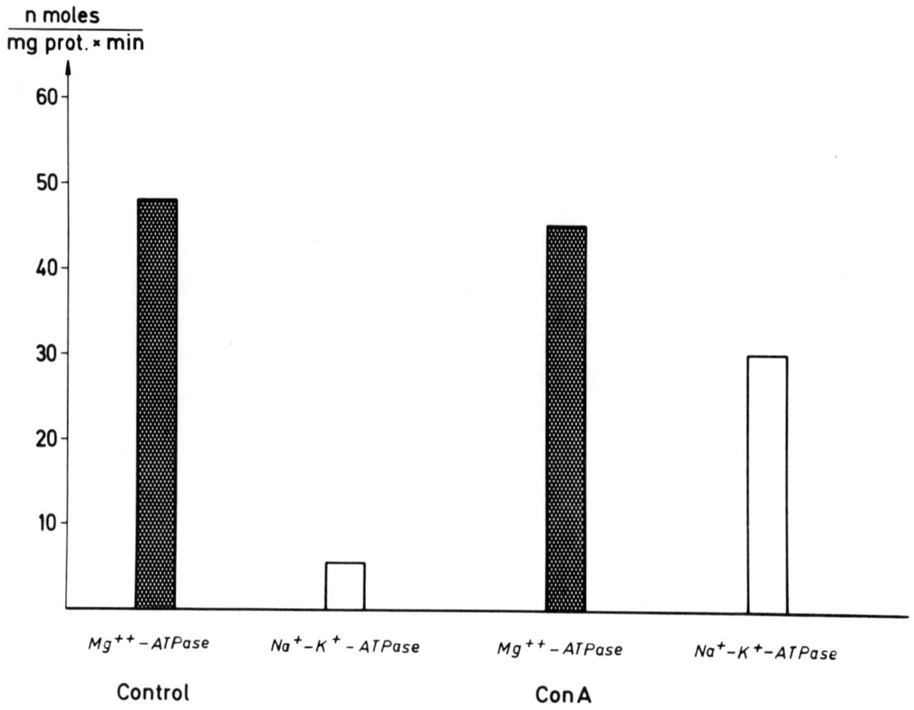

Fig. 12. Activation of ouabain-sensitive Na^{+}-K^{+}-ATPase in thymocyte microsomes by Con A. The Mg^{++}-ATPase is uneffected. The mixture contained 10 μg/ml Con A and 5 x 10^7 cells/ml EM-4. Incubation was for 45 min at 37°.

the state of its fluidity. Ferber et al. (8) have shown that in lipids extracted from thymocytes incubated with 5 μg/ml for 60 min, the fluorescence depolarization of perylene was complete at 20°; whereas in the lipids from controls this state was first reached at 40°. An incubation period of, however, less than 60 min brought about no apparent changes in lipid fluidity.

Because Con A also activates the Na^{+}-K^{+}-ATPase immediately upon binding, the lysolecithin acyltransferase is not exclusive in its prompt response to this mitogen. The cell surface membrane appears therefore to integrate a number of functions simultaneously upon mitogen contact; the nature of all these functions and how they relate to cell division will indeed be the subject of future research.

REFERENCES

1. ANDERSSON, J. and MELCHERS, F., Proc. Nat. Acad. Sci. USA 70 (1973) 416.
2. ATKINSON, D., HATHAWAY, T. and SMITH E., J. Biol. Chem. 240 (1965) 2682.
3. BETEL, I. and VAN DEN BERG, K., Eur. J. Biochem. 30 (1972) 571
4. CHANGEUX, J. P., Cold Spring Harbor Symosia Quant. Biol. 28 (1963) 497.
5. CUATRECASAS, P., Biochem. 12 (1973) 1312.
6. DAVIS, B., DULBECCO, R. EISEN, H., GINSBERG, H. and WOOD, B. Microbiology, Harper and Row, New York (1969) 370.
7. DUGAL, B. and LOUIS, B., FEBS Letters 28 (1972) 275.
8. FERBER, E., REILLY, C., DEPASQUALE, G. and RESCH, K., Lymphocyte Recognition and Effector Mechanisms (Lindahl-Kiessling, K. and Osoba, D. eds.) Academic Press, New York (1973) 529.
9. FERBER, E. and RESCH, K., Biochim. Biophys. Acta 296 (1973) 335.
10. FISHER, D. and MUELLER, G., Proc. Nat. Acad. Sci. USA 60 (1968) 1396.
11. GREEVES, M. and BAUMINGER, S., Nature New Biol. 235 (1972) 67.
12. HADDEN, T., HADDEN, E., HADDOX, M. and GOLDBERG, N., Proc. Nat. Acad. Sci. USA 69 (1972) 3024.
13. HUBER, A., Doctoral Dissertation, Medical Faculty, University of Freiburg (1974).
14. LAUF, P., Biochim. Biophys. Acta 415 (1975) 173.
15. LAWRENCE, H. and LANDY, M. (eds), Mediators of Cellular Immunity. Academic Press, New York (1969).
16. MUELLER, C. and LE MAHIEU, M., Biochim. Biophys. Acta 174 (1966) 100.
17. NOSSAL, G. J. V. and ADA, G. L., Antigens, Lymphoid Cells and the Immune Response, Academic Press, New York (1971).
18. NOVOGRODSKY, A., Biochim. Biophys. Acta 266 (1972) 343.
19. NOVOGRODSKY, A. and KATCHALSKI, E., Biochim. Biophys. Acta 228 (1971) 579.
20. NOWELL, P., Cancer Res. 20 (1960) 462.
21. PERLMANN, P., PERLMANN, H. and HOLM, G., Science 160 (1968) 306.
22. QUASTEL, M. and KAPLAN, J., Exp. Cell Res. 63 (1970) 714.
23. RESCH, K. and FERBER, E., Euro. J. Biochem. 27 (1972) 153.
24. SELL, S. and GELL, P., J. Exptl. Med. 122 (1965) 423.
25. SINGER, S. and NICOLSON, G., Science 175 (1972) 720.
26. SMITH, J. W., STEINER, A. L. and PARKER, C. W., J. Clin. Invest. 50 (1971) 442.
27. WYMAN, T., Advan. Prot. Chem. 19 (1964) 223.

COOPERATIVE REGULATION OF HORMONE BINDING AFFINITY FOR CELL SURFACE RECEPTORS

Pierre DE MEYTS

Diabetes Branch, National Institute of Arthritis
Metabolism and Digestive Diseases
National Institutes of Health, Bethesda, Maryland
20014 USA

Many enzymes in key metabolic pathways are subject to "regulation," i.e. their level of activity (K_m and/or V_{max}) is modulated by the substrate itself, by end-products of the chain of biochemical reactions or by substrate-unrelated effectors (42, 59). In the majority of cases, this regulation is effected through cooperative conformational transitions in a polymeric enzyme molecule, which are either induced (33) or stabilized (41) by the substrate or effector. Cooperative or "allosteric" properties are not an exclusivity of enzymes. Hemoglobin also exhibits strong cooperative interactions upon binding of oxygen (1, 5, 27, 44, 50, 60), and it is quite possible that some antibodies are endowed with similar properties when they bind antigens (40, 58) or complement (56).

It is natural to anticipate that a key step in the action of polypeptide hormones, their binding to a specific receptor at the surface of their target cell, would be also subject to regulation, especially in the light of the extraordinary biological potency of hormones, active at 10^{-12} - 10^{-9} M, and with circulating levels susceptible to rather large and rapid fluctuations.

Singer and Nicolson, considering the fluid mosaic nature of cell membranes, (54, see also Singer, elsewhere in this volume) and the fact that surface proteins are mobile and can aggregate into "clusters," proposed in 1972 that surface receptors would exhibit cooperative properties. The first evidence that this was indeed the case for polypeptide hormone receptors arose from the non-linearity of binding data at steady-state when plotted according to Scatchard (3, 29, 51). For most peptide hormone receptors, the curve obtained was concave upwards (Figure 1) suggesting that the system disobeyed the law of mass action. The early interpretation of such curves,

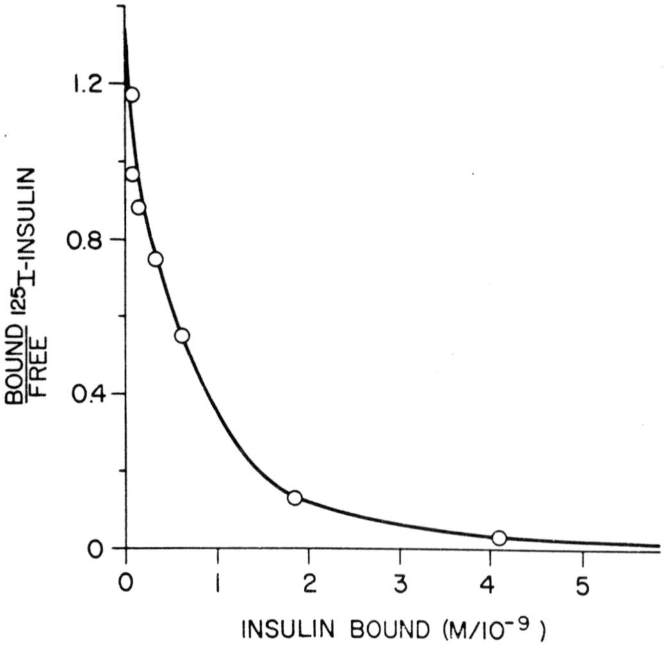

Fig. 1. Curvilinear Scatchard plot for insulin binding to cultured human lymphocytes (IM-9). [^{125}I]-Insulin (7 x 10^{-12} M) was incubated with 10^7 cells at 15° for 90 min. with a range of concentrations of unlabeled insulin. The bound/free of [^{125}I]-insulin is plotted as a function of the insulin bound. Non-specific binding has been subtracted.

by analogy with theories proposed for radioimmunoassay curves (3, 20) and steroid receptors (47), was that the receptor population was inhomogeneous and comprised at least two classes with discrete affinities (30, 34). An alternative explanation would be that the binding of the hormone induces site-site interactions which lower the affinity of the receptor sites, a phenomenon called negative cooperativity (7, 36). Assuming that the fall in affinity would be at least partially due to an increased dissociation rate, we proposed a method based on studying alterations in dissociation kinetics to demonstrate the presence of site-site interactions among insulin receptors (19). We have extensively developed the theoretical basis and experimental methodology elsewhere (10, 13).

In summary, a tracer amount of radioactively labeled hormone is allowed to bind to the receptors and is chosen so that only a

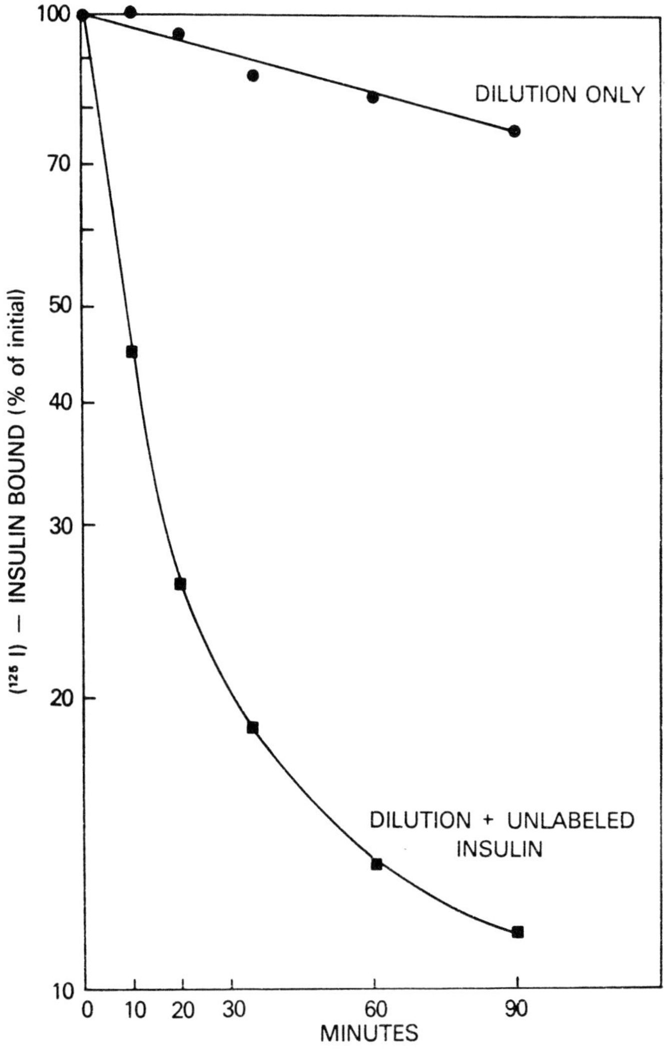

Fig. 2. Dissociation of [^{125}I]-insulin from cultured human lymphocytes in a 100-fold dilution in the absence (dilution only) and in the presence (dilution and unlabeled insulin) of 1.7 x 10^{-7} unlabeled insulin. [^{125}I]-insulin (1.7 x 10^{-11} M) was incubated for 30 min at 15° with 5 x 10^7 cells/ml after which the cells were diluted 100-fold in the absence (dilution only) and in the presence (dilution and unlabeled insulin) of unlabeled insulin.

small minority of receptor sites is occupied. After separating the unbound labeled hormone, the receptors are then diluted to an extent sufficient to prevent rebinding of the dissociating label, This is usually done for insulin receptors on cells or membranes by diluting the system 100-fold. Half of the cells are dissociated in hormone-free medium ("dilution only"), and half in medium containing an excess of unlabeled hormone ("diluted and cold hormone"). clearly, in the absence of hormone-induced site-site interactions, the dissociation rate of the label should be independent of the extent of occupancy of sites by the unlabeled hormone. So, in non-cooperative systems (whose steady-state binding yields linear Scatchard plots), the dissociation rates by "dilution only" and "dilution and cold" should be identical. This was indeed the case for growth hormone receptors on human cultured lymphocytes (7). In contrast, insulin receptors on the same cells showed a dissociation rate which was markedly accelerated in the presence of unlabeled insulin (Figure 2), demonstrating that occupancy of empty sites with unlabeled hormone affects sites occupied with label (site-site interactions).

While this experimental finding cannot be explained by heterogeneous populations of sites alone, it is not totally unequivocal since aside from site-site interactions some other models can yield the same accelerated dissociation (e.g. unstirred layers (43), retention effects (52), or ligand-ligand interactions (9, 12, 13)). We have accumulated much evidence which tends to exclude such models, and make it likely that only site-site interactions explain the modulation of the affinity of insulin receptors (10-14, 18, 19). The best argument for a very specific mechanism is the high degree of structural specificity of the cooperativity. In a study of about 30 insulin structural analogs, we have shown that the negative cooperativity is highly dependent on the integrity of a well defined portion on the surface of the insulin monomer. Both insulin and the receptor contain a "cooperative site" which is distinct from the surface residues responsible for triggering biological effects. The cooperative site appears to be present on insulin as early as in birds (15, 53) and fish (15), and on the receptor in birds (26)[1] and frogs.[2] The characteristics of the negative cooperativity of insulin receptors are summarized in Table I. Data suggesting that this negative cooperativity is not due to ligand-ligand interactions (insulin dimerization) (9, 12, 13) are summarized in Table II.

A phenomenological interpretation of the negative cooperativity observed in insulin receptors postulates that the receptor sites exist as a homogeneous population with high affinity when unoccupied

---

[1] Ginsberg, B., Kahn, C. R., and Roth, J., submitted for publication.

[2] Muggeo, personal communication.

TABLE I

Characteristics of the Negative Co-operativity of Insulin Receptors

Due to an accelerated dissociation rate
Due to site-site interactions
Significant with physiological insulin levels (2 x 10^{-10} - 10^{-9} M)
Maximal at 100-1000 ng/ml insulin (∿ 2 x 10^{-8} - 2 x 10^{-7} M)
Induced upon binding of the insulin monomer
Not induced by insulin dimers
Induced by most insulin analogs in direct proportion to their affinity for the receptor itself, and unrelated to their ability to dimerize
Not induced by desalanine-desasparagine and desoctapeptide (DAA and DOP) insulins
A discrete "cooperative" site, distinct from the bioactive site, is present on insulin and the receptor; this site is buried in the dimer and partially deleted in DAA and DOP

("high affinity," "slow dissociating" conformation), but are switched to a "low affinity," "fast dissociating" conformation upon occupancy with insulin (13). This phenomenon is dependent upon pH, temperature, ionic milieu, and other factors as summarized in Table III.

Negative cooperativity has been demonstrated kinetically not only with insulin receptors in a dozen different tissues, but also with receptors for several other peptide hormones and growth factors (Table IV). In the case of nerve growth factor, Pulliam and coworkers have excluded ligand dimerization as the cause of negative cooperativity by showing that the dissociation of a cross-linked dimeric labeled NGF is accelerated by unlabeled ligand exactly like the monomer (46, see also Bradshaw et al., elsewhere this volume). Majerus et al. have recently extended these methods and concepts to non-hormonal systems by demonstrating that thrombin acts on platelets by a hormone-like interaction with a surface receptor, which exhibits a negative cooperativity strikingly similar to that of insulin receptors (39).

It would be premature at this point to postulate any detailed microscopic mechanisms for the negative cooperativity; these could

TABLE II

Evidence that the Negative Cooperativity of Insulin Receptors is Unrelated to Insulin Dimerization

Insulin Derivative	Ability to Dimerize	Ability to Induce Negative Cooperativity
Pork Insulin up to 10^{-7} M (Mostly monomeric)	Normal ($K_a = 2.22 \times 10^5\ M^{-1}$ at pH 8)	Normal
Pork Insulin > 10^{-7} M Mostly dimeric	Dimer	Decreased in proportion of % dimers
Guinea-Pig Insulin	Abolished	Normal
Tetranitro-Insulin	Abolished	Normal
DAA-insulin	Decreased but present ($K_a = 1.74 \times 10^3\ M^{-1}$ at pH 8)	Abolished

TABLE III

"Two-State" Model for Insulin Receptors

"Slow-dissociating state" favored by	"Fast-dissociating state" favored by
Low fractional saturation with insulin	Increased fractional saturation with insulin
Alkaline pH (8-9)	Acid pH (5-6)
Low temperature (4°)	High temperature (37°)
High concentration of Ca^{++} - Mg^{++}	Low concentration of Ca^{++} - Mg^{++}
Concanavalin A (20 μg/ml)	------------
----------	Urea

TABLE IV
Kinetic Demonstration of Negative Cooperativity in Cell Surface Receptors for Hormones, Neurotransmitters, Growth Factors

Receptor for	Tissue Preparation	Reference
Insulin	Human cultured lymphocytes	De Meyts, Roth, Neville, Gavin & Lesniak, (19)
	Human blood monocytes	Bianco, Schwartz & Handwerger, (4)
	Human blood granulocytes	Fussgänger, Kahn, Roth & De Meyts, (24)
	Turkey erythrocytes	Ginsberg, Kahn & Roth, (26)
	Turkey erythrocyte membranes	Ginsberg, Kahn & Roth,[1]
	Rat liver membranes	De Meyts, Bianco & Roth, (13)
	Mouse liver membranes	Soll, Kahn & Neville, (55)
	Chicken liver membranes	De Meyts, Kahn, Ginsberg & Roth, (15)
	Guinea-pig liver membranes	De Meyts, Kahn, Ginsberg & Roth, (15)
	Human cultured placental cells	Podskalny, Chou & Rechler, (45)
	Human placental membrane	Cuatrecasas & Hollenberg, (9)
	Cultured human fibroblasts	Rechler & Podskalny (submitted for publication)
	Mouse 3T3 fibroblasts	Thomopoulos, (personal communication)
	Rat mammary adenocarcinoma R 3230 AC	Harman & Hilf, (personal communication)
	Mouse macrophage tumor line P38801	Bar & Korne, (personal communication)
Nerve Growth Factor	Sympathetic and dorsal root ganglia	Frazier, Boyd & Bradshaw, (22)
	Embryonic heart & brain	Frazier, Boyd, Pulliam, Szutowicz & Bradshaw (23)
Epidermal Growth Factor	Human placental membrane	Cuatrecasas & Hollenberg, (9)
Thyrotropin (TSH)	Thyroid membranes	Lee, Winand & Kohn, (submitted for publication)
	Retro-orbital membranes	Lee, Winand & Kohn, (submitted for publication)
Thyroliberin (TRF)	Anterior pituitary membranes	Boss, Vale & Grant, (6)
β-Adrenergic	Frog erythrocyte membranes	Limbird, De Meyts & Lefkowitz, (37)

include a major membrane reorganization, receptors clustering (29, 35) or only minute changes in the conformation of individual subunit structures. Some findings however tend to restrict the possible role of the membrane:

1) drugs which modify the microtubules and microfilaments do not affect the cooperative properties of insulin receptors (57).

2) detergent-solubilized insulin receptors still exhibit negative cooperativity with a dependency on insulin concentration superimposable to the one found with intact cells.[3]

Hormone receptors are not yet available in highly purified form accessible to chemical analysis. Understanding the mechanism underlying the cooperativity is important since it would give us some clues to the structure of receptors. However, irrespective of the molecular nature of this event, its biological relevance lies in its regulatory nature, i.e. in the fact that the affinity of the receptor is inversely related to ambient hormone concentration.

In the presence of cooperativity, quantitative analysis of the binding data is complex (8, 21, 28, 31, 32). For example, the slopes of the Scatchard plot do not represent any more the intrinsic affinity constants of the sites, but a complex function of these. If we had some data on the stoichiometry of the hormone-receptor interaction (e.g., number of subunits/receptor) we could analyze the Scatchard slopes in terms of "stepwise equilibrium constants" with some chance of obtaining physically meaningful parameters (8, 21, 28, 31, 32 38).[3] We have recently proposed an alternative method of graphical analysis (16, 17).

To overcome the problems caused by the unknown stoichiometry, we consider in analyzing the data not the "stepwise" equilibrium constants but an "average affinity: ($\bar{K}$) related to occupancy of the sites, and calculated as $(B/F)/(R_o - B)$ where B = concentration of bound hormone, F = concentration of unbound hormone and R_o = total concentration of binding sites (from the abcissa intercept of the Scatchard plot).[3] The limit of K for $B \Rightarrow 0$ is $\bar{K}_e$, the affinity of "empty sites," and for $B \Rightarrow R_o$ is $\bar{K}_f$, the affinity of "filled sites."

$\bar{K}$ can be directly plotted as a function of log $\bar{Y}$, the fractional occupancy of the sites of B/R_o. The negative cooperativity is immediately apparent (Figure 3). $\bar{K}$ starts falling when as little as 1-5% of the sites are occupied, and $\bar{K}_f$ is reached usually when only 20-50% of the sites are occupied.

---

[3] Ginsberg, B., Kohn, R., Kahn, C. R., and Roth, J., submitted for publication.

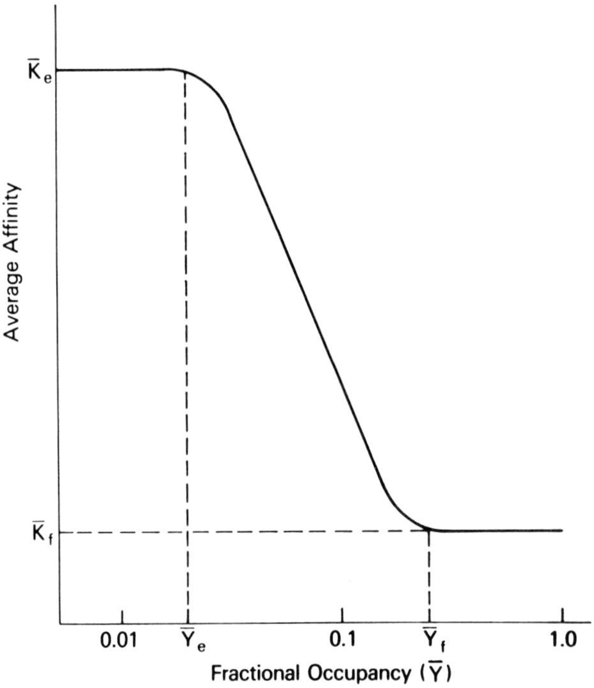

Fig. 3. The "average affinity" profile. $\bar{K}$, the average affinity, calculated as $(B/F)/(R_o - B)$, is plotted as a function of log $\bar{Y}$ = log (B/R_o). $\bar{Y} = 0$ when all sites are empty, and = 1 when all sites are occupied. $\bar{K}_e$ = affinity of the "empty sites" conformation, $\bar{Y}_e$ = threshold for measurable site-site interactions and $\bar{Y}_f$ = fractional occupancy at which site-site interactions are maximal (all sites in $\bar{K}_f$), $\bar{K}_f$ = affinity of the "filled sites" concentration.

The biological relevance and the regulatory nature of negative cooperativity is also strikingly displayed by the plot in Figure 3. The system is highly sensitive at low hormone concentration, but the affinity starts falling when a minimal proportion of the sites is occupied, thus buffering the system against sudden variations of hormone levels. This is an acute type of regulatory mechanism. In situations where insulin levels are chronically altered, a second type of mechanism comes into play: a "down regulation" of receptor concentration on cells directly related to ambient hormone levels (2, 25, 48, 49, 55).[4]

---

[4] Bar, R. S., Gorden, P., Roth, J., Kahn, C. R. and De Meyts, P. Submitted for publication.

ACKNOWLEDGEMENTS

I am extremely grateful to Dr. J. Roth for his advice and support, and to Dr. C. R. Kahn for helpful discussions and careful review of the manuscript. This work was done under a PHS International Postdoctoral Fellowship F05-TW1918 and the 1975 Solomon A. Berson Research and Development Award of the American Diabetes Association.

REFERENCES

1. ADAIR, G. S., J. Biol. Chem. 63 (1925) 529.
2. ARCHER, J. A., GORDEN, P. and ROTH, J., J. Clin. Invest. 55 (1975) 166.
3. BERSON, S. A. and YALOW, R. S., J. Clin. Invest. 38 (1959) 1996.
4. BIANCO, A. R., SCHWARTZ, R. H. and HANDWERGER, B. S., Diabetologia 10 (1974) 359.
5. BOHR, C. Zentralb für Physiol. 17 (1903) 682.
6. BOOS, B., VALE, W. and GRANT, G., in Biochemical Actions of Hormones (G. Litwak ed.) (1975) 87.
7. CONWAY, A. and KOSHLAND, D. E., JR., Biochemistry 7 (1968) 4011.
8. CORNISH-BOWDEN, A. and KOSHLAND, D. E., JR., Biochemistry 9 (1970) 3325.
9. CUATRECASAS, P. and HOLLENBERG, M., Biochem. Biophys. Res. Commun. 62 (1975) 31).
10. DE MEYTS, P., in Methods in Receptor Research (M. Blecher ed.) M. Dekker, N. Y., (1976) in press.
11. DE MEYTS, P., Proceedings of the Ninth Miles Symposium, (1976) in press.
12. DE MEYTS, P., J. Supramolecular Struct. (1975) in press.
13. DE MEYTS, P., BIANCO, A. R. and ROTH, J., J. Biol. Chem. (1976) in press.
14. DE MEYTS, P., GAVIN, J. R., III, ROTH, J. and NEVILLE, D. M., JR., Diabetes 23: suppl. 1 (1974) 355.
15. DE MEYTS, P., KAHN, C. R., GINSBERG, B. and ROTH, J., Diabetes 24: suppl. 2 (1975) 393.
16. DE MEYTS, P. and ROTH, J., The Endocrine Society, 57th Annual Meeting, Abstracts (1975) 51.
17. DE MEYTS, P. and ROTH, J., Biochem. Biophys. Res. Commun. 66 (1975) 1118.
18. DE MEYTS, P., ROTH, J., NEVILLE, D. M., JR. and FREYCHET, P., The Endocrine Society, 56th Annual Meeting, Abstracts (1974) 189.
19. DE MEYTS, P., ROTH, J., NEVILLE, D. M., JR., GAVIN, J. R., III and LESNIAK, M. A., Biochem. Biophys. Res. Commun. 54 (1973) 154.
20. FELDMAN, H. A., Anal. Biochem. 48 (1972) 317.

21. FLETCHER, J. E., SPECTOR, A. A. and ASHBROOK, J. D., Biochemistry 9 (1970) 2580.
22. FRAZIER, W. A., BOYD, L. F. and BRADSHAW, R. A., J. Biol. Chem. 249 (1974) 5513.
23. FRAZIER, W. A., BOYD, L. F., PULLIAM, M. W., SZUTOWICZ, A. and BRADSHAW, R. A., J. Biol. Chem. 249 (1974) 5918.
24. FUSSGÄNGER, R. D., KAHN, C. R., ROTH, J. and DE MEYTS, P., Diabetes 24: suppl. 2 (1975) 393.
25. GAVIN, J. R., III, ROTH, J., NEVILLE, D. M., JR., DE MEYTS, P. and BUELL, D. N., Proc. Nat. Acad. Sci. USA 71 (1974) 84.
26. GINSBERG, B., KAHN, C. R. and ROTH, J., Endocrinology (1976) in press.
27. HILL, A. V., J. Physiol. 40 (1910) IV.
28. HUNSTON, D. L., Anal. Biochem. 63 (1975) 99.
29. KAHN, C. R. in Methods in Membrane Biology, Vol. 3. (E. D. Korn ed.) Plenum Publishing Co., New York (1975) 81.
30. KAHN, C. R., FREYCHET, P., ROTH, J. and NEVILLE, D. M., JR., J. Biol. Chem. 249 (1974) 2249.
31. KLOTZ, I. M., Accounts Chem. Res. 7 (1974) 162.
32. KLOTZ, I. M. and HUNSTON, D. L., J. Biol. Chem. 250 (1975) 3001.
33. KOSHLAND, D. E., JR., NEMETHY, G. and FILMER, D., Biochemistry 5 (1966) 365.
34. LEFKOWITZ, R., ROTH, J. and PASTAN, I., Ann. N. Y. Acad. Sci. 185 (1971) 195.
35. LEVITZKI, A., J. Theor. Biol. 44 (1974) 307.
36. LEVITZKI, A. and KOSHLAND, D. E., JR., Proc. Nat. Acad. Sci. USA 62 (1969) 1121.
37. LIMBIRD, L., DE MEYTS, P. and LEFKOWITZ, R. J., Biochem. Biophys. Res. Comm. 64 (1975) 1160.
38. MAGAR, M. E., in Data Analysis in Biochemistry and Biophysics, Academic Press, (1972) 401.
39. MAJERUS, P. W., TOLLEFSEN, D. M. and SHUMAN, M. A., in Cellular Reactions of Blood Platelets, (J. L. Gordon ed.), ASP Biological Press, B. V., (1976) in press.
40. METZGER, H., Adv. in Immunology 18 (1974) 169.
41. MONOD, J., WYMAN, J. and CHANGEUX, J. P., J. Mol. Biol. 12. (1965) 88.
42. NEWSHOLME, E. A. and START, C., Regulation in Metabolism, J. Wiley & Sons, N. Y. (1973).
43. NOYES, A. A. and WHITNEY, W. R., Z. Physik. Chem. 23 (1897) 689.
44. PERUTZ, M. F., Nature (London) 228 (1970) 726.
45. PODSKALNY, J., CHOU, J. and RECHLER, M. M., Arch. Biochem. Biophys. 170 (1975) 504.
46. PULLIAM, M. W., BOYD, L. F., BAGLAN, N. C. and BRADSHAW, R. A., Biochem. Biophys. Res. Commun. 67 (1975) 1281.
47. ROSENTHAL, H. E., Anal. Biochem. 20 (1967) 525.
48. ROTH, J., KAHN, C. R., DE MEYTS, P., GORDEN, P., and NEVILLE, D. M., JR., in Insulin and Metabolism (J. S. Bajaj ed.), Elsevier Excerpta Medica, North Holland Publishers (1976) in press.

49. ROTH, J., KAHN, C. R., LESNIAK, M. A., GORDEN, P., DE MEYTS, P., MEGYESI, K., NEVILLE, D. M., JR., GAVIN, J. R., III, SOLL, A. H., FREYCHET, P., GOLDFINE, I. D., BAR, R. S. and ARCHER, J. A., in Recent Progress in Hormone Research 31 (1975) 95.
50. SAROFF, H. A., Biopolymers 12 (1973) 599.
51. SCATCHARD, G., Ann. N. Y. Acad. Sci. 51 (1949) 660.
52. SILHAVY, T. J., SZMELCMAN, S., BOOS, W. and SCHWARTZ, M., Proc. Nat. Acad. Sci. USA 72 (1975) 2120.
53. SIMON, J., FREYCHET, P., ROSSELIN, G. and DE MEYTS, P., Endocrinology (1976) in press.
54. SINGER, S. J. and NICOLSON, G. L., Science 175 (1972) 720.
55. SOLL, A. H., KAHN, C. R. and NEVILLE, D. M., JR., J. Biol. Chem. 250 (1975) 4702.
56. THOMPSON, J. J. and HOFFMAN, L. G., Immunochemistry 11 (1974) 431.
57. VAN OBBERGHEN, E., ROTH, J. and DE MEYTS, P., Diabetes (1976) in press.
58. WEINTRAUB, B. D., ROSEN, S. W., MC CAMMON, J. A. and PERLMAN, R. L., Endocrinology 92 (1973) 1250.
59. WHITEHEAD, E., Progr. Biophys., Mol. Biol. 21 (1970) 323.
60. WYMAN, J., Adv. Prot. Chem. 4 (1948) 407.

SPECIFIC INTERACTION OF NERVE GROWTH FACTOR WITH RECEPTORS IN THE CENTRAL AND PERIPHERAL NERVOUS SYSTEMS

Ralph A. BRADSHAW, Morris W. PULLIAM, Ing Ming JENG, Roger Y. ANDRES, Andrjez SZUTOWICZ†, William A. FRAZIER, Ruth A. HOGUE-ANGELETTI* and Robert E. SILVERMAN

Department of Biological Chemistry, Division of Biology and Biomedical Sciences, Washington University
St. Louis, MO 63110 (USA)

Nerve growth factor (NGF), a protein originally discovered as a diffusable substance produced by two mouse sarcomas (180 and 37) (28), stimulates marked neurite outgrowth by embryonic sympathetic and sensory ganglia both *in vivo* and *in vitro*. The latter phenomenon provided the basis for the bioassay (27). Antiserum to NGF blocks this effect, and if given to neonatal animals, produces a profound immunosympathectomy (26). Subsequent workers have noted wide-spread effects of NGF and its antisera on degeneration and regeneration of adrenergic fibers in both the central and peripheral nervous systems in embryonic and adult animals (3, 4). Its biological effects in the sensory nervous system, however, have been demonstrated only during a brief time span in chick embryonic development (27). To date no biological effects on the motor nervous system have been demonstrated. However, effects of the hormone on the development of neuroblastoma and pheochromocytoma cells in culture have been reported (22, 24, 31, 32, 39). In addition to the pronounced morphological changes, i.e. neurite production, several anabolic effects have been noted (7, 25, 27, 34). The sum of these findings suggests a hormone-like mode of action, pleiotypic in in character, that is important in regulating the development of peripheral neurons during embryonic development, and in some cases, in the mature animal.

†Present Address: Dept. of Biochemistry, Institute of Clinical Pathology, Gdansk, Poland

*Present Address: Division of Neuropathology, University of Pennsylvania School of Medicine, Philadelphia, PA 19185 (USA)

Although initial attemps at isolation and purification were performed with sarcomas (10), the molecular properties of NGF were determined for the mouse submaxillary protein. The reason for the high concentrations of NGF in adult male submaxillary glands and snake venoms is still obscure, despite the fact that these are by far the richest sources of the protein. Mouse NGF is isolated as a large polymeric species, designated 7S, consisting of 3 types of polypeptide chains, or as a smaller form representing only one of the 3 subunits (5, 41, 42). This latter form, depending on the method of isolation, has been called β or 2.5S NGF (the latter designation referring to its sedimentation coefficient). The 7S form probably represents the molecule as it is stored in the submaxillary gland, whereas the smaller form, which represents the only active subunit of the complex (40), results from pH induced dissociation of the complex during isolation.

The 2.5S molecule is a dimer of two subunits associated by non-covalent forces (2). Although in the gland the two polypeptides are identical, due to proteolysis during isolation, one subunit contains 118 amino acids (A chain) and the other, 110 amino acids (B chain) (1). They differ only in the additional 8 amino acids at the amino terminus of the A chain. For this reason the preferred notation of this species is NGF(AB) to distinguish it from β NGF (the other commonly used form of NGF) in which both chains are 118 amino acids in length and can be denoted NGF(A_2). In some 5 to 10% of the chains, the carboxyl terminal arginine residue is also missing. As with the amino terminal octapeptide, this modification has no effect on biologic activity (34).

The amino acid sequence of NGF(AB) was noted to bear statistically significant similarities to the structure of insulin and proinsulin, suggesting an evolutionary relationship arising from a common ancestral gene (16, 17). This observation raised the possibility that both substances induced their cellular effects in similar ways. Indeed, it has been shown that insulin covalently bound to Sepharose beads could still elicit biological responses, indicating a cell surface effect (11). Considerable additional evidence has now accumulated to support the view that insulin does indeed produce its biological effects by interaction with a membrane bound cell surface receptor. When similar experiments were carried out with NGF covalently bound to Sepharose, the typical biological response, as manifested by neurite production in chick embryonic dorsal root and sympathetic ganglia, was obtained (18). Thus, at least a portion of the NGF response as well appears to be elicited through cell-surface receptors.

BINDING OF [^{125}I]-NGF TO RESPONSIVE NEURONS

The properties of the interaction of NGF and its plasma membrane receptors of responsive neurons was studied using [^{125}I]-NGF. As shown in Figure 1, the binding is non-saturable up to a concentration

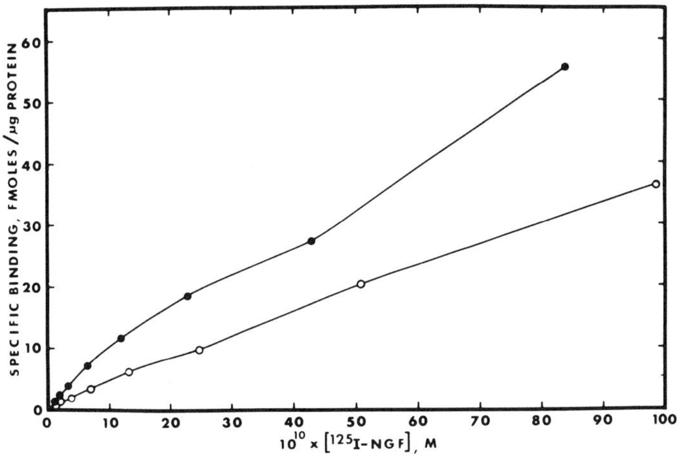

Fig. 1. The specific binding of ^{125}I-NGF to sympathetic (O) and dorsal root ganglia (●) tissue suspensions (2 µg of tissue protein per 300 µl of assay volume) as a function of the concentration of [^{125}I]-NGF. Non-saturating binding curves were obtained in a total of 20 experiments with ganglia tissue suspensions, all with different preparations of [^{125}I]-NGF. Nonspecific binding was determined at each point and subtracted from the data shown. (19).

of 10^{-8} M [^{125}I]-NGF (19). Expression of the data in the form of a Scatchard plot, as seen in Fig. 2, indicates at least two apparent classes of sites of quite different affinity and number. The affinity of the sites binding the hormone at low concentrations (early portion of plot) is 9.0 x 10^9 liters per mole (l/mol) for both tissues. The lower affinity binding seen at higher concentrations of [^{125}I]-NGF is 7.8 x 10^6 l/mol for sympathetic and 1.4 x 10^7 l/mol for dorsal root ganglia. The apparent number of high affinity sites is 6 x 10^8 receptors per µg of tissue protein with the number of low affinity sites being some 200 times greater.

While such results are consistent with the presence of 2 or more distinct classes of binding sites, other explanations are possible. One such explanation, that has been proposed by De Meyts et al. (15) to explain similar results with insulin binding (see elsewhere in this volume), is that of negative cooperativity. With this interpretation, the curvilinear Scatchard plots of the present data for NGF represent a range of association constants from 10^6 to 10^9 liters per mole for a single receptor class. This concept as applied to the hormone-receptor interaction means that the binding of a molecule of hormone (ligand) with a receptor molecule induces an interaction with other receptor molecules resulting in a decrease in the affinity of the receptor molecules for the hormone. This phenomenon has been described in other hormone-receptor systems (13, 14, 29) (see also Limbird and Lefkowitz, elsewhere this volume).

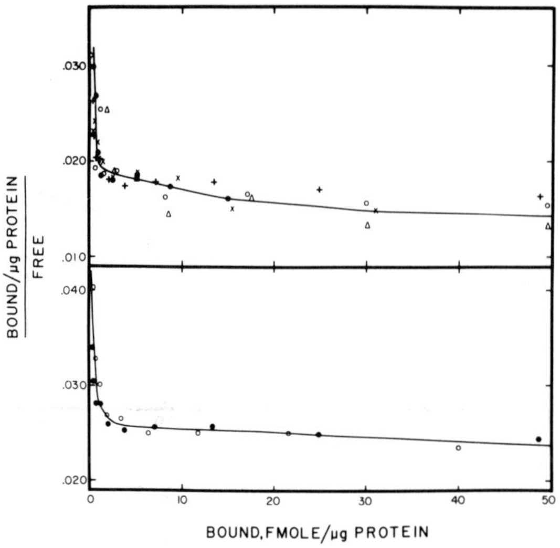

Fig. 2. Scatchard plots of specific [^{125}I]-NGF binding data for sympathetic (upper plot, five experiments) and dorsal root ganglia (lower plot, two experiments). The concentration of tissue protein in all experiments was from 1.9 to 2.2 μg/300 μl of assay volume. Nonspecific binding was determined at each point and subtracted from the data (19).

The utility of such a phenomenon is that it permits sufficient hormonal effect through hormone receptor interaction at relatively low hormone concentrations while serving as a protective mechanism at high hormone concentrations. While a non-linear Scatchard plot, as above, is consistent with negative cooperativity, the most meaningful experiment to establish its presence in a given system is a measurement of the kinetics of dissociation. If binding of additional hormone results in progressively decreasing affinity of the receptor for the hormone, then the dissociation of bound labeled hormone should be accelerated by dilution with buffer containing native hormone over that produced by dilution with buffer alone. Such results had already been reported with the binding of insulin to lymphocytes (15).

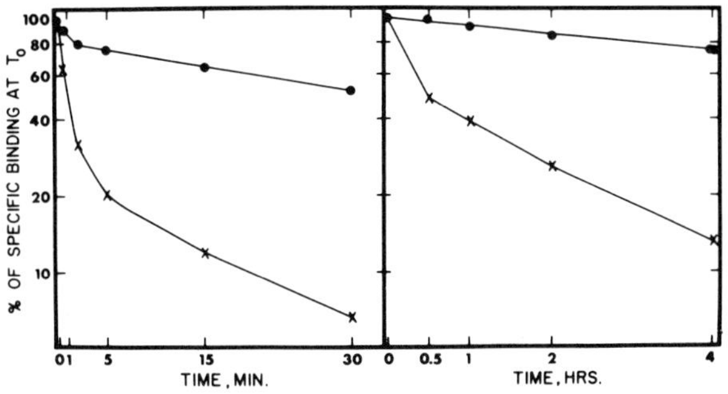

Fig. 3. (Left) Time course of dissociation of specifically bound [^{125}I]-NGF (2×10^{-9} M, equilibrated at 24° for 2 hours) from sympathetic tissue suspensions (2.4 μg/300 μl) at 24°. One set of samples (4 tubes) was diluted with 33 volumes of Hank's salt solution with 1 mg of albumin per ml at 24° (●) and another set was diluted in the same way except that the diluent contained 2×10^{-7} M native NGF (X). At the times indicated the samples were transferred to 0.8 μm Nuclepore filters under reduced pressure and washed twice with 5 ml of ice cold Hank's salt solution with 1 mg of albumin per ml. The process required 12 s. Nonspecific binding was determined under the conditions of the experiment in quadruplicate and subtracted from each data point. (Right) Time course of dissociation of specifically bound ^{125}I-NGF (2×10^{-9} M) from sympathetic tissue suspensions at 0°. The experimental design was identical to that of the experiment shown on left except that the diluent was at 0° and the tubes were maintained in an ice bath for the times indicated (19).

As shown in Fig. 3, at both 24°(left) and 0° (right), the rate of dissociation of bound [^{125}I]-NGF produced by 1:34 dilution with buffer containing only bovine serum albumin (BSA) (upper lines) is significantly accelerated by the same dilution containing BSA plus 2×10^{-7} M native NGF. At 24° the half-time of dissociation produced by dilution alone is 30 min while being about 1 min when induced by dilution plus native hormone, or a 30-fold acceleration. These values correspond to first order rate constants at 24° of 3.8×10^{-4} s^{-1} and 1.2×10^{-2} s^{-1}, respectively. Thus, there is a clear acceleration of dissociation produced by native hormone, consistent with the concept of negative cooperativity. However, other more complex interpretations of the data, such as dimerization of [^{125}I]-NGF at the receptor site are possible (12). If this dimer has a lower affinity for the receptor than the monomer, the data observed could result.

To investigate this possibility, NGF(AB) was covalently cross-linked with dimethylsuberimidate according to the method of Stach and Shooter (36). Gel electrophoresis in sodium dodecyl sulfate

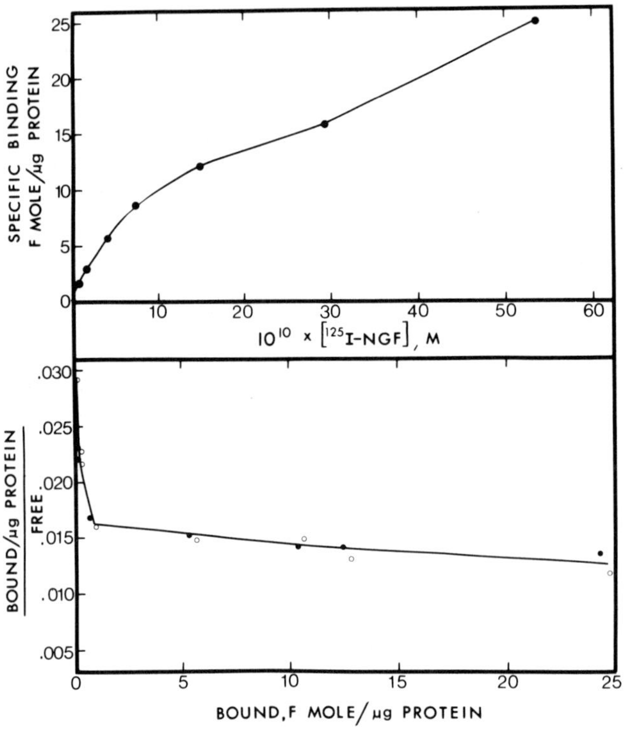

Fig. 4. Upper plot: Specific binding of [^{125}I]-NGF$(AB)_{XL}$ to suspensions of 8 day chick embryo dorsal root ganglia (2 μg tissue protein/ 300 μl of assay volume) as a function of increasing concentrations of [^{125}I]-NGF$(AB)_{XL}$. Nonspecific binding was determined at each point using a 500-fold excess of native NGF(AB) and subtracted to obtain the data shown. Lower plot: Data from A expressed in the form of a Scatchard plot. Specific activity of the [^{125}I]-NGF$(AB)_{XL}$ was 3300 cpm/fmole of dimer (65% counting efficiency). Assays were performed in triplicate. The data shown represent two separate experiments (35).

confirmed that the cross-linked dimer (NGF$(AB)_{XL}$) had a molecular weight of 26,000 (twice the size of the monomer). As with the NGF$(A_2)_{XL}$, prepared by Stach and Shooter (35), NGF$(AB)_{XL}$ was biologically active in the dorsal root ganglia assay at an optimum concentration of 2 ng/ml (the same molar concentration as that of monomer giving maximal biological response).

The specific binding of [^{125}]-NGF$(AB)_{XL}$ to a suspension of 8 day dorsal root ganglia is shown in Fig. 4 (top) (35). When graphed in the Scatchard form (Fig. 4, bottom), a biphasic curve was also obtained. The limiting slope of the data at low concen-

TABLE I

Comparison of Constants for Specific Binding of ^{125}I-NGF(AB) and ^{125}I-NGF(AB)$_{XL}$ to Chick Embryo Sympathetic and Dorsal Root Ganglia

	Binding Constant (K_A)			Dissociation (24°)	
	High Affinity	Low Affinity	Association (K_1) (24°)	Dilution (k_{-1_I})	Dilution + Native ($k_{-1_{II}}$)
	L^1 M^{-1}		M^{-1} s^{-1}	s^{-1}	
NGF(AB)	9.0×10^9	1.4×10^7	7.5×10^6	3.8×10^{-4}	1.2×10^{-2}
NGF(AB)$_{XL}$	8.4×10^9	4.7×10^7	5.1×10^6	3.6×10^{-4}	1.5×10^{-2}

tration of dimer yields a class of apparent high affinity sites with an association constant (K_A) of 8.4 x 10^9 l/mol and a class of low affinity sites with a K_A of 4.7 x 10^7 l/mol. As can be seen in Table I, this compares closely with the K_A seen in dorsal root ganglia with NGF (AB) that has not been cross-linked.

The kinetic analyses of the association and dissociation of [^{125}I]-NGF(AB)$_{XL}$ were indistinguishable from native [^{125}I]-NGF(AB) including the enhancement of the dissociation by the unlabeled hormone. Thus, it is apparent that the binding parameters of NGF(AB) are not appreciably different from those of NGF(AB)$_{XL}$, disproving dimerization of the hormone as the reason for the apparent negative cooperativity observed.

BINDING OF NGF TO END ORGANS AND BRAIN

Although the presence of specific binding of NGF to responsive neurons was not unexpected, the observation of such binding in homogenates of peripheral tissues and brain of 13 day embryonic chick and neonatal rats was (21). As shown in Table II, that while sympatheic ganglia bind the most NGF(AB) per μg protein, there is clearly a significant amount of specific binding in all tissues tested with the exception of red blood cells. In addition, animals that had been rendered sympathectomized by administration of guanethidine at the neonatal stage showed essentially the same pattern of binding in their tissues as untreated animals. These results seem to indicate that, in all tissues, a significant portion of the total specific binding is to structures other than sympathetic nerve terminals.

The properties of the binding to embryonic heart (as a representative peripheral tissue) and brain were found to be very similar to those observed for dorsal root and sympathetic ganglia (20). In Fig. 5, it can be seen that the specific binding of [^{125}I]-NGF(AB) to suspensions of 13 day embryonic chick heart and brain, up to a concentration of 10^{-8} M, is a non-saturable process. If these data are represented in the form of a Scatchard plot, as seen in Figs. 6 and 7, it is clear that the lines are not linear and are consistent with a minimum of two apparent classes of binding sites, or, as noted above, with a negatively cooperative interaction between the sites. Examination of the initial portion of the curves corresponding to the high affinity binding yields an association constant of 2.9 x 10^9 l/mol for heart and brain, respectively. The low affinity binding sites in both organs have apparent association constants of about 10^7 l/mol.

The association and dissociation kinetics of labeled NGF were identical to those found for the responsive neural tissue. Both tissues showed enhancement of dissociation by unlabeled hormone,

TABLE II

Binding of ^{125}I-NGF to Tissues of Newborn Rat and 13 Day Chick Embryo[a]

Tissue	Specific Binding (cpm/μg protein) Chick[b]	Rat[c]
Sympathetic ganglia	2,050	1,790
Heart	687	215
Brain	442	93
Liver	309	45
Adrenal[d]	n.d.	399
Red blood cells[d]	n.d.	2

[a]Specific activity of this preparation was 1,374 cpm/fmole.

[b]9.8×10^{-10} M ^{125}I-NGF.

[c]3.9×10^{-10} M ^{125}I-NGF.

[d]n.d., not determined.

consistent with a negatively cooperative interaction between the receptor molecules on the cell surface. In Table III, results are compared to those previously reported for sympathetic ganglia. Clearly, these similarities in binding characteristics support the hypothesis that the receptor molecule is quite similar, if not the same, in all the tissues examined.

Although no immediate explanations for these results in terms of physiological function are possible for heart (and other end organs), a somewhat clearer picture of the role of NGF in brain emerges from the subcellular fractionation of the receptors of this tissue (37, 38). Using a procedure modelled after that described by

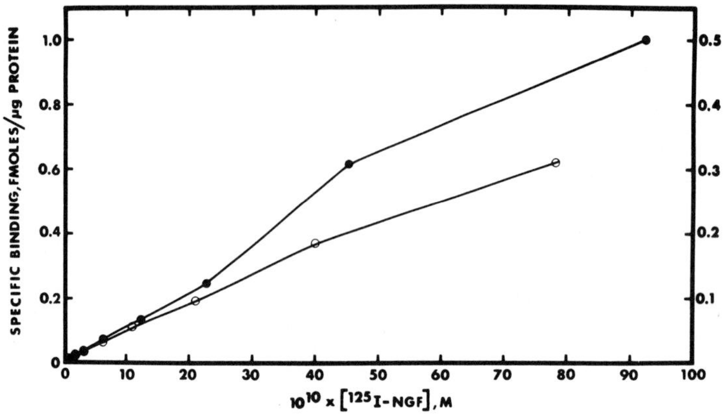

Fig. 5. The specific binding of ^{125}I-NGF to 13 day chick embryo heart (●) and brain (O) tissue suspensions (30 to 50 μg of tissue protein per 300 μl of assay volume) as a function of the concentration of ^{125}I-NGF. Nonsaturable binding curves were obtained in a total of 15 experiments with heart and brain with different preparations of ^{125}I-NGF. Nonspecific binding was determined at each point and subtracted to produce the data points shown (20).

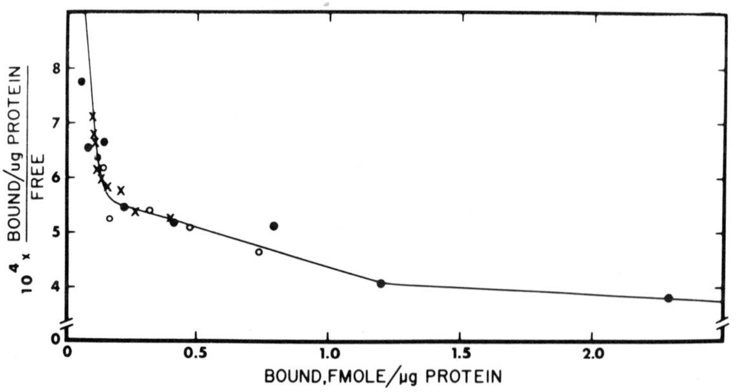

Fig. 6. Scatchard plot of specific ^{125}I-NGF binding data for heart. The concentration of tissue protein in all experiments was from 30 to 50 μg/300 μl of assay volume. Nonspecific binding was determined at each point and subtracted from the data (20).

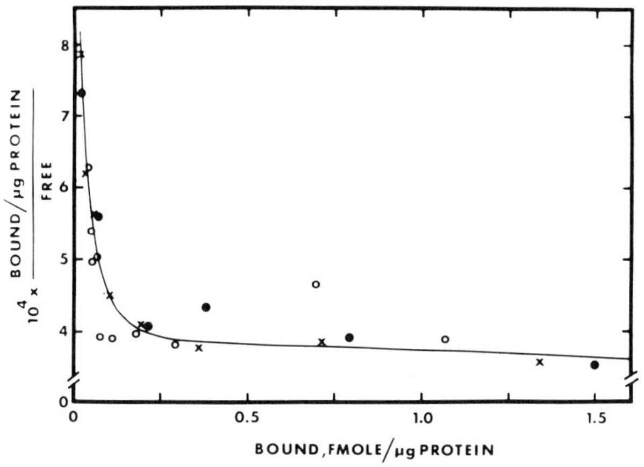

Fig. 7. Scatchard plot of specific ^{125}I-NGF binding data from brain. Details as in Fig. 6 (20).

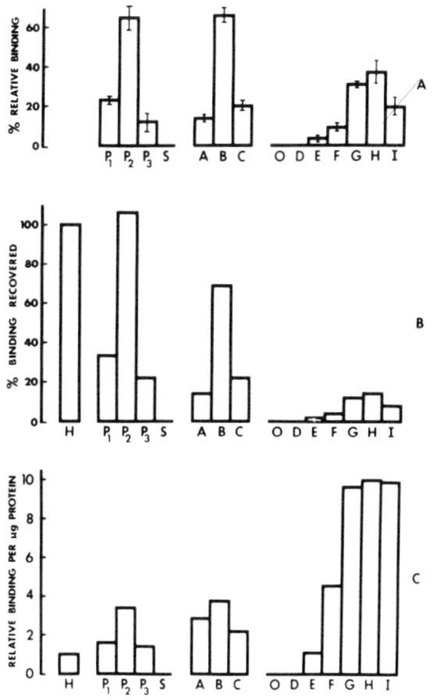

Fig. 8. Graphic representation of the levels of specific binding of ^{125}I-nerve growth factor in subcellular fractions of 13 day chick embryo brain. (A) Binding data presented as the sum of each set of fractions (P_1 through S, A through C and D through I) equal to 100%. Error bars indicate the standard error of the mean. (B) Binding activity of each fraction presented as the actual percentage of the binding activity of the homogenate (H) recovered in each fraction. (C) Binding data presented as ratio of recovered activity to that found in the homogenate (relative specific binding per μg protein (37)).

TABLE III

Comparison of Equilibrium and Kinetic Constants for Specific Binding of ^{125}I-Nerve Growth Factor to Chick Embryo Sympathetic Ganglia, Heart, and Brain at 24°

Constant	Sympathetic Ganglia	Heart	Brain
Kinetic constants			
k_1 association (M^{-1} s^{-1})	7.5×10^6	7.0×10^6	7.0×10^6
k_{-1_I} dissociation (s^{-1})	3.8×10^{-4}	7.0×10^{-4}	6.6×10^{-4}
$k_{-1_{II}}$ (+ Native NGF) (s^{-1})	1.2×10^{-2}	4.4×10^{-3}	4.2×10^{-3}
Association constants (liters/mole)			
Limit K_A			
High affinity	9.0×10^9	2.9×10^9	2.5×10^9
Low affinity	7.6×10^6	5.7×10^6	8.2×10^6

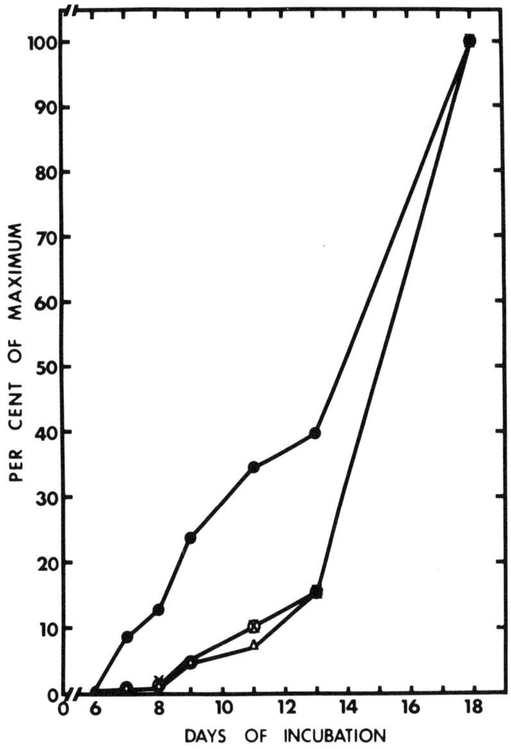

Fig. 9. Detailed developmental time course of the appearance of specific [^{125}I]-NGF binding (●) in the washed particulate fraction of whole chick embryo brain and specific binding of [^{125}I]-α-bungarotoxin (X), choline acetyltransferase activity (O) and dopamine-β-hydroxylase activity (Δ) in whole brain homogenate. Data are presented as activity per one brain compared to the value of 18 days (set at 100%) (38).

Whittaker (43) for adult brain, 13 day chick embryonic brain was separated into various enriched subcellular fractions and examined for specific [^{125}I]-NGF binding. As shown in Fig. 8, the majority of this binding was found to reside in the P_2 (crude synaptosomal) fraction. Further purification of P_2 on a discontinuous sucrose density gradient concentrated binding in the "purified" synaptosomal fraction, B. When these synaptosomes, which result from the shearing off and resealing of synaptic terminals, are lysed by hypoosmotic shock and further separated on a more complex sucrose density (37), the NGF-binding is found in the denser fractions, G, H and I. The recovery of binding following this procedure (as shown in panel B of Fig. 8) is less than 50%. However a ten-fold purification of

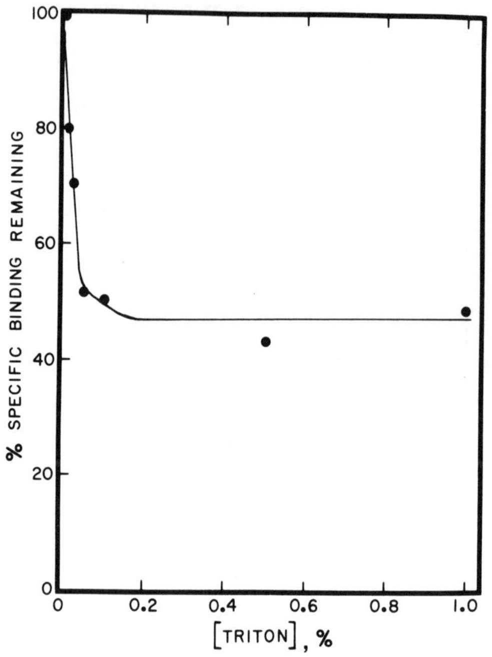

Fig. 10. The effect of triton X-100 on the specific binding of [^{125}I]-NGF to dorsal root ganglia. Dorsal root ganglia (2/400 μl incubation solution) were incubated with various concentrations of triton X-100 for 1 hr. The particulate fractions were removed and washed twice with ice cold Hank's salt solution containing 1 mg/ml bovine serum albumin. The binding activity remaining was measured as described previously (19).

specific binding over that found in the homogenate is achieved (Panel C, Fig. 8). The validity of the assignment of subcellular entities and organelles was checked with a number of appropriate macromolecular and enzymatic markers (37). In all cases these were found to correspond very closely to the distribution found by Whittaker (43) for adult brain.

Although these results support the concept that the majority of NGF binding in brain is located on synaptosomal structures, an examination of the developmental time course of these receptors that this location is not exclusive (38). As shown in Fig. 9, specific [^{125}I]-NGF binding appears at detectable levels at day 6 of embryonic development, the earliest day that it is practical to measure it. However the synaptosomal markers, dopamine-β-hydroxylase, [^{125}I]-α-bungarotoxin binding and choline acetyltransferase, which exactly parallel each other, do not appear significantly before day 8. Subsequently, all activities (including NGF-binding) rise in a fashion identical with synaptogenesis. Prior to this, however, there is significant binding that is apparently not associated with synaptosomal structures (38). It seems probable that these structures are associated with the plasma membrane receptors of cell bodies and modulate such activities as cellular adhesive specificity demonstrated in the optic tectum (30) (see also Glaser *et al.*, elsewhere in this volume).

It should be noted that the binding of NGF in brain (as well as the tectal activity) may in fact reflect an analog activity of NGF (30). Nonetheless, even if this is true, the correct physiological substance, which would have to strongly resemble NGF in both structure and function, would also have an apparent bimodal functionality (*vide infra*).

EFFECT OF TRITON X-100 ON NGF RECEPTORS IN CHICK DORSAL ROOT GANGLIA

The demonstration that NGF could exert its neurite-proliferation activity apparently solely through interactions with plasma membrane receptors (18) (*vide supra*) suggested that these molecules might be amenable to further characterization following solubilization by detergent. Thus, chick embryonic DRG were incubated with different concentrations of triton X-100 for one hour, centrifuged at 2,000 x g for 20 minutes, and washed twice with 2 ml ice-cold Hank's salt solution containing 1 mg per ml bovine serum albumin (BSA) to remove the triton. Surprisingly, not all of the NGF-specific binding was solubilized by this treatment. As shown in Fig. 10, about 50% of the specific binding remained in the pellet. Since, in general, 1% triton effectively solubilizes the receptors for other hormones, this finding raised the question whether the apparent multiple affinities for the specific NGF binding, reported by Frazier *et al.*, (19) were not in fact the result of two types of receptors present in the plasma membrane. To explore this possibility, the pellet remaining after extraction of chick DRG by triton X-100 was subfractionated by discontinuous sucrose density gradient centrifugation. Virtually all of the binding was found in the densest fraction which contained primarily nuclei. That these are, in fact, NGF-specific receptors associated with the nucleus was confirmed by the isolation of chromatin from these cells. This material showed high affinity binding ($\sim 10^9$ M^{-1}) that, in contradistinction to that observed for whole cell preparations (19), was saturable. These results clearly indicate that there is found in DRG neurons a class of binding sites not seen before in studies using whole cell preparations of dorsal root and sympathetic ganglia (19).

The receptors solubilized by the triton X-100 were detected on Sephadex G-150 gel filtration by specific, reversible interaction

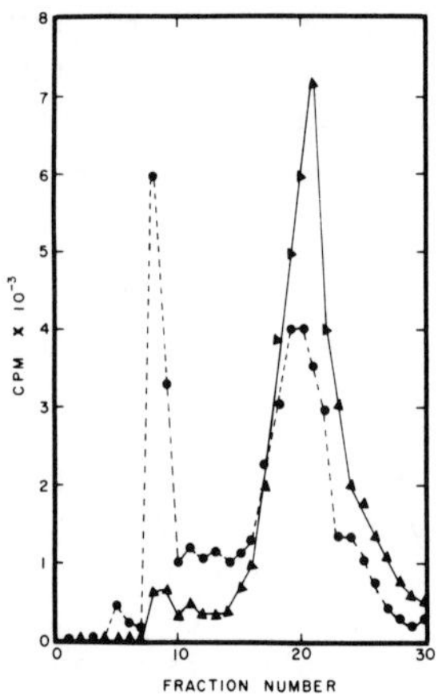

Fig. 11. Elution profile on Sephadex G-100 of the Triton X-100 solubilized binding components for [^{125}I]-NGF isolated from chick dorsal root ganglia. The dialyzed soluble fractions obtained from the Triton X-100 (0.1%) extraction of 250 dorsal root ganglia was incubated with [^{125}I]-NGF at room temperature for 1 hour. Following cooling to 4°, the sample was fractionated in a column of Sephadex G-100 (0.8 x 22 cm) equilibrated in 0.02% Triton X-100, 50 mM Tris-HCl, pH 7.4, in 1 mg/ml bovine serum albumin; (▲) and (●) indicates profiles of samples incubated in the presence and absence of excess unlabeled NGF.

with [^{125}I]-NGF. As shown in Fig. 11, when the dialyzed soluble fraction of 250 DRG was incubated with [^{125}I]-NGF and analyzed by gel filtration on Sephadex G-100 at 4°, two peaks, one eluting at the void volume and one in the position of native iodinated NGF were observed. However, if the sample was pre-treated with native NGF prior to loading, the void volume peak was eliminated. When this high molecular weight peak was pooled and reanalyzed, the pattern in Fig. 12A was obtained. Clearly, this complex is only very slightly dissociated by this treatment. However, if the isolated pool was incubated with unlabelled NGF and passed through the G-100 column again (Fig. 12B), most of the complex was dissociated. Thus, the Triton-solubilized receptor is specific for and reversibly bound to NGF.

BIFUNCTIONAL ACTIVITY OF NGF

It seems clear from these results that NGF may exert its physiological action on responsive tissues by two, probably independent, actions (6). During embryonic development, NGF, synthesized by as yet unidentified sources (see, however, Young _et al_., elsewhere in

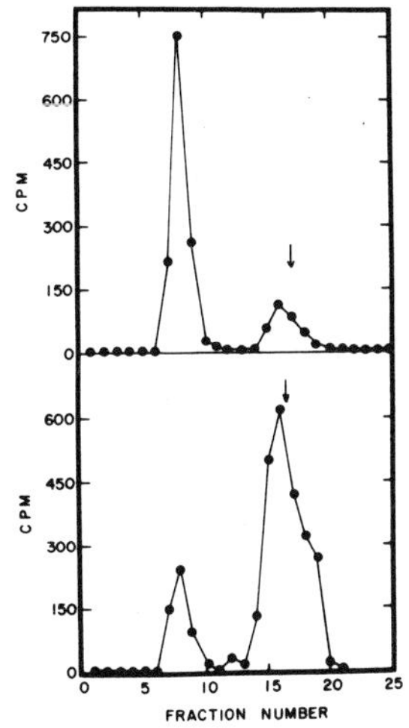

Fig. 12. (Left) Elution profile of the rechromatography of the void volume peak (see Fig. 11) of the separation of the triton X-100 solubilized receptor of dorsal root ganglia. (Right) Same sample as in left panel preincubated with 20 μg/ml unlabeled NGF for 30 min prior to analysis.

this volume), interacts with responsive neurons of dorsal root and superior cervical ganglia to elicit the neurite proliferation for which NGF is most well known. This trophic activity of NGF is apparently mediated through cell surface receptors which show non-saturability and negative cooperativity. Subsequently, when these neurons have matured, and axon proliferation and synapse formation are complete, a new role for NGF appears - namely the messenger between synapse and cell body. NGF, presumably synthesized by the post-synaptic cell, crosses the synaptic junction where it is complexed by membrane-bound receptors which may be related to or identical to the plasma membrane receptors active during embryonic development. However, in contrast to these moieties which are most likely coupled with an as yet unidentified effector, the synaptosomal receptors act as portals for the internalization of NGF which is transported to the perikaryon by retrograde axoplasmic transport (23). When it reaches the cell body, this NGF finds its way to the nucleus where it mediates its effects through complexing with the nuclear receptors which differ markedly from the plasma membrane receptors in their solubility in detergent and binding properties.

Although this scheme is clearly hypothetical, it is based on a large number of established observations. What remains to be determined is to what extent this scheme is mimicked by other hormones or growth factors. While the embryonic behavior of NGF is quite insulin-like, the internalized role seems unique. On the other hand, other non-NGF responsive neurons must also have such messengers, the interruption of which would produce chromatolysis and, eventually, cell death. Whether these are also protein factors related to NGF remains to be established.

ACKNOWLEDGEMENT

This work was supported by a U.S.P.H.S. research grant from the National Institutes of Health, NS 10229. M.W.P. was supported by a N.I.H. postdoctoral fellowship, NS 00757. R.Y.A. is a fellow of the Swiss National Fund, and R.E.S. is a predoctoral trainee of N.I.H. General Medical Scientists Training Program GM 02016.

REFERENCES

1. ANGELETTI, R. H., and BRADSHAW, R. A., Proc. Nat. Acad. Sci. USA 68 (1972) 2417.
2. ANGELETTI, R. H., BRADSHAW, R. A. and WADE, R. D., Biochemistry 10 (1971) 463.
3. BJERRE, B., WIKLUND, L., and EDWARDS, D. C., Brain Res. 92 (1975) 257.
4. BJORKLUND, A., BJERRE, B. and STENEVI, U., in: Dynamics of Degeneration and Growth in Neurons (K. Fuxe, L. Olson, and Y. Zotterman eds.), Pergamon Press, Oxford and New York (1974) 389.
5. BOCCHINI, V., and ANGELETTI, P. U., Proc. Nat. Acad. Sci. USA 64 (1969) 787.
6. BRADSHAW, R. A., FRAZIER, W. A., PULLIAM, M. W., SZUTOWICZ, A., JENG, I., HOGUE-ANGELETTI, R. A., BOYD, L. F. and SILVERMAN, R. E., Proc. 6th Int. Cong. Pharm., Vol. 2, Neurotransmission (1975) 231.
7. BURNHAM, P. A., SILVA, J. A. and VARON, S., J. Neurochem. 23 (1974) 689.
8. COHEN, S., Proc. Nat. Acad. Sci. USA 46 (1960) 302.
9. COHEN, S. and LEVI-MONTALCINI, R., Proc. Nat. Acad. Sci. USA 42 (1956) 571.
10. COHEN, S., LEVI-MONTALCINI, R. and HAMBURGER, V., Proc. Nat. Acad. Sci. USA 40 (1954) 1014.
11. CUATRECASAS, P., Proc. Nat. Acad. Sci. USA 63 (1969) 450.
12. CUATRECASAS, P. and HOLLENBERG, M. D., Biochem. Biophys. Res. Commun. 62 (1975) 31.

13. DE MEYTS, P., J. Supramol. Struc. 4 (1975) 241.
14. DE MEYTS, P. and ROTH, J., Biochem. Biophys. Res. Commun. 66 (1975) 1118.
15. DE MEYTS, P., ROTH, J., NEVILLE, D. M., JR., GAVIN, J. R. III and LESNIAK, M. A., Biochem. Biophys. Res. Commun. 55 (1973) 154.
16. FRAZIER, W. A., ANGELETTI, R. H. and BRADSHAW, R. A., Science 176 (1972) 482.
17. FRAZIER, W. A., ANGELETTI, R. H., SHERMAN, R. and BRADSHAW, R. A., Biochemistry 12 (1973) 3281.
18. FRAZIER, W. A., BOYD, L. F. and BRADSHAW, R. A., Proc. Nat. Acad. Sci. USA 70 (1973) 2931.
19. FRAZIER, W. A., BOYD, L. F. and BRADSHAW, R. A., J. Biol. Chem. 249 (1974) 5513.
20. FRAZIER, W. A., BOYD, L. F., PULLIAM, M. W., SZUTOWICZ, A. and BRADSHAW, R. A., J. Biol. Chem. 249 (1974) 5918.
21. FRAZIER, W. A., BOYD, L. F., SZUTOWICZ, A., PULLIAM, M. W. and BRADSHAW, R. A., Biochem. Biophys. Res. Commun. 57 (1975) 1096.
22. GOLDSTEIN, M. N. in: Cell Differentiation (R. Harris, P. Allin and D. Viza eds.) Munksgaard, Copenhagen (1972) 131.
23. HENDRY, I. A., STOECKEL, K., THOENEN, H. and IVERSON, L. L., Brain Res. 68 (1974) 103.
24. KOBLER, A. R., GOLDSTEIN, M. N. and MOORE, B. W., Proc. Nat. Acad. Sci. USA 71 (1974) 4203.
25. LARRABEE, M. G., in: Prog. in Brain Res. (K. Akbert and P. G. Waser eds.), Elsevier Publishing Co., Amsterdam (1969) 95.
26. LEVI-MONTALCINI, R. and ANGELETTI, P. U., Pharm. Rev. 18 (1966) 619.
27. LEVI-MONTALCINI, R. and ANGELETTI, P. U., Physiol. Rev. 48 (1968) 534.
28. LEVI-MONTALCINI, R. and HAMBURGER, V., J. Exp. Zool. 123 (1953) 233.
29. LIMBIRD, L. E., DE MEYTS, P. and LEFKOWITZ, R. J., Biochem. Biophys. Res. Commun. 64 (1975) 1160.
30. MERRELL, R., PULLIAM, M. W., RANDONO, L., BOYD, L. F., BRADSHAW, R. A. and GLASER, L., Proc. Nat. Acad. Sci. USA 72 (1975) 4270.
31. MONARD, D., SOLOMON, F., RENTSCH, M. and GYSIN, R., Proc. Nat. Acad. Sci. USA 70 (1973) 1894.
32. MONARD, D., STOECKEL, K., GOODMAN, R. and THEONEN, H., Nature 258 (1975) 444.
33. MOORE, J. B., JR., MOBLEY, W. C. and SHOOTER, E. M., Biochemistry 13 (1974) 833.
34. PARTLOW, L. M. and LARRABEE, M. G., J. Neurochem. 18 (1971) 2101.
35. PULLIAM, M. W., BOYD, L. F., BAGLAN, N. C. and BRADSHAW, R. A., Biochem. Biophys. Res. Commun. 67 (1975) 1281.

36. STACH, R. W. and SHOOTER, E. M., J. Biol. Chem. 249 (1974) 6668.
37. SZUTOWICZ, A., FRAZIER, W. A. and BRADSHAW, R. A., J. Biol. Chem. 251 (1976) 1516.
38. SZUTOWICZ, A., FRAZIER, W. A. and BRADSHAW, R. A., J. Biol. Chem. 251 (1976) 1524.
39. TISCHLER, A. S. and GREENE, L. A., Nature 258 (1975) 341.
40. VARON, S., NOMURA, J. and SHOOTER, E. M., Proc. Nat. Acad. Sci. USA 59 (1967) 1782.
41. VARON, S., NOMURA, J. and SHOOTER, E. M., Biochemistry 6 (1967) 2202.
42. VARON, S., NOMURA, J. and SHOOTER, E. M., Biochemistry 7 (1968) 1296.
43. WHITTAKER, V. P., in: Handbook of Neurochemistry, (Lajtha, A. ed.), Plenum Press, New York 2 (1969) 327.

STUDIES ON THE MOLECULAR PROPERTIES OF NERVE GROWTH FACTOR AND ITS CELLULAR BIOSYNTHESIS AND SECRETION

Michael YOUNG, Richard A. MURPHY, Judith D. SAIDE, Nicholas J. PANTAZIS, Muriel H. BLANCHARD and Barry G. W. ARNASON

Departments of Biological Chemistry, Medicine and Neurology, Harvard Medical School and the Massachusetts General Hospital, Boston, Massachusetts 02114 (USA)

In the preceding chapter, Bradshaw *et al.* have summarized much of the historical background information arising from the discovery of and early studies on nerve growth factor (NGF) by Bueker and by Levi-Montalcini, Hamburger and their colleagues. Here we should only like to draw attention to a brief list of those historical aspects which are most pertinent to the studies to be presented below.

1. NGF was first discovered as a soluble diffusible factor present in 2 mouse sarcomas. This factor has never been isolated and purified, and thus its chemical properties are unknown.

2. Male adult mouse submandibular glands contain large amounts of NGF, whose covalent structure *is* known, and male glands contain much more than female glands.

3. Little, if any, NGF is present in the glands of newborn mice--a perplexing observation since the factor is believed to play a central role in the *embryonic* development of the autonomic and sensory nervous systems.

4. Removal of mouse submandibular glands appears to have no obvious deleterious effect upon the animal--and NGF has not been detected in the salivary glands of other mammals.

5. Treatment of newborn mice with antibody to NGF results in selective destruction of the sympathetic nervous system without other obvious pathologic changes.

6. Small amounts of a nerve growth-promoting activity have been found in experimentally-induced granulation tissue, serum, and many peripheral tissues--for reasons unknown.

Against this background, let us turn now to the problem of the biosynthesis of NGF and its cellular sources of origin. For reviews of the early and more recent NGF literature, we refer the reader to (21) and (6, 7), respectively.

Experimental Procedures

Preparation of mouse submandibular gland NGF by the methods of Bocchini and Angeletti (4), demonstration of its electrophoretic and immunoelectrophoretic homogeneity, and preparation of monospecific antibody have been previously described (26).

Throughout these studies, two different immunoassays have been used to measure NGF. One is a radioimmunoassay, the details of which are presented elsewhere (25). A second type of assay employing NGF covalently linked to bacteriophage T_4 has also been employed. The basic feature of this method is that infectivity of the NGF-phage conjugates (for *E. coli*) can be blocked by antibody to NGF. Free, unconjugated NGF competes with the phage-NGF conjugate in the antigen-antibody reaction and this forms the basis of the immunoassay (26) (see also (18) for other references to this method). Within experimental error both phage and radioimmunoassays yield the same results.

Biological assays for nerve growth factor activity were performed with chick embryo sensory ganglia explanted onto collagen-coated coverslips by minor modifications (25) of the methods of Levi-Montalcini *et al.* (24). All other experimental details including methods used for cell cultures, are given in the legends to the figures and tables or in the original papers cited.

Biosynthesis and Secretion of NGF

The first evidence that certain cells in culture can synthesize and secrete NGF or an NGF-like factor was obtained from the studies with mouse L cells in 1973 (26). The L cell line was originally derived in 1943 from explanted mouse C3H subcutaneous and adipose tissue which had been treated *in vitro* for several months with 20-methylcholanthrene (10) and it has since been cloned (31). Fig. 1 reveals that when L cells (clone 929) are placed in culture together with sensory ganglia, intense neurite outgrowth occurs after 18 hr, and this effect is indistinguishable from that produced by NGF itself. The biologic effect depicted in Fig. 1 is abolished when NGF antiserum is included in the ganglion assay system (26). Moreover, when serum-free L cell culture supernatant fluids are analyzed, they

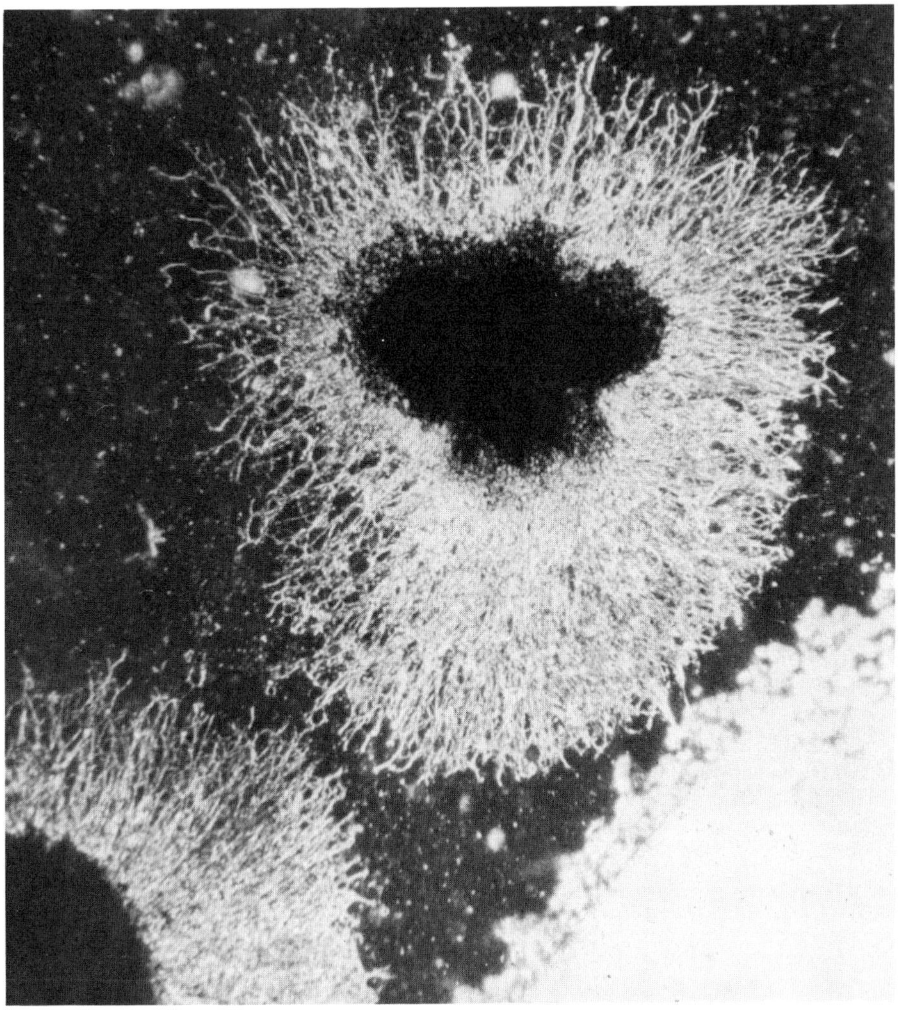

Fig. 1. Stimulation of ganglionic neurite extension by L cells. Chick embryo dorsal root ganglia (10-14 days) were placed on a collagen-coated coverslip together with about 10^6 L-929 cells. Photomicrograph (X 100) taken after 18 hr in culture. Neurite outgrowth can be seen to be directed predominantly toward the L cells which lie within the right lower corner (26).

were found to contain considerable quantities of a nondialyzable, biologically active protein which was immunochemically similar to NGF itself (26). Fig. 2. presents results of phage immunoassays of L cell culture fluids for NGF as a function of time in culture, and it will be seen that the levels of immunoreactive material rise to a maximum at about 4 days and then decline. The reason for the decline is not

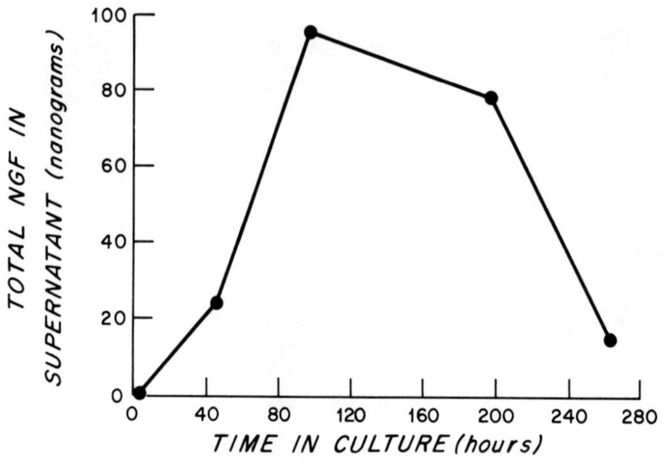

Fig. 2. Production of NGF by L cells. L cells were maintained in the presence of Eagle's minimal essential medium in plastic tissue culture flasks, each containing about 3 x 10^7 cells. Samples were withdrawn over time and analyzed for NGF by the bacteriophage method (26).

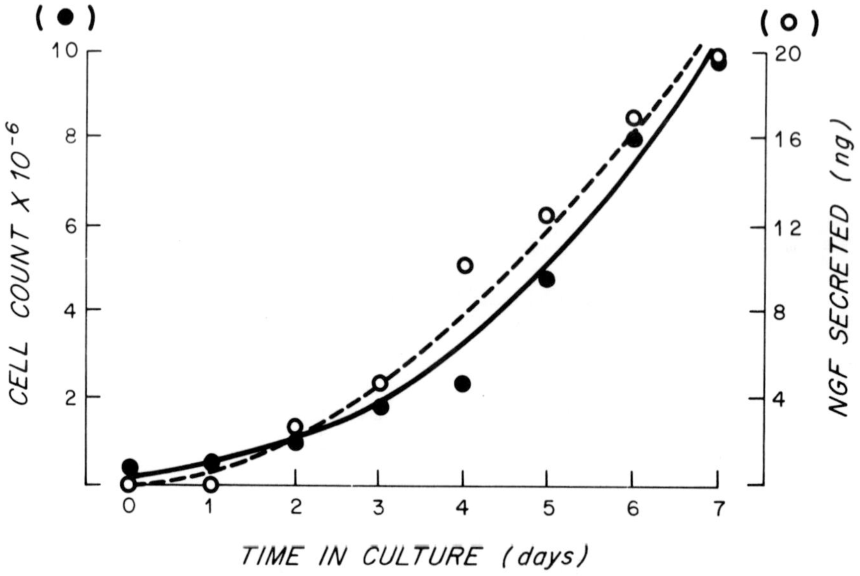

Fig. 3. Secretion of NGF by L-cells in logarithmic growth phase. Cultures were grown in serum free medium and analyzed for cell number and NGF (bacteriophage method) over time.

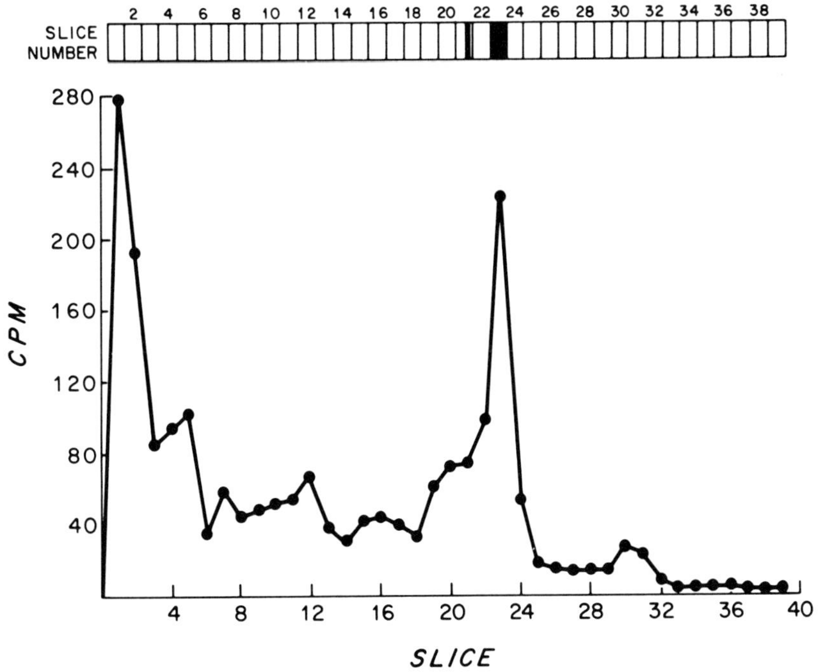

Fig. 4. Polyacrylamide gel electrophoretic pattern of partially purified radioactive L-cell culture medium. Solvent: 6 M urea, 0.9 M acetic acid. Prominent stained band is carrier NGF and the minor stained component arises from the L-cell culture medium (27).

clear. It may be that cell secreted proteases are responsible, but it is also possible that production declines after 4 days in culture and that the cells somehow subsequently utilize the factor. This latter notion is pertinent to the observation that L cells are a source of so-called "conditioned medium" (32). That is, these cells secrete factors which stimulate the proliferation of other cells. The relationship among these factors if any, is not known, but an NGF-like molecule is clearly one of them and it could be responsible for some of the conditioning effects observed.

To be sure that the L cell factor was actually being secreted and that it did not arise from dead and ruptured cells, L cells were seeded at such a density that they would enter a logarithmic growth phase. Fig. 3 presents a plot of NGF production and cell number as a function of time. It will be seen that the concentration of the growth factor in the culture supernatant solution closely parallels the increase in number of cells.

To further characterize the L cell factor, L cells were grown in the presence of ^{14}C-labeled amino acids in the absence of serum.

After 4 days, the culture medium was removed, mouse NGF was added as a carrier, and this mixture was subjected to carboxymethyl cellulose chromatography was described by Bocchini and Angeletti (4). Fractions containing carrier NGF were pooled, concentrated and examined by gel electrophoresis (27). Fig. 4 reveals that L cell culture fluid contains a radioactive component which migrates with the carrier protein. Clearly, further chemical studies are required to establish whether the radioactive component depicted in Fig. 4 is in fact identical to mouse submandibular gland NGF.

L cells injected into newborn C3H mice produce sarcomas, and it will be recalled that nerve growth factor activity was first detected many years ago in mouse sarcomas 180 and 37 (8, 22, 23). Thus, it occurred to us that secretion of NGF by L cells could be a counterpart in culture of its presence in the two mouse sarcomas. However, although the early NGF literature consistently refers to these tumors as sarcomas, they did not originate as such. Sarcoma 180 was originally described in 1914 as an axillary carcinoma of a male mouse and sarcoma 37 arose in 1908 as an adenocarcinoma of mouse mammary gland (34). Thus the precise nature of these tumors is unclear.

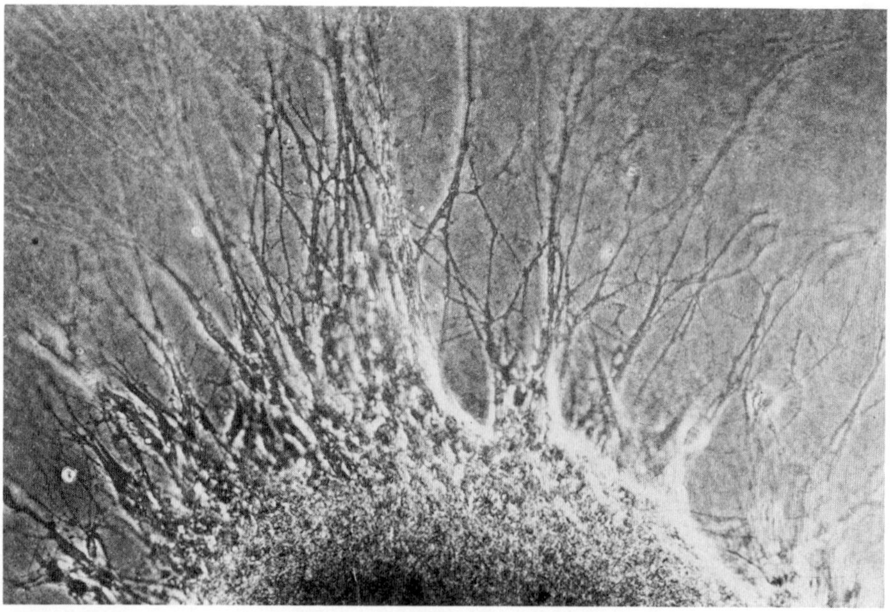

Fig. 5. Stimulation of ganglionic neurite extension by chick embryo fibroblast culture medium. The ganglion was treated with a 50-fold concentrated culture fluid removed from cells after 120 hr in culture. (X 260) (36).

TABLE I

Bacteriophage Immunoassays of Chick Fibroblast Culture Supernatant Solutions as a Function of Time in Culture

The NGF values refer to the amounts of immunoreactive material (± standard error of the mean) per 120 ml serum free-medium, based upon a standard assay for mouse NGF (36).

Time in Culture (hours)	NGF (ng)
0	0
36	17.4 ± 1.7
96	25.8 ± 2.7
120	38.5 ± 4.0
144	29.3 ± 3.0

Secretion of NGF by L cells could be simply a characteristic of the malignant state and not a property of "normal" or untransformed cells. To resolve this problem, we turned to primary chick embryo fibroblast cultures (36). Fig. 5 presents results of a sensory ganglion bioassay of concentrated serum-free culture fluid removed from cells after 120 hr in culture and Table I summarizes results of immunoassays of culture fluids as a function of time in culture. The data reveal that these chick embryo fibroblasts also secrete a factor which is biologically active in producing neurite outgrowth and which is immunologically similarly to mouse submandibular gland NGF. Now it should be recalled that Levi-Montalcini and Angeletti demonstrated many years ago that experimentally-induced granulation tissue displayed NGF-like biological activity _in vitro_ (20). Since granulation tissue is rich in fibroblasts, we began to speculate at this time that secretion of NGF (or a molecule similar to it) might be a general property of fibroblasts and that NGF might play some role in the biological function of granulation tissue (36). One of these functions is wound healing. Yet, secretion of NGF is not solely a property of fibroblasts since many other cells in culture also have the capacity to secrete it.

Table II presents a list of those cells which so far have been shown by the combined criteria of biological activity, bacteriophage immunoassay, and radioimmunoassay to produce an NGF-like factor in culture. Several features emerge from examination of Table II.

TABLE II

Cells Which Secrete NGF in Culture

Cell Type	Source
L-cells (26)	Mouse
3T3 cells (26)	Mouse
SV40 3T3 cells (26)	Mouse
Primary fibroblasts (36)	Chick
Neuroblastoma (25)	Mouse
Melanoma[a]	Mouse
Myoblasts[b]	Rat
Glioma (2)	Rat
Glioblastoma (2)	Human
Primary skin fibroblasts (30)	Human
Primary synovial fibroblasts (30)	Human

[a] Murphy, R. A., Krane, S. M., Arnason, B. G. W., and Young, M., unpublished observations.

[b] Murphy, R. A., Singer, R., Saide, J. D., Arnason, B. G. W., and Young, M., manuscript in preparation.

First, both biological activity and immunoreactivity are preserved over a wide range of species. Second, primary as well as transformed cells synthesize and secrete an NGF-like molecule. Third, this property is not confined to a particular cell type. Taken together, the information of Table II, while not extensive, suggests the possibility that many (and conceivably all) cells have the capacity to secrete NGF.

Fig. 6. Effects of neuroblastoma cell conditioned medium on embryonic sensory ganglia. Top: Ganglion incubated in Eagle's minimal essential medium plus 10% normal rabbit serum. Middle: Ganglion incubated in serum-free medium conditioned by neuroblastoma cells. Bottom: The same conditioned medium plus 10% rabbit NGF-immune serum. (X260) (25).

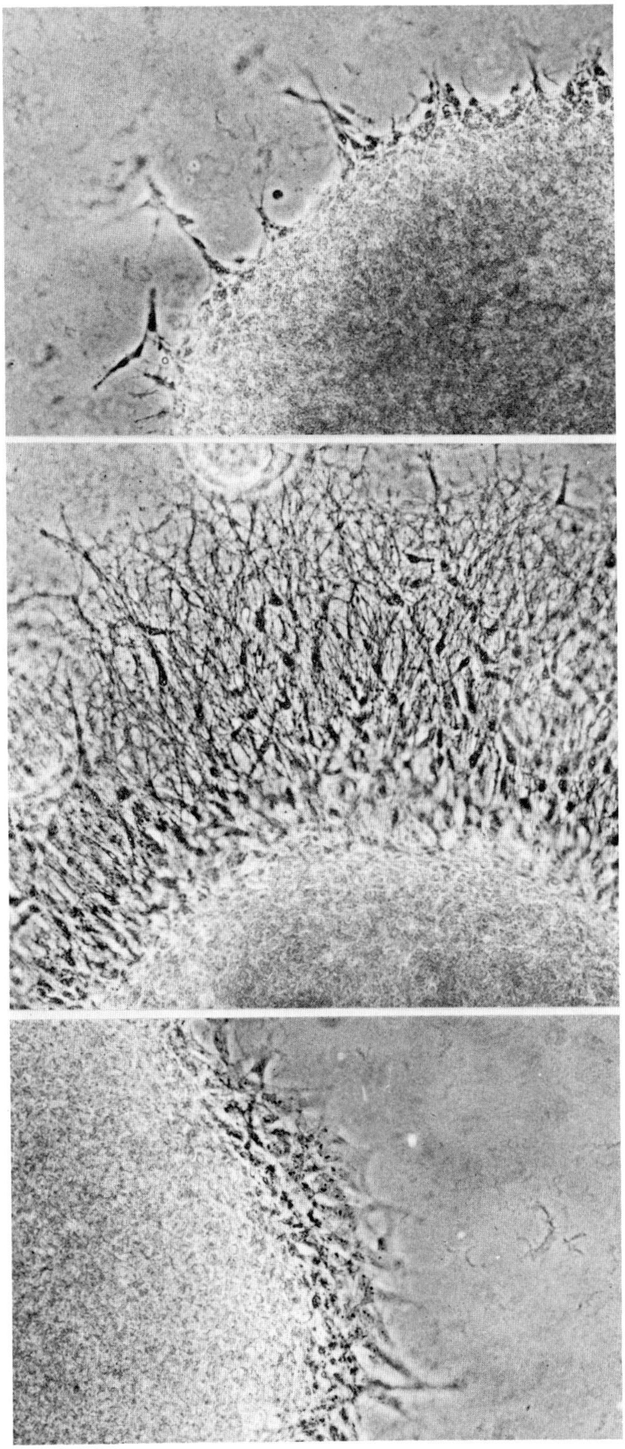

Fig. 6

TABLE III

Bacteriophage and Radioimmunoassays of Concentrated Neuroblastoma Culture Solutions

Cells were grown for 4 days in serum-free medium and the supernatant solution was removed and concentrated. Values represent nanograms of immunoreactive material per ml of original unconcentrated culture fluid (± SEM) (25).

Method	NGF (ng/ml)
Radioimmunoassay	0.15 ± .02
Bacteriophage	0.18 ± .03

We should like to call attention to the mouse neuroblastoma line listed in Table II. Several lines of evidence indicate that both mouse (14) and human (19) neuroblastoma cells can respond biologically to NGF *in vitro* and that mouse neuroblastoma cells possess cell surface receptors for the protein (5, 28, 29). When these features were taken together with the observation that glial cells, (Table II) also secrete NGF, we predicted that perhaps one function of glial cells, in their intimate association with neurons *in vivo*, might be to secrete NGF for the purpose of neuronal cell growth or maintenance. If so, then perhaps primary neuronal cell cultures might not produce NGF. Whether they do or not is not known, but Fig. 6 and Table III reveal that mouse C1300 neuroblastoma cells do secrete a biologically active, immunoreactive factor (25). Also depicted in Fig. 6 is the fact that antibody to NGF abolishes the biological activity of the neuroblastoma NGF. Moreover, Fig. 7 reveals that these cells contain considerable intracellular amounts of the factor. After 4 days in culture, appreciable NGF has been secreted into the medium with little net change in the intracellular concentration--an effect which is blocked by cycloheximide. These results demonstrate that the neuroblastoma nerve growth factor is actively synthesized and then secreted into the culture fluid (25).

The observation that neuroblastoma cells can respond to mouse submandibular gland NGF, and the fact that they secrete a factor closely similar to it, raises some interesting possibilities. One is that certain cells within the clone secrete an NGF-like molecule to which other cells respond. Alternatively, a neuroblastoma cell may secrete the factor at one stage of the cell cycle and respond to it at another stage--perhaps as part of some kind of growth auto-regulatory mechanism. In this connection, evidence from studies with synchronized neuroblastoma cells suggests that binding of NGF to the cell surface is maximal during the late G_1 and early

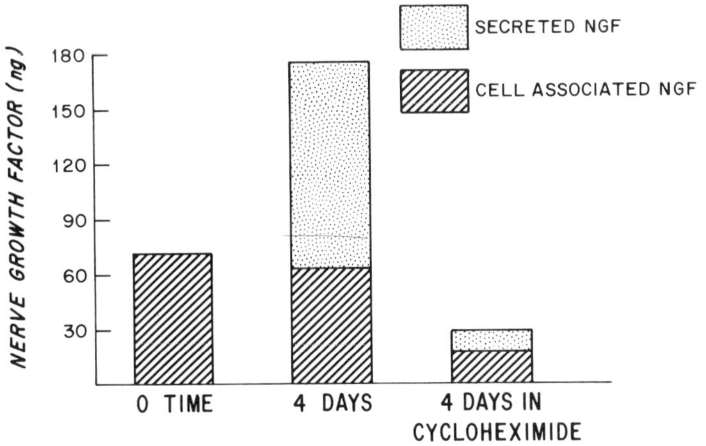

Fig. 7. Synthesis and secretion of NGF by neuroblastoma cells. Three identical groups of confluent cultures were incubated in serum-free medium for the time periods indicated. NGF levels in the culture medium as well as in cell extracts from each group were measured by radioimmunoassay.

S phases of the cell cycle (28). Whether cells other than neuroblastoma which secrete NGF also functionally respond to it is not known, but it is known that a variety of nonneuronal tissues can bind NGF with a high association constant (12, 13). The meaning of these findings is not clear, but it could be that these nonneuronal NGF-binding tissues contain functional receptors for NGF which are unrelated to their sympathetic innervation.

NGF in Sera of Mouse and Man

It has been recognized for a long time that the sera of a variety of species are capable of stimulating neurite outgrowth from embryonic ganglia in culture (21). The factor(s) responsible has never been isolated nor has it been shown that the nerve growth stimulating activity arises from the presence in serum of a molecule which is chemically similar to NGF.

Using a radioimmunoassay for NGF, Hendry (16) has measured levels of about 30 ng/ml and 4 ng/ml in the plasma of adult male and female mice, respectively, and this result was interpreted as a reflection of the fact that male mouse submandibular glands contain much higher levels of NGF than female glands. In a subsequent study, Hendry and Iverson (17) observed somewhat different plasma values--11.5 ng/ml for male mice and 2.6 ng/ml for females. Following bilateral removal of the submandibular glands, these values fell to

TABLE IV

Immunoassays of Adult Mouse Sera

Method	Sex	Strain	NGF-equivalents (ng/ml)
Bacteriophage	M	Swiss	32
	F	Swiss	50
	F	Swiss	42
	F	Swiss	42
	M	C-57	60
	M	C-57	47
	M	C-57	47
			(Mean = 45 ± SEM)
Radioimmunoassay	M	C-57	42
	M	A/HeJ	44
	M	A/HeJ	38
	M	A/HeJ	29
	M	A/HeJ	27
	M	Swiss	54
	M	Swiss	44
	M	Swiss	52
	M	Swiss	49
	M	Swiss	54
			(Mean = 43 ± 3)

a minimum of 15% of control levels over a period of about 30 days and then rose again to normal over a period of 2 months. Our own studies[1] on mouse serum levels are in reasonable agreement with those reported in the earlier paper by Hendry (16) although we have been unable to detect any significant differences in serum levels between the sexes. Table IV represents results of both bacteriophage and radioimmunoassays of adult male and female mice sera of several strains. Both

[1] Chelmicka, E., Arnason, B. G. W., Blanchard, M. H., and Young, M., unpublished observations.

TABLE V

Radioimmunoassays of Sera of Control, Sham Operated, and Sialoadenectomized Swiss Mice as a Function of Time after Surgery.[a]

Time after operation (weeks)	Control[a] (ng/ml)	Sham operated[a] (ng/ml)	Sialoadenectomized[a] (ng/ml)
1	43 ± 3	25 ± 5	37 ± 3
2	40 ± 4	23 ± 10	29 ± 1
3	34 ± 6	37 ± 5	25 ± 6
4	32 ± 2	46 ± 4	32 ± 1

[a]Three to four mice comprise each group.

assays yield closely similar values of immunoreactive material (about 44 ng/ml) and this number is close to that reported by Hendry (16) for male mice.

We have also examined the effect of removal of the submandibular glands upon the serum NGF-immunoreactive levels of male mice. Table V presents results of radioimmunoassays of sera of control, sham operated, and sialoadenectomized animals as a function of time after surgery.[1] Over a period of 5 weeks, serum values of NGF-immunoreactive material are not significantly different among the 3 groups of animals. We are unable to offer any explanation for the differences between our results and those of Hendry and Iverson (17). On the other hand, both studies reveal that the mouse submandibular glands cannot be the whole (or perhaps even the most important) source of NGF *in vivo*. In this regard, we note that serum levels of epidermal growth factor (also found in high concentration in the mouse submandibular gland) are closely similar in males and females and do not decrease following removal of the glands (9).

Recently we have begun to measure serum levels of NGF-immunoreactive material in human serum by radioimmunoassay. The ultimate purpose of these studies is to isolate the growth factor from human serum, to characterize it chemically and to see whether serum levels are altered in disease. We wish to emphasize that the immunologic results to be presented here are based upon assays with mouse NGF as the standard. Consequently, depending upon the degree of cross-reactivity of mouse NGF and the immunoreactive material in human serum, the human values obtained could be falsely low.

Table VI summarizes levels of a mouse NGF cross-reactive substance measured by assays of human serum from individuals free of

TABLE VI

Radioimmunoassays of Normal Adult Human Serum[a]

Male (ng/ml)[a]	Female (ng/ml)[a]
46	41
78	21
42	41
33	90
26	21
43	66
46	37
50	58
16	65
	62
Mean = 42 + 6 (SEM)	Mean = 50 + 7 (SEM)

[a]Values are expressed as NGF-equivalents based upon mouse NGF standards.

known disease. It will be seen that the human serum levels are similar to those of mouse serum, and that no significant sex differences are detected.

In our laboratory, we have found that biological assays (chick sensory ganglion system) of human serum yield erratic results. Sometimes neurite outgrowth is observed but more often, no biological activity can be detected and the results are difficult to correlate with those obtained by quantitative radioimmunoassay. These findings are similar to those of Banks _et al_. (3), who observe nerve growth-promoting activity only when serum was partially purified by chromatographic methods. Consequently we turned to the Cohn fractions of human serum, with the result that fraction III contains the highest level of immunoreactive material (30). Moreover, biological assays of this fraction yield consistently positive results. Fig. 8 illustrates the neurite-outgrowth stimulating activity of Cohn fraction III as well as the effect of antiserum to mouse NGF in preventing outgrowth.

The observation that fibroblasts, both mouse and human, secrete an NGF-like factor in culture has led us to measure by radioimmunoassay the levels of immunoreactive material in sera of patients with

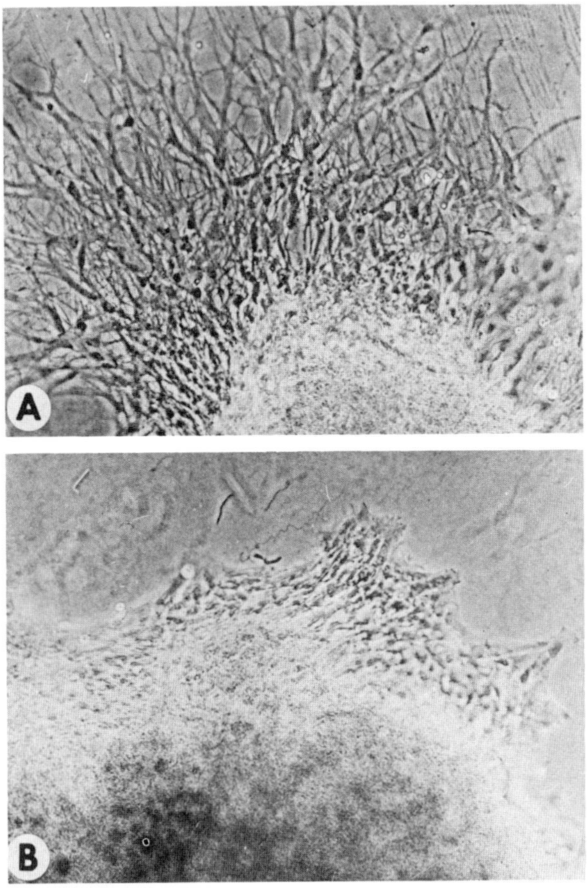

Fig. 8. Neurite-outgrowth stimulating activity of Cohn fraction III of human serum. A: Cohn fraction III. B: Cohn fraction III plus antiserum to mouse NGF.

Paget's Disease of Bone. This disease is characterized by intense metabolic activity and cellular proliferation in affected bones. Our reasoning was that perhaps cells in an active state of proliferation might secrete increased amounts of the NGF-like factor and that this could be reflected by higher serum levels of the factor. Table VII presents values for NGF-equivalents of immunoreactive material in the sera of several patients with this disease (30). By comparison with the values given in Table VI, it will be seen that all patients display significantly elevated levels of this factor. Of

TABLE VII

Radioimmunoassay Measurements of NGF-Immunoreactive Material in Sera of Patients with Paget's Disease of Bone before and after Treatment

Serum	Pre-therapy	Post-therapy[a]
1	182	91
2	239	87
3	390	206
4	283	117
5	223	85
6	175	93
7	182	98
8	163	95
9	186	136
10	220	119
11	190	97
12	206	158
13	410	134

[a]Post-treatment values were measured after 3 months therapy with 20 mg/kg/day of disodium ethane-1-hydroxy-1,1-diphosphonate (30). Values refer to NGF-equivalents in ng/ml based upon mouse NGF standards.

particular interest is the observation that prolonged treatment of these patients with diphosphonates produced a fall in serum levels, in some cases to within the "normal" range.

The significance of these findings, either for the function of NGF or for Paget's Disease, is not clear. But it could be that enhanced cell proliferation in this disease, as well as perhaps in the malignant state, could be accompanied by increased secretion of NGF or a factor immunologically similar to it. Detailed chemical studies on the human factor are required to approach this problem.

On the Molecular Structure of NGF in Dilute Solution

In the course of studies on the chemical properties of NGF of NGF in serum, we happened to observe that the gel filtration proper-

TABLE VIII

Summary of Sedimentation and Gel Filtration Data for an NGF Monomer-Dimer Equilibrium Reaction

Values for the sedimentation coefficient ($S_{20,w}$) and gel filtration partition coefficient ($\bar{\sigma}_w$) were calculated as a function of the initial NGF concentration (C_o). K is the apparent association constant calculated for each C_o, with 0.1M potassium phosphate, pH 7.0, as solvent (37).

Method	C_o (μg/ml)	$S_{20,w}$(S) or $\bar{\sigma}_w$	K(M^{-1} x 10^{-6})
Sedimentation	1200	2.38 ± .04	--
	82	2.33 ± .04	9.1
	59	2.32 ± .04	9.9
	9.0	2.28 ± .06	8.7
	6.2	2.14 ± .03	8.9
	6.2	2.05 ± .20	3.9
	1.11	1.90 ± .05	5.9
Gel Filtration	3.6	0.456	7.1
	.0615	0.510	22.0

ties of dilute solutions of pure mouse submandibular gland NGF were inconsistent with those of a dimer of molecular weight close to 26,000 (37). Specifically, when NGF at initial concentrations on the order of 1 μg/ml was applied to Sephadex, the protein emerged with a partition coefficient consistent with a molecular weight much less than 26,000. Now evidence from several laboratories clearly demonstrates that NGF exists as a dimer at concentrations of 1 mg/ml and above (1, 4). Consequently, the available information suggested that we were observing the existence of a monomer ⇄ dimer equilibrium at the lower concentrations. A combined study of the sedimentation equilibrium, sedimentation velocity, and gel filtration properties of NGF demonstrated that this is indeed the case (37). For the sedimentation velocity work, the photoelectric absorption scanning system of the analytical ultracentrifuge was used with light of wavelength as low as 220 nm to measure the sedimentation coefficient of the protein at concentrations as low as 6 μg/ml. Table VIII reveals that the weight-average sedimentation

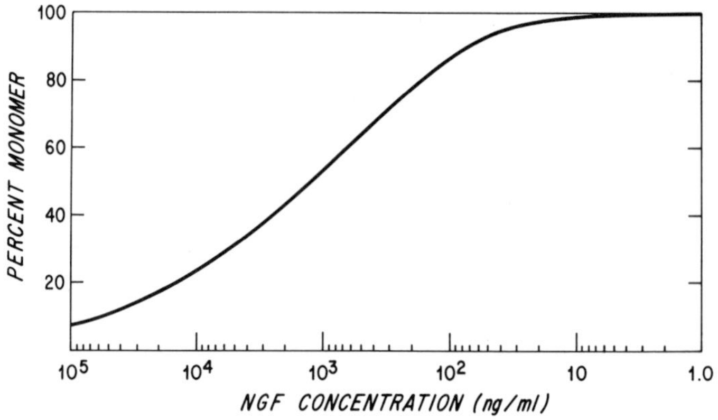

Fig. 9. Plot of percent monomer as a function of total NGF concentration for a monomer $\leftrightarrows$ dimer equilibrium reaction (37).

coefficient falls as the protein concentration is lowered and that the weight-average partition coefficient (σ_W) of NGF on columns of G-75 Sephadex is also a function of protein concentration. Both the sedimentation and gel filtration data can be used to calculate the association constant (K) for the monomer $\leftrightarrows$ dimer reaction, and when all data are taken together a mean value of $K = 9.4 \times 10^6\ M^{-1}$ is obtained. Using this number we have calculated a theoretical curve for the dissociation reaction. Fig. 9 illustrated a plot of percent monomer in the equilibrium mixture as a function of total NGF concentration. It will be seen that an NGF solution of concentration 1.4 µg/ml consists of an equal mixture of monomer and dimer. At a total concentration of 1 ng/ml, the mixture contains greater than 99% monomer, and this is 20 to 30 times the NGF concentration which has been shown to be biologically active (15).

The question arises as to the possible biological significance of this equilibrium reaction. First of all, the data indicate that at concentrations of NGF usually employed in the ganglion bioassay (1-10 ng/ml), the monomer is the biologically active species. Several lines of evidence indicate that the monomer $\leftrightarrows$ dimer reaction is a chemically specific one. For example, polymerization appears to stop at the dimer stage (See (37) for a summary of the evidence). Moreover, crystals of NGF have now been prepared and they were grown at sufficiently high protein concentrations where dimer predominates (35). Consequently, the packing arrangement between monomer chains in the crystal must be highly stereospecific.

One possible explanation for the existence of a specific monomer $\leftrightarrows$ dimer reaction should not be overlooked and that is that monomer is biologically active and dimer inactive. In this regard, evidence from several laboratories indicates that both by morphologic and biochemical criteria, there is a maximum in the biological dose-response curve of NGF, with apparent neurite growth-promoting activity decreasing sharply at concentrations greater than 1 μg/ml where formation of the dimer is beginning to predominate (See (37) for a summary of the evidence). If the monomer is biologically active and dimer inactive, this could serve to explain why an optimum in dose-responsiveness to NGF is observed, provided that monomer and dimer compete for cellular NGF binding sites. Evidence that dimer can bind arises from the studies of Frazier *et al.* (11) (see also Bradshaw *et al.*, elsewhere in this volume) who measured binding of ^{125}I-NGF to sympathetic ganglia. Multiple binding affinities were observed and binding was unsaturable at concentrations of NGF as high as 10^{-8} M (about 0.26 μg/ml). At this concentration, solutions of NGF contain considerable amounts of dimer (Fig. 9). Moreover, it may be that the multiple binding affinities observed by Frazier *et al.* (11) arise, at least in part, from different binding affinities of the monomer and dimer.

The question of whether the NGF dimer is biologically active has been examined by Stach and Shooter (33) who prepared NGF covalently crosslinked with dimethylsuberimidate. This preparation was found to be indistinguishable from native NGF in the sensory ganglion assay. While this result argues against the idea that the dimer may be inactive, there is another possible explanation, since it is not known how extensively (in a 3-dimensional sense) the two chains were cross-linked. That is, it remains possible that at high dilution, the interchain noncovalent bonds can still be broken with concomitant unfolding of the cross-linked dimer and exposure of biologically active segments of individual monomer chains. In this connection, Stach and Shooter (33) also observed a maximum in the dose-response curve for cross-linked dimer which was closely similar to that of native NGF.

In concluding, we should like to point out that a monomer $\rightleftarrows$ polymer equilibrium system (monomer active, polymer inactive) could be a rather simple control mechanism for regulating the biological activity of a molecule *in vivo* as a function of its local concentration at its site of action.

ACKNOWLEDGEMENTS

We wish to express our gratitude to Dr. Joël Oger for his early and important contributions to this work and for many subsequent conversations with him. Professors Harold Amos and John T. Potts, Jr. have contributed significantly to the development of our ideas. This work was supported by NIH grants CA 17137 and AM 09404 (M.Y.) and by NINDS 06021 (B.G.W.A.). R.A.M. and J.D.S. were supported by NIH Posdoctoral Fellowships NS 01828 and HL 05079 respectively.

REFERENCES

1. ANGELETTI, R. H., BRADSHAW, R. A. and WADE, R. D. , Biochemistry 10 (1971) 463.
2. ARNASON, B. G. W., OGER, J., PANTAZIS, N. J. and YOUNG, M., J. Clin. Invest. 53 (1974) 2a.
3. BANKS, B. E. C., BANTHORPE D. V., CHARLWOOD, K. A., PEARCE, F.L., VERNON, C. A. and EDWARDS, D. C., Nature 246 (1973) 503.
4. BOCCHINI, V. and ANGELETTI, P. U., Proc. Nat. Acad. Sci. USA 64 (1969) 787.
5. BOSMAN, C., REVOLTELLA, R. and BERTOLINI, L., Cancer Res. 35 (1975) 896.
6. BOYD, L. F., BRADSHAW, R. A., FRAZIER, W. A., HOGUE-ANGELETTI, R. A., JENG, I., PULLIAM, M. W. and SZUTOWICZ, A., Life Sciences 15 (1974) 1381.
7. BRADSHAW, R. A. and YOUNG, M. , Biochem. Pharm. (1976) in press.
8. BUEKER, E. D., Anat. Rec. 102 (1948) 369.
9. BYYNY, R. L., ORTH, D. N., COHEN, S. and DOYNE, E. S., Endocrinology 95 (1974) 776.
10. EARLE, W. R. , J. Nat. Cancer Inst. 4 (1943) 165.
11. FRAZIER, W. A., BOYD, L. F. and BRADSHAW, R. A. , J. Biol. Chem. 249 (1974) 5513.
12. FRAZIER, W. A., BOYD, L. F., PULLIAM, M. W., SZUTOWICZ, A. and BRADSHAW, R. A. , J. Biol. Chem. 249 (1974) 5918.
13. FRAZIER, W. A., BOYD, L. F., SZUTOWICZ, A., PULLIAM, M. W. and BRADSHAW, R. A., Biochem. Biophys. Res. Commun. 57 (1974) 1096.
14. GOLDSTEIN, M. N., BRODEUR, G. M. and ROSS, D., Anat. Rec. 175 (1973) 330.
15. GREENE, L. A. , Neurobiology 4 (1974) 286.
16. HENDRY, I. A., Biochem. J. 128 (1972) 1265.
17. HENDRY, I. A. and IVERSON, L. L., Nature 243 (1973) 500.
18. HURWITZ, E., DIETRICH, F. M. and SELA, M., Eur. J. Biochem. 17 (1970) 273.
19. KOLBER, A. R., GOLDSTEIN, M. N. and MOORE, B. W., Proc. Nat. Acad. Sci. USA 71 (1974) 4203.
20. LEVI-MONTALCINI, R. and ANGELETTI, P. U., Regional Neurochemistry (Kety, S. S. and Elkes, J. eds.) Pergamon Press, Oxford (1961) 362.
21. LEVI-MONTALCINI, R. and ANGELETTI, P. U., Physiol. Rev. 48 (1968) 534.
22. LEVI-MONTALCINI, R. and HAMBURGER, V., J. Exp. Zool. 116 (1951) 321.
23. LEVI-MONTALCINI, R. and HAMBURGER, V., J. Exp. Zool. 123 (1953) 233.
24. LEVI-MONTALCINI, R., MEYER, H. and HAMBURGER, V., Cancer Res. 14 (1954) 49.
25. MURPHY, R. A., PANTAZIS, N. J., ARNASON, B. G. W. and YOUNG, M., Proc. Nat. Acad. Sci. USA 72 (1975) 1895.
26. OGER, J., ARNASON, B. G. W., PANTAZIS, N. J., LEHRICH, J., and YOUNG, M., Proc. Nat. Acad. Sci. USA 71 (1974) 1554.

27. PANTAZIS, N. J., OGER, J., ARNASON, B. G. W. and YOUNG, M., Fed. Proc. 33 (1974) 1343.
28. REVOLTELLA, R., BERTOLINI, L., PEDICONI, M. and VIGNETTI, E., J. Exp. Med. 140 (1974) 437.
29. REVOLTELLA, R., BOSMAN, C. and BERTOLINI, L., Cancer Res. 35 (1975) 890.
30. SAIDE, J. D., MURPHY, R. A., CANFIELD, R. E., SKINNER, J., ROBINSON, D. R., ARNASON, B. G. W. and YOUNG, M., J. Cell Biol. 67 (1975) 376a.
31. SANFORD, K. K., EARLE, W. R. and LIKELY, G. D., J. Nat. Cancer Inst. 9 (1948) 229.
32. SHODELL, M., Proc. Nat. Acad. Sci. USA 69 (1972) 1455.
33. STACH, R. W. and SHOOTER, E. M., J. Biol. Chem. 249 (1974) 6668.
34. STEWARD, H. L., SNELL, K. C., DUNHAM, L. J. and SCHLYEN, S. M., Transplantable and Transmissible Tumors of Animals (Armed Forces Institute of Pathology, Sec. 12, F40 Washington, D. C.) (1956) 243, 324.
35. WLODAWER, A., HODGSON, R. O. and SHOOTER, E. M., Proc. Nat. Acad Sci. USA 72 (1975) 777.
36. YOUNG, M., OGER, J., BLANCHARD, M. H., ASDOURIAN, H., AMOS, H. and ARNASON, B. G. W., Science 187 (1975) 361.
37. YOUNG, M., SAIDE, J. D., MURPHY, R. A. and ARNASON, B. G. W., J. Biol. Chem. 251 (1976) 459.

SPECIFIC BINDING AND ABILITY OF VASOACTIVE INTESTINAL OCTACOSA-PEPTIDE (VIP) TO ACTIVATE ADENYLATE CYCLASE IN ISOLATED PANCREATIC ACINAR CELLS FROM THE GUINEA PIG

Jean CHRISTOPHE*, Patrick ROBBERECHT*,
Thomas P. CONLON and Jerry D. GARDNER

Section on Gastroenterology, Digestive
Diseases Branch, National Institute of
Arthritis, Metabolism, and Digestive Diseases
Bethesda, Md. 20014 (USA)

This is probably the first instance in which the specific binding of a gastrointestinal hormone to a digestive cell other than the hepatocyte has been studied. The vasoactive intestinal octacosapeptide (VIP) was recently isolated from porcine upper small intestine (19) and its amino acid sequence has been determined (4, 12).

Nine amino acids out of 28 in the VIP molecule occur in the same position as they do in secretin (Fig. 1.) It is therefore not surprising that when injected intravenously into cats, VIP stimulates bicarbonate and water secretion from the pancreas in a secretin-like fashion (19).

Upon intravenous injection of VIP in rats, a transient doubling in cyclic AMP levels is observed in the pancreas (13). *In vitro*, four-fold increases in cyclic AMP levels are observed in rat pancreatic fragments incubated with 10^{-6} M VIP (8). Both *in vivo* and *in vitro*, the potency of VIP appears to be ten times lower than that of secretin.

If the pancreas is undoubtedly a target for VIP, further studies are complicated by the heterogeneity of the intact tissue. Indeed the coexistence of centroacinar and ductular cells with acinar cells makes interpretation of the preceding data on intact

---

* Recipients of a scientific mission award from the Fonds National de la Recherche Scientifique of Belgium; permanent address: Department of Biochemistry and Nutrition, University of Brussels Medical School, 1000 Brussels, Belgium.
Abbreviation: VIP: vasoactive intestinal polypeptide.

	1	2	3	4	5	6	7	8	9	10	11	12	13	14	
VIP	H-HIS	SER	ASP$^-$	ALA	VAL	PHE	THR	ASP$^-$	ASN	TYR	THR	ARG$^+$	LEU	ARG$^+$	
SECRETIN	H-HIS	SER	ASP$^-$	GLY	THR	PHE	THR	SER	GLU$^-$	LEU	SER	ARG$^+$	LEU	ARG$^+$	
GLUCAGON	H-HIS	SER	GLN	GLY	THR	PHE	THR	SER	ASP$^-$	TYR	SER	LYS$^+$	TYR	LEU	
	15	16	17	18	19	20	21	22	23	24	25	26	27	28	29
VIP	LYS$^+$	GLN	MET	ALA	VAL	LYS$^+$	LYS$^+$	TYR	LEU	ASN	SER	ILE	LEU	ASN(NH_2)	
SECRETIN	ASP$^-$	SER	ALA	ARG$^+$	LEU	GLN	ARG$^+$	LEU	LEU	GLN	GLY	LEU	VAL(NH_2)		
GLUCAGON	ASP$^-$	SER	ARG$^+$	ARG$^+$	ALA	GLN	ASP$^-$	PHE	VAL	GLN	TRP	LEU	MET	ASN	THR

Fig. 1. Comparison of the amino acid sequences of porcine hormones VIP (vasoactive intestinal octacosapeptide), secretin and glucagon. The amino acid identities with VIP are underlined twice and similar polarities once.

tissue hazardous. Fortunately, there is now a method available for isolating exocrine acinar cells (1), and with this tool at hand it was recently shown in this laboratory that VIP can activate adenylate cyclase in a cell suspension which is made of about 98% acinar cells.[1] In liver and fat cells VIP also binds to plasma membrane and stimulates cyclase activity (2, 6, 7, 9). With this in mind, and taking into consideration the relative easiness of radioiodinating a peptide with two tyrosyl residues, we decided to investigate the specific binding of ^{125}I-VIP to plasma membrane receptors in isolated intact acinar cells. In addition, we compared the effects of VIP, secretin, and secretin fragments on adenylate cyclase activity in the same preparation.

EXPERIMENTAL PROCEDURE

Pancreatic acinar cells from the guinea pig pancreas were isolated following the procedure of Amsterdam and Jamieson (1) using crude collagenase and hyaluronidase, EDTA and mild shearing forces. These cells were viable and metabolically active. Since at least 95% of such suspensions were made of acinar cells (1), our basic assumption is that these cells were responsible for the observed effects.

VIP was stoichiometrically monoiodinated by a modification of the chloramine-T method of Hunter and Greenwood (10). We aimed at

[1] Gardner, J. D. *et.al.* (1975) *in press*.

incorporating no more than 0.27 iodine atoms per trysyl residue, i.e., 0.54 iodine atoms per molecule in the case of VIP. Most of the various steps of this method were conducted as described by Rodbell _et al_. (15), and Bataille _et al_. (2). The importance of limiting the specific activity aimed at and of adding chloramine-T in stepwise amounts, i.e. with a systematic control of organification progress has been stressed recently (17). ^{125}I-VIP had a calculated specific radioactivity of 246 μCi/μg (937 Ci/nmol) and was easily purified by adsorption over cellulose powder. This material was precipitable to a 95% extent by 5% trichloroacetic acid and could be kept at -20° for at least two months without obvious damage other than the normal reduction in specific activity.

Unless otherwise specified in the Figures and Tables, the assay for binding of ^{125}I-VIP was performed at 37° in a Krebs-Ringer bicarbonate buffer equilibrated with 90% O_2 and 5% CO_2 to pH 7.4. This medium contained 0.5 mM Ca and was supplemented, as suggested by Amsterdam and Jamieson (1), with 1% (w/v) bovine serum albumin, 14 mM glucose, a diluted L-amino acid supplement, and 0.1 mg/ml of chromatographically purified soybean trypsin inhibitor. Basically, for a binding assay, the enriched Krebs-Ringer bicarbonate buffer contained $1.66 \pm 0.14 \times 10^7$ cells/ml (mean ± SEM; n = 53) and approximately 2.6×10^{-11} M ^{125}I-VIP. After 10 min, duplicate 100 μl samples were layered on top of 300 μl ice-cold buffer in plastic microcentrifuge tubes. This Krebs-Ringer bicarbonate buffer (pH 7.4) was enriched with 2.5 mM Ca and 1% (w/v) albumin. The cells were immediately washed by centrifugation for 15 seconds at 10,000 x g with a Beckman Model 153 microfuge. The bulk of the supernatant fluid was discarded by aspiration. The pellet was agitated with a Vortex and washed twice more with 300 μl fresh buffer and centrifugation. The washed pellet was then dispersed in 200 μl of 5% (v/v) triton X-100. The microcentrifuge tube was poured upside down and agitated in a vial containing 20 ml of a liquid scintillation solution made of 15 parts toluene, 5 parts triton X-100 and 1 part Liquifluor (New England Nuclear Corp.) before liquid scintillation counting. Data were corrected for non-specific binding of ^{125}I-hormone to pancreatic acinar cells by performing parallel incubations in which unlabeled hormone in large excess was added prior to the iodinated peptide.

To estimate adenylate cyclase activity in intact cells [8-^{14}C] adenine (100 μCi) was added during the second digestion period of the isolation procedure (1). Washed cells were resuspended in the medium previously described. Samples of cells (0.5 ml) were distributed in polystyrene tubes containing the appropriate agents. After 15 min at 37° incubations were terminated by addition of 10 ml of ice-cold Krebs-Ringer bicarbonate buffer containing 5 mM theophylline. After vigorous vortexing and rapid centrifugation at 1000 x g for 2 min at 4°, the supernatant was discarded and the cell pellet was treated with 0.5 ml of 5% perchloric acid. [^{3}H]cyclic AMP

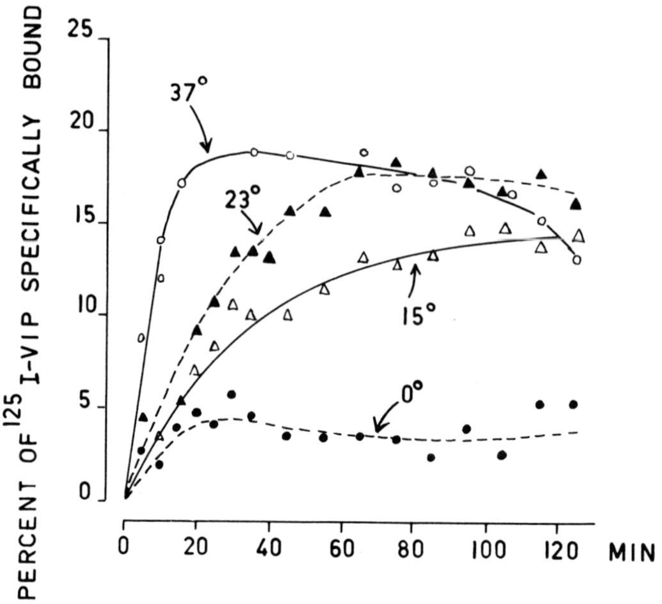

Fig. 2. Time course of specific binding of ^{125}I-VIP to pancreatic acinar cells as a function of temperature. Cells (3.8 x 10^6 per ml) were incubated with 2.6 x 10^{-11} M ^{125}I-VIP. The standard medium composition and the cell washing procedure are described in the text. The temperature of the incubation medium was 37° (O——O), 23° (▲——▲), 15° (Δ——Δ) or 0° (● — — ●). At various times duplicate aliquots (100 µl) were withdrawn. Each experimental point was corrected for nonspecific binding observed in simultaneous incubations conducted in the presence of 4 x 10^{-7} M unlabeled VIP.

(5000 cpm) was added in order to monitor recovery. After more vortexing and 5 min standing over ice the mixture was neutralized with 0.5 ml 0.8 M Tris. Each tube was centrifuged at 1000 x g for 5 min and cyclic AMP was isolated from the supernatant using the procedure of Salomon et al. (20).

RESULTS AND DISCUSSION

Specific Binding of ^{125}I-VIP. The binding of ^{125}I-VIP to pancreatic acinar cells was a time, temperature and pH-dependent process. At 37° and at tracer concentration, specific binding was moderately rapid, 50% of maximum binding occurring within the first 7 min of incubation. An apparent equilibrium of binding was achieved in 10-20 min. Non-specific binding at that time was only about 15%

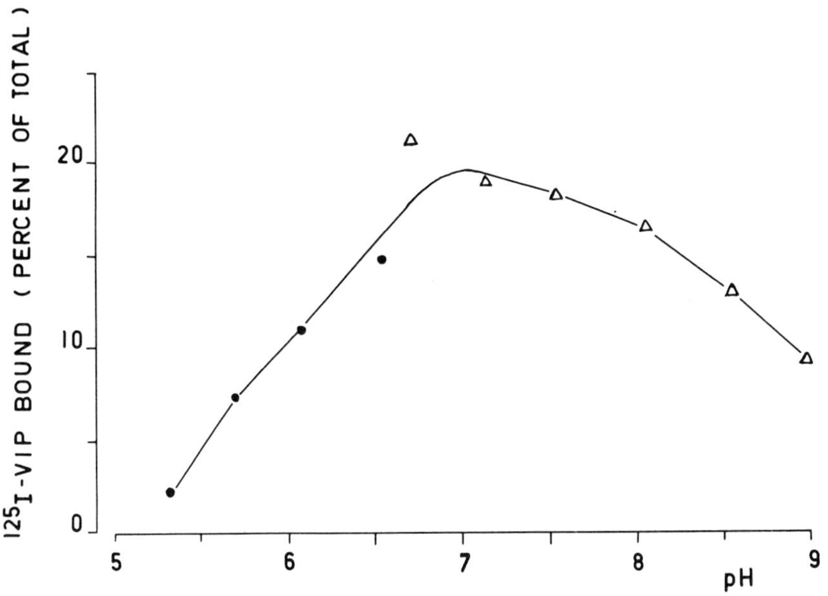

Fig. 3. Effect of pH on specific binding of ^{125}I-VIP to pancreatic acinar cells. Cells (3.8×10^7 per ml) were incubated at 37° with 2.6×10^{-11} M ^{125}I-VIP for 10 min a buffer system made of one portion of Krebs-Ringer bicarbonate (in which a concentrated cell suspension had been prepared) and four portions of 0.154 M of two N-substituted taurines. The first one was MES (● : 2-1^{N}-morpholinoethanesulfonic acid) with a pK_a of 6.5 and the other one was HEPES (Δ : N-2-hydroxyethylpiperazine-N'-2-ethanesulfonic acid) with a pK_a of 7.55. The pH of the organic buffers was adjusted with HCl or NaOH and the final pH of the mixture was measured. Data were corrected for unspecific binding as shown in the legend of Fig. 2.

of the total radioactive VIP bound (data not shown). Reducing the incubation temperature to 23° and 15° produced a progressive slowing of the binding reaction (Fig. 2). The specific binding at steady-state occurred over a relatively large range of pH. It was optimal between pH 6.7 and 7.6. The declines above pH 8.0 and below pH 6.5 (Fig. 3) may reflect increased dissociation (vide infra).

The presence of EDTA resulted in a significant 30% increase in the percentage of ^{125}I-VIP bound to acinar cells for 10 min at 37° (Table I). On the other hand, a concentration of 20 mM $CaCl_2$ and $MgCl_2$ decreased specific binding by approximately 50%. Since the dissociation of ^{125}I-VIP determined from semilogarithmic plots (Fig. 4) was not modified by EDTA or EGTA (data not shown), it may

TABLE I

Effects of EDTA, EGTA, calcium, magnesium, ionic strength and detergents on the specific binding of ^{125}I-VIP to pancreatic acinar cells. Cells (1.5 to 2.0 x 10^7 per ml) were incubated at 37° with 2.6 x 10^{-11} M ^{125}I-VIP for 10 min. Results are given in percentage of control values observed with the standard Krebs-Ringer bicarbonate buffer containing 0.5 mM Ca^{3+} and 1.2 mM Mg^{2+} (Expts 1 and 3), or with a 0.154 M buffer (pH 7.4) made of one portion of the standard buffer and 4 portions of 0.154 HEPES buffer (Expt. 2). Each condition was tested once except for 5 mM EDTA when the experiment was repeated 7 times (percent ± SEM). Experimental values were corrected for unspecific binding observed in simultaneous incubations conducted in the presence of 1 x 10^{-7} M unlabeled VIP. All determinations were made in duplicate.

Experiment 1	Molarity of chelator or divalent cation	2.5 M	5M	10 M	15 M	20 M
	EDTA		130 ± 3			
	EGTA		119			
	Ca^{2+}	100	81	75	60	69
	Mg^{2+}	96	77	72	80	56
Experiment 2	Total ionic strength*:	0.154 M	0.123 M	0.093 M	0.062 M	0.031 M
		100	90	59	49	26
Experiment 3	Detergent† (in % w/v):			0.010%		0.025%
	Digitonin			9		16
	Triton X-100			55		68
	Sodium dodecylsulfate			84		100

*The lowest ionic strength (0.031 M) was obtained by a 5-fold dilution of the usual Krebs-Ringer bicarbonate buffer with distilled water. Increasing ionic strength was achieved by diluting the same buffer with increasing proportions of 0.154 M HEPES buffer (pH 7.4) at the expense of distilled water.

†Detergents were added 5 min before the addition of ^{125}I-VIP.

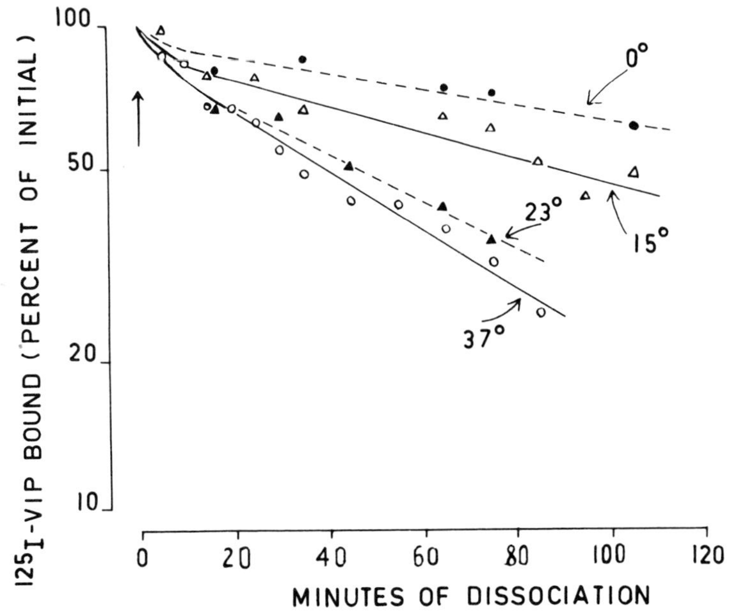

Fig. 4. Effect of temperature on time course of dissociation of specifically bound ^{125}I-VIP from pacreatic acinar cells. ^{125}I-VIP (2.6×10^{-11} M) was first incubated with acinar cells (7.6×10^6 cells/ml) at 37°. After 10 min the temperature was adjusted at four levels (37° O——O; 23° ▲- - -▲; 15° Δ——Δ ; and 0° ●- - -●) and natural VIP was added (↑) to give a final concentration of 5×10^{-8} M. Data were corrected for unspecific binding determined in parallel in tubes where unlabeled VIP (1×10^{-7} M) had been added during the 10 min preincubation period. Results are expressed as a percentage of ^{125}I-VIP specifically bound to cells at zero time (the moment of addition of unlabeled VIP).

well be that the effect of these cationic chelators was due to the unmasking of new VIP binding sites resulting from a change in plasma membrane configuration and/or from a better peeling of basement membrane remnants (1).

The hydrophobic nature of the binding process is suggested by the finding that ^{125}I-VIP binding was markedly reduced by moderate concentrations of surface active agents (Table I). The most effective one, on a weight basis, was digitonin, followed by triton X-100 and sodium dodecylsulfate. The hydrophobic C-terminal moiety of VIP (Fig. 1) may explain why the binding component of the system behaved as a lipoprotein in the presence of detergents. However other explanations for these effects are conceivable such as a perturbation of membrane lipids or a partial solubilization of the receptors.

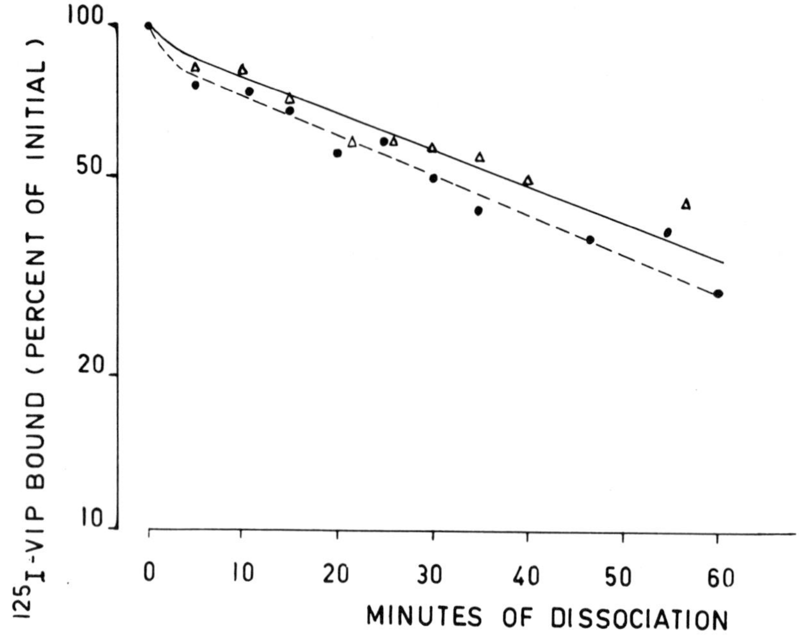

Fig. 5. Comparison of the effects of unlabeled VIP addition and acinar cell dilution on the dissociation of specifically bound ^{125}I-VIP. ^{125}I-VIP (2.6×10^{-11} M) was incubated with acinar cells (4.0×10^{7} cells/ml) at 37°. Data were corrected for unspecific binding determined in parallel in tubes where unlabeled VIP (1×10^{-7} M) had been added. After 10 min the dissociation reaction was initiated in one tube by adding 1×10^{-7} M unlabeled VIP and the procedure described in Fig. 4 was followed. In another tube cells were centrifuged after the 10 min preincubation period. The pellet was resuspended in a similar volume of peptide-free medium. One hundred µl aliquots of this cell suspension were further diluted into 10 ml fresh peptide-free medium and tested at various time intervals. The initial ^{125}I-VIP concentration was diluted more than 5000-fold by this procedure. The amount of bound ^{125}I-VIP was measured after centrifuging the 10 ml suspension at 1000 x g for 2 min. Results are expressed as a percentage of ^{125}I-VIP specifically bound to cells at zero time (the moment of addition of unlabeled VIP (●- - -●) or final cell dilution (Δ——Δ)).

Dissociation of Bound ^{125}I-VIP. The VIP-cell complex dissociated by a process which was also time, temperature and pH-dependent. Fig. 4 illustrates the reversible binding of ^{125}I-VIP when the concentration of the labeled peptide was diluted 2×10^{3}-fold by the addition of unlabeled VIP after 10 min preincubation. ^{125}I-VIP was released more rapidly during the next 10-15 min and this was followed by the first order kinetics expected for simple dissociation of VIP from high affinity binding sites. At 37°, 50% of the

TABLE II

Equilibrium constants, kinetic constants, and binding capacities for the interaction of ^{125}I-VIP with pancreatic acinar cells at 37°.*

Affinity (association) constant:	$K_a = 2.36 \pm 0.29 \times 10^9\ M^{-1}$	(n = 13)
Dissociation constant:	$K_d = 4.24 \pm 0.53 \times 10^{-10}\ M$	(n = 13)
Rate of dissociation:	$k_{-1} = 2.88 \pm 0.33 \times 10^{-4} sec^{-1}$	(n = 7)
Rate of association (calculated):	$k_1 = 6.80 \times 10^5\ M^{-1}\ sec^{-1}$	
Binding sites/cell:	8,900 ± 1,500	(n = 13)
Low affinity sites		
Affinity (association) constant:	$K_a = 1.00 \pm 0.54 \times 10^7\ M^{-1}$	(n = 8)
Dissociation constant:	$K_d = 1.00 \pm 0.54 \times 10^{-7}\ M$	(n = 8)
Binding sites/cell:	53,000 ± 38,000	(n = 8)

*The values indicated are the means ± SEM (number of experiments in parenthesis). The integrated denaturation of ^{125}I-VIP was estimated to be 20% of the original concentration after 10 min incubation in order to correct the bound/free ratio of ^{125}I-VIP in Scatchard plots. The rate of dissociation k_{-1} was determined from semilogarithmic plots of the dissociation of ^{125}I-VIP bound to acinar cells as a function of time (Fig. 4). The rate of association k_1 was calculated from the product $K_a \times k_{-1}$.

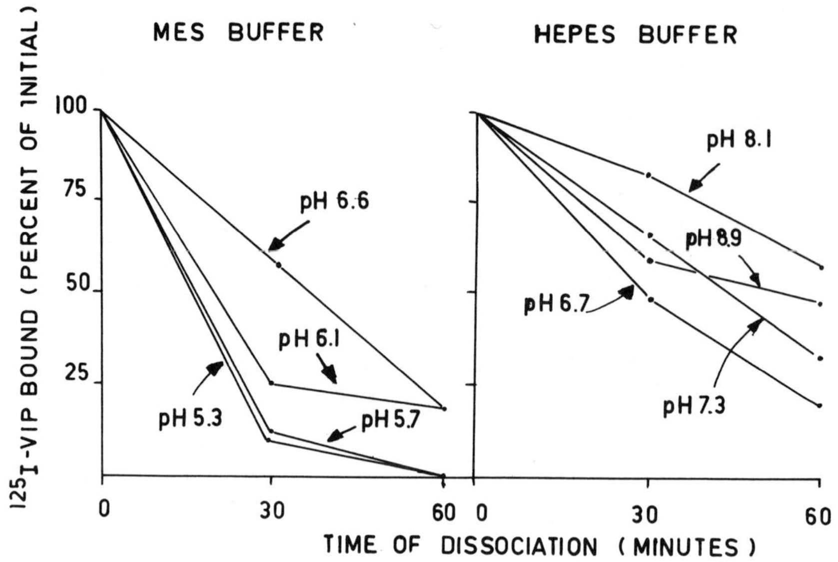

Fig. 6. Effect of pH on time course of dissociation of specifically bound ^{125}I-VIP from pancreatic acinar cells. Cells (7.6×10^6/ml) were preincubated with ^{125}I-VIP (2.6×10^{-11} M) at 37° in Krebs-Ringer bicarbonate buffer. After 10 min the cells were centrifuged and resuspended in 0.154 M MES buffer (to explore the effects of pH 5.3 to 6.6: left panel) or 0.154 HEPES buffer (to examine the effects of pH 6.7 to 8.9 : right panel) in the presence of 1×10^{-7} M VIP. Data were corrected for unspecific binding determined in parallel in tubes where unlabeled VIP (1×10^{-7} M) had been added during the preincubation period. Results are expressed as a percentage of ^{125}I-VIP specifically bound at zero time (the moment of pH adjustment).

VIP originally bound to acinar cells was released into the medium in 40 ± 5 min (mean ± SEM of 7 experiments: cf Table II). Fig. 5 indicates that similar results were obtained when the dissociation reaction was initiated by adding a 1×10^5-fold molar excess of non-radioactive VIP or by reducing the free ^{125}I-VIP concentration by the rapid washing and resuspension of labeled cells in a fresh peptide-free medium.

The rate of dissociation of specifically bound ^{125}I-VIP was influenced by the pH, being minimal at pH 8.1, increasing slightly at pH 8.9 and sharply below pH 6.1 (Fig. 6). Decreasing temperature reduced the rate of dissociation of ^{125}I-VIP (Fig. 4). It is possible that temperature dependence of dissociation and of association of the VIP-cell complex reflect alterations in the fluidity of the plasma membrane as well as in the structure of binding sites.

We feel that Arrhenius plots of equilibrium and kinetic constants would be meaningless in our case considering the difficulty in attaining true steady-state levels at each considered temperature (Fig. 2). However the slower binding observed when lowering temperature (Fig. 2) suggests that the decrease in the rate of association k_1 was disproportionately greater than the decrease in the rate of dissociation (Fig. 4) and the decline of degradation of unbound ^{125}I-VIP (Fig. 7; *vide infra*).

The Fate of ^{125}I-VIP Incubated with Acinar Cells. Two reactions were observed: one, already discussed, involved binding VIP by forces partially reversible with acidic treatment; the other reaction involved a relatively rapid degradation of VIP, the inactivated hormone being unable to bind to specific sites (Fig. 7). From 5 experiments repeated under similar conditions it can be extrapolated that 24 ± 4% (mean ± SEM) of the ^{125}I-VIP originally TCA-precipitable lost this property after 10 min incubation at 37° when exposed to a concentration of 1.7 x 10^7 cells/ml. Large concentrations of unlabeled VIP, bacitracin, and Trasylol (the kallikrein inhibitor) were unable to protect ^{125}I-VIP from degradation. On the other hand, increasing the concentration of bovine serum albumin from 0.1 to 4.0% progressively reduced the relative rate of degradation from 31% to 17% after 30 min (data not shown).

The rate of degradation as illustrated by TCA-precipitability was also markedly temperature-dependent, decreasing by 38% after 20 min when the temperature in the incubation medium was reduced to 15° (Fig. 7).

The biological characterization of free ^{125}I-VIP was tested in parallel by studying the ability of this peptide to bind to fresh acinar cells (Fig. 7, lower panel). The degradation of free ^{125}I-VIP tested by this criterion was also temperature-dependent. In addition, Fig. 7 indicates that bound ^{125}I-VIP was more stable in its capacity to remain bound than unused free ^{125}I-VIP in its binding ability. It appears therefore that the inactivation of VIP is independent of binding and that the rapid change in VIP concentration incurred by the inactivation process may, in part, be responsible for the fall in the rate of binding (Fig. 2).

Analysis of Specific Binding at Steady-State, as a Function of Unlabeled VIP Concentration. Unlabeled VIP readily inhibited the specific binding of ^{125}I-VIP observed after 10 min at 37° (Fig. 8). This period of time was utilized as a good compromise since it corresponded to the beginning of a plateau of binding (Fig. 2) and since degradation of free ^{125}I-VIP in the medium was still moderate (Fig. 7).

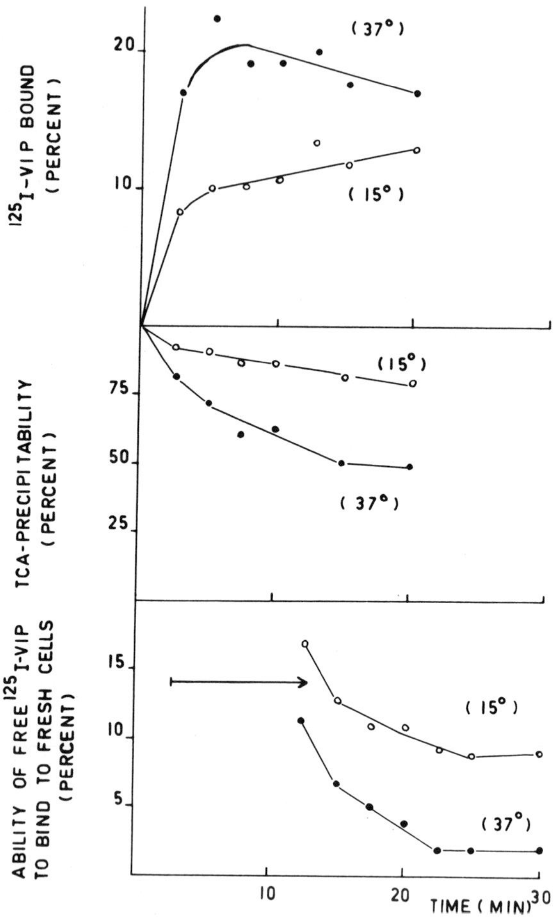

Fig. 7. Effect of temperature on the time course of degradation of ^{125}I-VIP with respect to TCA-precipitability and binding ability. Pancreatic acinar cells (8.3 x 10^6 cells/ml) were incubated with ^{125}I-VIP (2.6 x 10^{-11} M). Incubations were at 37° (●——●) or 15° (O——O) for the indicated times. At seven time intervals specific binding of ^{125}I-VIP was estimated as described in the text (data were corrected for unspecific binding as shown in the legend of Fig. 2). In addition, 200 µl aliquots from each incubation mixture were transferred to microfuge tubes and centrifuged for 15 seconds at 10,000 x g. Duplicate 20 µl aliquots from each supernatant were tested for TCA-precipitability (middle panel) while a larger aliquot (100 µl) was added to 200 µl of fresh acinar cells to give a final cell concentration of 5.5 x 10^{-6} ml. After incubation for 8 min, bound ^{125}I-VIP was assayed (lower panel) as previously described. The horizontal arrow in the panel indicates the time required between samplings and the corresponding estimation of binding ability.

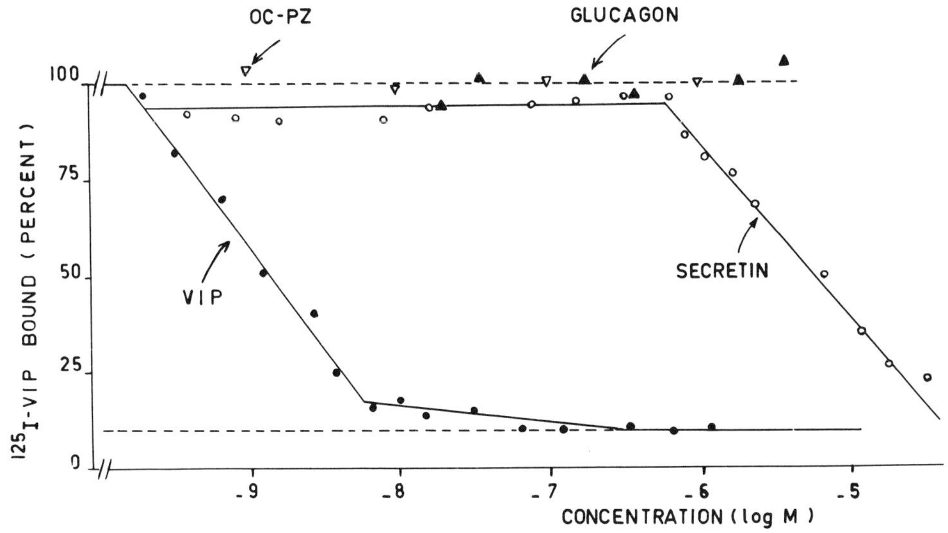

Fig. 8. Displacement of ^{125}I-VIP as a function of the concentration of unlabeled VIP (●———●), synthetic secretin (O———O), glucagon (Δ———Δ) and C-terminal octapeptide of pancreozymin (OC-PZ: ∇- - -∇). Pancreatic acinar cells (1.8 x 10^7 cells/ml) were incubated for 10 min at 37° with 2.6 x 10^{-11} M ^{125}I-VIP and increasing concentrations of unlabeled peptide in the assay buffer as described in the text. The total radioactivity bound to cells (as a percentage of maximal binding) is plotted as a function of unlabeled peptide concentration. Each point is the mean for 2 or 3 experiments. All determinations were made in duplicate.

The concentration of VIP at which half-maximal binding occurred was about 1 x 10^{-9} M. In addition the broken line suggests the presence of more than one class of binding sites. Scatchard plots for specific binding of ^{125}I-VIP were drawn on several occasions. The bound/free ratio of ^{125}I-VIP was plotted as a function of the VIP bound. The ratio was corrected for VIP inactivation (20% in 10 min) and nonspecific binding was subtracted. These plots were curvilinear with a concavity upward (Fig. 9). From these broken plots, two best-fitting lines could be drawn. This quantitative analysis (Table II) suggests the existence of at least two types of binding sites: one of high affinity with an apparent dissociation constant for the binding process of 4 x 10^{-10} M and another one of lower affinity (apparent K_d of 1 x 10^{-7} M). The corresponding maximal number of binding sites per cell were 8,900 and 53,000, respectively, indicating a relatively lower capacity for the high affinity binding site. Thus, there are good indications that pancreatic

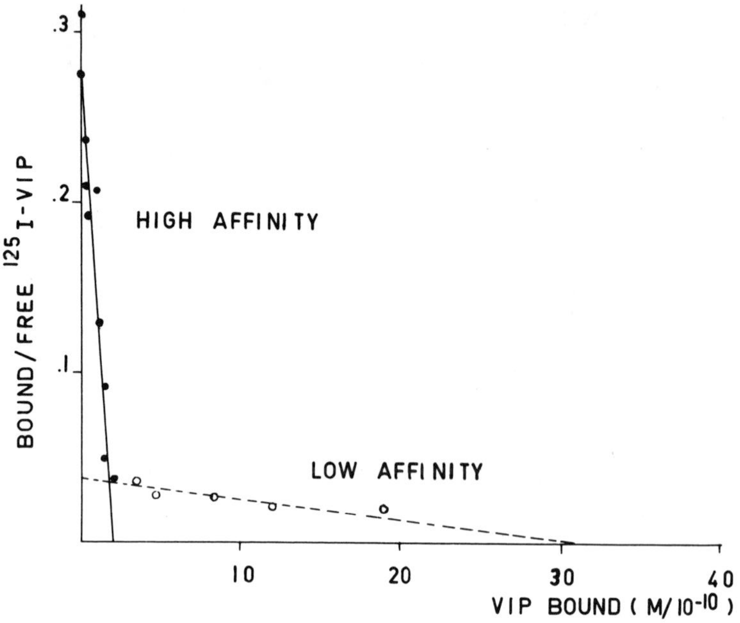

Fig. 9. A representative Scatchard analysis for specific binding of ^{125}I-VIP to pancreatic acinar cells. The bound/free ratio of ^{125}I-VIP was plotted as a function of the VIP bound. Non-specific binding was subtracted. From the broken plot two best-fitting lines corresponding to apparent high (●———●) and low (O- - -O) affinity could be drawn.

acinar cells possess two functionally distinct classes of VIP binding sites. However in view of recent evidence with insulin binding (5), it is recognized that Scatchard plots alone do not establish the existence of two populations of sites for VIP binding and that a negatively cooperative model involving site: site interaction might also explain the curvilinearity observed. Evidence favoring the first hypothesis follows.

Characteristics of High-Affinity VIP Binding Sites. The chemical specificity required for VIP binding to high affinity binding sites was explored in conjunction with the corresponding ability to stimulate the catalytic unit of the adenylate cyclase system.

VIP, glucagon and secretin are similar in their chemical structure (Fig. 1) and in their activation of a number of adenylate cyclase systems. However, glucagon, at concentrations as high as 3×10^{-5} M, failed to inhibit binding of ^{125}I-VIP (Fig. 8). Glucagon is also neither an agonist nor an antagonist of VIP and secretin on the adenylate cyclase activity of rat pancreatic fragments (8) and

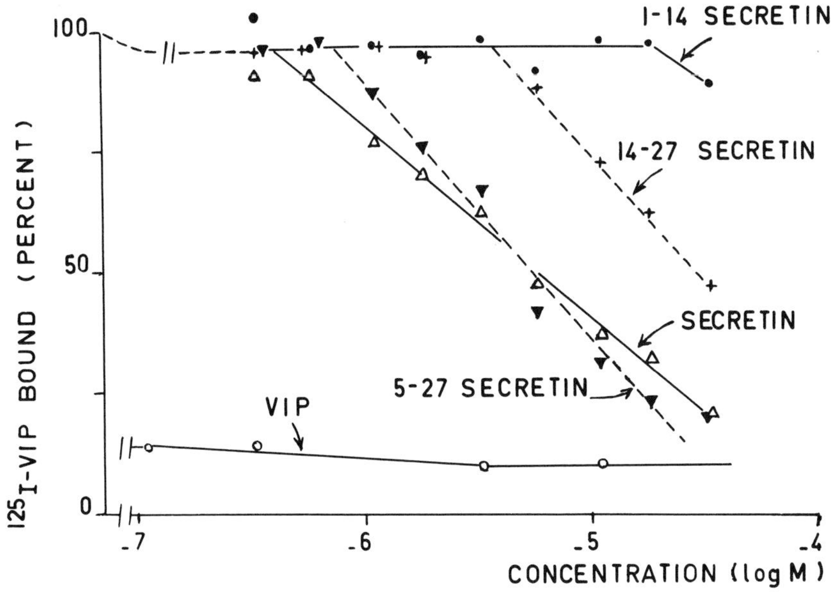

Fig. 10. Displacement of ^{125}I-VIP as a function of the concentration of unlabeled VIP, secretin, and secretin fragments. Conditions were comparable to those in Fig. 8. VIP, O———O; secretin, Δ———Δ; secretin fragment [1-14], ●———●; secretin fragment [14-17] +- - -+; and secretin fragment [5-27], ▼- - -▼. Each point is the mean for 2 experiments. All determinations were made in duplicate.

of acinar cells and acinar cells extracts from the guinea pig pancreas.[1] Even in tissues responsive to glucagon, such as the liver and adipose tissue, there is strong evidence that binding sites for glucagon are distinct from binding sites for VIP and secretin (6, 9, 14).

The concentration of synthetic secretin giving half-maximal displacement of the ^{125}I-VIP-acinar cell complex was 5 x 10^{-6} M as compared to 1 x 10^{-9} M VIP on the same experimental day (Fig. 8). Thus secretin was capable of inhibiting binding of ^{125}I-VIP to pancreatic acinar cells. However high-affinity VIP binding sites had a 4000-fold lower affinity for secretin and are therefore poor candidates for the physiological effects of secretin.

No VIP fragments were available at the time of our study but we could test three fragments of synthetic secretin for structure-function relationship in the binding of ^{125}I-VIP and the activation of adenylate cyclase. Secretin 5-27 was as potent as native secretin in its ability to bind to VIP receptors on the basis of competitive inhibition (Fig. 10). By itself, this peptide fragment had no

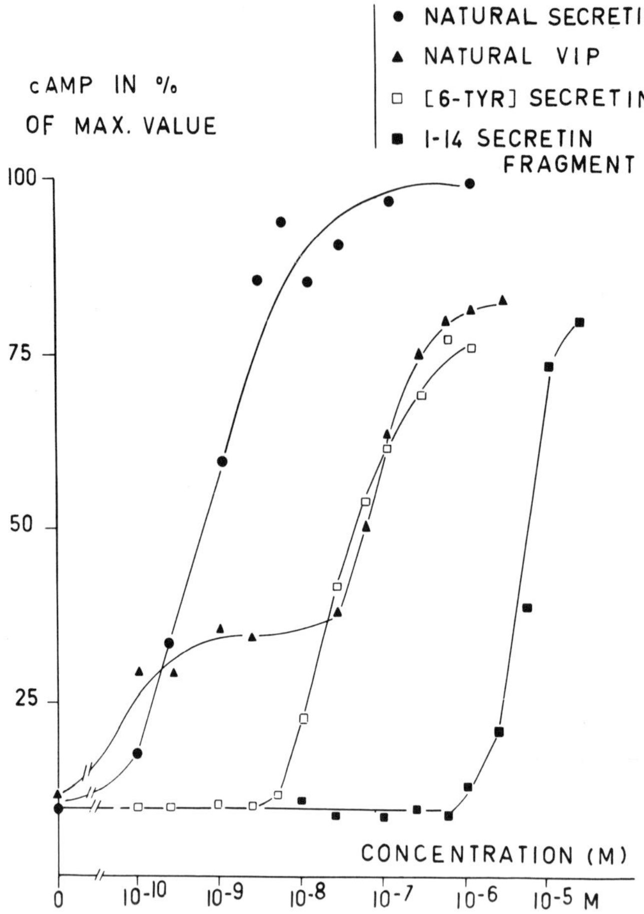

Fig. 11. Accumulation of $[^{14}C]$ cyclic AMP in isolated acinar cells from the guinea pig pancreas as a function of peptide concentration. Cellular cyclic AMP was determined at the end of a 15 min incubation period at 37°. Results are expressed in % of the maximal value obtained with 10^{-6} M secretin. Each curve is the mean of at least 2 experiments made in duplicate.

effect on adenylate cyclase activity (data not shown). These results suggest that the N-terminal [1-4] portion of the present family of hormones is important for adenylate cyclase activation but not for binding. The N-terminal portion of secretin, [1-14], failed to inhibit ^{125}I-VIP binding (Fig. 10) and was 10^4 times less potent than secretin in stimulating adenylate cyclase (Fig. 11). The fact that activation was possible suggests that the N-terminal [1-14] moiety, in addition to provoking chemical excitation, contains an amino acid sequence allowing minimal binding. The comparison of the primary structure of VIP with secretin (Fig. 1) indicates that eight identical residues are among the first 14 and one among the remaining residues and it is clear that most of these amino acids must play an important role. Only the substitution of phenylalanine by tyrosine in the sixth position of secretin, i.e. the use of [6-Tyr] secretin for example, decreased the efficiency of the peptide 100 times (Fig. 11).

The C-terminal moiety of secretin [14-27] was relatively efficient in inhibiting the binding of ^{125}I-VIP (Fig. 10) while being inefficient on adenylate cyclase activity (data not shown). This suggests that the C-terminal region functions to enhance the binding capacity. Five to six hydrophobic amino acids are clustered in the C-terminal moiety of VIP and secretin. It is tempting to consider that this region contributes to the binding of VIP and secretin through hydrophobic forces.

These data demonstrate the existence of 2 receptors, one having a high affinity for VIP and a low affinity for secretin while the second one has a low affinity for VIP but is a good candidate for binding secretin with high affinity (vide infra). It is therefore tempting to assign their relative specificity to their capacity to distinguish the C-terminal portion of each peptide. If we neglect the 9 similarities in addition to the 9 identities of VIP and secretin, the major differences are 1) the presence of an extra hydrophilic residue at position 28 and 2) variations in charge distribution, i.e. when VIP is compared to secretin, there is a negative charge at 8 rather than at 9, a positive charge at 20 rather than at 18, and perhaps more conspicuously a positive charge instead of a negative one at 15 (Fig. 1).

Characteristics of Low-Affinity VIP Binding Sites. We suggest that there exists a second population of VIP binding sites having a low affinity for VIP and a high affinity for secretin on the basis of five lines of evidence:

1) The curvilinearity of Scatchard plots (Fig. 9) was observed systematically. The shallow second slope was 42-fold lower than the first one. Fifty % of the low affinity binding sites were occupied at a VIP concentration of 1×10^{-7} M (Table II) and the dose-effect curve for low affinity binding corresponds to the second part of the dose-effect curve describing adenylate cyclase activation in response to VIP (Fig. 11).

The number of low affinity binding sites may be greater than that estimated for the other class of sites (Table II). However, precise measurement was difficult due to the interference of nonspecific binding (Figs. 8 and 9).

2) A constant rate of release of ^{125}I-VIP bound to acinar cells was observed during the major part of the time-course of dissociation, not only after cell dilution but also after adding a large excess of nonradioactive VIP (Fig. 5). This observation precludes negative cooperativity among VIP receptors (5) and indicates that the shallow slope in Scatchard plots (Fig. 9) was indeed characteristic of a second, independent class of binding sites. The affinity of these sites for VIP (Table II) was probably responsible for the abnormally rapid release of bound VIP observed during the first 15 min of the dissociation process (Figs. 4 and 5).

3) There is reason to believe that low-affinity binding sites may in fact be high-affinity secretin binding sites. Indeed, concentrations of secretin as low as 4 x 10^{-10} M were able to reduce moderately (by 6%) the total specific binding of ^{125}I-VIP (Fig. 8). In addition, a concentration of secretin too low to release ^{125}I-VIP from high-affinity binding sites could reduce ^{125}I-VIP remaining bound in the presence of a concentration of unlabeled VIP which still allows binding to low affinity receptors (data not shown). In other terms, low-affinity VIP-binding sites appear to have an affinity for secretin higher than that for VIP.

4) The activity of secretin to interact with pancreatic acinar cells was studied directly by using ^{125}I-synthetic secretin (Table III). The binding of ^{125}I-secretin was competitively inhibited by synthetic and natural secretin. The proportion of unspecific ^{125}I-secretin binding represented 30% of the total as compared to 14% for ^{125}I-VIP. This was probably due to the lower quality of the ^{125}I-secretin preparation. However, one important point seems valid: a 1 x 10^{-8} M unlabeled secretin concentration was high enough to inhibit 50% of ^{125}I-secretin binding. The concentration of unlabeled VIP inducing the same extent of binding inhibition was 1000-fold higher.

5) The second part of the dose-response curve of VIP-stimulated adenylate cyclase (vide infra).

VIP Binding and Adenylate Cyclase Activation. In intestinal mucosa (21), liver (6) and fat cells (6, 9) VIP activation of adenylate cyclase activity has been demonstrated. In isolated pancreatic acinar cells the concentration dependence of VIP binding was closely similar to that of hormone-stimulated adenylate cyclase. Indeed, the dose-response curve of VIP-stimulated adenylate cyclase (Fig.11) indicates the presence of two functionally distinct classes of receptors having affinities similar to those illustrated for VIP binding. The first part of the dose-effect curve for high affinity VIP-binding corresponds to the first part of the dose-effect curve describing adenylate cyclase activity.

The stimulation by secretin of adenylate cyclase activity, previously reported in the intact pancreas (11, 18) was detected in

TABLE III

Comparative binding of ^{125}I-secretin and ^{125}I-VIP on the same pancreatic acinar cell preparation, in the absence and in the presence of unlabeled peptides.*

Radio-iodinated peptide	Addition of unlabeled peptides						
	none	natural secretin		synthetic secretin		natural VIP	
		10^{-5} M	10^{-8} M	10^{-5} M	10^{-8} M	10^{-5} M	10^{-8} M
^{125}I-synthetic secretin	8.41 (100)	2.33 (28)	3.67 (44)	2.65 (32)	5.29 (62)	4.10 (49)	7.64 (91)
^{125}I-natural VIP	21.4 (100)	4.6 (21)	20.6 (96)	9.7 (45)	20.9 (98)	3.0 (14)	5.1 (24)

*The natural hormones utilized were provided by Dr. Mutt and the synthetic product was a gift of Dr. Wünsch. Binding was estimated after 10 min incubation at 37° and expressed as percent of total radioactivity offered, and percent of the radioactivity bound in the absence of unlabeled hormones (in parentheses). The specific activity of secretin was 66 μCi/μg and that of VIP 246 μCi/μg. The final radioactivity/ml incubation medium was reduced four-fold with secretin in order to compare similar concentrations (2-3 x 10^{-11} M) of both peptides.

isolated acinar cells from guinea pig pancreas at concentrations as low as 10^{-10} M (Fig. 11). The fact that the stimulation of cellular cyclic AMP by secretin and VIP is not additive (8)[2] suggests that both of these peptides activate the same adenylate cyclase. This evidence indicates that a careful distinction must be made between the activation of a common catalytic subunit and the sharing of identical receptors. A common site of action for secretin and VIP is also advocated in both liver and fat cell plasma membranes (2, 6).

There was no linear relationship between the time course of receptor occupation by VIP and enzyme activity: maximal binding was reached after 10-20 min only (Fig. 2) and in contrast the activity of adenylate cyclase was maximal within 3 min (data not shown). Under similar circumstances, with glucagon and the plasma membrane of rat liver, the concept of additional binding sites over and above those required for stimulation of the enzyme has been advocated (3) and refuted (16). A selective damage of the catalytic unit during the isolation of pancreatic acinar cells might disrupt the coupling existing *in vivo*. More probably, allosteric transitions were involved as indicated by the interactions between hormones and ethanol on the adenylate cyclase activity in intact acinar cells, and between hormones, ethanol and guanylylimidodiphosphate on the same enzyme in acinar cell extracts.[3]

In conclusion, in isolated guinea pig pancreatic acinar cells, our results suggest that VIP (vasoactive intestinal polypeptide) and secretin share two distinct receptor binding sites with overlapping affinity. The first receptor class shows high VIP affinity and low secretin affinity and low secretin affinity, whereas the second class shows opposite affinities. In addition, the binding of both hormones does not require the N-terminal [1-4] portion but activation of adenylate cyclase only occurs in the presence of this extremity.

ACKNOWLEDGEMENTS

We thank Dr. V. Mutt for supplying pure natural VIP and secretin, Dr. E. Wünsch for synthetic secretin, and Dr. M. Ondetti for synthetic secretin fragments and the C-terminal octapeptide of pancreozymin. The helpful suggestions and stimulating interest of Dr. P. DeMeyts are gratefully acknowledged.

---

[2] Robberecht, P. *et al*, submitted for publication.

[3] Robberecht, P., Olinger, E., and Gardner, J., manuscript in preparation.

REFERENCES

1. AMSTERDAM, A. and JAMIESON, J. D., J. Cell Biol. 63 (1974) 1037, 1057.
2. BATAILLE, D., FREYCHET, P. and ROSSELIN, G., Endocrinology 95 (1974) 713.
3. BIRNBAUMER, L. and POHL, S. L., J. Biol. Chem. 248 (1973) 2056.
4. BODANSZKY, M., KLAUSNER, Y. S. and SAID, S. I., Proc. Nat. Acad. Sci. USA 70 (1973) 382.
5. DE MEYTS, P., ROTH, J., NEVILLE, D. M., GAVIN, J. R. and LESNIAK, M. A., Biochem. Biophys. Res. Commun. 55 (1973) 154.
6. DESBUQUOIS, B., LAUDAT, M. H. and LAUDAT, P., Biochem. Biophys. Res. Commun. 53 (1973) 1187.
7. DESBUQUOIS, B., Eur. J. Biochem, 46 (1974) 439.
8. DESCHODT-LANCKMAN, M., ROBBERECHT, P., DE NEEF, P., LABRIE, F. and CHRISTOPHE, J., Gastroenterology 68 (1975) 318.
9. FRANDSEN, E. K. and MOODY, A. J., Horm. Metab. Res. 5 (1973) 196.
10. HUNTER, W. M., and GREENWOOD, F. C., Nature 194 (1972) 495.
11. MAROIS, C., NORISSET, J. and DUNNIGAN, J., Rev. Can. Biol. 31 (1972) 253.
12. MUTT, V. and SAID, S. I., Eur. J. Biochem. 42 (1974) 581.
13. ROBBERECHT, P., DESCHODT-LANCKMAN, M., DE NEEF, P., BORGEAT, P. and CHRISTOPHE, J., FEBS Letters 43 (1974) 139.
14. RODBELL, M., BIRNBAUMER, L. and POHL, S. L., J. Biol. Chem. 245 (1970) 718.
15. RODBELL, M., KRANS, H. M. J., POHL, S. L. and BIRNBAUMER, L., J. Biol. Chem. 246 (1971) 1861.
16. RODBELL, M., LIN, M. C. and SALOMON, Y., J. Biol. Chem. 249 (1974) 59.
17. ROTH, J., Metabolism 22 (1973) 1059.
18. RUTTEN, W. J., DE PONT, J. J. H. H. M. and BONTING, S. L., Biochim. Biophys. Acta 274 (1972) 201.
19. SAID, S. I. and MUTT, V., Eur. J. Biochem. 28 (1972) 199.
20. SALOMON, Y., LONDOS, C. and RODBELL, M., Anal. Biochem. 58 (1974) 541.
21. SCHWARTZ, C. J., KIMBERG, D. V., SHEERIN, H. E., FIELD, M. and SAID, S. I. (1975) in press.

α-MELANOTROPIN RECEPTORS: NON-IDENTICAL HORMONAL MESSAGE SEQUENCES (ACTIVE SITES) TRIGGERING RECEPTORS IN MELANOCYTES, ADIPOCTYTES AND CNS CELLS*

Alex EBERLE and Robert SCHWYZER

Institute of Molecular Biology and Biophysics
Swiss Federal Institute of Technology (ETH)
CH-8049 Zürich (Switzerland)

α-Melanotropin (α-melanocyte-stimulating hormone, α-MSH), a pituitary tridecapeptide, exhibits strong effects on a variety of tissues. The best known is certainly the dramatic impact on the color change of cold-blooded vertebrates where α-MSH causes darkening of the skin by dispersion of the melanophore pigment granules. α-MSH seems also to control--at least partially--the pigmentation of mammals and men (35). Only a small number of extrapigmentary effects of α-melanotropin will be mentioned here as e.g. the lipolytic activity in rabbit adipose tissue (7, 45), the effects on learning and the central nervous system (11, 13, 46, 47, 58), and the activity on sebum secretion in the rat (49, 60).

α-Melanotropin is structurally related to β-melanotropin and adrenocorticotropin (Fig. 1). They belong to the class of polypeptide hormones with sychnological organization (56). This means that discrete sequences of adjacent amino acids ("continuate words") are responsible for different components of the biological activity (51-54). The common biological activity of the mammalian peptides on melanocytes is claimed to be due to the common heptapeptide sequence, -Met-Glu-His-Phe-Arg-Trp-Gly- (21). The same sequence is

---

*Abbreviations and nomenclature according to E. Wünsch (63) (three-letter symbolism) and Margaret O. Dayhoff (10) (one-letter symbolism). The two systems correlate as follows: Ala = A, Arg = R, Asn = N, Asp = D, Cys = C, Gln = Q, Glu = E, Gly = G, His = H, Ile = I, Leu = L, Lys = K, Met = M, Phe = F, Pro = P, Ser = S, Thr = T, Trp = W, Tyr = Y, Val = V; CNS: central nervous system.

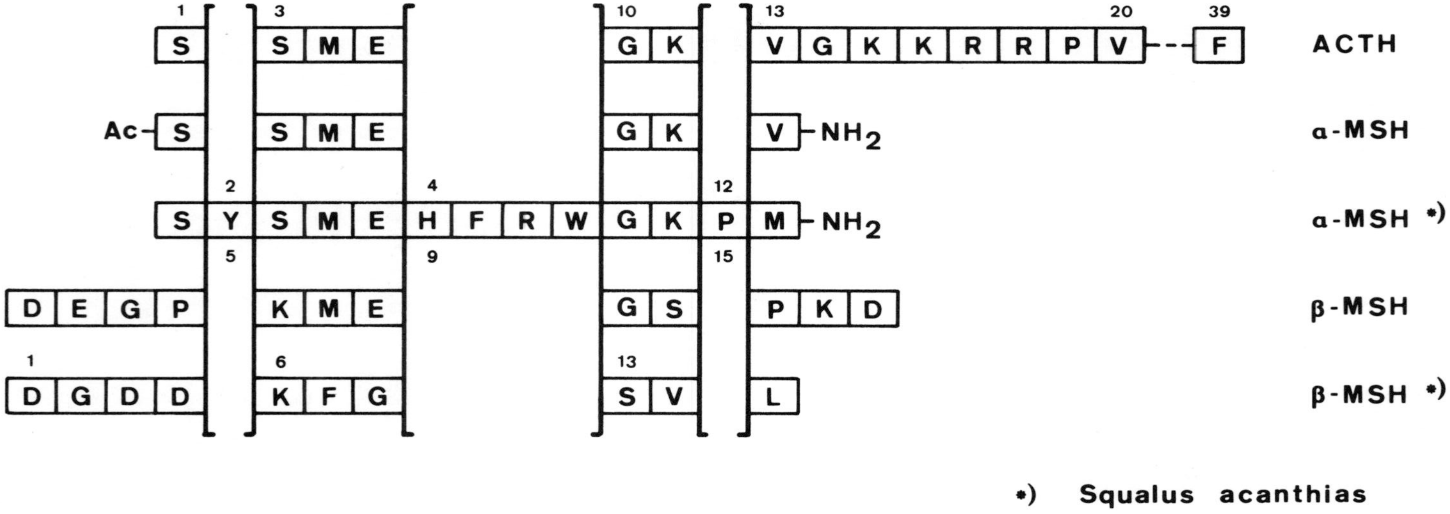

Fig. 1. Primary structures of melanotropic and corticotropic hormones: α-MSH, β-MSH (pig), ACTH, and non-mammalian α- and β-MSH of the dogfish *Squalus acanthias.*

also capable of eliciting the adipose lipolytic response (in the rabbit, but not in the rat where the C-terminal binding sequence of ACTH is indispensable) (8), and the adrenal steroidogenic response typical for ACTH (55), and furthermore of affecting behavior in the same way as MSH and ACTH (13). This sequence has therefore been called the hormonal "active site" (25) or "message sequence" (56).

Little is known about the α-melanotropin receptors of the different target cells. There is much evidence that the mode (or one of the modes) of action of this hormone results - as in the case of most other hormones--in the stimulation of adenylcyclase (1, 3, 4, 40), but the localization of the receptor itself remains to be carried out.

Since hormonal activity is dependent on the amino acid sequence and on the chemical topography of the molecule (53), one basic information about the molecular mechanism by which the hormone acts upon its receptor can be gained from the relationship between primary structure and biological activity, namely some indications of the "active site" topology of the hormone-receptor complex. Comparison of the minimal structural requirements for triggering the receptors of different target cells results in knowledge about the possible identity or non-identity of these different receptors.

Synthesis, Bioassay and Melanotropic Potencies of 32 Synthetic Peptides Related to α-MSH

Synthesis of peptides was carried out by a classical approach and is described elsewhere (17).

Bioassay: The melanotropic activity of peptides was determined *in vitro* using essentially the reflectometric assay of Shizume, Lerner and Fitzpatrick (57) as improved by Geschwind and Huseby (19) and modified to be statistically more pleasing (15). All assays were performed with specimens of *Rana pipiens*, 7.5 - 9 cm long, which had been kept under constant illumination for at least 48 hours preceding the experiment. The values of melanotropic activities are the means of the measurements of twelve different skins, four skins each, at three different concentrations.

Results: Table I lists the agonistic melanotropic potencies of peptides related to the heptapeptide sequence, -Met-Glu-His-Phe-Arg-Trp-Gly-, (I), and to the C-terminal tripeptide sequence of α-MSH, -Lys-Pro-Val-NH_2, (II), and of peptides covering both sequences (III). The values are expressed in units (U) per millimole: examples from the literature were included and repeated to test the consistency of our methods.

TABLE 1

Melanotropic activity of synthetic peptides related to the first message sequence of α-MSH, β-MSH, and ACTH (I), to the C-terminal second message sequence of α-MSH (II), and covering both message sequences (III)

		Melanotropic activity (in vitro) U/mmol	Ref.
(I)	Ac-Trp-Gly-OH	0	(16)
	H-Phe-Arg-Trp-Gly-OH	$6 \cdot 10^{2}$	(15)
	Ac-Phe-Arg-Trp-Gly-OH	$6 \cdot 10^{3}$	(15)
	H-Phe-Arg-Leu-Gly-OH	$3 \cdot 10^{3}$	(15)
	H-Pfp-Arg-Trp-Gly-OH	0	(16)
	H-Phe-Arg-OH	$6 \cdot 10^{2}$	(15)
	H-His-Phe-Arg-Trp-OH	$6 \cdot 10^{3}$	(41)
	H-His-Phe-Arg-Trp-Gly-OH	$2 \cdot 10^{4}$	(27,50)
	H-Glu-His-Phe-Arg-Trp-Gly-OH	$1//2 \cdot 10^{5}$	(15)//(36)
	Ac-Glu-His-Phe-Arg-Trp-Gly-OH	$4 \cdot 10^{5}$	(15)
	H-Gln-His-Phe-Arg-Trp-Gly-OH	$2 \cdot 10^{5}$	(32)
	H-Gly-His-Phe-Arg-Trp-Gly-OH	$2 \cdot 10^{5}$	(49)
	H-Met-Glu-His-Phe-Arg-Trp-Gly-OH	$1 \cdot 10^{6}$	(37)
	H-Met-Gln-His-Phe-Arg-Trp-Gly-OH	$3 \cdot 10^{5}$	(31,33)
	H-Ser-Met-Gln-His-Phe-Arg-Trp-Gly-OH	$7 \cdot 10^{5}$	(23)
	Ac-Ser-Met-Glu-His-Phe-Arg-Trp-Gly-OH	$4 \cdot 10^{6}$	(15)
	H-Ser-Tyr-Ser-Met-Glu-His-Phe-Arg-Trp-Gly-OH	$3 \cdot 10^{6}$	(15)
	Ac-Ser-Tyr-Ser-Met-Glu-His-Phe-Arg-Trp-Gly-OH	$1 \cdot 10^{7}$	(15,51)
(II)	H-Pro-Val-NH_2	0	(15)
	H-Lys-Pro-NH_2	$2 \cdot 10^{3}$	(16)
	H-Lys-Pro-Val-NH_2	$3 \cdot 10^{4}$	(15)
	Ac-Lys-Pro-Val-NH_2	$8 \cdot 10^{4}$	(15)
	Ac-Lys-NH_2	$8 \cdot 10^{2}$	(15)
	H-Gly-Lys-NH_2	$8 \cdot 10^{2}$	(16)
	H-Gly-Lys-Pro-Val-NH_2	$5 \cdot 10^{4}$	(16)
	Ac-Trp-Gly-Lys-Pro-Val-NH_2	$6 \cdot 10^{5}$	(15)
(III)	Ac-Phe-Arg-Trp-Gly-Lys-Pro-Val-NH_2	$5 \cdot 10^{6}$	(15)
	H-His-Phe-Arg-Trp-Gly-Lys-Pro-Val-NH_2	$8 \cdot 10^{6}$	(26)
	Ac-Glu-His-Phe-Arg-Trp-Gly-Lys-Pro-Val-NH_2	$5 \cdot 10^{8}$	(15)
	Ac-Met-Glu-His-Phe-Arg-Trp-Gly-Lys-Pro-Val-NH_2	$3 \cdot 10^{9}$	(15)
	Ac-Nle-Glu-His-Phe-Arg-Trp-Gly-Lys-Pro-Val-NH_2	$2 \cdot 10^{9}$	(15)
	Ac-Ser-Met-Glu-His-Phe-Arg-Trp-Gly-Lys-Pro-Val-NH_2	$7 \cdot 10^{9}$	(15)
	Ac-Tyr-Ser-Met-Glu-His-Phe-Arg-Trp-Gly-Lys-Pro-Val-NH_2	$2 \cdot 10^{10}$	(15)
	Ac-Ser-Tyr-Ser-Met-Glu-His-Phe-Arg-Trp-Gly-Lys-Pro-Val-NH_2	$4 \cdot 10^{10}$	(15)
	Ac-Ala-Tyr-Gly-Nva-Glu-His-Phe-Arg-Trp-Gly-Lys-Pro-Val-NH_2	$2 \cdot 10^{9}$	--
	Ac-Ser-Tyr-Ser-Met-Glu-His-Phe-Arg-Pmp-Gly-Lys-Pro-Val-NH_2	$2 \cdot 10^{10}$	(61)
	Ac-Ser-Tyr-Ser-Met-Glu-His-Phe-Arg-Phe-Gly-Lys-Pro-Val-NH_2	$2 \cdot 10^{10}$	(61)
	Ac-Ser-Tyr-Ser-Met-Glu-His-Phe-Arg-Trp-Gly-Lys-Pro-NH_2	$2 \cdot 10^{9}$	--
	Ac-Ser-Tyr-Ser-Met-Glu-His-Phe-Arg-Trp-Gly-Lys-NH_2	$4 \cdot 10^{8}$	--

Pfp Pentafluorophenylalanine Pmp Pentamethylphenylalanine

Relationship Between Chain Length and Melanotropic Activity

α-MSH is the most potent melanin-dispersing agent. Chain shortening or chain elongation both lead to a gradual decrease of melanotropic activity (Fig. 2). Sequence extentions 13→1 and 10→1 show for both cases a high increase of activity in the range 9→4 and a much smaller (only 10-fold) by adding the N-terminal tripeptide, -Ser-Tyr-Ser. Although this part seems to contribute only little to the full melanotropic potency of the hormone, changing its hydrophilic nature to a more hydrophobic one leads to a marked loss of activity (see below). Sequence extension from Gly^{10} to the C-terminus results in a much steeper slope of the curve than at the N-terminal part indicating a much greater importance of this tripeptide amide.

In all fragments N-terminal acetylation produces an approximately 2 to 10-fold rise in activity. Whether this may reflect increased stability to exopeptidase action or whether it is due to a closer binding of the hormone (hydrophobic "stickiness") to its receptor remains to be established.

Modifications Within the Central Pentapeptide [5-9], the First Message Sequence of α-MSH

As already mentioned the pentapeptide -Glu-His-Phe-Arg-Trp- has been found to be invariant in adrenocorticotropins, melanotropins and lipotropins of most species sequenced to date [2, 22, 38, 44], ant it is regarded as "message sequence" (56). Concerning biological activity it seems to be very sensitive to alterations. A number of synthetic peptides related to this sequence have been tested for malanotropic activity by various authors: in the case of the pentapeptide [6-10] the all-D-isomer (65), the D-His^{6}-isomer (64) and peptides with modifications at the $arginine^{8}$-residue (ornithine, citrulline, D-arginine) (6, 34) have no or less activity than the peptide with the natural sequence. On the other hand, peptides with D-Phe^{7} and D-Trp^{9} exhibit a higher activity (48, 64).

The glutamic acid residue at position 5 does not appear to be specifically essential in small peptides containing only the sequences [5-10], [4-10] and [3-10] (23, 31-33, 49); however its replacement by glutamine in the tridecapeptide is reported to attenuate the activity to 20% (1), or to leave it unaltered (24).

Our own results support the idea that the "active core" of this first message sequence is centered around -Phe-Arg-, because this dipeptide was found to be the shortest fragment with small but definite melanotropic activity. Furthermore, Arg^{8} is very sensitive to alterations, and replacement of Phe^{7} by pentafluorophenylalanine (Pfp) in the tetrapeptide Phe-Arg-Trp-Gly leads to a complete loss of melanotropic potency, whereas Trp^{9} can be replaced by leucine without adverse effects. Modifications at this same position within the whole

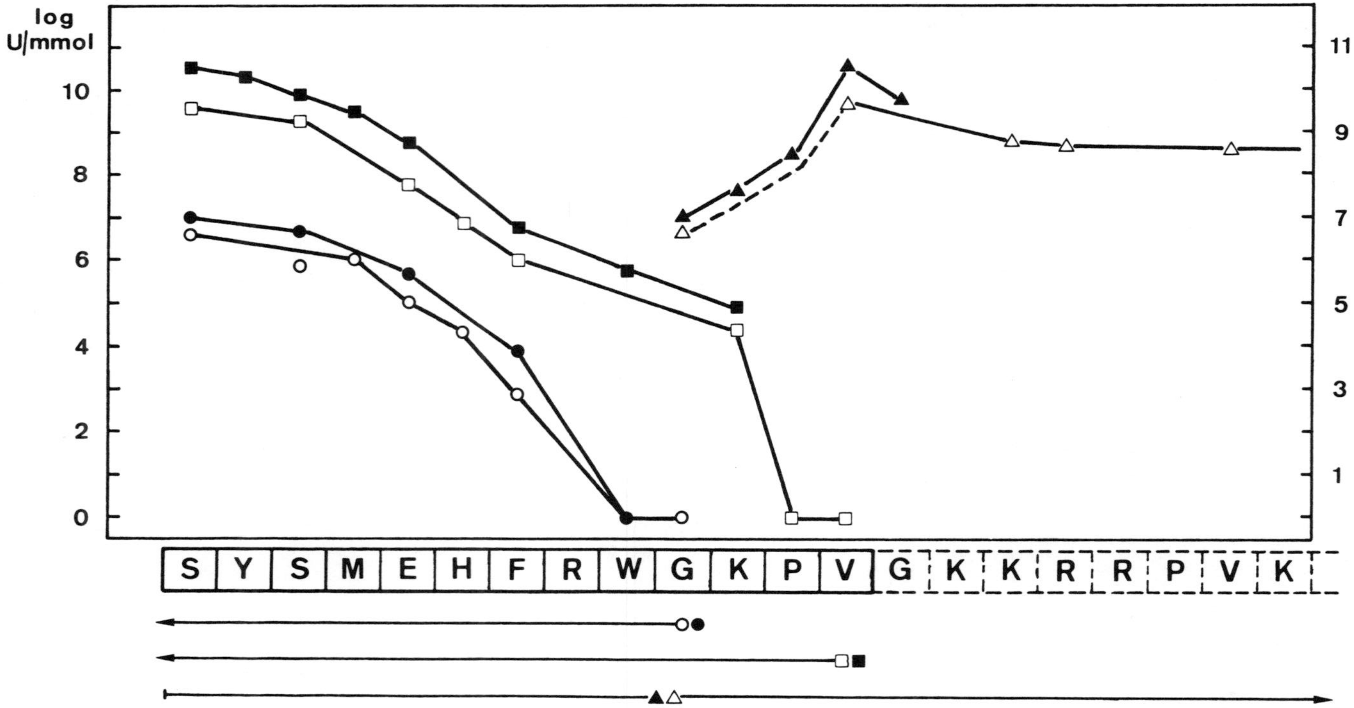

Fig. 2. Melanotropic activity in vitro of peptides and peptide derivatives obtained by sequence extensions 10 → 1 [o,●], 13 → 1 [□ ■], and 10 → 20 [△ ▲]. Open symbols indicate unsubstituted, dark symbols acetylated N-terminal amino groups.

TABLE II

Influence of Varying the Side Chain of Lysine[11] on the Melanotropic Activity of α-MSH

	Position 11		MELANOTROPIC ACTIVITY (IN VITRO) U/MMOL
Ac·Ser-Tyr-Ser-Met-Glu-His-Phe-Arg-Trp-Gly-	—	-Pro-Val-NH_2	$1.5 \cdot 10^{7}$
	Gly		$3 \cdot 10^{8}$
	Nle		$2 \cdot 10^{9}$
	Lys		$4 \cdot 10^{10}$
	Lys (Form)		$4 \cdot 10^{10}$
	Lys (Msoc)		$4 \cdot 10^{9}$
	Lys (Dansyl)		$4 \cdot 10^{8}$

tridecapeptide by phenylalanine or pentamethylphenylalanine (Pmp) lowers the MSH-activity only by a factor of 2-3 (62). We could also confirm the finding of Ramachandran (43) that [Nps-Trp^9] α-MSH exhibits the same potency as the natural hormone. Thus, as Trp^9 is not an essential requirement for hormonal activity, no charge transfer properties at this position are needed for triggering the stimulus. His^6, on the other hand, seems to be an important potentiating factor.

The C-Terminal Part of α-MSH: a Second Message Sequence

The C-terminal tripeptide [11-13] has hitherto been regarded as inactive *per se*; it appears to potentiate the biological activity of the central heptapeptide 4000-fold. The lysine^{11}-side chain has been thought to contribute very little in the hormonal activity, since Hofmann *et al.* (24) have found that N^{ε}-formyl protected lysine^{11} does *not* alter the melanotropic potency of α-MSH.

Table II clearly demonstrates the importance of lysine^{11}-side chain. [Nle^{11}] α-MSH exhibits only 5% melanotropic activity compared to the intact hormone. The deleterious effect is even larger if lysine^{11} is replaced by glycine: potency less than 1%. The α-MSH-dodecapeptide with complete omission of lysine^{11} exhibits only a slightly higher activity than the N-terminal decapeptide, namely

0.04%. On the other hand the melanotropic potency decreases also with larger blocking groups than the formyl group, e.g. with the methylsulfonylethoxycarbonyl (MSOC)-group to about 10% and with the dansyl group even to about 1%.

These results suggest that the C-terminal tripeptide has not only the purpose of binding to the receptor site, it even plays a role in triggering the stimulus: the C-terminal pentapeptide Ac-Trp-Gly-Lys-Pro-Val-NH_2 is almost equally active as the central heptapeptide sequence, -Met-Glu-His-Phe-Arg-Trp-Gly- (Table I) compared to the dipeptide Ac-Trp-Gly-OH which is inactive *per se*. The C-terminal tetra- and tripeptides still possess melanotropic activities in the range of the minimal sequences of α-MSH so far known to elicit MSH-activity: His-Phe-Arg-Trp(40) and His-Phe-Arg-Trp-Gly (50). (The activity of the tripeptide [11-13] may depend somewhat on the origin of the test animals.)

Thus, it appears that the melanocyte α-MSH receptor contains two message-recognizing sites, one for -(Glu-)His-Phe-Arg-Trp-, the N-terminal or *first message sequence*, and one for -Gly-Lys-Pro-Val-NH_2, the C-terminal or *second message sequence* (16). The two sites can--in the experimental situation--operate either alone or in combination to trigger melanin dispersion. In combination, they have a multiplicative, "cooperative" effect. Whether they produce melanin dispersion by the same or by different mechanisms (cAMP, cGMP, other) remain to be elucidated.

The important area of the second "active site" of α-MSH seems to be centered around -Lys-Pro-; Val^{13}-amide can be replaced by, also lipophilic Met^{13}-amide as in the α-MSH of the dogfish *Squalus acanthias* (2) (see Fig. 1). Maybe a residue with strong hydrophobicity is required at this position. (It is worth mentioning that also the steroidogenic activity of ACTH-peptides may be effected by replacing $valine^{13}$, as e.g. in the $[Gly^{13}]ACTH_{1-17}$-amide: activity only 3% compared to the Val^{13}-analogue (42)).

Alterations of the N-Terminal Sequence Part of α-MSH

The N-terminal tetrapeptide Ac-Ser-Tyr-Ser-Met is completely devoid of melanotropic activity; it potentiates the biological activity of the central hexapeptide [5-10] and the C-terminal nonapeptide [5-13] about 100-fold whereas the tripeptide -Ser-Tyr-Ser- does only about 10-fold. The subdivision into a hydrophilic and a hydrophobic part within this tetrapeptide seems to be very important. Thus, methionine can be replaced by norleucine without significant loss of activity (15), but when it is oxidized to the sulfoxide analogue or replaced by α-amino-butyric acid (shorter side chain), the activity of α-MSH drops to 2.5% (51) and 1% (28) respectively. On the other hand, modifying the hydrophilic N-terminal tripeptide towards a more hydrophobic one as in the case of [D-Ala[1], Gly[3], Nva[4]]α-MSH has about the same effect as oxidizing methionine[4]: activity 5%. Conclusion: like the C-terminal valinamide[13], methionine[4] possesses strong hydrophobic binding properties to the receptor. This effect is intensified by hydrophilic (hydrogen-bonding?) interactions of the N-terminal -Ser-Tyr-Ser- with the receptor.

Comparison of α-MSH Message Sequences Triggering Stimuli in Melanocytes, Adipocytes and CNS-Cells

The finding of two non-overlapping, biologically active sequences in α-melanotropin which independently trigger melanin-dispersion is not restricted to melanocyte-receptors. De Wied (13) has recently shown that the tripeptide [4-6] and the tetrapeptide [7-10] both possess behavioral activity in the rat pole jumping test (Fig. 3). The former had a relative potency of 30%, and the latter of 10% compared to the parent heptapeptide, Met-Glu-His-Phe-Arg-Trp-Gly. This heptapeptide exhibits the same behavior activity as ACTH,α-MSH or β-MSH(12, 20). Thus, it represents the message sequence of these hormones regarding their effects on CNS-cells. Shortening at the C-terminus of this peptide by two residues (i.e. [des-Arg[8], des-Trp[9]]-analogue) or replacement of these amino acids by lysine[8] and phenylalanine[9] did not alter the activity (13). In the same way Phe[7] could be replaced by leucine, tryptophan or pentamethylphenylalanine (slightly increased activity), but changing the configuration at this residue led to reversal of the behavioral effect (extinction of the avoidance response) (13). Oxidation of Met[4] increases the behavioral potency about 10-fold (in contrast to the melanotropic potency). From his results, De Wied concluded that the essential requirements for the behavioral effect of melanotropin and adrenocorticotropin analogues may not be restricted only to the tetrapeptide sequence -Met-Glu-His-Phe- (which is equally active as the longer peptides), but that there may be a second "active site", -Phe-Arg-Trp-, which may contain the information in a "dormant form" and needs potentiating modifications (e.g. chain elongation) to become expressed. There is a close analogy between the second message sequence for α-melanotropin-receptors in melanocytes and CNS-cells.

It appears that there may also be significant structural differences between hormone receptors of melanocytes and adipocytes, and equally of rat and rabbit adipocytes: α-MSH and [Nps-Trp[9]]ACTH exhibit a higher lipolytic potency in isolated rabbit fat cells than natural ACTH (39). However both are inactive in rat adipocytes. Whether this is due to non-identical receptors or to MSH-specific receptors which may be present only in rabbit adipocytes (8), remains to be established. Much work has been done in elucidating the structural requirements of the message sequence, -Met-Glu-His-Phe-Arg-Trp-Gly-, in melanotropins and adrenocorticotropins triggering the

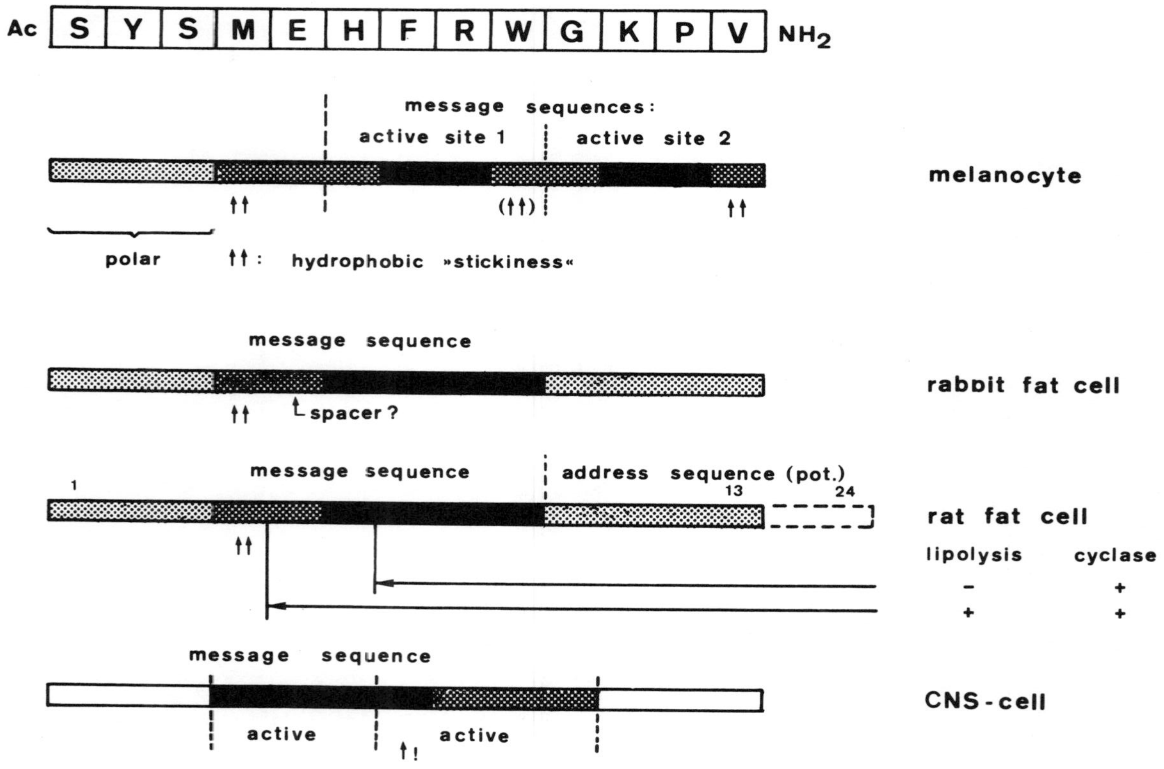

Fig. 3. Comparison of α-MSH message sequences triggering stimuli in melanocytes, adipocytes and CNS-cells.

steroidogenic and/or the lipolytic response. It seems that the first of these amino acids, methionine, possesses strong hydrophobic binding properties to the receptor as in the case of the action on melanocytes (18). Glu^5 appears to play an important "spacer role" within the peptide--at least in the case of rabbit adipocytes --but the biological activity does not depend on the nature of its side chain (14). The tetrapeptide sequence [6-9], -His-Phe-Arg-Trp-, is more sensitive to alterations. This is especially true for tryptophan^9 and arginine^8 where replacement by phenylalanine^9 and/or lysine^8, ornithine^8 dropped the activity about 100-fold (9, 29, 59, 62). The same is true for modifications at the position 7 (Phe), whereas the reduction of activity by replacing the imidazolyl- by a pyrazolyl-residue of His^6 is nearly negligible (30). On the other hand, replacing it by a phenyl-residue is very unfavorable (30).

A very interesting relation between lipolytic, adenylate cyclase stimulation and chain length of ACTH-peptides in rat isolated fat cells has been found by Dr. Ursula Lang in our laboratory, namely that a hormone analogue lacking the N-terminal hexapeptide [1-6] (i.e. $ACTH_{7-24}$) is lipolytically inactive, but a partial agonist of ACTH for cyclase-stimulation. Chain extension by two residues (Glu^5 and His^6) leads to a partial agonist also for lipolysis.[1]

In conclusion, the important "cores" of the message sequence of α-melanotropin and related peptides are not identical for triggering receptors in melanocytes, adipocytes and CNS-cells. Therefore, the complementary regions of the receptors involved in interacting with these sequence parts of the hormone may be structurally different. Apparently the information of a hormonal polypeptide is used only partially to produce one single stimulus. Thus, the peptide seems to possess several "information units" which can cause a biological response in different tissues by triggering different receptor sites. Some of these receptors are closely related to one another like melanocyte and rabbit adipocyte α-melanotropin receptors, or rat fat cell and adrenal cell receptors for ACTH. Others appear to be very different like CNS-cell and adrenal cell receptors.

ACKNOWLEDGEMENTS

We thank the Swiss National Foundation for financial support, and Mr. Willi Hübscher for his excellent technical assistance.

[1] U. Lang and R. Schwyzer, unpublished results.

1. ABE, K., GOBISON, G. A., LIDDLE, G. W., BUTCHER, R. W., NICHOLSON, W. E. and BAIRD, C. E., Endocrinology 85 (1969) 674.
2. BENNETT, H. P. J., LOWRY, P. J., MC MARTIN, C. and SCOTT, A. P., Biochem. J. 141 (1974) 439.
3. BITENSKY, M. W. and BURSTEIN, S. R., Nature 208 (1965) 1282.
4. BITENSKY, M. W., DEMOPOULOS, H. B. and RUSSELL, V., Genesis and Biologic Control (V. Riley, ed.), Appleton-Century-Crofts, New York (1972) 247.
5. BLAKE, J. and LI, C. H., Biochemistry 11 (1972) 3459.
6. BODANSZKY, M., ONDETTI, M. A., RUBIN, B., PIALA, J. J., FRIED, J., SHEEHAN, J. T. and BIRKHEIMER, C. A., Nature 194 (1962) 485.
7. BRAUN, T., HECHTER, O. and LI, C. H., 52nd Meeting Endocrine Soc., (1970) 48.
8. BRAUN, T. and HECHTER, O., Adipose Tissue, Regulation and Metabolic Functions (B. Jeanrenaud & D. Hepp, eds.), G. Thieme, Stuttgart, (1970) 11.
9. CHUNG, D. and LI, C. H., J. Am. Chem. Soc. 89 (1967) 4208.
10. DAYHOFF, M. O., Atlas of Protein Sequence and Structure, National Biomedical Research Foundation, Silver Spring, Maryland, Vol. 5 (1972).
11. DEMPSEY, G. I., KASTIN, A. J. and SCHALLY, A. V., Horm. Behav. 3 (1972) 333.
12. DE WIED, D., Frontiers in Neuroendocrinology (W. F. Ganong and L. Martini, ed.), Oxford University Press, London (1969) 97.
13. DE WIED, D., WITTER, A. and GREVEN, H. M., Biochem. Pharm. 24 (1975) 1463.
14. DRAPPER, M. W., RIZACK, M. A. and MERRIFIELD, R. B., Biochemistry 14 (1975) 2933.
15. EBERLE, A. and SCHWYZER, R., Helv. Chim. Acta 58 (1975) in press.
16. EBERLE, A. and SCHWYZER, R., J. Clin. Endocrine., Suppl. (1975) in press.
17. EBERLE, A., FAUCHÈRE, J.-L., TESSER, G. I. and SCHWYZER, R., Helv. Chim. Acta 58 (1975) in press.
18. FUGINO, M., NISHIMURA, O. and HATANAKA, C., Chem. Pharm. Bull. 18 (1970) 1291.
19. GESCHWIND, I. I. and HUSEBY, R. A., Endocrinology 79 (1966) 97.
20. GREVEN, H. M. and DE WIED, D., Progress in Brain Research (E. Zimmermann, W. H. Gispen, B. H. Marks & D. DeWied eds.) Elsevier, Amsterdam, Vol. 39 (1973) 430.
21. HARRIS, J. I., Brit. Med. Bull. 16 (1960) 189.
22. HECHTER, O. and BRAUN, T., Structure Activity Relationship of Protein and Polypeptide Hormones (M. Margoulies & F. C. Greenwood, eds.) Excerpta Medica Intern. Congress Ser. No. 241 (1971) 212.)
23. HOFMANN, K., THOMPSON, T. A. and SCHWARTZ, E. T., J. Am. Chem. Soc. 79 (1957) 6087.

24. HOFMANN, K., YAJIMA, H. and SCHWARTZ, E. T., J. Am. Chem. Soc. 82 (1960) 3732.
25. HOFMANN, K., Brookhaven Symp. Biol. 13 (1960) 184.
26. HOFMANN, K. and YAJIMA, H., J. Am. Chem. Soc. 83 (1961) 2289.
27. HOFMANN, K. and LANDE, S., J. Am. Chem. Soc. 83 (1961) 2286.
28. HOFMANN, K. and YAJIMA, H., Recent Progr. Horm. Res. 18 (1962) 41.
29. HOFMANN, K., WINGENDER, W. and FINN, F. M., Proc. Nat. Acad. Sci. USA 67 (1970) 829.
30. HOFMANN, K., ANDREATTA, R., BOHN, H. and MORODER, L., J. Med. Chem. 13 (1970) 339.
31. KAPPELER, H. and SCHWYZER, R., Helv. Chim. Acta 43 (1960) 1453.
32. KAPPELER, H., Helv. Chim. Acta 44 (1961) 476.
33. KAPPELER, H. and SCHWYZER, R., Experientia 16 (1960) 415.
34. LANDE, S., Diss., University of Pittsburgh (1960).
35. LERNER, A. B., in Biology of Normal and Abnormal Melanocytes (T. Kawamura, B. Fitzpatrick & M. Seiji, eds.), University of Tokyo Press (1971) 3.
36. LI, C. H., GORUP, B., CHUNG, D. and RAMACHANDRAN, J., J. Org. Chem. 28 (1963) 178.
37. LI, C. H., SCHNABEL, E., CHUNG, D., LO, T.-B., Nature (Lond.) 189 (1961) 143.
38. LI, C. H., DANHO, W. O., CHUNG, D. and RAO, A. J., Biochemistry 14 (1975) 947.
39. MOYLE, W. R., KONG, Y. C. and RAMACHANDRAN, J., J. Biol. Chem. 248 (1973) 2409.
40. NOVALES, R. and NOVALES, B. J., Pigment Cell: Mechanisms in Pigmentation (V. J. McGovern & P. Russell eds.) S. Karger, Basel, (1973) 188.
41. OTSUKA, H. and INOUYE, K., Bull. Chem. Soc. Jap. 37 (1964) 1465.
42. RAMACHANDRAN, J., Hormonal Proteins and Peptides, (C. H. Li, ed.) Academic Press, New York (1973) 1.
43. RAMACHANDRAN, J., Biochem. Biophys. Res. Comm. 41 (1970) 353.
44. RINIKER, B., SIEBER, P., RITTEL, W. and ZUBER, H., Nature New Biol. 235 (1974) 114.
45. RUDMAN, D., BROWN, S. J. and MALKIN, M. F., Endocrinology 72 (1963) 527.
46. SANDMAN, C. A., DENMAN, P. H., MILLER, L. H. and KNOLT, J. R., J. Comp. Physiol. Psychol. 76 (1973) 103.
47. SANDMAN, C. A., KASTIN, A. J. and SCHALLY, A. V., Physiol. Behav. 6 (1971) 45.
48. SCHNABEL, E. and LI, C. H., J. Am. Chem. Soc. 82 (1960) 4576.
49. SCHNABEL, E. and LI, C. H., J. Biol. Chem. 235 (1960) 2010.
50. SCHWYZER, R. and LI, C. H., Nature 182 (1958) 1669.
51. SCHWYZER, R., Ergebnisse der Physiologie 53 (1963) 1.
52. SCHWYZER, R., Journal de Pharmacie 3 (1968) 254.
53. SCHWYZER, R., Proceedings of the International Symposium on Protein and Polypeptide Hormones, Liege, Excerpta Medica Intern. Congress Ser. No. 161 (1968) 201.

54. SCHWYZER, R., Proceedings 4th International Congress on Pharmacology, Schwabe, Rasel/Stuttgart, Vol. 5, (1970) 196.
55. SCHWYZER, R., SCHILLER, P., SEELIG, S. and SAYERS, G., FEBS Letters 19 (1971) 229.
56. SCHWYZER, R., Peptides 1972, Proceedings 12th European Peptide Symposium (H. Hanson & H. D. Jakubke, eds.), North-Holland Publ. Co., Amsterdam (1973) 424.
57. SHIZUME, K., LERNER, A. B. and FITZPATRICK, T. B., Endocrinology 54 (1954) 553.
58. STRATTON, L. O. and KASTIN, A. J., Horm. Behav. 5 (1972) 149.
59. TESSER, G. I. and RITTEL, W., Rec. Trav. Chim. Pays-Bas 88 (1969) 553.
60. THODY, A. J. and SHUSTER, S., J. Endocrin. 64 (1975) 503.
61. VAN NISPEN, J. W., Diss., Catholic University of Nijmegen, Netherlands.
62. VAN NISPEN, J. W. and TESSER, G. I., Intern. J. Peptide Protein Res. 7 (1975) 57.
63. WÜNSCH, E., Synthese von Peptiden, Vol. 15, Part I of Houben-Weyl, Methoden der Organischen Chemi (E. Müller, ed.) G. Thieme, Stuttgart (1974).
64. YAJIMA, H. and KUBO, K., Biochim. Biophys. Acta 97 (1965) 596.
65. YAJIMA, H. and KUBO, K., J. Am. Chem. Soc. 87 (1965) 2039.

ENDOTHELIAL PROLIFERATION FACTOR

B. HABER, R. L. SUDDITH, H. T. HUTCHISON and P. J. KELLY

Division of Neurobiology, The Marine Biomedical Institute, and Departments of Neurology, and Human Biological Chemistry and Genetics, University of Texas Medical Branch, Galveston, Texas 77550 (USA)

The viability of the brain is predicated upon its blood supply. Nowhere is the process of vascularization of the brain more dramatically visualized than in the expansion of the capillary bed within and around tumors of the CNS (6). Recent evidence suggests that this endothelial proliferation (or neovascularization as it is often called) is mediated by a soluble factor (12) released by the growing tumor cells as proposed earlier by Algire (1). The new vessels are recruited by the tumor from the already existent host vasculature in the region of the tumor (1, 26).

The importance of neovascularization for continued tumor growth is shown by Tannock's demonstration (22, 23) that the rate of tumor growth is directly related to its vascular supply. Further, Gimbrone (9) has shown that in the absence of neovascularization a solid tumor becomes restricted in size to a small spherical body 2 to 3 mm in diameter. Thus, the initiation of vascular growth is a critical rate-limiting step in the growth of a solid tumor.

The stimulation of capillary growth by the tumor has been suggested to be mediated by a diffusible or humoral factor (1, 4, 26). Direct evidence for such a factor was first provided by Greenblatt (12). In these experiments capillary growth was induced in host tissues separated from tumor cells by a Millipore filter. Several investigators have confirmed this observation in a variety of host tissues (2, 5, 7, 9, 10). Further support for this concept was the isolation by Folkman (7) of a diffusible factor named tumor angiogenesis factor (TAF) from animal and human solid tumors, which is capable of inducing vasoproliferation in a variety of *in vivo* bioassay systems.

The vasoproliferative response elicited by TAF is not restricted to the capillary endothelium, but also involves the pericytes and surrounding connective tissues as demonstrated by ^{3}H-thymidine autoradiography. Cavello (2, 3) has speculated that this broad mitogenic response reflects an impure TAF fraction containing several mitogenic factors or alternatively is a manifestation of the wound healing process. The non-specificity of the TAF mitogenic response, and the inability to adequately quantitate the mitogenic response _in vivo_ prompted us to develop an _in vitro_ assay system employing cultured human endothelial cells as the target of a trophic factor or factors elaborated by cells of tumor origin.

The development of the _in vitro_ assay was made possible by procedures developed for harvesting viable endothelial cells from the intimal lining of large blood vessels. Significant progress in such culture techniques was obtained by buffered collagenase digestion of the umbilical vein which consists of a layer of endothelial cells resting on a collagen-like basement membrane (11, 13). The advantage of collagenase is due in part to the fact that it strips off the endothelia from the basement membrane, and unlike trypsin, does not damage cell membranes. The endothelia used in these experiments were obtained from fresh umbilical cords by collagenase digestion, as described in detail (14, 20, 21). In this _in vitro_ assay, the proliferative response of endothelia is measured either radioautographically, or as acid-precipitable counts following a 72-hour pulse with ^{3}H-thymidine. This _in vitro_ assay for EPF, which is described in detail elsewhere (14, 20, 21) has permitted the demonstration that cells of tumor origin (clonal tumor cell lines and primary cultures of tumor biopsy specimens) _in vitro_ produce a factor which stimulates endothelial cells to incorporate thymidine.

Table I shows the endothelial proliferation response as measured by the thymidine labeling index (TI) following ^{3}H-thymidine pulsation in the presence of media conditioned by fibroblasts obtained from a biopsy scar, a rat (C6) and human astrocytoma, and a human (IMR-32) and mouse (NB-41) neuroblastoma. The endothelia exhibit a six to ten-fold increase in thymidine labeling index in response to a soluble factor elaborated by the clonal tumor cell lines. Figure 1 is a graphic display of endothelial proliferation response to media conditioned by primary cultures of biopsy specimens from several CNS tumors. In addition, a portion of the data shown in Table I is presented by easy comparison. These data show that a soluble factor capable of stimulating endothelial cell proliferation is elaborated by these relatively short-term cultures obtained from biopsy specimens of tumors. Media which has been conditioned by cells of non-tumor origin, such as fibroblasts, lymphocytes, or mouse neonate brain, does not, however, elicit the endothelial cell

TABLE I

Proliferation Response of Human Endothelial Cells in the Presence of Media Conditioned by Clonal Cell Lines of Tumor Origin *

CONTROLS		
Endothelia with:		
Fresh media		2.1%
Amniotic fluid		1.1%
Fibroblast conditioned media		4.5%
GLIA TUMOR CELLS		
Endothelia with conditioned media from:		
C6	(A)	98.3%
C6	(A)	97.6%
C6 JDV	(B)	92.6%
C6 JDV	(B)	83.3%
C6-CR1-107	(C)	25.7%
C6-CC1-107	(C)	36.7%
C6-CC1-107	Co-culture	28.7%
Human astrocytoma-primary		30.1%
NEUROBLASTOMA TUMOR CELLS		
Endothelia with conditioned media from:		
NB-41		90.0%
IMR-32		78.7%

* The proliferation response is measured as the thymidine labeling index (TI) of endothelial cells following a 72-hour pulse with ^3H-thymidine in presence of media conditioned by cell lines of tumor origin.

$$TI = \frac{\text{number of labeled cells}}{\text{total number of cells}} \times 100$$

The TI of endothelial cells in the presence of media conditioned by cells of non-tumor origin (fibroblasts), fresh media and amniotic fluid is given as control values.

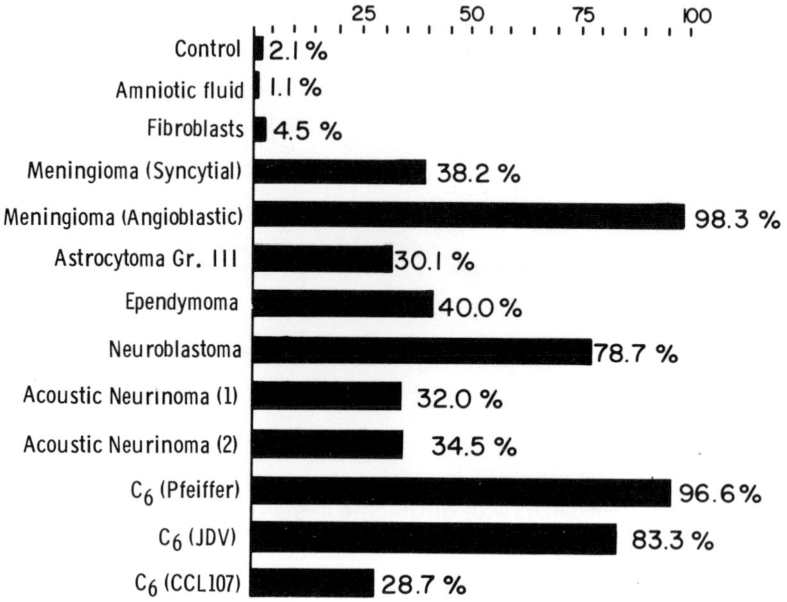

Fig. 1. Thymidine labeling index of endothelial cells in the presence of media conditioned by human and experimental CNS tumor cells in culture. Graphic display of the proliferation response of endothelial cells to media which has been conditioned by primary cultures of cells obtained from CNS tumor biopsy specimens. For comparison, the TI of endothelial cells in the presence of control (non-tumor cell) conditioned media and clonal tumor cell line (C6) is also presented.

proliferation response. Figure 2 shows the intense labeling localized over the chromosomes of an early metaphase endothelial cell, as well as, over the nuclei of several endothelial cells. Thus, it appears that a soluble factor which is capable of eliciting an endothelial cell proliferation response is elaborated by cells of tumor origin in culture. We have called this factor, "endothelial proliferation factor" or EPF. It is clear that production of this mitogenic factor or factors is not the property of rapidly proliferating cells, *per se*, but rather that of tumor cells. The production of EPF by tumor cells in culture makes conditioned media an attractive starting vehicle for the purification of EPF.

We have made some preliminary attempts to estimate the molecular weight of EPF. The material used was F-10 medium, conditioned by C6 astrocytoma cells and shown to be active in the *in vitro* assay. The medium was concentrated by ultrafiltration through an Amicon PM-10 membrane, yielding two fractions with material of molecular weights greater and lesser than 10,000. The ultrafiltrate

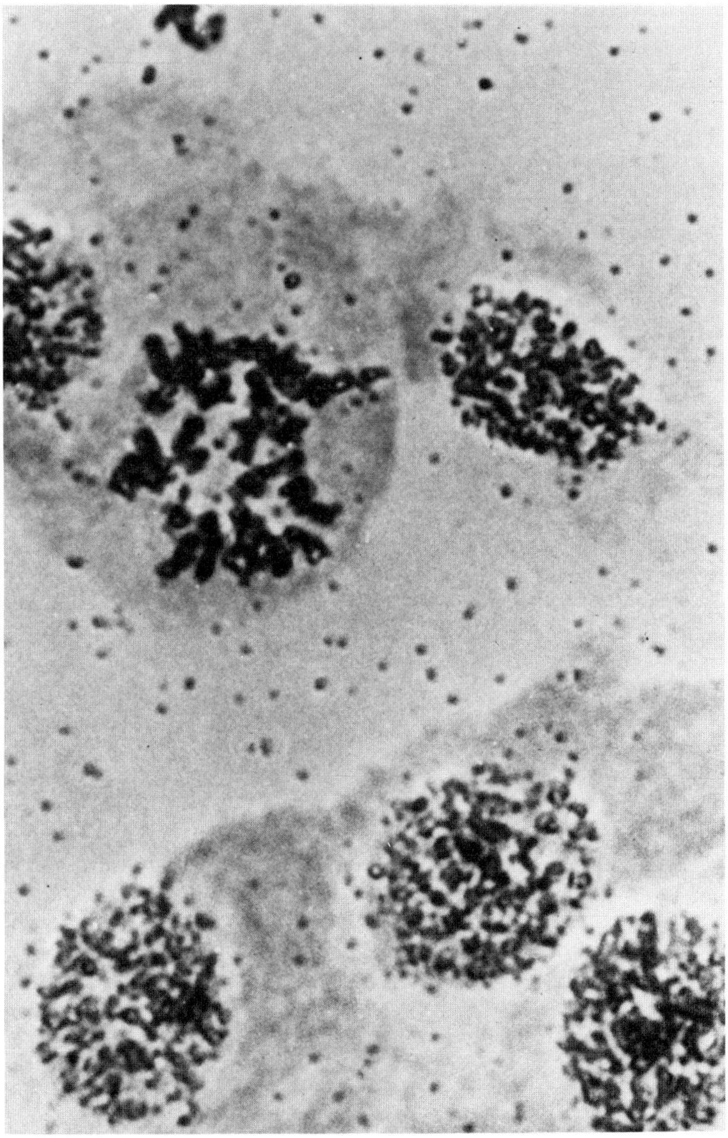

Fig. 2. Endothelial cells showing the localization of silver grains over the nuclei and early metaphase chromosomes. The endothelial cells were pulse-labeled with ^{3}H-thymidine in the presence of media conditioned by C6 astrocytoma cells.

(less than 10,000) was shown to have no EPF activity and was discarded. The second fraction, containing only moieties of MW greater than 10,000, was then fractionated on a calibrated Sephadex G-100

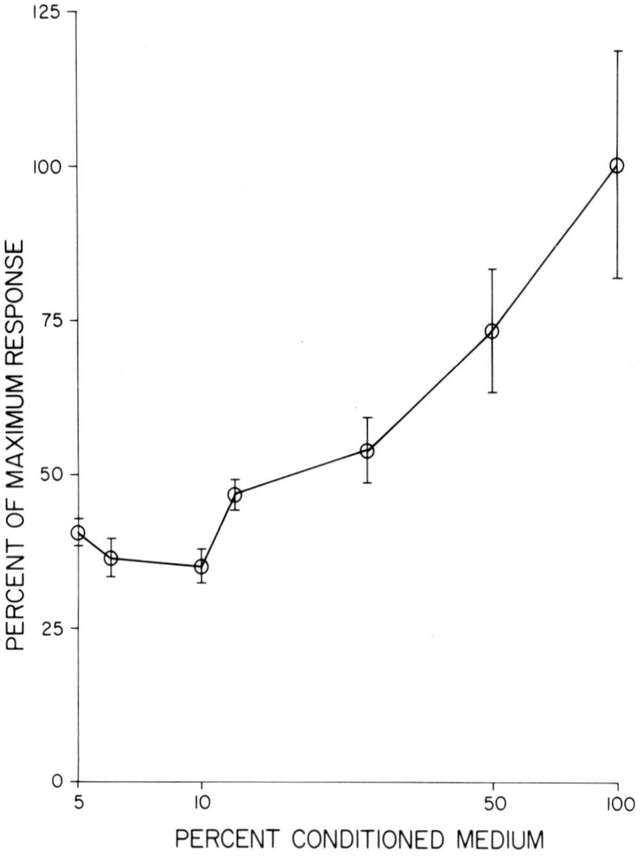

Fig. 3. Dose response curve of endothelial cells to media conditioned by the B16 rat melanoma cell line. The response of the cultured endothelial cells to tumor conditioned media was measured by determining the number of TCA-precipitable counts following a 72-hour pulse of ^{3}H-thymidine in the presence of conditioned medium. Each point shown represents the mean of four determinations at that concentration of conditioned medium. The bars represent the standard error of the mean. The half maximal response was obtained with approximately 30% concentration of this medium.

column at 4°, using 0.05 M phosphate buffer, pH 6.8 for elution. Pooled fractions were concentrated by pressure dialysis, reconstituted in F-10 medium containing heat inactivated serum, and tested in the *in vitro* assay. We estimate that the MW of the EPF moiety is roughly 150,000. The proliferative response of the various fractions was quantitated by a direct counting procedure and verified by radiography. Whether this EPF is a protein of MW about 150,000, or a low molecular polypeptide bound to a high MW protein cannot be answered at present.

Recently, a variety of growth factors has been described These include nerve growth factor (NGF) which has been sequenced and whose biological role is described in some detail in this volume. It has been shown recently that L cells, as well as glial tumor cells in culture, elaborate nerve growth factor (NGF), which is antigenically indistinguishable from 7S NGF (16)(and Young et al. elsewhere in this volume). NGF, molecular weight 131,750 (18) elaborated by glial tumors in culture (15), and astrocytoma conditioned culture medium induces a marked differentiation in IMR-32, a human neuroblastoma (17). Because of the above findings, we obtained some authentic mouse salivary gland NGF from Dr. Regino Perez-Polo at U. T. Austin, which was shown to be active in the chick dorsal root ganglia bioassay. The possible "EPF" activity of the NGF preparation was tested in the endothelial assay, at dosages of 1, 10 and 100 biological units/ml, with negative results. These findings suggest that EPF is different from NGF. However, any similarities or differences to NGF and other growth factors will be more rigorously tested as purified EPF fractions become available.

To determine the efficiency of our purification, it has become necessary to develop a means of quantitating the endothelial proliferation response. The first step is to demonstrate that the proliferation response is dosage-dependent. Figure 3 shows the dosage response curve of endothelial cells from a single umbilical vein to medium conditioned by rat melanoma cell line (B16). The half maximal response was obtained with approximately 30% concentration of this medium. These results were obtained by determining the number of acid-precipitable counts present following a 72-hour pulse with ^{3}H-thymidine in the presence of conditioned media. Using this assay system, we have initiated the purification of EPF by selective precipitation of an active fraction from conditioned medium using ammonium sulfate. So far, we have found that the active fraction is precipitated by a final concentration of 33% ammonium sulfate.

During the initial screening of EPF activity of tissue culture media conditioned by a variety of cell lines, we tested those media from normal and transformed human lymphocytes, with negative results. It was of interest to examine whether tissue culture supernatants shown to have EPF activity might alter the normal or mitogen induced blastogenic responses of human lymphocytes.

As shown in Table II, supernatants from various tumor culture lines have a variety of effects on human lymphocytes. Many of the samples tested caused increased ^{3}H-TdR uptake of unstimulated human lymphocytes, while causing a marked inhibition of the responses to PHA-P and Con A. The response to PWM was not inhibited and was somewhat enhanced in two cases. This is meaningful when it

TABLE II

Effect of Supernatants from Various Cell Cultures on the Responses of Human Lymphocytes to Mitogen

Supernatant from	Effect on control*	Effect on Mitogen Response*		
		PHA-P	Con A	PWM
Fibroblast	**	-19%	**	**
C6PF-29 cloned astrocytoma	1057%	-92%	-88%	**
C6PF-35 cloned astrocytoma	4006%	-87%	-87%	+138%
Endothelial cells	486%	-27%	+42%	**
Primary astrocytoma	77%	**	**	**
NB 41-4 neuroblastoma	5002%	-31%	**	+564%
Wilson's conditioned media	1559%	-57%	-65%	**
Human lymphoblasts	391%	**	**	**

$$*\text{Effect} = \frac{{}^{3}\text{H-TdR uptake experimental} - {}^{3}\text{H-TdR uptake control}}{{}^{3}\text{H-TdR uptake control}} \times 100$$

** Not significantly different from control ($p < 0.05$)

is realized that PHA-P and Con A stimulate T-cells (thymic derived lymphocytes)(19), which are responsible for cell mediated immunity whereas PWM also stimulates B-cells (25) (thymic-independent lymphocites) which are the anti-body producing cells. Tumors that produce T-cell inhibitors would have a much better chance of escaping the immune system's ability to recognize and destroy aberrant cells. Also, tumors that increase B-cell activity might have a greater probability of inducing the formation of blocking factors that interfere with the ability of T-cells to react with the cancerous cells. The preliminary data in Table II shows that most of the supernatants from tumor cells inhibited the T-cell responses, whereas none inhibited the B-cell responses.

The effects of the supernatants on non-stimulated controls indicates that the supernatants contain factors capable of causing lymphocytes to undergo DNA synthesis. It is not yet clear if this observation has direct meaning in vivo. Histological examination of tumor tissue often reveals a greater than normal infiltration of the malignant tissues with lymphocytes. This may be due to the cancerous tissue sequestering lymphocytes from the circulation as is generally assumed, or alternatively, as our preliminary data suggests, due to the blastogenic effects of factors produced by the cancerous tissue in situ. Lymphocytes isolated from tumor tissue do have a significantly higher spontaneous uptake of ^{3}H-TdR than do autologous peripheral blood lymphocytes, but it is unclear if this is due to a reaction of the lymphocytes to the tumor antigens or an effect on the lymphocytes of various factors produced by the cancerous cells.

Folkman (7, 8) obtained TAF in large quantities by the disruption with nitrogen of Walker 256 ascites carcinoma cells as well as a variety of other tumor cells. Sephadex G-100 column chromatography of the delipidated and trypsinized cytoplasm yielded an active TAF fraction having a molecular weight of approximately 100,000. The fraction contained approximately 25% RNA, 10% protein and 50% carbohydrate. Tuan et al., however, found similar TAF activity associated with the chromatin fraction of the tumor cell nucleus. Specifically, the activity was in the non-histone protein pool. This fraction contained only 3% RNA and a minimum of 20 proteins as shown by acrylamide gel electrophoresis. Similar TAF activity has recently been obtained from a cold saline wash of Walker carcinoma cells in tissue culture (7, 8). This fraction contains up to 10 proteins and has a nucleic acid content of 20%. Based on these analyses, Folkman (8) has concluded that TAF resembles a non-dialyzable ribonucleoprotein of approximately 100,000 molecular weight.

Neither the EPF described by us or the TAF described by Folkman are adequately characterized. Both EPF and TAF appear to be proteins, are produced by tumor cells, and have a MW of 100,000 - 150,000. EPF differs from TAF in that EPF is ribonuclease resistant, is inactivated by proteolytic digestion with pronase and stimulates the proliferation of endothelial cells, but not of fibroblasts. However, the in vivo assays of TAF activity may measure the proliferative responses of all the cellular components of the vascular bed, whereas the in vitro EPF assay measures only the responses of endothelia in the absence of other cell types. Understanding of whether these differences are apparent or real must await further characterization of EPF and TAF. Furthermore, the issue of specificity of EPF, and its possible homology to other growth factors, such as NGF, etc., is as yet unsettled and must await its characterization.

ACKNOWLEDGEMENTS

The collaboration of Dr. G. B. Thurman, Ms. S. Monroe and Ms. K. Werrbach is gratefully acknowledged. Supported in part by PHS grants NS 11255, 11344, Welch Grant H-504 and a grant from the Muscular Dystrophy Association of America. R. L. Suddith is a MDAA Post-doctoral Fellow.

REFERENCES

1. ALGIRE, G. H. and CHALKLEY, H. W., J. Nat. Cancer Inst. 6 (1945) 73.
2. CAVALLO, T., SADE, R., FOLKMAN, J. and COTRAN, R. S., J. Cell. Biol. 54 (1972) 408.
3. CAVALLO, T., SADE, R., FOLKMAN, J. and COTRAN, R. S., Am. J. Pathol. 70 (1973) 345.
4. CHALKEY, H. W., Comments on ALGIRE, G. H. and LEGALLAIS, F. Y., Special Publications. N. Y. Acad. Sci. 4 (1948) 164.
5. EHRMANN, R. L. and KNOTH, M., J. Nat. Canc. Inst. 41 (1968) 1329.
6. FEIGEN, I., ALLEN, L. B. and LIPKIN, L., Cancer 11 (1958) 264.
7. FOLKMAN, J., MERLER, E. and ABERNATHY, C., J. Exp. Med. 133 (1971) 275.
8. FOLKMAN, J., Adv. in Cancer Res. 19 (1974) 331.
9. GIMBRONE, M. A., LEAPMAN, S. B., COTRAN, R. S. and FOLKMAN, J., J. Exp. Med. 136 (1972) 261.
10. GIMBRONE, M. A., LEAPMAN, S. B., COTRAN, R. S. and FOLKMAN, J., J. Nat. Canc. Inst. 50 (1973) 219.
11. GIMBRONE, M. A., COTRAN, R. S. and FOLKMAN, J., J. Cell. Biol. 60 (1974) 673.
12. GREENBLATT, M. and SHUBI, K. P., J. Nat. Canc. Inst. 41 (1968) 111.
13. JAFFE, E. A., NACHMAN, R. L., BECKER, C. G. and MINICK, C. R., J. Clin. Invest. 52 (1973) 2745.
14. KELLY, P. J., SUDDITH, R. L., HUTCHISON, H. T., WERRBACH, K. and HABER, B., J. Neurosurgery (1976) in press.
15. LONGO, A. and PENHOET, E., Proc. Nat. Acad. Sci. USA 71 (1974) 2347.
16. OGER, J., ARNASON, B. G. W., PANTAZIS, N., LEHRICH, J. and YOUNG, M., Proc. Nat. Acad. Sci. USA 71 (1974) 1554.
17. REYNOLDS, P. and PEREZ-POLO, J. R., Neuroscience Letters (1975) in press.
18. SHINE, H. D., Master's Thesis. Univ. of Texas, Austin (1975).
19. STOBO, J. D., ROSENTHAL, A. and PAUL, W. D., J. Immunol. 108 (1972) 1.
20. SUDDITH, R. L., KELLY, P. J., HUTCHISON, H. T. and HABER, B., Fed. Proc. 34 (1975) 731.
21. SUDDITH, R. L., KELLY, P. J., HUTCHISON, H. T., MURRAY, E. A. and HABER, B., Science 190 (1975) 682.

22. TANNOCK, I. F., Br. J. Cancer 22 (1968) 258.
23. TANNOCK, I. F., Cancer Res. 30 (1970) 2470.
24. TUAN, D., SMITH, S., FOLKMAN, J. and MERLER, F., Biochem. 12 (1973) 159.
25. THURMAN, G. B., SILVER, B. B., HOOPER, J. A., GIOVANELLA, B. C. and GOLDSTEIN, A. L., Proceedings of the 1st International Workshop on Nude Mice, Gustav Fischer Verlag, Stuttgart (1974).
26. WARREN, B. A. and SHUBIK, P., Lab Invest. 15 (1966) 464.

EVIDENCE FOR A MACROMOLECULAR EFFECTOR OF CELL DIFFERENTIATION IN DICTYOSTELIUM DISCOIDEUM AMOEBAE

Michel DARMON, Claudette KLEIN and Philipe BRACHET

Unité de Différenciation Cellulaire
Département de Biologie Moléculaire
Institut Pasteur, Paris (France)

Upon depletion of their food supply, myxamoebae undertake their developmental program, the first stage of which is marked by the aggregation of cells toward central collection points. Aggregation follows a period of starvation called interphase, during which cells differentiate into aggregation-competent amoebae(7): they attain an elongated morphology, move chemotactically toward their neighboring cells (4), and form specific cell-cell contacts (6) which appear to require membrane components (contact sites A) present specifically on starved cells (3). Movement of amoebae is directed by an acrasin (4), cAMP (2, 14), which starved cells rhythmically release into the media (18). The response of cells to the chemotactic signal appears to involve surface cAMP-binding receptors (15) which probably function in the stimulation of cAMP synthesis and emission (17, 19, 20).

Although it has been shown that cAMP, applied as pulses which simulate the normally-arising chemotactic signal, can induce cell differentiation (5), and that this is accompanied by an increase in the number of cAMP-binding sites (9), the number of contact A sites (8), cAMP-phosphodiesterase activity (9)[1] and chemotaxis (5), it is not known if the cyclic nucleotide is the primary or unique regulator of cell differentiation or if another mediator(s) is involved. In a variety of systems, macromolecules which function at the cell surface appear to regulate various aspects of development, including cell growth (11), differentiation (12), and cell recognition and adhesiveness (1, 16). In light of the general nature of this control mechanism, it was of interest to determine if a macromolecular com-

[1] Klein, C. and Darmon, M., submitted for publication.

ponent is involved in the differentiation of Dictyostelium discoideum amoebae. This possibility was examined by testing the ability of supernatants from starved cells to enhance the aggregation of amoebae starved below the critical density for aggregation. Preliminary evidence for such a component(s) is presented and possible mechanisms of action are discussed.

Test Conditions

In order to detect any factor(s) which would stimulate cell differentiation of Ax-2 amoebae, cells were starved in the presence of added compounds under conditions in which little or no aggregation normally occurs. Therefore, cells were incubated in 17 mM phosphate buffer at very low densities on Falcon Petri dishes. The technique provides two important advantages:

1. No random cell contacts or clumping can occur.

2. The progression of morphological events characteristic of differentiating amoebae--cell elongation, contact and stream formation--can easily be monitored microscopically without perturbing the culture conditions.

Cell densities at which minimal development of aggregation-competence occurs were determined by plating various dilutions of amoebae onto Petri dishes and observing the morphological changes which occurred after 20-25 hours of starvation. When plated at a density of 3.3×10^3 cells/cm^2, amoebae retained the rounded appearance of undifferentiated cells and did not form any clouds (18) or specific intercellular contacts. With 6.6×10^3 cells/cm^2, cell elongation was observed to varying degrees and in some experiments limited cell streaming occurred. Tests for differentiation-promoting compounds were performed using these two cell concentrations. At higher densities, significant cell differentiation occurred which made assessment of any stimulatory factor difficult.

Preparation and Quantitation of Added Factors

Amoebae were starved at a density of 10^7 cells/ml in spinner suspensions as described by Beug et al. (3). At various times, aliquots were taken and cells removed by centrifugation at 1000 x g for 5 min. The supernatants were removed and filtered sterilely through 0.45 Millipore filters in order to assure total elimination of cells. Duplicate dilutions of the supernatants were added to the media of cells plated onto Petri dishes at the two densities indicated above. Cells were starved in a final volume of 5 ml. After approximately 20 hours, the number of organized territories (clouds, centers, aggregates) in four randomly chosen areas of the Petri dishes were counted. Control cells did not receive any supernatant. Preparations were tested immediately, and in some experiments, were also tested after storage at 4°.

Effect of Supernatants

Wild-type cells: When the supernatant of Ax-2 cells starved at high density for 3-4 hours was added to diluted cell populations, a slight stimulation of cell aggregation was observed: an increase in the number of clouds and varying extents of cell streaming occurred when cells were incubated directly in the supernatants, but this effect was generally lost if supernatants were diluted more than 2-fold. Supernatants from cells starved less than 3 hours were less effective, and preparations from cells starved for more than 5 hours often lead to a loss of activity. The stimulatory effect was retained when supernatants were dialyzed prior to their addition, indicating that a macromolecular component(s) is responsible for this stimulation. The factor(s) responsible will be referred to as DSF.

Mutant cells: The excretion of DSF by various mutants of Ax-2 amoebae isolated in this laboratory was examined in hopes of finding an "overproducer" which would induce more dramatic effects on cell aggregation and facilitate the eventual characterization of the active component. Of the mutants tested, only Agip 53 produced supernatants of elevated activity (Table I). Mutants Agip 55, 45 and 43, produced effects comparable to wild-type cells, while no effect was observed with mutants Agip 20 or 72. As with the wild-type Agip 53, supernatants from cells starved for 3 hrs provoked maximal responses and its effect was retained after dialysis. Frozen preparations were stable for at least 10 days. At room temperature, stimulatory activities of preparations were stable for the time period tested - 3 days.

Figure 1 shows the response of cells incubated with 1 ml of Agip 53 supernatant (i.e. 1-5 dilution) compared to control cells. The higher magnified image of an induced aggregate (Fig. 2) shows that the cell clusters formed under these conditions arise from a normal aggregation process, marked by typical end-to-end contacts and streams, and are not the result of enhanced random movements and adhesiveness.

A plot of the number of territories formed per fixed surface area versus the amount of Agip 53 supernatant added is shown in Fig. 3. In this experiment, the responses of cells incubated at two different concentrations are shown. At the lower density, minimal cell aggregation occurred in the absence of added DSF (number of territories of control cell = 1). Increasing amounts of supernatant provoked significantly greater numbers of aggregates. No effect was observed when less than 0.1 ml of supernatant was added. (This corresponds to over a 50-fold dilution of the supernatant.

TABLE I

Comparison of DSF and extracellular phosphodiesterase in various cell-types derived from Ax-2 amoebae

Cells	Activity	
	DSF	Phosphodiesterase (Extracellular)
Wild-type Ax-2	2	1
Agip 53	50	0.6
Agip 55	2	0.6
Agip 45	2	0.1
Agip 20	-	0.1
Agip 43	-	0.1
Agip 72	-	0.5
Wild type Ax-2 + $4x10^{-4}$M cAMP	2	10
Agip 53 + $4x10^{-4}$M cAMP	50; 50	1.8; 1.2

All activities are considered on a scale relative to wild-type cells. Phosphodiesterase was measured as previously described (10). DSF activity is referred to as the maximum dilution of supernatant which provoked a significant increase in the number of organized territories relative to control cells.

The same experiment using cells capable of undergoing some aggregation (control = 8) shows that higher amounts of supernatant must be added before any stimulatory effect is observed. This is not surprising since cells which can undergo aggregation would be expected to produce a certain amount of DSF and only when a proper excess is added would a response be observed. The effect of DSF does vary in different experiments. This seems to be due to changes in the sensitivity of diluted Ax-2 amoebae to the factor and may be related to the changes in the rate at which this strain completes

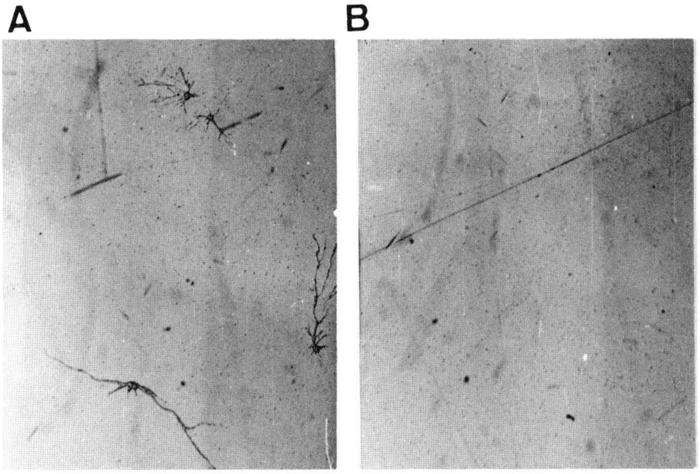

Fig. 1. Effect of DSF on cell aggregation. Cells were starved in the presence of DSF excreted by Agip 53 amoebae as described in the text. 1A shows response of cells incubated in the presence of DSF. 1B shows, in the same surface area, cells which have been starved without addition of DSF.

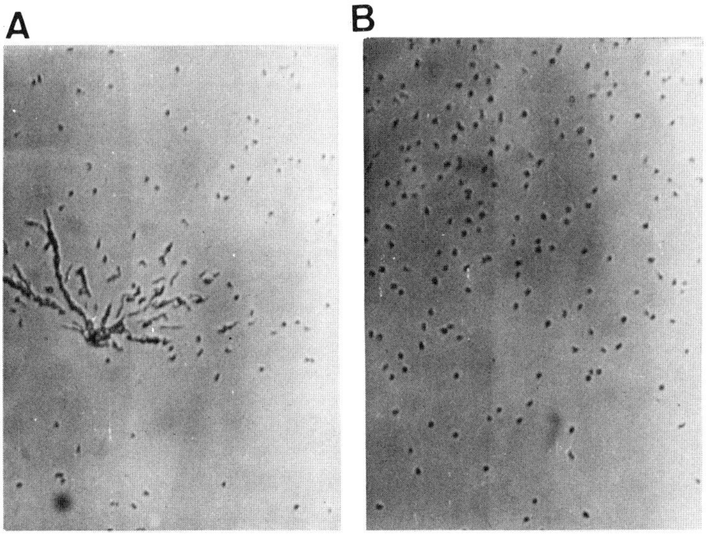

Fig. 2. Same experiment as Fig. 1 but showing cells at a 4-fold magnification:

2A: Cells incubated with DSF; 2B: No DSF added.

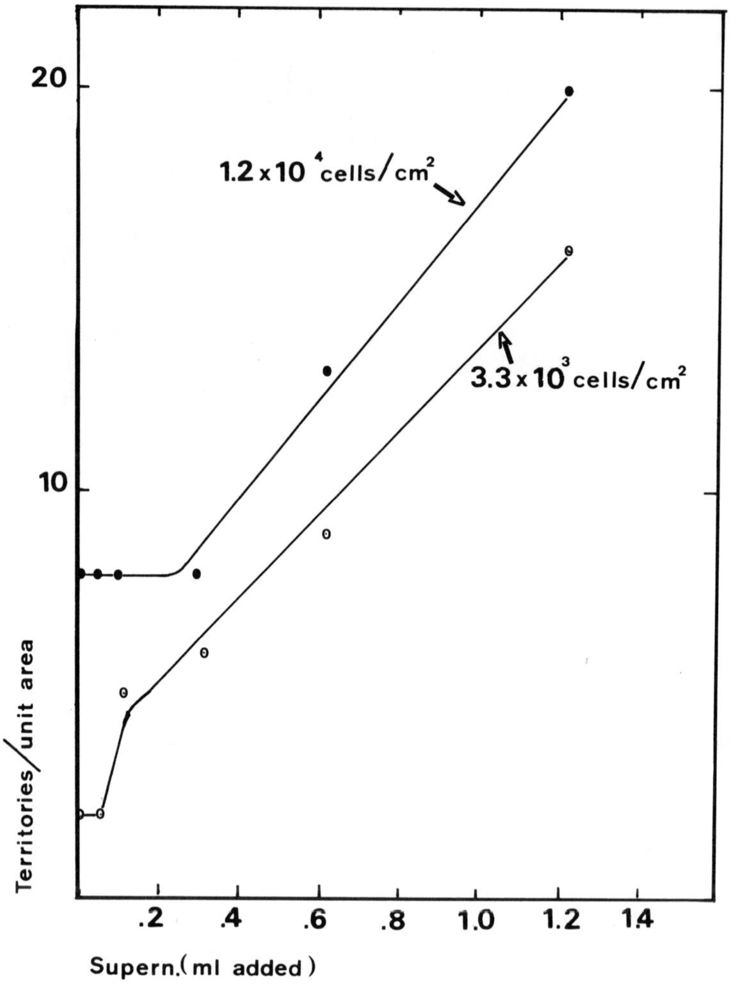

Fig. 3. Number of organized territories versus the amount of DSF added. Cells were incubated for 20 hours at the indicated densities with increasing amounts of DSF. Control cells were starved without added DSF. Quantitation of territories was performed as described in the text.

its aggregation process under optimal conditions. A maximal effect of DSF was generally observed with 1-1.5 ml of added supernatant. Higher amounts, in some cases, were found to be inhibitory. When dilute populations of cells were starved overnight prior to the addition of DSF, they still exhibited a stimulatory response to the factor and formed aggregates. Therefore, during the first 24 hours, the inability of cells to aggregate when starved at low densities is not due to their decreased viability or the loss of any unstable cell constituents.

Comparison of DSF and Phosphodiesterase

It has been reported that the aggregation competence of cells incubated at low densities in spinner suspensions is stimulated by phosphodiesterase.[2] Comparison of the phosphodiesterase activities excreted by various cell types with the aggregation-promoting effects of their supernatants (Table I) suggests that this enzyme activity is not equivalent to DSF. The most stimulatory supernatant (Agip 53) contains only two-thirds the enzymatic activity of wild-type cells while Agip 55, which excretes the same phosphodiesterase activity as Agip 53, produces minimal DSF (approximately equivalent to wild-type). Induction of phosphodiesterase in wild-type and Agip 53 cells with high concentrations of cAMP (13) did not result in an increase in DSF activity. In the two experiments using Agip 53, the minimal amount of supernatant necessary to provoke a response was determined and found to be the same using preparations from induced and uninduced cells. Further evidence that DSF is not phosphodiesterase was obtained by comparing the stability of these activities at different temperatures and pH's. The differences in the percent residual activity of phosphodiesterase and DSF in supernatants heated at various temperatures for 10 min are shown in Figure 4. Heating at 50° destroys DSF activity by approximately one-half, but does not affect phosphodiesterase activity (Fig. 4A). Treatment at 60° totally eliminates DSF activity, while one-half of that of the phosphodiesterase remains. The heat-sensitivity of DSF also indicates that it is not the phosphodiesterase inhibitor which is also excreted during starvation (10). The resistance of DSF and phosphodiesterase to changes in pH are shown in Figure 4B. In these experiments, supernatants were adjusted to the indicated pH with either HCl or KOH. DSF activity was determined after samples were re-adjusted to pH 6.2. DSF appears to be very sensitive to changes of pH, showing substantial loss of activity above or below pH 6.2. Maximal phosphodiesterase activity, however, is observed at the more alkaline pH's.

DISCUSSION

In this communication, evidence has been presented for the existence of a macromolecular factor excreted by Dictyostelium discoideum which stimulates the differentiation of these amoebae. Evidence for such an effector was obtained by adding the media of amoebae incubated

[2] Alcantara, F. and Bazill, Q., submitted for publication.

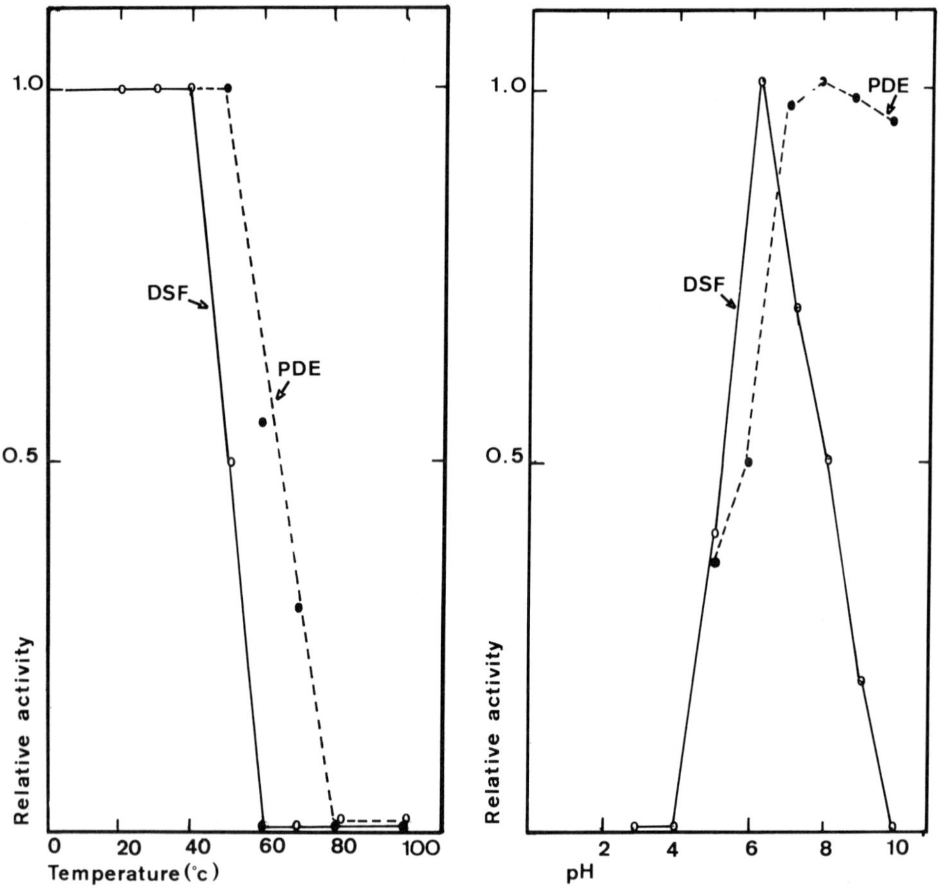

Fig. 4. Stability of DSF and phosphodiesterase. 4A: Aliquots of supernatant were heated at various temperatures for 10 min, cooled, and then assayed for DSF and phosphodiesterase activity. 4B: Aliquots were adjusted to the indicated pH's and tested for DSF and phosphodiesterase activity. In all cases activities are considered on a scale of 0-1 relative to peak activity.

at high densities in buffer to cells starved at very low densities, generally 3-6 x 10^3 cells/cm^2, on Petri dishes. In the absence of any added compounds, these low densities result in minimal cell differentiation and are more than an order of magnitude lower than the concentrations previously used to study the parameters involved in cell aggregation (5-70 x 10^4 cells/cm^2). Quantitation of the differentiation stimulating factor (DSF) was performed by comparing the number of organized territories which arose when low density amoebae were starved in the absence or presence of factor. Territories, as defined, consist of clouds of aggregation-competent cells, groups of cells streaming toward central collection points, and/or cell aggregates. The number of territories formed was proportional to the quantity of DSF present when added in amounts above a critical level. It is not known if DSF stimulates oriented cell movement, which then results in the formation of cell aggregates, or if this factor induces "centers", which in diluted populations would statistically be less likely to form, to which cells may converge.

DSF was shown to be excreted during the first few hours of cell starvation, reaching a maximum at three hours. The factor is thermosensitive and maximally stable between pH 6 and 7. The experiments presented here indicate that DSF is not one of the extracellular macromolecules previously described (phosphodiesterase or its inhibitor) which have been implicated in cell aggregation.

Alcantara and Bazill[2] have observed a stimulation of aggregation when cells are starved at 5-7 x 10^6/ml in spinner suspensions incubated in the presence of supernatants from AX-2 amoebae starved at higher concentrations. The fact that *D. discoideum* phosphodiesterase co-purified with their stimulatory factor (AF) and that rat brain phosphodiesterase was equally effective in stimulating aggregation under their conditions lead these authors to conclude that AF was probably phosphodiesterase.[2] However, in spinner suspensions, cells even at densities as low as 10^5/ml do not remain as individual amoebae.[3] Therefore, it is possible that the stimulatory effect of phosphodiesterase is an event secondary to the formation of cell clusters. This hypothesis is consistent with our observation that, under the conditions described here, in which no cell contacts normally occur, beef heart phosphodiesterase did not stimulate cell aggregation.[3] However, preliminary evidence suggests that increasing amounts of phosphodiesterase could advance the completion of aggregation in preparations containing a fixed DSF activity.

The level at which DSF exerts its effect is not known. Stimulation of cell aggregation naturally would occur if this factor enhanced either the reception or relay of the chemotactic signal, for example, by unmasking cell surface receptors for cAMP. In that case, DSF may actually prove to be the primary effector of the differentiation of cells to aggregation-competence. This is consistent with our observation that a stimulation of cell differentiation by DSF is only observed when cells are starved at low densities. Since with high cell concentrations DSF is no longer produced in limiting

---

[3] Darmon, M. and Klein, C., unpublished observations.

quantities, the differentiation of cells would then appear to be solely under the control of cAMP.

We would also like to advance the hypothesis that the effect of DSF on cell aggregation may be only a reflection of a more general effect of this compound in the life cycle of Dictyostelium discoideum. In this respect, DSF may play a role in inducing, or stabilizing, the metabolic state of cells vital to the successful undertaking or completion of their development program.

ACKNOWLEDGEMENTS

This research was supported by grants from the Centre National de la Recherche Scientifique. CK is a Chercheur Associé of the CNRS.

REFERENCES

1. BALSAMO, J., and LILIEN, J., Proc. Nat. Acad. Sci. USA 71 (1974) 727.
2. BARKLEY, D. S., Science 165 (1969) 1133.
3. BEUG, H., KATZ, F. E. and GERISCH, G., J. Cell. Biol. 56 (1973) 647.
4. BONNER, J. T., BARKLEY, D. S., HALL, E. M., KONIJN, T. M., MASON, J. W., O'KEIFE, G. and WOLF, P. B., Develop. Biol. 20 (1968) 72.
5. DARMON, M., BRACHET, P. and PEREIRA DA SILVA, L. H., Proc. Nat. Acad. Sci. USA (1975), in press.
6. GERISCH, G., Exptl. Cell Res. 25 (1961) 535.
7. GERISCH, G., Current Topics in Developmental Biology 3 (1968) 157.
8. GERISCH, G., FRONM, H., HUESGEN, A. and WICK, U. Nature 255 (1975) 547.
9. GERISCH, G., MALCHOW, D., HUESGEN, A., NANJUNDIAH, V., ROOS, W. and WICK, U., Proceedings of the 1975 ICN-UCLA Symposium on Developmental Biology (D. McMahon and C. F. Fox, eds) W. A. Benjamin (1975).
10. GERISCH, G., MALCHOW, D., RIEDEL, V., MÜLLER, E. and EVERY, M., Nature New Biology 235 (1972) 90.
11. GILBERT, L. I., The Hormones (G. Pincus, K. V. Thimann and E. B. Ashwood, eds) Vol. 4 (1964) 67.
12. HARRIS, R. J. C. (ed) Biological Organization at the Cellular and Supercellular Level. Academic Press, N. Y. (1963).
13. KLEIN, C., J. Biol. Chem. (1975) in press.
14. KONIJN, T. M., VAN DE MEENE, J. G. C., BONNER, J. T. and BARKLEY, D. S., Proc. Nat. Acad. Sci. USA 58 (1967) 1152.
15. MALCHOW, D. and GERISCH, G., Proc. Nat. Acad. Sci. USA 71 (1974) 2423.

16. MOSCONA, A. A., J. Cell Comp. Physiol. 60 (1962) 65.
17. ROOS, W., NANJUNDIAH, V., MALCHOW, D. and GERISCH, G., FEBS Letters 53 (1975) 139.
18. SCHAFFER, B. M., Advances in Morphogenesis 2 (1962) 109.
19. SCHAFFER, B. M., Biochem. Biophys. Res. Comm. 65 (1975) 364.
20. SCHAFFER, B. M., Nature 255 (1975) 549.

Section IV

Cell-Small Molecule Interactions

CONFORMATIONAL CHANGES OF THE CHOLINERGIC RECEPTOR PROTEIN FROM *TORPEDO MARMORATA* AS REVEALED BY QUINACRINE FLUORESCENCE†

Hans-Heinrich GRÜNHAGEN* and Jean-Pierre CHANGEUX

Institut Pasteur, Service de Neurobiologie Moleculaire
Paris (France)

In chemical synapses the arriving pulse induces on the presynaptic side the release of a transmitter, which diffuses through the synaptic cleft and binds on the post-synaptic side to a receptor. Subsequent events involving an ionophore finally lead to the opening of ionic channels and a passive flow of ions results in a local discharge of the potential across the membrane (6). Electrophysiological experiments have shown that the depolarizing effects of physiological transmitters can be mimicked by structurally related compounds (agonists), whereas a different class of compounds has the ability to inhibit reversibly the depolarizing action of agonists (antagonists). In recent years biochemical investigations have focussed on the acetylcholine receptor for which nicotine is an agonist and d-tubocurarine is an antagonist (3, 11). This receptor is found, for example, in the neuromuscular junction and in large quantities in the electric organ of some fish species. It has been found that snake α-toxins, which block *in vivo* the function of the postsynaptic membrane, bind with high affinity and specificity to biochemical preparations of the receptor (2). The use of these toxins has made possible considerable progress in the field of preparation of the receptor, and one can now study the interesting questions concerning the functional events in the receptor system.

---

*Recipient of a long-term fellowship from EMBO (1974/75) and of a fellowship from the Deutsche Forschungsgemeinshaft (1975/76).

†Abbreviations used: Quin : quinacrine; Carb : carbamylcholine; Hexa : hexamethonium; Flax : flaxedil; ACh : acetylcholine; Pril : prilocaine; PTA : phenyltrimethylammonium.

For this physicochemical approach quinacrine has been chosen as a fluorescence probe. Due to the optical properties of the acridine system in this compound its fluorescence depends on the local environment (9). On the other hand, there is a structural similarity between quinacrine and the group of local anesthetics containing a tertiary amine in the side chain. These local anesthetics at low concentrations increase the affinity of the receptor for agonists and at high concentrations compete with agonists for the binding to the receptor (4). An interaction of quinacrine with the receptor was therefore likely and a fluorescence response to functional events in the receptor system could be expected.

Fluorescence Spectra of Quinacrine Bound to the Receptor System

The physicochemical experiments have been carried out with DL-quinacrine and receptor-rich membrane fragments from *Torpedo marmorata* (5) in *Torpedo* physiological saline solution (150 mM NaCl, 5 mM KCl, 4 mM $CaCl_2$, 2 mM $MgCl_2$, 5 mM sodium phosphate, pH 7). The fluorescence spectra were obtained at 20° with a Fica differential spectrofluorimeter. The spectra in Fig. 1 show that quinacrine binds to receptor rich membrane fragments and responds in the membrane-bound state to the addition of cholinergic effectors. Part a) represents the emission spectra of quinacrine in the bound state, which are slightly redshifted compared to aqueous quinacrine (8). They have been obtained with the supernatant of the membrane suspension as the reference. The fluorescence has been excited at 350 nm, an isosbestic point of the pH-dependent absorption spectra. The quantitative analysis shows that the amount of quinacrine bound to the membrane is approximatively a linear function of the quinacrine concentration in the range of 10^{-7} to 2×10^{-5} M. For a typical receptor concentration of 10^{-7} to 10^{-6} M about 1 to 10% of the total quinacrine is bound to the membrane in this phase equilibrium.

Part b) of Fig. 1 shows that the fluorescence intensity increases when quinacrine binds to the membrane. The spectra have been obtained with an equal total concentration of quinacrine in sample and reference, membrane fragments being added to the sample only. It can be calculated that the quantum yield changes from about 0.05 in the aqueous phase (9) to about 0.2 in the membrane-bound state. The upper trace in this part of the figure shows an additional increase of the fluorescence after the addition of carbamylcholine to the sample solution. The amplitude of this effect has been used to determine the equilibrium binding constants of cholinergic agonists and antagonists. Table I demonstrates the agreement of these data with data determined by the toxin binding method (12).

TABLE I

Dissociation Constants of Nicotinic Effector-Receptor Complexes as Determined by Quinacrine Fluorescence *

Effector	Solution	pH	C_{50} [M] (Quinacrine-fluorescence)	K_p [M] (Protection against [^3H]-α-toxin binding) (a)
Carbamylcholine	Ringer	7	(4±2) x 10^{-7}	5 x 10^{-7}
Carbamylcholine	Ca-and Mg-free Ringer	8	(4±2) x 10^{-7}	
Carbamylcholine	Ca-and Mg-free Ringer	7	(6±2) x 10^{-7}	
Decamethonium	Ringer	7	(6±2) x 10^{-7}	7 x 10^{-7}
Hexamethonium	Ringer	7	(6±3) x 10^{-5}	4 x 10^{-5}
Flaxedil	Ringer	7	(2±1) x 10^{-5}	8 x 10^{-6}

*λ_{Ex} = 350 nm, λ_{Em} = 515 nm, quinacrine 10^{-5} M, C_{50} : concentration of effector causing the half-maximum fluorescence change.

(a) data from Weber and Changeux (12).

In the presence of saturating concentrations of, for example, carbamylcholine the amplitude of the additional increase is a function of the quinacrine concentration and is saturated at about 5 x 10^{-6} M quinacrine for a receptor concentration of 4 x 10^{-7} M. This additional fluorescence increase has been observed for all agonists and competitive antagonists tested with the exception of α-toxin. There is no increase of the fluorescence after addition of α-toxin, but incubation with α-toxin inhibits the additional fluorescence increase caused, for example, by carbamylcholine.

In part c) of Fig. 1 the fluorescence technique of energy-transfer from proteins (excited at 290 nm) to quinacrine has been applied. Sample and reference contain identical concentrations of membrane fragments and quinacrine. It can be seen that the addition of carbamylcholine to the sample causes an increase of the fluorescence. This effect is observed with all agonists tested and also for some antagonists. α-Toxin does not cause itself fluorescence changes, but again it inhibits the fluorescence effect of carbamylcholine.

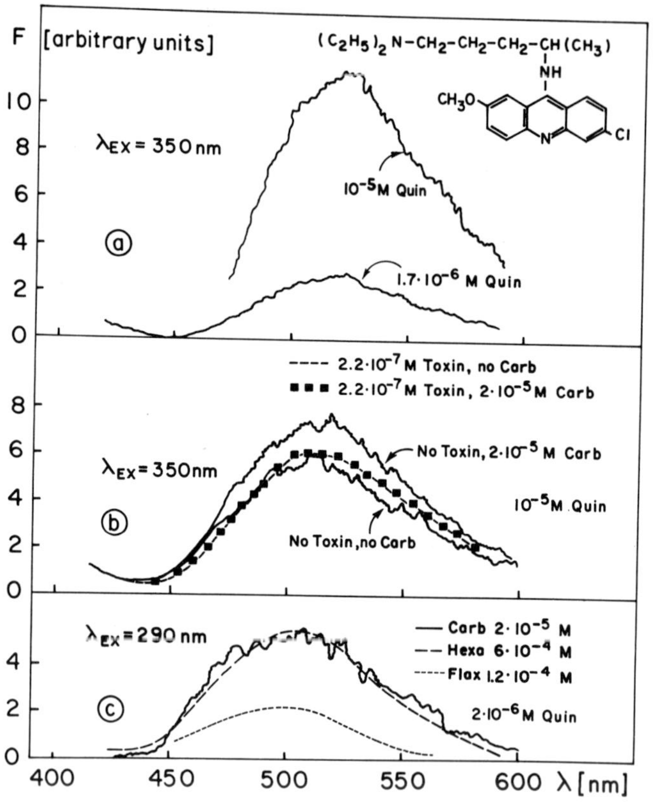

Fig. 1. Fluorescence Spectra of Membrane-bound Quinacrine.

a) and b) : Quinacrine directly excited.

a) Fluorescence spectrum of membrane-bound quinacrine. (Difference measurement; samples : quinacrine solution in the presence of membrane fragments (1.5 x 10^{-7} M toxin binding sites, 0.18 mg/ml protein); reference : supernatant of this solution (60 min, 100,000 x g)).

b) Effect of carbamylcholine. (Difference measurement; samples : quinacrine 10^{-5} M, membrane fragments (1.2 x 10^{-7} M toxin binding sites, 0.4 mg/ml protein; reference : quinacrine 10^{-5} M: reagent added to both cuvettes).

Fig. 1 (Cont'd.)

c) Effect of carbamylcholine, hexamethonium and flaxedil on energy-transfer from protein to quinacrine.
(Difference measurement; samples : quinacrine 2 x 10^{-6} M, membrane fragments (1.5 x 10^{-7} M toxin binding sites, 0.11 mg/ml protein, effector concentration as indicated; references : as samples, but without effector).

Pharmacological Characterization of Quinacrine

The fluorescence effects demonstrate an interaction between quinacrine and the receptor system. The nature of this interaction has been characterized by the following experiments : on *Electrophorus electricus* electroplaques quinacrine has no depolarizing activity. It blocks, however, non-competitively the depolarizing effect of agonists. The concentration required for half maximum of this antagonistic effect is about 10^{-5} M. In *Torpedo* membrane fragments quinacrine increases the affinity of acetylcholine for the receptor at low concentrations around 10^{-6} M, whereas it inhibits the binding of acetylcholine at high concentrations above 10^{-5} M. These effects resemble the effects of local anesthetics with a tertiary amine group (4). According to fluorescence investigation of quinacrine, binding to the membrane reveals a competition with typical local anesthetics such as prilocaine and quotane (Fig. 2). Both local anesthetics cause a strong decrease of quinacrine fluorescence in the concentration range where they increase the affinity of acetylcholine for the receptor (4). This suggests that the interaction of quinacrine with the receptor qualitatively corresponds to the effect of local anesthetics. Nevertheless a quantitative analysis indicates a local anesthetic effect of quinacrine less pronounced than found for a typical member of this group.

Quinacrine as a Probe for Conformation Changes of the Receptor

The fluorescence technique of energy transfer has been used to study the response of the receptor system to the addition of cholinergic effectors. In addition to the equilibrium values, the transient equilibration phenomena were recorded to gain further insight into the processes leading from one state to another. To facilitate the discussion of the experimental data, a schematic model of elementary states of the receptor-ionophore-system is presented in Fig. 3. The R-state may be related to the physiological resting state (closed channel), the A-state to the activated (open channel) and the D-state to the desensitized state (channel closed after prolonged application of an agonist). In the absence of effectors, under physiological conditions, the intrinsic equilibria

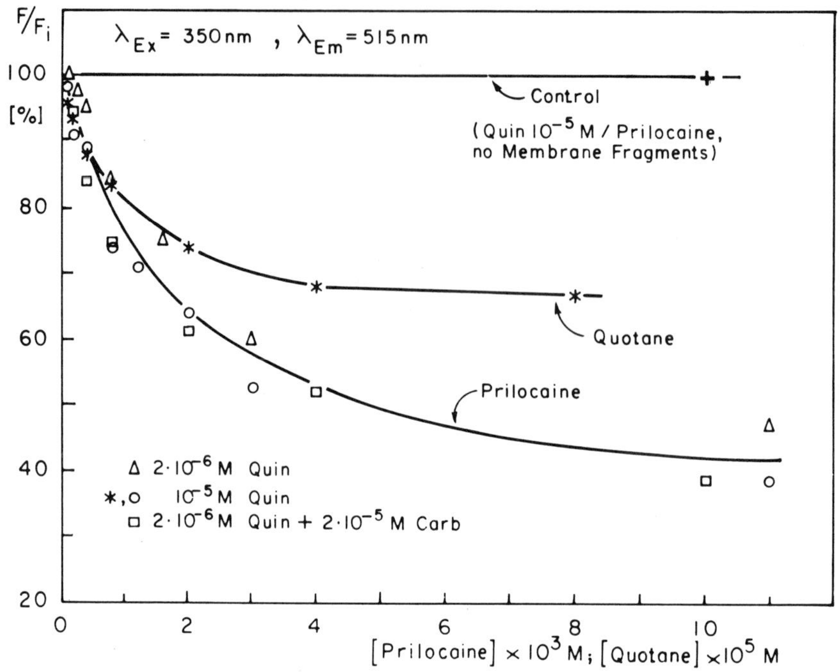

Fig. 2. Effect of Prilocaine and Quotane on Fluorescence of Membrane-bound Quinacrine. Differential measurement; sample solution : quinacrine as indicated, membrane fragments (3×10^{-7} M toxin binding sites, 0.4 mg/ml protein for Δ and O; 4×10^{-7} M toxin binding sites, 0.3 mg/ml for x and ■ ; carbamylcholine where indicated; prilocaine and quotane added as indicated (maximal dilution 3%); reference solution : as sample, but without membrane fragments. F is the fluorescence increase due to the presence of membrane fragments (and carbamylcholine where indicated); F_i, its initial value in the absence of prilocaine or quotane. In the case of the control experiment, the ratio F/F_i indicates the absolute flourescence of the corresponding quinacrine solutions in the absence of membrane fragments (test for absorption and quenching effects).

in this system favor the resting state. They can be shifted after addition of an effector E. Since after binding of an agonist, the desensitized state is attained, it is assumed that the D-state has an affinity for agonists higher than that of the R-state. The A state may have an intermediate affinity for agonists or one slightly higher than the R-state.

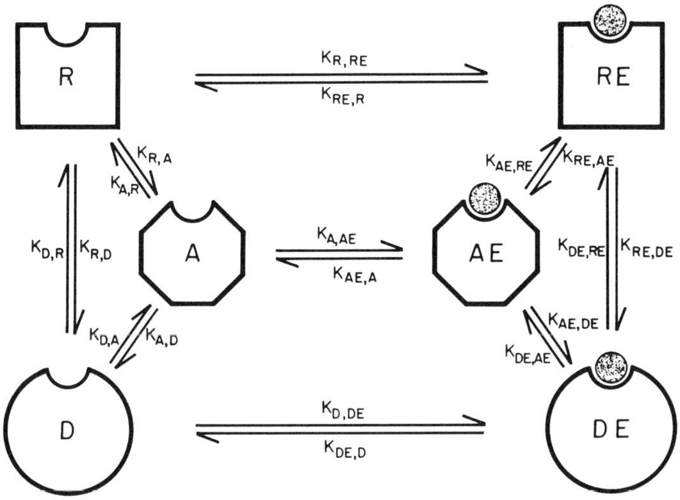

Fig. 3. Schematic representation of structural and binding states of the receptor.

Agonists. Fig. 4 shows the equilibration process after addition of carbamylcholine. If immediately a high concentration is added, a fast increase of the fluorescence intensity occurs, the time course of which cannot be resolved with this technique. Afterwards a slow decrease leads to the final state. If, however, small concentrations,of the order of the known equilibrium constant, are added, only slow upward signals are observed, before the same saturated state is reached. For all tested agonists (carbamylcholine, acetylcholine, phenyltrimethylammonium) this qualitative difference between concentrations of the order of the dissociation constants and concentrations considerably higher is observed.

If one assumes tentatively, 1) that quinacrine responds to the conformation changes between the R-, A-, and D-state and 2) that the fluorescence intensity is lowest in the R-state, intermediate in the D-state and highest in the A-state, the following conclusions may be drawn : at high concentrations agonists initially bind largely to the low affinity R-state, thereby causing a transition from RE to AE (fast fluorescence increase). Afterwards, the receptor transforms slowly to the high affinity DE state (slow fluorescence decrease). Low concentrations are not sufficient to bind

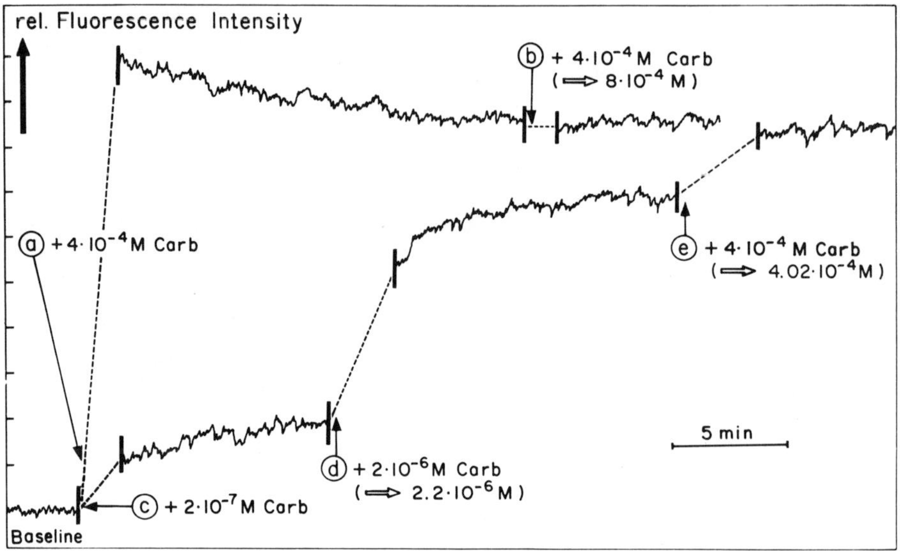

Fig. 4. Equilibration processes after addition of carbamylcholine. Receptor-rich membrane fragments labeled with quinacrine. Change of energy-transfer fluorescence. Baseline-conditions, sample and reference : quinacrine 2 x 10^{-6} M, membrane fragments (2 x 10^{-7} M toxin binding sites). Additions to 2.5 ml sample (reference) : a) 10 µl Carb, 0.1 M (10 µl H_2O); b) as a); c) 5 µl Carb, 10^{-4}M (5 µl H_2O); d) 0.5 µl Carb, 10^{-2} M (-); e) 10 µl Carb, 10^{-1} M (10 µl H_2O).

substantially to the low affinity R-state and instead of a prompt fluorescence increase, reflecting the transition to AE, only the equilibration towards DE is observed (slow fluorescence increase). To verify these hypothetical explanations the equilibration processes of other classes of cholinergic effectors were investigated.

Antagonists. Under physiological conditions antagonists by definition do not cause a transition from the R- to the A-state. In Fig. 5, the effects following the addition of antagonists (hexamethonium and flaxedil) are compared with those of the agonist phenyltrimethylammonium. In spite of their high concentration (cf Table I) the antagonists do not provoke the fast fluorescence increase assigned to the transition from the RE to the AE state. Their addition causes only a slow saturable fluorescence increase, the amplitude of which differs within the groups of antagonists.

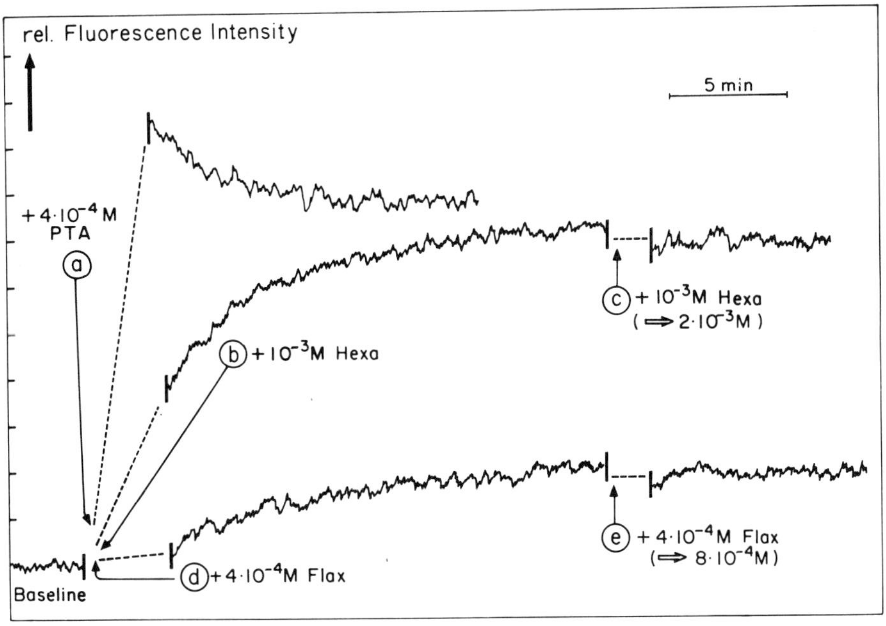

Fig. 5. The effect of antagonists in comparison with the effect of an agonist. Receptor-rich membrane fragments labeled with quinacrine. Change of energy-transfer fluorescence. Baseline conditions as in Fig. 4. Additions to 2.5 ml sample (reference): a) 10 µl PTA, 10^{-1} M (10 µl NaCl 10^{-1} M); b) 25 µl Hexa, 10^{-1} M (25 µl NaBr 10^{-1} M; c) 25 µl Hexa, 10^{-1} M (25 µl NaBr 10^{-1} M); d) 10 µl Flax, 10^{-1} M (10 µl NaCl 0.3 M); e) 10 µl Flax, 10^{-1} M (10 µl NaCl 0.3 M).

The antagonists, tetraethylammonium and α-toxin, do not change the energy transfer fluorescence of quinacrine at all. In contrast the final level of fluorescence increase caused by the agonists carbamylcholine, choline, phenyltrimethylammonium and decamethonium has been found to be identical in a given preparation (the amplitude of the acetylcholine effect is smaller and depends on the applied esterase agent). The close similarity of the final state in this chemically heterogenous group of effectors strongly supports the previous assumption, that the energy transfer from protein to quinacrine responds to a conformation change rather than to direct interaction between effector and quinacrine.

The transformation of the receptor to a well-defined conformation after binding of agonists raises the question of whether the partial transformation (flaxedil) or complete transformation (hexamethonium) of the receptor in the presence of antagonists reflects

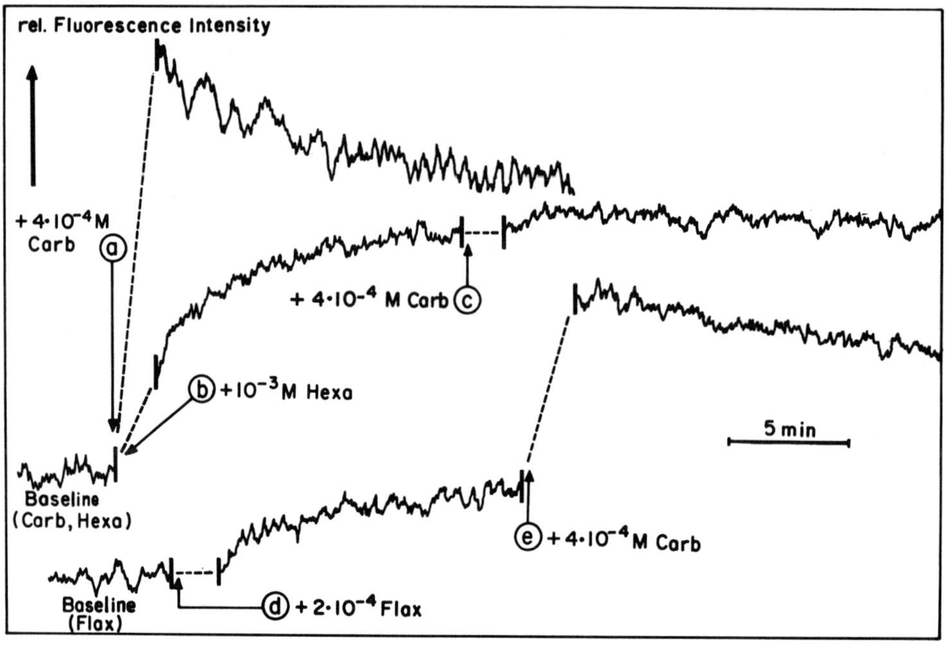

Fig. 6. Response to an agonist after preincubation with antagonists. Receptor-rich membrane fragments labeled with quinacrine. Change of energy-transfer fluorescence. Baseline conditions as in Fig. 4. Additions to 2.5 ml sample (reference): a) 10 µl Carb. 10^{-1} M 10 µl NaCl 10^{-1} M); b) 5 µl Hexa. 0.5 M (5 µl NaCl 0.5 M); c) 10 µl Carb. 10^{-1} M (10 µl NaCl 10^{-1} M; d) 5 µl Flax. 10^{-1} M (5 µl NaCl 0.3 M); e) 10 µl Carb. 10^{-1} M (10 µl NaCl 10^{-1} M. In contrast to the figures 4 and 5 and 7 and 8, where only one common baseline is used, in this figure the flaxedil experiment has a baseline differing from the other traces (as indicated).

the same type of conformation change as observed with agonists. The emission spectra (cf. Fig. 1c) are in agreement with the assumption of one common conformation change.

Further strong evidence comes from a kinetic experiment (Fig. 6). After a saturating dose of hexamethonium, an addition of carbamycholine does not cause any further change, not even a transient one. When carbamylcholine is added after a saturating dose of flaxedil, one observes a fast increase of the fluorescence to a level higher than the final one, followed by a slow decrease to the final level obtained in the absence of flaxedil. The amplitude of the transient over-shoot after preincubation with flaxedil is

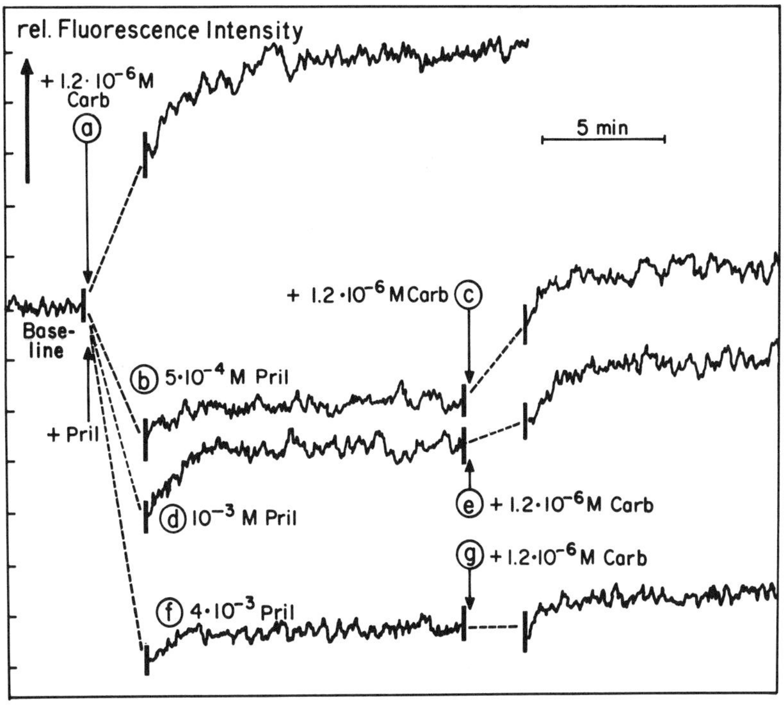

Fig. 7. Effect of prilocaine on equilibration processes. Receptor-rich membrane fragments labeled with quinacrine. Change of energy transfer fluorescence. Baseline conditions as in Fig. 4. Additions to 2.5 ml sample (reference): a) 3 µl Carb. 10^{-3} M (-); b) 12.5 µl Pril/Ringer, 10^{-1} M (12.5 µl Ringer); c) as a); d) 25 µl Pril/Ringer, 10^{-1} M (25 µl Ringer); e) as a); f) 100 µl Pril/Ringer, 10^{-1} M (100 µl Ringer); g) as a).

reduced compared to the immediate addition of carbamylcholine. These preincubation effects of antagonists are exactly those to be expected after pre-treatment with an agonist causing the corresponding fluorescence change. It has therefore to be concluded, that some antagonists can transform the receptor towards the DE state and that the effects of antagonists under physiological conditions may be due to two different molecular phenomena : competition for receptor sites in the prefunctional state (RE) and transformation of receptor units to the refractory D and DE state.

Local Anesthetics. The slow conformational change of the receptor from the R to the D state is accompanied by an increase of the affinity for agonists. Nevertheless, the ionophore is closed in the D state under physiological conditions. This phenomenon reveals an interesting analogy to the effect of local anesthetics which block the depolarizing activity of agonists although the affinity of the receptor for the agonists is found to be increased (4). Fig. 7 shows the effect of a typical local anesthetic, prilocaine, on the conformation of the receptor as revealed by quinacrine. The addition of prilocaine causes first a fast decrease of the fluourescence, which corresponds to the competition between prilocaine and quinacrine (cf Fig. 2). This is followed by a small but significant slow increase. The addition of carbamylcholine to this system causes a slow additional increase which however is reduced in amplitude compared to the pure carbamylcholine effect. At 4 x 10^{-3} M prilocaine, which causes the maximal increase of agonist binding to the receptor (4), the addition of carbamylcholine does not cause any further increase of the fluorescence. This suggests that prilocaine transforms the receptor from the R to the D state. Quantitative analysis of the amplitudes, taking into account the partial displacement of quinacrine from the receptor by prilocaine, is in agreement with this.

TABLE II

Effect of Prilocaine and Quinacrine on the Half-lifetime of the Slow Equilibration Process *

Quinacrine [M]	Addition [M]	Pretreatment [M]	$\tau_{1/2}$ [sec]
1 x 10^{-6}	Carbamylcholine 1.2 x 10^{-6}	-	129 ± 10
2 x 10^{-6}	1.2 x 10^{-6}	-	95 ± 10
2 x 10^{-6}	1.2 x 10^{-6}	Prilocaine 10^{-3}	34 ± 15
2 x 10^{-6}	Prilocaine 10^{-3}	-	45 ± 15
4 x 10^{-6}	Carbamylcholine 1.2 x 10^{-6}	-	70 ± 10

*Energy transfer fluorescence; λ_{Ex} = 290 nm, λ_{Em} = 515 nm.

Concerning the local anesthetic effect of quinacrine itself, the kinetics of the slow fluorescence signal corresponding to the conformation change towards the D state have been analyzed. Table II shows that increasing quinacrine concentration accelerates the response of the receptor to carbamylcholine addition, but obviously the acceleration effect is less prounced than with prilocaine at concentrations where it shows its pharmacological activity.

The following conclusions can be drawn from these kinetic data: 1) local anesthetics shift the intrinsic equilibria between the states of the receptor towards the D-state; 2) an increased rate of the forward reaction towards the D state contributes to this change of the thermodynamic constant; 3) quinacrine itself shows the corresponding effect, although to a considerably smaller extent than prilocaine. Finally, the acceleration effect of prilocaine on the fluorescence increase demonstrates that the kinetics are determined by the transition of the receptor and not by, for example, binding of quinacrine to the receptor as such.

α-Toxin. The interaction of the α-toxin from *Naja nigricollis* with the different receptor states is demonstrated in Fig. 8. The binding of α-toxin itself does not cause any change of the energy transfer signal, it blocks the effect of carbamylcholine and reverses any existing effect of carbamylcholine. This is also true after preincubation of the receptor with prilocaine. These effects can be explained as follows: the α-toxin has no obvious preference for binding to the R or D state, it does not affect the equilibria between these receptor states. Since its binding displaces any bound agonist (4), the receptor is no longer transformed to the D state and equilibrates towards the R state. On the other hand local anesthetics do not compete with the α-toxin for the acetylcholine binding site in a concentration range where they cause the increase of affinity (4). Therefore, the binding of the α-toxin cannot be expected to cause the displacement of the local anesthetic from the receptor and the equilibria between the R and D state should not be changed. The phenomenon, that the conformation change of the receptor affects the affinity for agonists and not for the competiviely binding α-toxin, may be accounted for by partial local exclusion instead of binding to exactly the same site.

Conclusions

The fluorescence signals revealed by quinacrine in receptor-rich membrane fragments make possible a physicochemical approach concerning the molecular basis of the physiological receptor function. The analysis of the equilibration processes after the addition of cholinergic effectors to the receptor revealed characteristic differences between the pharmacologically defined classes of ligands. The observed effects can be explained qualitatively by a model of the receptor protein which assumes the existence of three conformation states in reversible equilibria (cf Fig. 3).

This model combines the phenomenological description of the receptor function by Katz and Thesleff (7) with elements from the

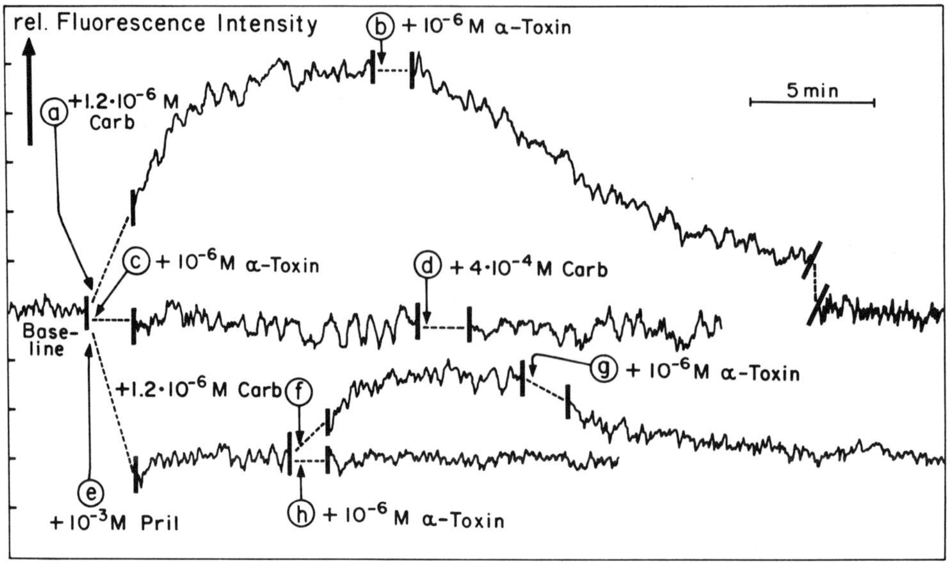

Fig. 8. Interaction of α-toxin with different receptor states. Receptor-rich membrane fragments labeled with quinacrine. Change of energy-transfer fluorescence. Baseline conditions as in Fig. 4. Additions to 2.5 ml sample (reference): a) 3 µl Carb/Ringer, 10^{-3} M (-); b) 25 µl toxin, 10^{-4} M (25 µl H_2O); c) 25 µl toxin, 10^{-4} M (25 µl H_2O); d) 10 µl Carb, 10^{-1} M (10 µl NaCl 10^{-1} M); e) 25 µl Pril/Ringer, 10^{-1} M (25 µl Ringer); f) 3 µl Carb/Ringer, 10^{-3} M (-); g) 25 µl toxin, 10^{-4} M (25 H_2O); h) as g).

theory of allosteric transitions developed by Monod, Wyman and Changeux (10). The meaning of the signals has been discussed in detail in the single classes of effectors. It can be summarized that the equilibria between the receptor states are shifted by agonists and some competitive antagonists as homotropic effectors and furthermore by local anesthetics as heterotropic effectors. The conformation changes can be classified as fast, reflecting the transition from the R to the A state, and as slow, corresponding to the equilibration towards the D state with its greater affinity for agonists. As a direct consequence it follows from these conclusions that the most interesting transition in the receptor-ionophore system, namely the activation of the ionophore by agonists, can only be studied with fast methods. Analyzing the fast transient increase of energy transfer fluorescence, which follows the addition of agonists to the receptor, may give insight into this phenomenon.

Further support for the assumption of reversible conformational changes comes from binding studies with α-toxin (Weber and Changeux) and ion flux experiments (Popot and Changeux) which were carried out with the same receptor-rich membrane fragments (1). With the toxin binding technique the affinity of the receptor for agonists has been observed to increase in the first minutes after the addition of the agonist. Furthermore, a cross preincubation effect of hexamethonium for an agonist has been observed as predicted on the basis of the fluorescence data (cf Fig. 6). The ion flux experiments with $^{22}Na^{+}$-charged microsacs revealed a desensitization phenomenon in the minute range. In agreement with the three state model and the low affinity of the prefunctional R state to agonists it has furthermore been found that the concentrations of agonists necessary to stimulate the ion flux are considerably higher than the known equilibrium binding constants.

BIBLIOGRAPHY

1. CHANGEUX, J. P., BENEDITTI, L., BOURGEOIS, J. P. BRISSON, A., CARTAUD, J., DEVAUX, P., GRÜNHAGEN, H. H., MOREAU, M., POPOT, J. L., SOBEL, A., WEBER, M., Cold Spring Harbor Symposia on Quantitative Biology (1975) in press.
2. CHANGEUX, J. P., KASAI, M., LEE, C. Y., Proc. Nat. Acad. Sci. USA 67, (1970) 1241.
3. COHEN, J. B. and CHANGEUX, J. P., Ann. Rev. Pharmacol. 15 (1975) 83.
4. COHEN, J. B., WEBER, M., and CHANGEUX, J. P., Mol. Pharmacol. 10 (1974) 904.
5. COHEN, J. B., WEBER, M., HUCHET, M., and CHANGEUX, J. P., FEBS Lett. 26 (1972) 43.
6. KATZ, B., Nerve, Muscle and Synapse. McGraw Hill, New York (1966)
7. KATZ, B. and THESLEFF, S., J. Physiol. (London) 138 (1957) 63.
8. KREY, A. K., and HAHN, F. E., Mol. Pharmacol. 10 (1974) 686.
9. MASSARI, S., DELL'ANTONE, P., COLONNA, R. and AZZONE, G. F., Biochemistry 13 (1974) 1038.
10. MONOD, J., WYMAN, J. and CHANGEUX, J. P., J. Mol. Biol. 12 (1965) 88.
11. RANG, H. P., Quart. Rev. Biophysics 7 (1975) 283.
12. WEBER, M. and CHANGEUX, J. P., Mol. Pharmacol. 10 (1974) 1, 15.

CHOLINERGIC SITES IN SKELETAL MUSCLE

Richard R. ALMON and Stanley H. APPEL

Department of Biology, University of California
La Jolla, California 92093, and
Division of Neurology, Duke University Medical Center
Durham, North Carolina 27710 (USA)

The physiological properties of mammalian skeletal muscle vary considerably as a function of innervation. Developmentally, the resting membrane potential, the cell surface distribution of acetylcholine sensitivity and the characteristics of the action potential all change (1, 2, 25). Experimental denervation causes the muscles to revert to a state similar to the one observed prior to innervation ontogenetically. In recent years considerable effort has been directed towards the correlation of molecular properties of the membrane with denervation induced changes in physiological properties of the muscle (3, 5, 8-11, 15-17, 19, 21, 22). This approach is pertinent to the understanding of both the molecular mechanisms of these physiological phenomena, and their significance to the innervation process.

The study of the molecular correlates of junctional and extrajunctional acetylcholine (ACh) sensitivity is a topic which may provide information pertinent not only to understanding receptor mechanisms, but also to understanding the role of the receptor as a surface indicator. Because of the relatively low number of cholinergic sites in skeletal muscle, as well as the variety of conditions, variables and tissues studied, several questions concerning cholingic sites in skeletal muscle require further clarification and elaboration. Among these are questions concerning the number of types of cholinergic sites and the possible identity between junctional and extrajunctional ACh receptors. Heterogeneity in the population of cholinergic sites as well as differences between normal junctional and denervated extrajunctional receptors have been indicated by a number of previous reports. Cholinergic sites derived from different sources, preparations and locations on the cell surface have been reported to vary in turnover rate (14), sensitivity to ACh (13, 23), sensitivity to cholinergic antagonists (18, 24), α-Bgtx binding (3), antigenicity (5), and isoelectric point (17).

Whether these data reflect differences in primary structure, conformation, or state of aggregation of the receptor has not yet been resolved. In addition, it is not clear that all types of binding observed relate to the biological sites responsible for muscle depolarization following interaction with ACh. The present report examines the equilibrium interactions between skeletal muscle cholinergic sites and several agents, and how they are affected by denervation.

Cholinergic Sites

A. Definition

The binding isotherm of diiodo-α-bungarotoxin (^{125}I-α-Bgtx) to detergent extracted preparations from skeletal muscle can be separated into two components. The first component is an unsaturable function representing that binding between 10^{-10} and 10^{-5} M ^{125}I- -Bgtx which cannot be blocked by d-tubocurarine (dTc) or by adding a 500-fold excess of unlabeled α-Bgtx. Similar unsaturable components have been reported in other receptor systems (12, 19, 20, 28). This unsaturable component is variously denoted as "non-biological" or "non-specific" binding. In those systems the unsaturable component is defined either by competition with an agonist or antagonist (28) or by adding a large excess of the unlabeled ligands (12, 19, 20).

Both methods (dTc and unlabeled α-Bgtx) have been employed to define an unsaturable component with comparable results. In the other systems, the unsaturable component has approximated linearity over the one to two orders of magnitude concentration range studied. In the present report, the unsaturable component also approximates linearity over a concentration range of one to two orders of magnitude. However, over the five orders of magnitude studied here, the component is best approximated by a power function in which the power (b) is less than unity. Considering the wide concentration range studied and the inclusion of a straight line in the class of power functions, the current formulation does not represent an extraordinary conceptual alteration. The most appropriate interpretation of these unsaturable components is that they represent a large variety of molecular association mediated by several different non-covalent binding mechanisms.

The second component of the isotherm represents the binding which can be blocked by dTc or made analytically insignificant by isotope dilution. This second component, therefore, defines one or more sets of sites which presumably interact as a specific index of their cholinergic definition.

B. Innervated Skeletal Muscle

Figure 1 is the isotherm for the association of ^{125}I-α-Bgtx to muscle preparations derived from normal innervated muscle. The bottom solid line is the power function empirically fit to the experimental points derived in the presence of dTc (closed circles). Statistical analysis indicates that these points clearly fit a power function (coefficient of determination 0.99). The open circles represent the experimental points in the absence of dTc. The difference between the upper and lower solid lines, therefore, represents interactions (5) of ^{125}I-α-Bgtx which are a function of a specific cholinergic character. This component indicates a discrete number of sites with an isotherm which progresses over five orders of magnitude (10^{-10} to 10^{-5} M ^{125}I-α-Bgtx). Two interpretations of these data are possible. The long progression of the isotherm can indicate either negative cooperativity or more than one set of sites. The former interpretation (cooperativity) would indicate that the binding of the toxin to each site decreases the probability of binding to each successive site. If the latter interpretation (two sets of sites) is employed, the data closely approximate the binding of the toxin to two separate sets of sites, the sites in each set being identical and independent. In the figure, the dashed line represents the sum of the power function plus a proposed high affinity set of sites ($K_A \cong 10^9$ l/mole). The upper solid line represents the sum of the power function, the proposed high affinity set ($K_A \cong 10^9$ l/mole) and a proposed low affinity set of sites ($K_A \cong 10^5$ l/mole). A definite distinction between the two interpretations cannot be made without additional information (29).

C. Effects of Denervation

The binding isotherms over the first ten day period following denervation were studied. Figure 2 is the isotherm of the binding of ^{125}I-α-Bgtx to the preparation derived from muscle on the tenth day following denervation. The bottom solid line is the power function derived from the experimental points in the presence of dTc (closed circles). The open circles are the experimental points in the absence of dTc. As in Figure 1 (innervated muscle), two interpretations of the dTc inhibited (specific) binding are possible. However, the two figures taken together suggest that the two site interpretation is most appropriate. More specifically, the data show that the high affinity area of the isotherm (10^{-10} -10^{-8} M ^{125}I-α-Bgtx) is affected much more by denervation than is the lower affinity area of the isotherm (10^{-7} -10^{-5} M ^{125}I-α-Bgtx). The simplest explanation of the changes in the isotherm over the denervation period is an alteration of N of two separate sets of sites rather than a complex alteration of a single set of sites. Therefore, this differential effect of denervation suggests independence of the two areas of the isotherm and supports an analysis of the isotherm

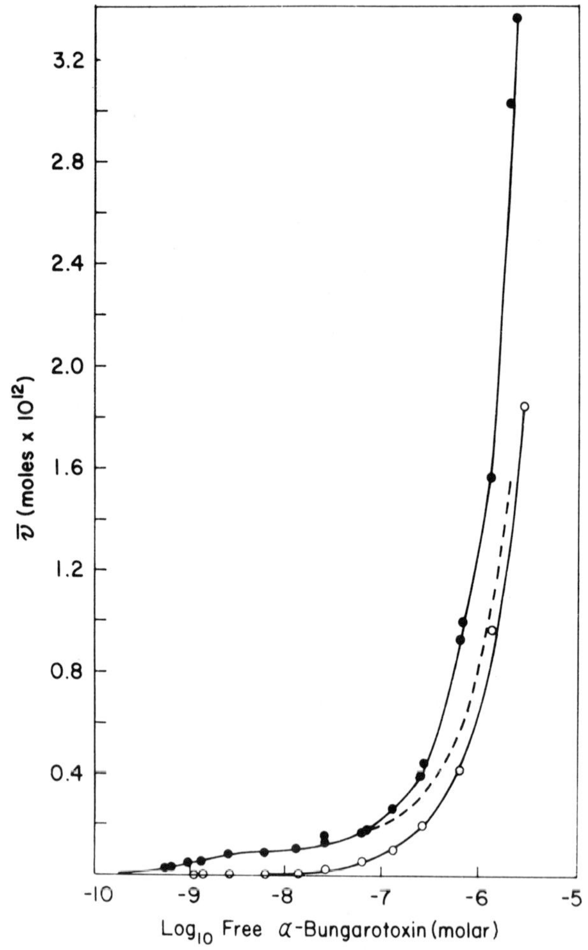

Fig. 1. The binding of ^{125}I-α-Bgtx to the triton X-100 extracted normal muscle preparation. The lower unbroken line represents the non-specific binding fitted to the power function $a[C]^b$. The open circles from which the power function is derived are experimental binding points in the presence of 4×10^{-3} Mol d-tubocurarine. The closed circles are experimental points in the absence of d-tubocurarine. The center dashed line represents the sum of the power function plus the high affinity set of binding sites.

$$\bar{\nu} = a[c]^b + \frac{N_1 K_1 [C]}{1 + K_1 [C]} + \frac{N_2 N_2 [C]}{1 + K_3 \ [C]}$$

The upper solid line represents the sum of the power function plus the high affinity set of sites and the low affinity sets of sites.

$$\bar{\nu} = a[c]^{b} + \frac{N_1 K_1 [C]}{1 + K_1 [C]}$$

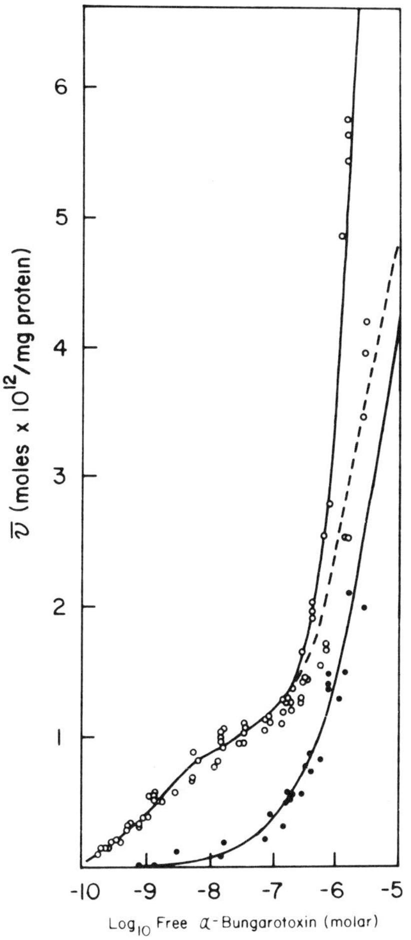

Fig. 2. Same as Figure 1 except that the triton X-100 preparation was extracted from ten day denervated muscle. (Closed circles are points derived in the presence of d-tubocurarine and open circles are points derived in the absence of d-tubocurarine).

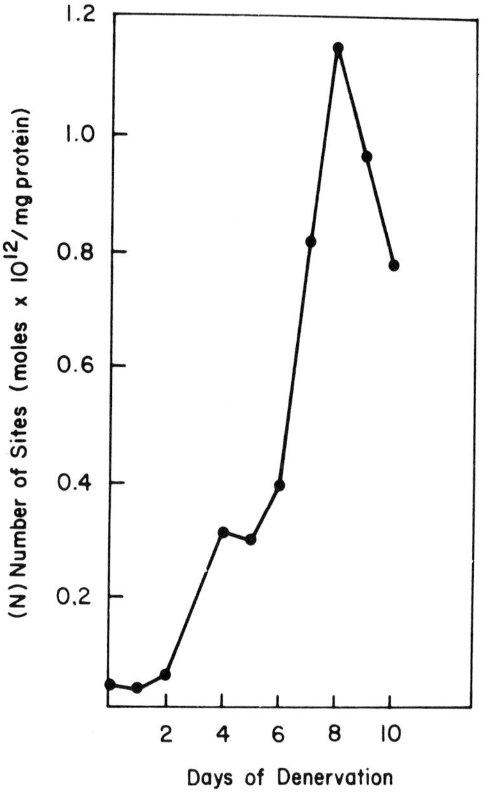

Fig. 3a. The change in the number of binding sites in the high affinity set (N_1) over the first ten day period following denervation. The points are derived from an analysis of the isotherms from preparations from each day following denervation.

as the sum of two sets of binding sites ($K_{A_1} = 10^9$ l/mole, $K_{A_2} = 10^5$ l/mole). The dashed line in this figure represents the sum of the power function and a high affinity set. The upper solid line represents the sum of the power function, a high affinity set and a low affinity set.

Based on a two site analysis, Figures 3a and 3b are graphical presentations of the change in the number of sites in the high affinity and low affinity sets of the sites respectively during the 10 day denervation period. The increase in the number of sites in the high affinity set begins at 3 days following denervation and progresses

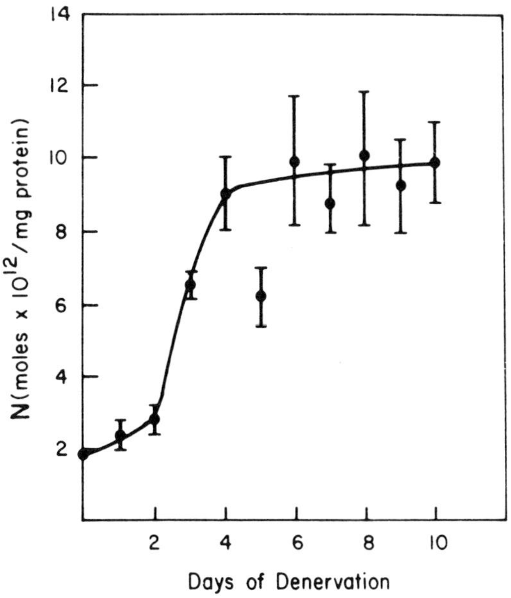

Fib. 3b. The change in the number of binding sites in the low affinity set (N_2) over the first ten day period following denervation. The points are derived from an analysis of the isotherms from preparations from each day following denervation.

to a peak at eight days (28-fold) then begins to decline. The increase in the number of sites in the low affinity set begins almost immediately following denervation but becomes most apparent at three days. The increase continues to a new level (four-fold) at about six days and remains rather constant to day ten.

Two factors suggest that the high affinity set represents the cholinergic sites responsible for muscle depolarization following its combination with acetylcholine (Ach receptor). First, the extraordinary increase in the number of sites in the set (28-fold) which probably reflects denervation supersensitivity. Second, the dissociation constant ($\sim 10^{-9}$ M) is comparable to the pharmacological activity of α-Bgtx. The biological definition of the low affinity set of sites is not clear and requires further study.

D. Interaction with Concanavalin A

Concanavalin A (Con A) which binds to α-D-mannopyranosyl or α-D-glucopyranosyl residues blocks the binding of ^{125}I-α-Bgtx to both sets of cholinergic sites defined above. Figure 4 shows the

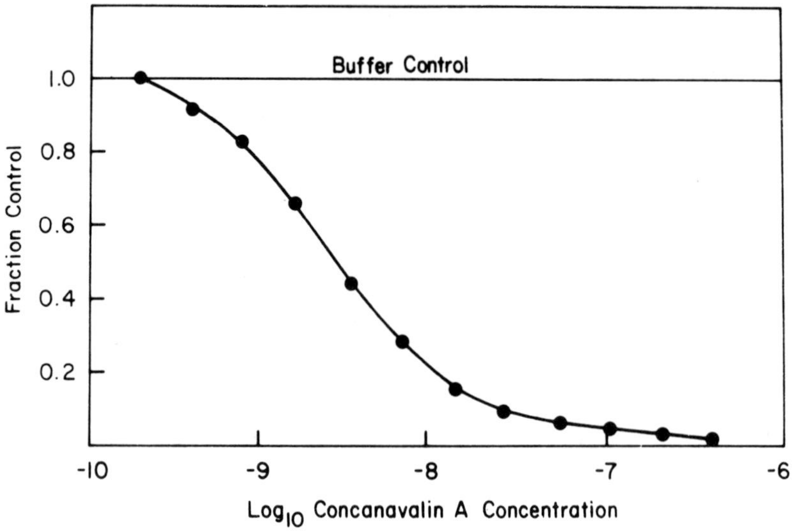

Fig. 4. A titration of the effect of concanavalin A in the binding of ^{125}I-α-Bgtx to a ten day denervated muscle preparation. The α-Bgtx concentration is 2 x 10^{-8} M.

inhibition of the high affinity set of sites by Con A. Since the inhibition is taking place in a crude skeletal muscle preparation, 10^{-9} M Con A represents an upper limit of the concentration necessary to block one half of the sites. As discussed above, this set most likely contains the sites which are responsible for ion translocation following combination with a cholinergic agonist. Similar experiments studying putative ACh receptors purified from *Electrophorus electricus* (26) and *Torpedo californica* (27) yield comparable results. In a previous report from this laboratory, preliminary data were presented indicating an interaction between Con A and the muscle ACh receptor (6). In addition, Brockes and Hall provide data demonstrating that Con A binds to a macromolecule containing a high affinity cholinergic site purified from both normal and denervated rat diaphragm (17).

The effect of Con A on the binding isotherm of ^{125}I-α-Bgtx to the high affinity set of sites has also been studied. Whereas in the absence of Con A, the isotherm of the binding of α-Bgtx to the set of high affinity cholinergic sites closely approximates the interaction of a homogeneous ligand population with a single set of identical, independent sites; in the presence of Con A, the isotherm is complex. The isotherm in the presence of 3 x 10^{-9} M Con A indicates either heterogeneity or cooperativity (data not presented). Since cooperativity is essentially a ligand-induced heterogeneity, the distinction between these two interpretations is not possible from the data. However, considering the specificity of Con A, the most likely interpretation is a steric block mediated by several carbohydrate chains in the area of the ACh site.

The apparent glycoprotein nature of the macromolecular complex(es) containing the sets of cholinergic sites has also been used to effect a partial purification. The data indicate that more than 99% of the α-Bgtx binding activity which can be inhibited by d-tubucurarine absorbs to Con A Sepharose resin. However, unlike previous reports detailing glycoprotein purifications by Con A chromatography, 20-50 mg/ml methyl-α-D-glucoside or methyl-α-D-mannoside does not remove the cholinergic binding activity. The removal of the binding activity required both a high salt concentration and the appropriate sugars. The most probable explanation of the difficulty in removal of the cholinergic binding activity from the column is that the site associated macromolecule(s) contain several carbohydrate chains with the appropriate terminal sugars. Since removal from the column will be a function of NK (N = number of sites; K = affinity of binding) rather than the single binding affinity of the individual sites (K), the requirement for having several sites free simultaneously would explain the difficult elution. The observation that Dextran Blue 2000 provides a comparably difficult elution from the Con A Sepharose column supports this conclusion.

THE ACh RECEPTOR

A. Interaction with Myasthenic IgG

In previous reports from this laboratory, differences were described in the high affinity set of sites from normal and 10 day denervated muscle preparations with respect to the interactions with the IgG fraction of sera from certain patients with the human muscle disorder myasthenia gravis (4). Using a developed radioimmunoassay method, it was observed that the IgG bound to the macromolecular complex associated with the high affinity set in denervated muscle preparations but did not bind to the comparable complex in normal muscle preparations. This binding was also reflected by the ability of myasthenic IgG to block the binding of ^{125}I-α-Bgtx to the high affinity set of sites in the 10 day denervated muscle preparations. The inhibition of the α-Bgtx binding by the myasthenic IgG did not affect the affinity of binding, but simply reduced the number of available sites (5). In ten day denervated muscle preparations the maximal extent of this inhibition was between 44-48% of the total number of sites in the high affinity set. The characteristics of the association of the myasthenic IgG with the macromolecular complex and the characteristics of the inhibition of

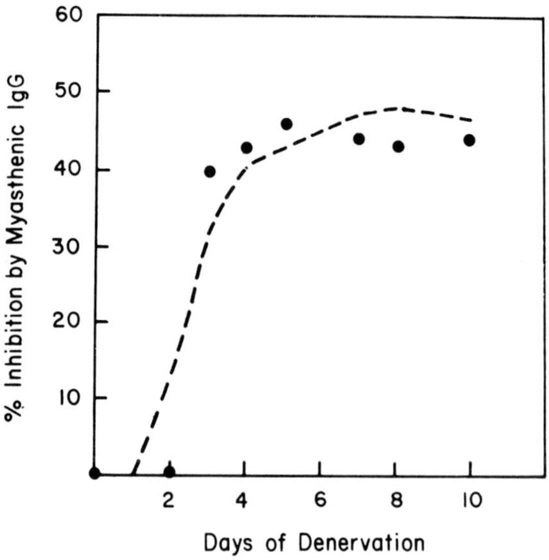

Fig. 5. The change in maximum amount of inhibition of ^{125}I-α-Bgtx to the high affinity set by IgG derived from the serum of patients T.S. and J. S. The points are derived from a titration of the IgG effect in muscle preparation from each day. The line is calculated based on the IgG producing a maximum inhibition of 50% of all new sites in the set observed following denervation.

α-Bgtx binding suggested that the myasthenic IgG associated with the newly appearing high affinity cholinergic sites in pairs. The results of this association was a steric block of toxin accessibility to one site in each pair (5, 7). In the present report, the transition from the totally insensitive normal muscle state over the 10 day denervated period is analyzed. These data are presented in Figure 5. The data closely approximate the results expected if 50% of all newly appearing sites were blocked by the myasthenic IgG. More specifically, the data show that sensitivity to the myasthenic IgG appears concurrently with the increase in sites between two and three days. The fractional inhibition progresses as would be expected if 50% of the new sites were blocked. Since the number of residual normal muscle sites become relatively insignificant by 3-5 days following denervation, the data appear to plateau between 40-50% inhibition.

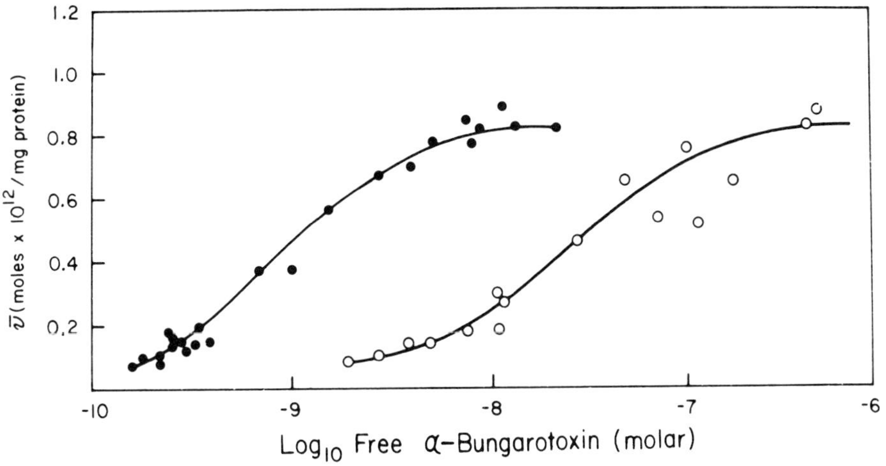

Fig. 6. The binding of ^{125}I-α-Bgtx to the high affinity set in a triton X-100 preparation from ten day denervated muscle. The control isotherm is indicated by the closed circles (●). The experimental isotherm in the presence of 8.7 x 10^{-6} M d-tubocurarine chloride is indicated by the open circles. The lines are calculated from the mass law equation for a homogeneous ligand population interacting with a single set of identical independent sites. The values (N) and (K) are derived from a Scatchard analysis of these data (2B).

B. Interaction with d-Tubocurarine (Antagonist)

Analysis of the effect of d-tubocurarine on the binding of ^{125}I-α-Bgtx to the high affinity ($K_A \cong 10^9$ l/mole) set of cholinergic sites in skeletal muscle indicates that this cholinergic antagonist is a competitive inhibitor. Figures 6 and 7 present the data for ten day denervated muscle. In these figures the closed circles represent the control data in the absence of d-tubocurarine, and the open circles represent the experimental data in the presence of 8.7 x 10^{-6} M d-tubocurarine. The log plot of the data for denervated muscle preparations (Figure 6) shows that the experimental and control isotherms saturate at the same level. Therefore, these data indicate that the number of sites available to the ^{125}I-α-Bgtx are unchanged by d-tubocurarine. Similarly, both the control and experimental isotherms fit the form of the mass law equation for a homogeneous ligand population binding to a single set of identical independent sites. The only effect of the d-tubocurarine is observed in the apparent K_D of the ^{125}I-α-Bgtx binding. The data indicates that in the presence of 8.7 x 10^{-6} M d-tubocurarine, the toxin concentration at which half the sites are occupied by the labeled toxin

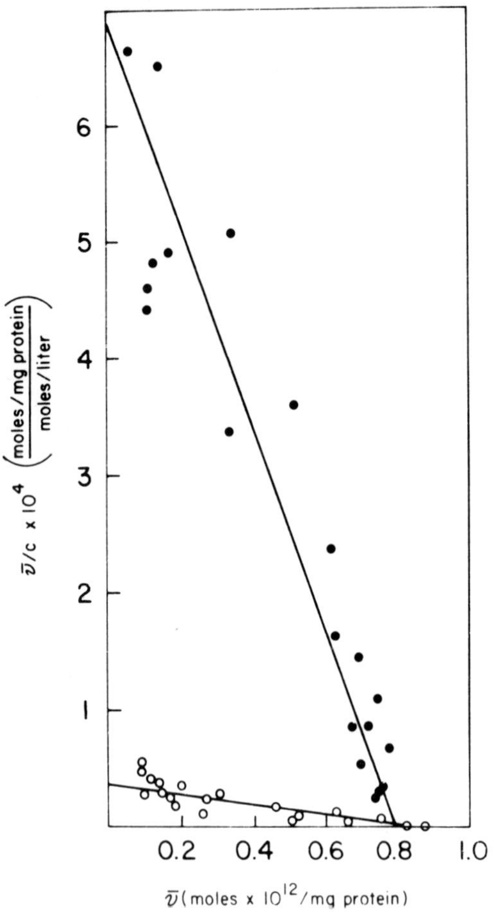

Fig. 7. Scatchard analysis of the binding ^{125}I-α-Bgtx to the high affinity set in a Triton X-100 preparation from ten day denervated muscle. The control data are indicated by closed circles and the experimental data in the presence of 8.7 x 10^6 M d-tubocurarine chloride is indicated by the open circles. The intercept on the (ν) axis is the number of available sites (N). The slope of the lines (least squares) is the affinity constant (K). The data demonstrate a lower apparent affinity in the presence of d-tubocurarine.

has shifted approximately one order of magnitude. In the absence of d-tubocurarine, the K_D is in the order of 10^{-9} M and in the presence of this competitive antagonist the K_D is in the order of 10^{-8} M α-Bgtx. These conclusions are reinforced by a Scatchard

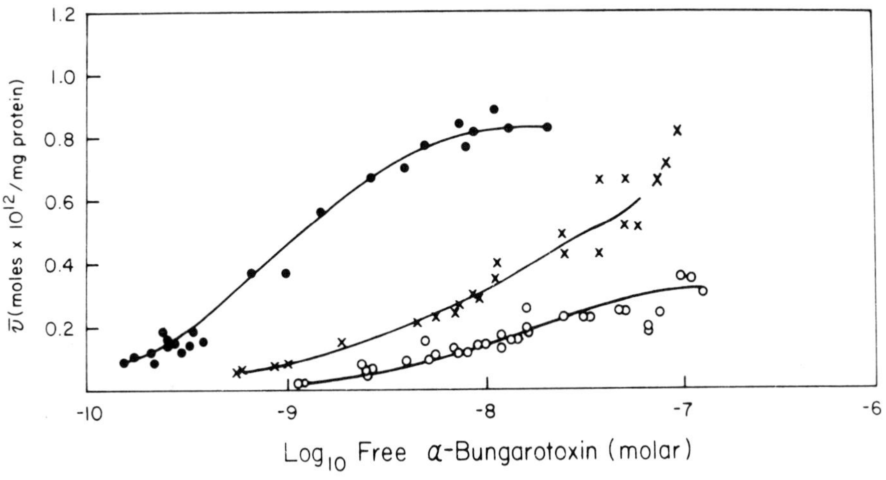

Fig. 8. The binding of ^{125}I-α-Bgtx to the high affinity set in a triton extracted preparation from ten day denervated muscle. The control isotherm is indicated by the closed circles (●). The experimental isotherm in the presence of 1.2 x 10^{-4} M carbamylcholine is indicated by open circles (o). The experimental isotherm in the presence of 3.2 x 10^{-5} M carbamylcholine is indicated by (X). The lines for the control isotherm (o) and the experimental isotherm in the presence of 1.2 x 10^{-4} M carbamylcholine (o) are calculated from the mass law equation for the interaction of a homogeneous ligand population interacting with a single set of identical independent sites. The line for the isotherm in the presence of 3.2 x 10^{-5} M carbamylcholine is approximated to the data.

analysis of the data. In the Scatchard plot a straight line indicates a homogeneous ligand population interacting with a single set of identical, independent sites. The intercept on the $\bar{\nu}$ axis indicates the number of available sites, and the slope of the line is the association constant (K_A) (the reciprocal of the dissociation constant, K_D). The analysis of the toxin binding shows that d-tubocurarine does not change the intercept of the $\bar{\nu}$ axis and that in both the absence and the presence of d-tubocurarine the data can be approximated by a straight line. On the other hand, the slope of the line in the presence of dTc is altered substantially. Similar results were obtained with normal muscle preparations. It can be concluded from these data that both antagonists (α-Bgtx and d-tubocurarine) associate competitively with this set of cholinergic sites.

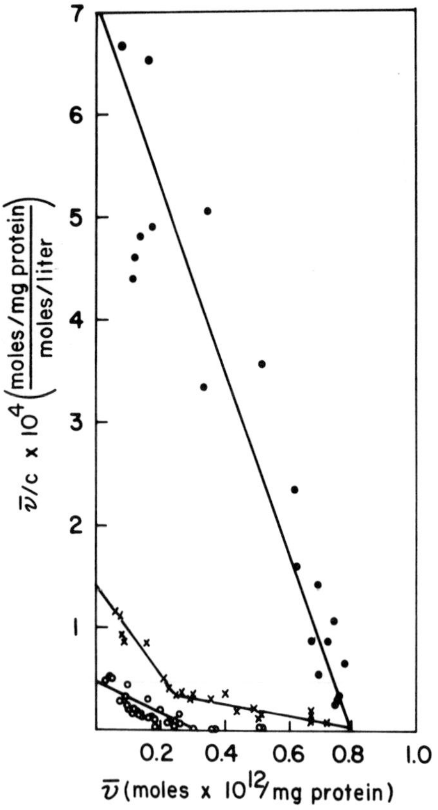

Fig. 9. Scatchard analysis of the binding of ^{125}I-α-Bgtx to the high affinity set of sites in a triton X-100 preparation from ten day denervated muscle. The control data are indicated by the closed circles (9). The experimental data in the presence of 1.2×10^{-4} M carbamycholine approximate a straight line indicating a single set of identical sites. The number of sites in the set are: control, 0.8 pmoles/mg protein, and in the presence of 1.2×10^{-4} M carbamycholine, 0.3 pmoles/mg protein. The experimental data in the presence of the lower carbamycholine concentration (3.5×10^{-5} M) have been analyzed as containing two sets of independent sites. The number of sites in the set with the highest apparent α-Bgtx binding affinity is 0.3 pmoles/mg protein. The number of sites in the set with the lowest apparent toxin binding is 0.5 pmoles/mg protein. It should be noted that a higher apparent toxin binding affinity indicates a lower carbamycholine binding affinity.

C. ACh Receptor: Interaction with Carbamylcholine (Agonist)

The interaction between carbamylcholine and the set of high affinity cholinergic sites was also investigated by analyzing the effect of this agonist on the α-Bgtx binding to the set. Figures 8 and 9 present those data. In these figures the closed circles represent control data, the (X)'s represent the addition of 3.2 x 10^{-5} M carbamylcholine, and the open circles represent the addition of 3.2 x 10^{-4} M carbamylcholine. The log plot (Figure 8) shows that the higher concentration of carbamylcholine reduces the number of available sites (N) and increases the apparent K_D (the toxin concentration at which 1/2 N is occupied). The magnitude of the range of toxin concentrations over which the residual set of sites saturate does not appear to be increased. However, in the presence of the lower concentration of carbamycholine, saturation is not apparent; K_D cannot be similarly analyzed and the magnitude of the range in toxin concentrations over which the isotherm extends is increased. The analysis of these data by the method of Scatchard (Figure 9) suggests that the appropriate interpretation is that two subsets of sites with different affinities for carbamylcholine exist. The data for both the control and experimental isotherms in the presence of the higher concentration of carbamylcholine approximate a straight line. The control (without dTc) intersects the $\bar{\nu}$ axis at 0.8 pmol/mg protein, and the experimental (with dTc) intersects at 0.3 pmol/mg protein. The experimental (high concentration) would, therefore, appear to reflect the complete reduction in the affinity of a second subset with an N of 0.3 pmol/mg protein. This interpretation is supported by the two site analysis of the isotherm in the presence of the lower carbamycholine concentration. Extrapolation of the two major portions of this curved line to the $\bar{\nu}$ axis would indicate two subsets which competitively bind carbamylcholine and toxin. The first has an (N) of 0.3 pmol/mg protein and a lower affinity for carbamylcholine (reflected by a higher apparent toxin binding affinity). The second has an (N) of 0.3 pmol/mg protein and a higher affinity for carbamylcholine (refleced by a lower apparent toxin binding affinity). Qualitatively similar results were obtained for normal muscle preparations.

The analysis of the effect of carbamylcholine in the equilibrium binding of ^{125}I-α-Btgx to both normal junctional and denervated extrajunctional receptors indicates that this agonist, unlike the antagonist (d-tubocurarine), does not effect a simple competitive inhibition of a α-Bgtx binding. The results show that although the sites in the set are identical and independent with respect to the antagonist (α-Bgtx and d-tubocurarine), they are not both identical and independent with respect to the agonist (carbamylcholine). Analysis of the results suggest that the high affinity set of antagonist binding sites contains two subsets of agonist binding sites with different affinities for the carbamylcholine. The significance of the differences between the agonist and the antagonist in their interactions with the ACh receptor is not clear. However, it is possible that these differences represent a reflection of the depolarization and/or desensitizing phenomena associated with the agonist binding.

Summary and Conclusions

Cholinergic interactions in systems derived from rat skeletal mixed muscle are detailed. The isotherms of the binding of a α-Bgtx over an extended range (10^{-10} mol to 10^{-5} mol toxin) suggests the presence of two sets of sites. The first is a high affinity set of sites ($K_A \simeq 10^9$ l/mol) corresponding to the set previously described in slow and fast muscle systems. The second is a lower affinity set ($K_D \simeq 10^6$ l/mol) in the area of the isotherm not previously studied. Based on interactions and purification with concanavalin A, both these sets of sites appear to be either part of or closely associated with carbohydrate containing macromolecules. Although the high affinity set of sites (ACh receptor site) is independent and identical with respect to toxin, it is either heterogeneous or cooperative with respect to cholinergic agonists. An examination of the period between 1 and 10 days denervation shows that the number of sites in the high affinity set drops slightly immediately following denervation (day 1, 0.03 pmol/mg then increases sharply (days 3-7) to the peak at day 8 (1.05 pmol/mg) and begins to decline (day 8 > 9 > 10). Similar results are obtained when sensitivity to the myasthenia antibody is examined. Sensitivity first appears at three days and progresses coincidentally with the increase in extrajunctional sites. In contrast, the number of sites in the low affinity set increases only about 4-fold following denervation. This increase also begins between 2 and 4 days. The definition of the low affinity set of sites remains to be determined. These studies indicate that the definition of molecular interactions in closed systems derived from dynamic biological systems (innervated muscle, denervated muscle, myasthenia gravis) can provide substantial information pertinent to the understanding of surface membrane receptors.

REFERENCES

1. ALBUQUERQUE, E. X., MC ISSAC R. T., Exp. Neurol. 26 (1969)183 183.
2. ALBUQUERQUE, E. X., WARNICK, J. E., TASSE, J. R., and SANSONE, F. M., Exp. Neurol. 37 (1972) 607.
3. ALMON, R. R., ANDREW, C. G., and APPEL, S. H., Biochemistry 13 (1974) 5522.
4. ALMON, R. R., ANDREW, C. G., and APPEL, S. H., Science 186 (1974) 55.
5. ALMON, R. R. and APPEL, S. H., Biochim, Biophys. Acta. 393 (1975) 66.

6. ALMON, R. R. and APPEL, S. H., Fed. Proc. 34 (1975) 307.
7. ALMON, R. R. and APPEL, S. H., Ann. N. Y. Acad. Sci. (1975) in press.
8. ALPER, R., LOWY, J. and SCHMIDT, J., FEBS Letters 48 (1974) 130.
9. ANDREW, C. G., ALMON, R. R., and APPEL, S. H., J. Biol. Chem. 249 (1974) 6163.
10. ANDREW, C. G., ALMON, R. R., and APPEL, S. H. , J. Biol. Chem. 250 (1975) 3972.
11. APPEL, S. H., ANDREW, C. G., and ALMON, R. R. , J. Neurochem. 23 (1974) 1077.
12. BANERJEE, S. P., SNYDER, S. H., CUATRECASAS, P., and GREEN, L. A., Proc. Nat. Acad. Sci., USA 70 (1973) 2519.
13. BERANEK, R. and VYSKOCIL, F., J. Physiol. 188 (1967) 53.
14. BERG, D. K. and HALL, Z. W. , Science 184 (1974) 473.
15. BERG, D. K., KELLY, R. B., SARGENT, P. B., WILLIAMSON, P., and HALL, Z. W. , Proc. Nat. Acad. Sci. USA 69 (1972) 147.
16. BROCKES, J. P. and HALL, A. W. , Biochemistry 14 (1975) 2092.
17. BROCKES, J. P. and HALL, Z. W., Biochemistry 14 (1975) 2100.
18. CHIU, T. H., LAPA, A. J., BARNARD, E. A., and ALBUQUERQUE, E. X., Exp. Neurol. 43 (1974) 399.
19. COLQUHOUN, D., RANG, H. P., and RICHIE, J. M., J. Physiol. 240 (1974) 199.
20. CUATRECASAS, P., J. Biol. Chem. 246 (1971) 6522.
21. EAKER, D., HARRIS, J. B., and THESLEFT, S., Europ. J. Pharmacol. 15 (1971) 254.
22. FAMBROUGH, D. M. Science 168 (1970) 372.
23. FELTZ, A. and MALLART, A., J. Physiol. 218 (1971) 85.
24. LAPA, A. J., ALBUQUERQUE, E. X., and DALY, J., Exp. Neurol. 43 (1974) 375.
25. MC ARDLE, J. J. and ALBUQUERQUE, E. X., J. Gen. Physiol. 61 (1973) 1.
26. MEUNIER, J. C., OLSEN, R. W., MENEZ, A., FROMAGEOT, P., BOQUET, P., and CHANGEUX, J. P. , Biochemistry 11 (1972) 1200.
27. MICHAELSON, D. M. and RAFTERY, M. A., Proc. Nat. Acad. Sci. USA 71 (1974) 4768.
28. MAKHERJEE, C., CARON, M. G., COVERSTONE, M., and LEFKOWITZ, R. J., J. Biol. Chem. 250 (1975) 4869.
29. SCATCHARD, G. , Ann. N. Y. Acad. Sci. 51 (1949) 660.

APPEARANCE OF SPECIALIZED CELL MEMBRANE COMPONENTS DURING DIFFERENTIATION OF EMBRYONIC SKELETAL MUSCLE CELLS IN CULTURE

Joav M. PRIVES

Neurobiology Unit, The Weizmann Institute of Science
Rehovot, (Israel)

The properties of cell membranes of skeletal muscle fibers are known to be strongly influenced by interaction with motor neurons. The subsynaptic areas of muscle surface membranes are distinct in organization and function from other membrane areas, having a unique convoluted morphology and containing a high density of acetylcholine receptor (AChR). These aspects of cell surface specialization are not initially present in embryonic muscle, appearing only during the period of formation of synaptic contacts with motor neurons (6,8). To study the mechanism by which such aspects of muscle cell phenotype may be regulated by nerve and the extent of these inductive effects, it is useful to compare the properties of muscle cells which have developed in the absence of neuronal influence with those of muscle *in situ*. Like muscle development in the intact embryo, the differentiation in culture of embryonic mononucleated myogenic cells to multinucleated cross-striated muscle fibers is well characterized (6, 21) and includes formation of excitable cell membranes and the biosynthesis of AChR and acetylcholinesterase (AChE) (3-5). The elaboration of these subsynaptic membrane components, which *in situ* mediate neuromuscular transmission of impulses, and the biosynthesis of membrane constituents responsible for excitability, are intrinsic aspects of muscle cell differentiation not requiring the presence of neurons. Homogeneous cultures of muscle cells thus constitute a highly suitable system for the study of cell membrane differentiation in the absence of direct neuronal influence. Moreover, muscle in culture can be made to differentiate more rapidly and with a higher degree of synchrony than occur in the embryo, making it possible to determine sequential relationships between developmental events. In the studies to be summarized here, the appearance of AChR, AChE, and other membrane compo-

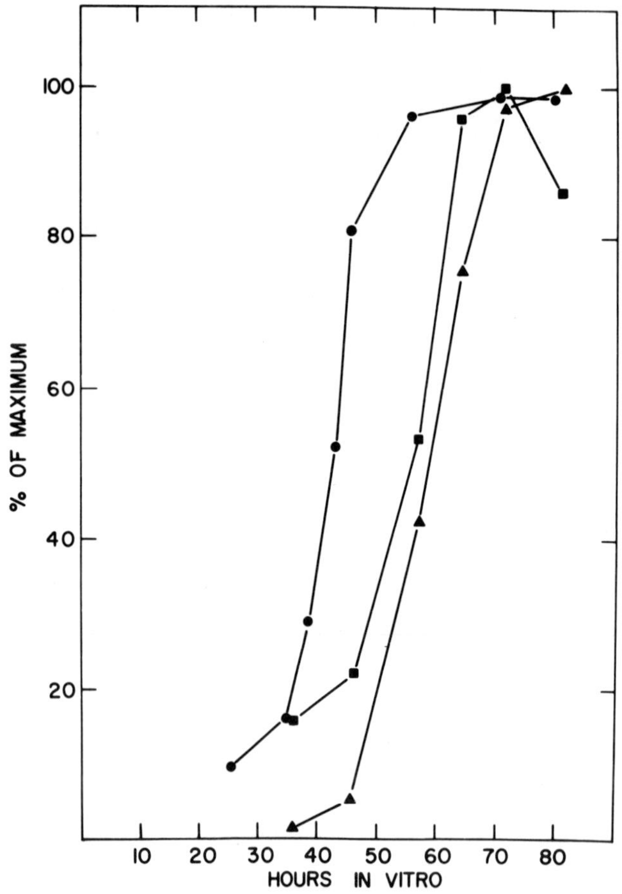

Fig. 1. Kinetics of fusion and relative rates of appearance of acetylcholine receptor and acetylcholinesterase during differentiation of muscle cultures. Plating density was 2 x 10^6 cells per dish. Fusion (●) was scored as described (12). Acetylcholine receptor (▲) was measured by specific binding of ^{125}I-α-bungarotoxin (12) and acetylcholinesterase (■) by the hydrolysis of [^{14}C] acetylcholine (15), using replicate dishes of each time point. Data are expressed as percentage of maximum. Maximal values were: fusion index, 65%; toxin binding, 3.6 x 10^{-13} moles per culture; acetylcholine hydrolysis, 1.8 x 10^{-7} moles/min per culture.

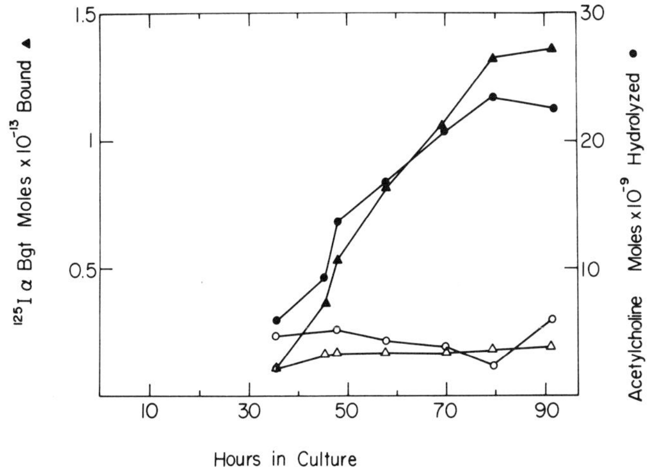

Fig. 2. Effects of BUDR on appearance of acetylcholine receptor and acetylcholinesterase. BUDR was added to cultures at a concentration of 10 μgm/ml at 24 hours post-plating. Receptor was measured by Bgt binding to control (▲) and BUDR-treated (Δ) cultures. In the same experiments, acetylcholinesterase was measured in control (●) and BUDR (O) cultures. Plating density was 1.5×10^6 per dish.

nents has been measured during development of embryonic chick skeletal muscle in cell culture, and these parameters have been used as markers to study the nature and sequence of events in development of specialized muscle surface membrane.

Myogenic cells dissociated from breast of 12 day chick embryos underwent rapid and synchronous differentiation with the plating densities and culture conditions used in these studies (12,14). These characteristics are evident from the time table of morphological development which was observed. The newly plated cells adhered to the collagen coated culture dishes, assumed an elongated shape, and carried out several cycles of cell division during the initial 25 or 30 hours *in vitro*. The myoblasts then withdrew from mitosis, became aligned and underwent rapid fusion; within 50 hours after plating 60% to 70% of muscle cell nuclei came to be situated in multinucleated myotubes. These ribbon-like cells increased in length and diameter and with continuing development acquired the cross-striated appearance characteristic of mature muscle fibers. Spontaneous contractility could be observed in cultures after approximately 150 hours of incubation. In synchronously differentiating cultures, the period subsequent to rapid fusion is characterized by sharp increases in levels of actin and myosin (13, 14). Similarly, the rapid appearance of several components of

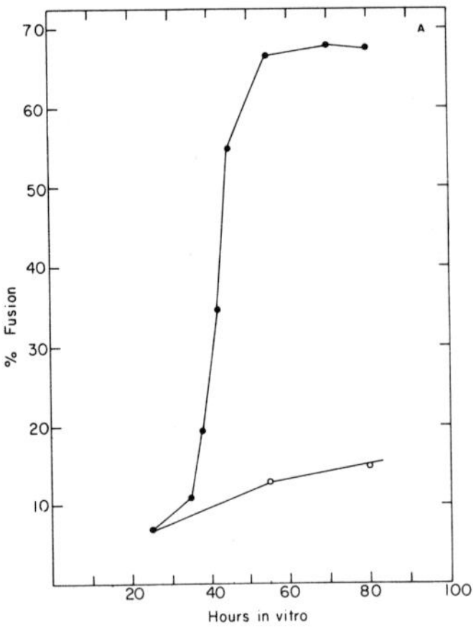

A

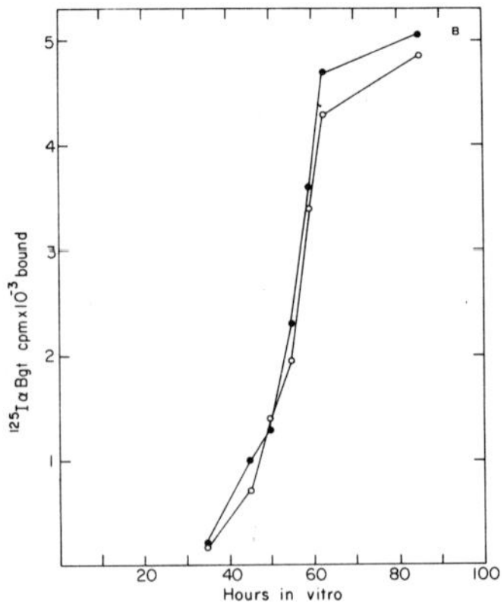

B

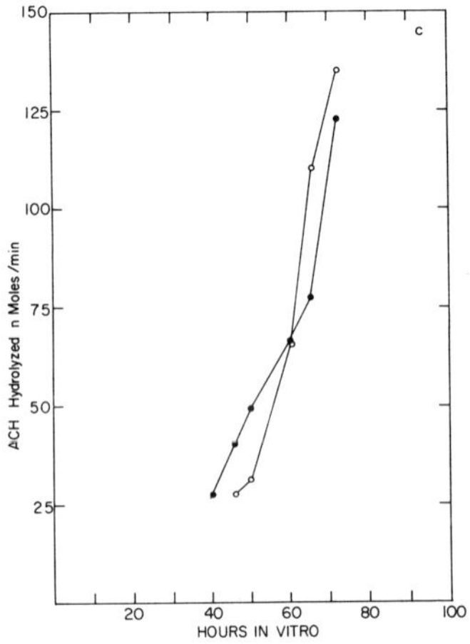

C

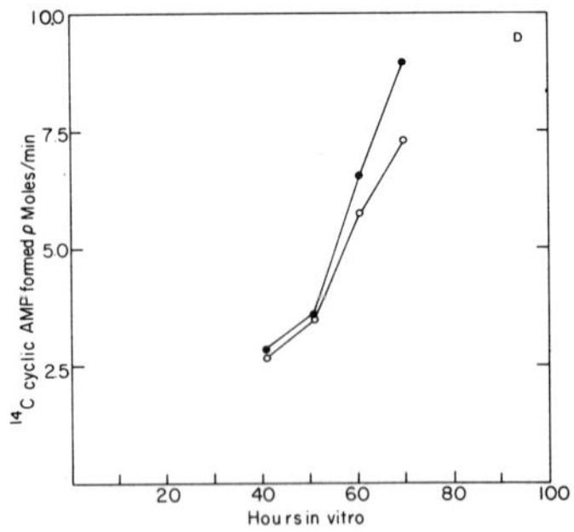

D

Fig. 3. A, B, C, D. (Fig. 3. E next page).

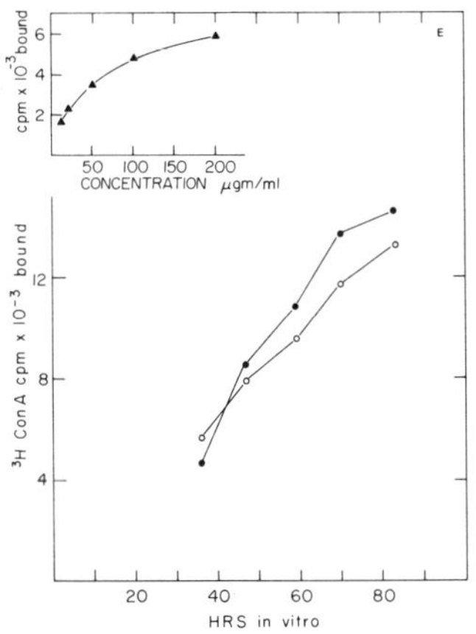

E

Fig. 3. A. Kinetics of fusion in control (●) and fusion-arrested (O) cultures. Fusion arrested cultures were changed to low calcium medium (25 μM Ca++) 24 hours after plating. B. Acetylcholine receptor appearance in control (●) and fusion-arrested (O) cultures C. Appearance of acetylcholinesterase activity in control (●) and fusion-arrested (O) cultures. D. Increase in adenylate cyclase activity in control (●) and fusion-arrested (O) cultures. Adenylate cyclase was assayed as described (15). E. Specific binding of ^{3}H-labeled concanavalin A to control (●) and fusion-arrested (O) cultures. Cultures were incubated with ^{3}H-Con A (0.2 mg/ml, specific activity 3 x 10^6 cpm/mg) for 60 min at 37°. The specific labeling was the component of total labeling eliminated by presence of α-methyl mannoside (50 mM) (20). Inset: saturation of specific binding (▲) with increasing ^{3}H-Con A concentration.

myotube cell membranes has been observed to occur shortly after fusion (15). As shown in Fig. 1., AChR, measured by the specific binding of α-bungarotoxin (Bgt) (9), and AChE both begin to increase at parallel rates 5 to 10 hours after the main period of fusion. The notion that the fusion process is a critical event in muscle development (22) is based on the observation that fusion precedes the appearance of proteins characteristic of differentiated muscle and, furthermore, that biosynthesis of such proteins is inhibited in cultures in which fusion has been prevented (11, 17). For example exposure of actively dividing myoblasts to the thymidine analogue 5-bromo-2-deoxyuridine (BUDR), results in failure of the cells to fuse and to synthesize myosin (18) as well as cytoplasmic enzymes characteristic of muscle (10, 18). In contrast, synthesis of those proteins required for viability and cell division is unaffected. The selective inhibition of muscle specific proteins extends to cell membrane constituents: neither AChR nor AChE appear in BUDR-blocked cultures, as shown in Fig. 2. It is also possible to eliminate myoblast fusion by markedly reducing the calcium concentration in the culture medium (15, 17) and this procedure, in contrast to BUDR treatment, distinguishes between the synthesis of cytoplasmic proteins and of cell membrane proteins. In cultures where fusion is eliminated by calcium depletion (Fig.3A) the synthesis of actin (13), myosin (14), or associated cytoplasmic enzymes (17) is strongly inhibited. In contrast, we have observed that elaboration of AChR and AChE is unaffected in muscle cultures where fusion has been blocked by lowered calcium level (15), as shown in Figs. 3B and 3C. An additional component of specialized muscle cell membranes, adenylate cyclase was shown to increase in activity in a manner similar to the cholinergic proteins in both fusing and non-fusing cultures (15), as can be seen in Fig. 3D. That the non-dependence on cell fusion may be a more general characteristic of elaboration of muscle cell membrane components is suggested by data shown in Fig. 3E which indicate that the specific binding of concanavalin A to sugar moieties on the cell membrane increases with the same timing during maturation of fused and fusion-arrested cultures. It is thus possible that a significant proportion of cell surface glycoproteins are elaborated under the same conditions and with similar kinetics as AChR, AChE, and adenylate cyclase. The results shown in Fig. 3 suggest two conclusions: firstly, that several components of muscle cell membranes are elaborated simultaneously during differentiation. This observation has recently been extended to an additional membrane constituent, Na^+, K^+-ATPase, which is seen in Fig. 4 to appear with highly similar kinetics as AChR (11). Secondly, the co-appearance of membrane proteins is not linked to cell fusion, unlike the synthesis of other muscle proteins. A general interpretation is that elaboration of these membrane components is regulated by a developmental program distinct from the mechanisms regulating cytoplasmic aspects of muscle differentiation.

The simultaneous appearance of these membrane constituents may result from their simultaneous biosynthesis and translocation from ribosomes to the cell surface. Alternately, these components may be synthesized at separate times and coordinated at the level of incorporation into the cell membrane. To distinguish between these two possibilities, we compared the rates of inhibition of elaboration of AChR and AChE resulting from interruption of protein synthesis. Cycloheximide was added to rapidly differentiating muscle cultures during the period when AChR and AChE were increasing at

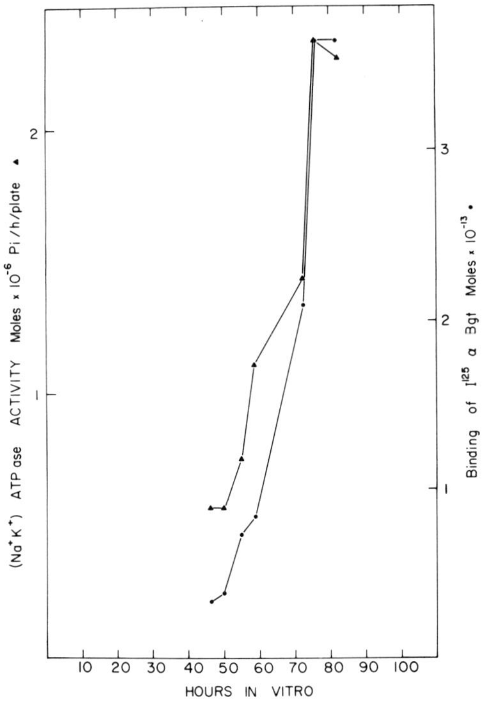

Fig. 4. Appearance of acetylcholine receptor (●) and Na^+, K^+-ATPase activity (▲). During differentiation of muscle cultures, binding of Bgt and Ouabain-sensitive ATPase activities were measured in replicate cultures in the same experiments.

maximal rates. Specific binding of Bgt and AChE activity were measured at increasing intervals after addition of the inhibitor. Cycloheximide blocked the incorporation of radioactive amino acids into proteins within 10 minutes. In contrast, AChR and AChE continued to appear at control rates for a considerably longer period - approximately 2 hours, after which the further appearance of both components ceased simultaneously. As shown in Fig. 5, the parallel inhibitory effects of cycloheximide on elaboration of the two cholinergic proteins are reversible: removal of the inhibitor is followed by renewed appearance on the cell surface of AChR and AChE at similar rates, about two hours after resumption of protein synthesis. In Table I, relative levels of AChR and AChE at various times after addition of cycloheximide are expressed as percentages of control values. As can be seen, elaboration of AChr and AChE was not appreciably diminished at the end of a 90 min period of exposure of the cultures to cycloheximide. However, upon longer exposure surface activities of both components decreased by similar

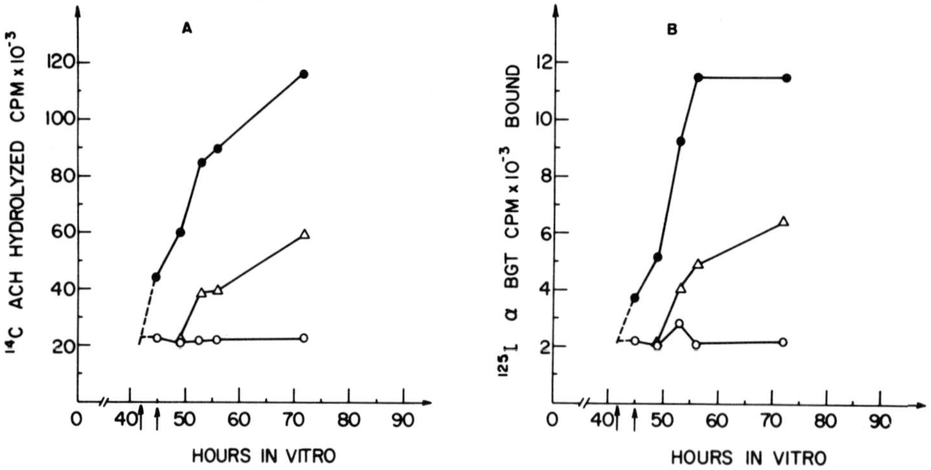

Fig. 5. Effects of cycloheximide on appearance of acetylcholinesterase and acetylcholine receptor. Acetylcholinesterase (A) and acetylcholine receptor (B) were measured in control cultures (●), in cultures to which cycloheximide (10 µgm/ml) was added (O) 42 hours after plating, and lastly, in cultures where cycloheximide was added at 42 hours, and removed at 46 hrs. (Δ)

TABLE I

Effect of cycloheximide on the appearance of acetylcholine receptor and acetylcholinesterase*

Exposure to Cycloheximide (hr)	Percent of Control	
	AChR	AChE
1.5	98.5	88.2
3	66.0	66.1
4.5	53.9	50.9
6	46	45.2
11	35.4	43.5

*Cycloheximide (10 µgm/ml) was added to differentiating muscle cultures 42 hours after plating. Acetylcholine receptor (AChR) and acetylcholinesterase (AChE) activities were determined in the same cultures at the specified intervals after cycloheximide addition and expressed as percent of values obtained with control cultures.

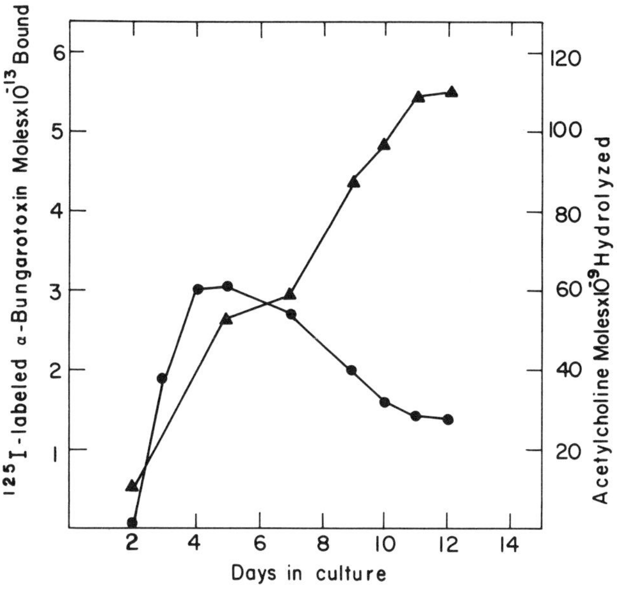

Fig. 6. Levels of acetylcholine receptor and acetylcholinesterase during muscle differentiation in culture. Acetylcholine receptor (●) is expressed as moles x 10^{-13} of ^{125}I-α-Bgt bound per plate. Acetylcholinesterase (▲) is expressed as moles x 10^{-9} hydrolyzed per plate during a 2 min incubation at 25°.

proportions. In contrast to cycloheximide, under these conditions actinomycin D had no effect on the rates of appearance of either AChR or AChE. These results support the possibility that the co-appearance of AChR and AChE is coordinated via their simultaneous biosynthesis directed by relatively stable messenger RNA species. The lag period preceding the inhibitory effect of cycloheximide may correspond to the interval between biosynthesis and incorporation into surface membrane. That this interval is similar for the two components suggests that AChR and AChE may utilize the same pathway from ribosomes to cell surface. Other membrane components shown to have highly similar kinetics of elaboration as the cholinergic proteins may be regulated in a similar manner.

As shown in Fig. 1, the period following shortly after myoblast fusion is marked by co-appearance of AChR and AChE at highly similar rates. Upon further differentiation of the muscle in culture an "uncoupling" is seen in cell surface levels of these two components It has recently been observed that, whereas AChE continues to increase, AChR decreases significantly during the second week in culture as seen in Fig. 6.[1] As Fig. 7 shows,the decline in AChR can be

---

[1]Prives, J. M., Silman, I., and Amsterdam, A., <u>in preparation</u>.

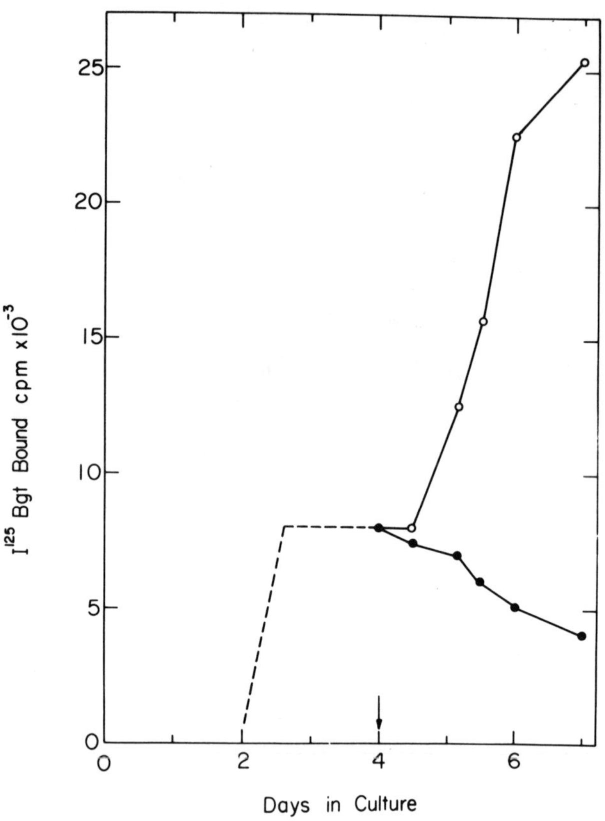

Fig. 7. Effect of D-600 on levels of acetylcholine receptor. D-600 (15 μg/ml) was added to muscle cultures 96 hrs after plating. Receptor was measured in cultures exposed to D-600 (O) and in control cultures (●).

reversed by D-600, an inhibitor of membrane conductance to calcium which also causes stimulation of receptor biosynthesis (16). The fact that muscle cultures in which levels of AChR are declining with maturation retain the capacity for renewed elaboration of this component suggests that the decrease in AChR is a regulated aspect of

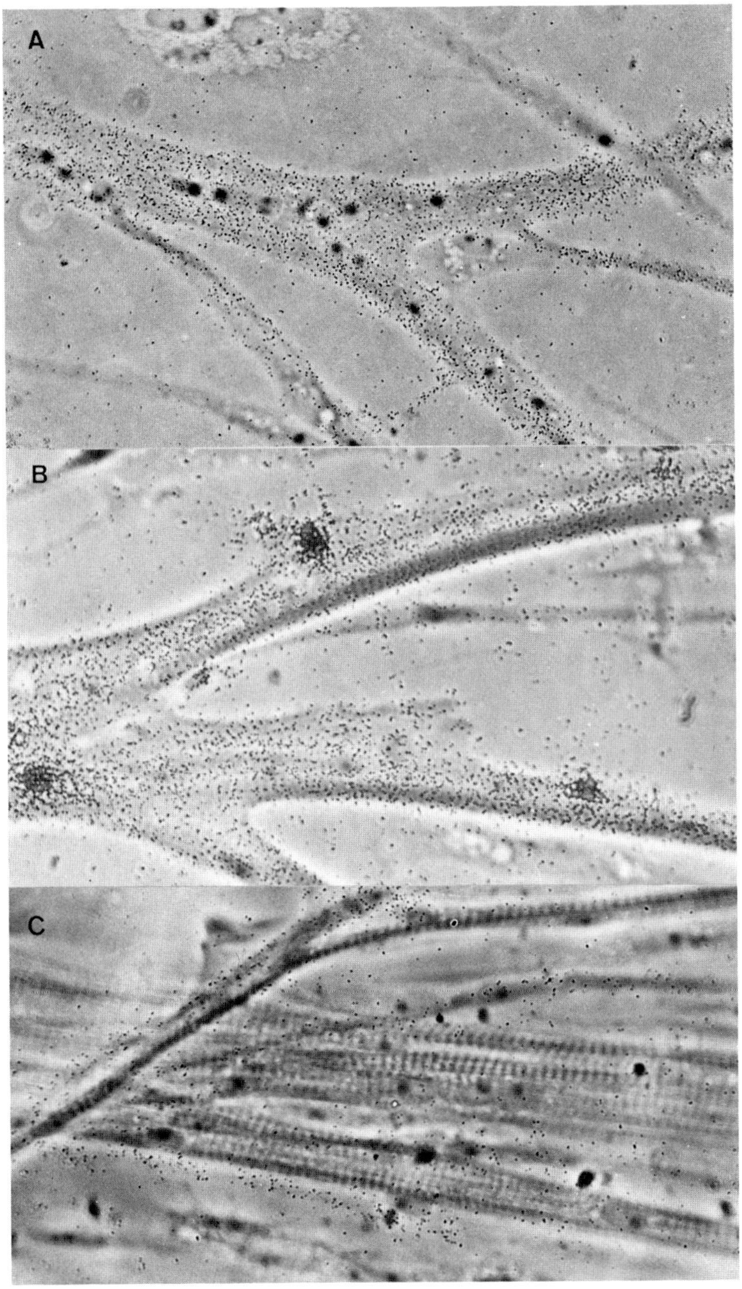

See next page for Fig. 8 legend.

Fig. 8. Autoradiograms of myotubes labeled with ^{125}I-αBgt: A. After 96 hrs in culture. Note the homogeneous distribution of silver grains on the myotube. B. After 7 days in culture. A large number of silver grains lying over the myotube are aggregated in clusters of 5-15 μm diameter. Note the presence of faint cross striations. C. After 13 days in culture. Highly developed cross striated myofibers are only lightly labeled.

differentiation *in vitro* rather than a consequence of deterioration. This notion was supported by studies in which changes in amount and distribution of AChR on surfaces of muscle fibers during development were visualized by radioautography of cultures labeled with ^{125}I-Bgt at various intervals after plating.[1] As shown in Fig. 8A, it was observed that the receptor was homogeneously distributed on the myotubes after 96 hr in culture, at the time when AChR levels were maximal. Fig. 8B shows that by 7 days in culture, AChR was no longer homogeneous but was aggregated into patches 5-15 μm in diameter, similar to the clusters reported by others (1, 19). These clusters have been shown to coincide with membrane loci sensitive to ACh (7) and have been postulated to participate in synapse formation with motor neurons *in situ* (19). However, in the rapidly differentiating muscle cultures which have been used, the clusters represent a transitory phenomenon. It has been observed that further development is accompanied by disappearance of AChR as detectable by binding of Bgt from the external surface of muscle fibers. As shown in Fig. 8C, in 13 day cultures, the density of AChR was drastically reduced in highly developed myotubes displaying sharp cross striation. In contrast, more poorly developed cells present in these same cultures which did not show cross striations can be seen to have retained a high density of AChR.

These results suggest that AChR elaboration, aggregation, and subsequent disappearance from the surface of myotubes are sequential events in the differentiation of muscle cell membranes in culture in the absence of neuronal elements. The appearance of AChR may be coordinated with elaboration of other specialized membrane components and regulated independently from the synthesis of the structural proteins and cytoplasmic enzymes characteristic of muscle. The aggregation of receptor to form clusters might be a prerequisite to the subsequent depletion of this component, possibly by the internalization of the AChR-rich membrane patches. Lastly, the disappearance of AChR from the surface of muscle cells, differentiated in culture, may be to some degree analogous with the marked decrease in receptor which occurs during development of embryonic skeletal muscle *in situ* during the period of formation of motor synapses (2).

BIBLIOGRAPHY

1. COHEN, S. and FISCHBACH, G., Science 181 (1973) 76.
2. DIAMOND, J. and MILEDI, R., J. Physiol. 162 (1962) 393.
3. ENGEL, W., J. Histochem. Cytochem. 9 (1961) 66.
4. FAMBROUGH, D., HARTZELL, H., POWELL, J., RASH, J., and JOSEPH, N., in Synaptic Transmission and Neuronal Interaction Raven Press, New York, (1974) 285. (1972) 75.
5. FAMBROUGH, D. and RASH, J., Dev. Biol. 26 (1971) 55.
6. FISCHMAN, D. in The Structure and Function of Muscle Vol. 1, Academic Press, New York,
7. HARTZELL, H. and FAMBROUGH, D., Dev. Biol. 30 (1973) 153.
8. JACOBSON, M. in Developmental Neurobiology, Holt, Rinehart and Winston (1970).
9. LEE, C. Y., Ann. Rev. Pharmacol. 12 (1972) 265.
10. MOCHAN, B., MOCHAN, E., and DE LA HABA, G., Biochim, Biophys. Acta 335 (1974) 408.
11. PAGLIN, S. and PRIVES, J., Israel J. Med. Sci. (1975) <u>in press</u>.
12. PATERSON, B. and PRIVES, J., J. Cell Biol. 59 (1973) 241.
13. PATERSON, B., ROBERTS, B. and YAFFE, D., Proc. Nat. Acad. Sci. USA. 71 (1974) 4467.
14. PATERSON, B. and STROHMAN, R. C., Dev. Biol. 29 (1972) 113.
15. PRIVES, J. and PATERSON, B. Proc. Nat. Acad. Sci. USA. 71 (1974) 3208.
16. SHAINBERG, A. and NELSON, P., Abstr. XXVI Congress of Physiological Sciences, Jerusalem Satellite Symposium on Mechanisms of Synaptic Action (1974) 53.
17. SHAINBERG, A., YAGIL, G. and YAFFE, D., Dev. Biol. 25 (1971) 1.
18. STOCKDALE, F., OKAZAKI, K., NAMEROFF, M. and HOLTZER, H., Science 146 (1964) 533.
19. SYTKOWSKI, A., VOGEL, Z. and NIRENBERG, M., Proc. Natl. Acad. Sci. USA 70 (1973) 270.
20. VLODAVSKY, I., INBAR, M. and SACHS, L., Proc. Nat. Acad. Sci. USA 70 (1973) 1780.
21. YAFFE, D., Current Topics Dev. Biol. 4 (1969) 37.
22. YAFFE, D., and DYM, H., Cold Spring Harbor Symp. Quant. Biol. 37 (1972) 543.

AFFINITY PARTITIONING OF RECEPTOR-RICH MEMBRANE FRACTIONS AND THEIR PURIFICATION

Steven D. FLANAGAN*, Samuel H. BARONDES and Palmer TAYLOR

Departments of Biology, Psychiatry and Medicine,
University of California, La Jolla, California 92037 (USA)

The study of the molecular properties of membrane bound receptors is greatly facilitated by their solubilization and purification. An excellent example of this is the nicotinic acetylcholine receptor, which can be solubilized by the use of nonionic detergents and purified by affinity chromatography (15, 20). Initial studies indicated that the binding properties of the nicotinic receptor remain intact in the presence of nonionic detergents. Such fortuitous circumstances do not occur in the study of other receptors. For example, the muscarinic acetylcholine receptor has proved rather labile to treatment with nonionic detergents (7), resulting in considerable loss of binding properties. Even when binding properties of a receptor are intact after solubilization, questions often remain concerning the *in situ* structure of the receptor and its associated activities. For example, functional coupling between receptor and enzymatic activities and ionophores may be disrupted or altered upon solubilization. Even in the case of the nicotinic receptor, differences in the binding properties of solubilized and membrane bound receptors have been detected (5).

An alternative to solubilization is the study of receptor binding and enzymatic activities in crude homogenates or partially purified membrane fractions. However, it is difficult to study the structural properties of receptors present in crude preparations. Thus it is often advantageous to purify fractions using methods that separate membrane fragments on the basis of their physical chemical properties. When the source of receptor is tissue containing homogeneous structures, then substantial purification may be achieved

*Present address: Friedrich Miescher-Institut, Postfach 273, CH-4002, Basel, Switzerland

using classical methods. In the case of tissue as specialized and hetergeneous as mammalian brain, it has proved extremely difficult to obtain membrane fractions of high purity using classical centrifugation methods. Even in cases where synaptic fractions obtained are highly purified on the basis of morphological criteria, it is likely that they are hetergeneous in neurotransmitter receptor specificities.

An attractive alternative to classical methods is the use of affinity chromatography to purify receptor-rich membrane fragments. Such purification of membrane fragments should, in theory, be straightforward and yield a high degree of purification in a few steps. However, recent re-evaluation of applications of affinity chromatography to the purification of soluble enzymes and receptors indicates that in many cases there are complicating nonspecific effects (14). Frequently, affinity ligands or the spacers used for their attachment to suitable matrices have ion exchange or hydrophobic properties. The problems encountered with affinity chromatography are further complicated in purifying multivalent membrane fragments. This may be because the stable binding of a particulate fraction to an affinity matrix requires the formation of several bonds, and is a cooperative process. Once initial specific bonds are formed, numerous additional bonds may form (both specific and nonspecific) resulting in almost irreversible binding to the affinity matrix. In practice it has proved difficult to bind particulate material to an affinity matrix unless the matrix contains high substitution levels of affinity ligand. Once bound, it has proved difficult to remove particulate fractions except in the presence of binding competitors at concentrations orders of magnitude above their known dissociation constants. In other applications elution even with high concentrations of specific competitor has proved impossible (18). These problems may be alleviated by the use of affinity ligands and matrices having low nonspecific sorption properties. In this connection it should be pointed out that many of the more potent binding agents (e.g., neurotransmitter congeners) have significant ion exchange and hydrophobic character.

In an effort to overcome the problems inherent in the application of affinity chromatography to the purification of particulate fractions, we have developed an alternative called affinity partitioning (8-11). Affinity partitioning is a modification (21, 22) of the phase partition method originally introduced by Albertsson and co-workers (1, 3, 12, 23). In the phase partition method, purification is achieved by extraction between water-rich phases formed by the addition of two water soluble polymers to an appropriate buffer. The most commonly used system contains 4% (w/w) poly(ethylene oxide) [6,000 mol. wt.] and 5% (w/w) dextran [500,000 mol. wt.] in a NaCl or phosphate buffer. The two phases are formed; one is poly(ethylene oxide) rich but contains substantial dextran content, and the other is dextran rich and contains poly(ethylene oxide).

Both phases contain more than 85% water. Such phase systems have proved extremely gentle and nondenaturing when applied to the purification of a number of enzyme activities (1, 3). Also, erythrocytes which are relatively sensitive to detergent activity are not lysed in dextran-poly(ethylene oxide) systems containing isotonic buffers (24). Purification can be achieved when components of a hetergeneous mixture distribute differently between the phases or between the interface and the phases. Complete purification may be achieved by several manual extractions or by the application of countercurrent distribution.

In our affinity partitioning method we increase the selectivity by the addition of derivatives of poly(ethylene oxide) that have unique affinity for receptor sites. In principle an affinity ligand conjugated to any molecule that distributes heavily into one of the phases (e.g. derivatives of dextran) can be employed as an affinity "puller." The poly(ethylene oxide) derivatives bind specifically to the soluble or membrane bound receptor and the resulting complex distributes more heavily into the poly(ethylene oxide) derivative rich phase. In the first application of affinity partitioning (9), it was shown that a dinitrophenyl derivative of poly(ethylene oxide) distributes quantitatively in the same manner as unsubstituted poly(ethylene oxide). Furthermore, replacement of a fraction of the unsubstituted poly(ethylene oxide) with the dinitrophenyl derivative changed the distribution of a myeloma protein that binds dinitrophenyl compounds. Immunoglobulins that do not bind the dinitrophenyl moiety are relatively unaffected by the replacement of the unsubstituted poly(ethylene oxide) with the dinitrophenyl derivative. To further demonstrate the utility of affinity partitioning in the purification of membrane fragments, we chose to study the acetylcholine receptor-rich membrane fragments from the electric organ of *Torpedo californica*. The success of this application of affinity partitioning was insured by three considerations: i) acetylcholine receptor may be assayed conveniently using ^{125}I-labeled cobra α-toxin (19), ii) the receptor is present on discrete membrane fragments after homogenization (6,16) and iii) the density of receptor is extremely high (13) -- 20,000 to 30,000 α-toxin binding sites per μ^2.

We synthesized several derivatives of poly(ethylene oxide) which should have various binding affinities for the receptor (Fig.1). All three are positively charged at neutral pH, thus it is necessary to control for the possibility that the addition of these derivatives causes selective changes in distribution simply as a result of coulombic effects (1, 17). Initial experiments were performed with membrane fragments rich in α-toxin binding activity purified by velocity and sucrose density gradient centrifigation (6, 16). Upon the replacement of various amounts of unsubstituted poly(ethylene oxide) with the various substituted poly(ethylene oxide) derivatives a selective and substantial change in the distribution of the

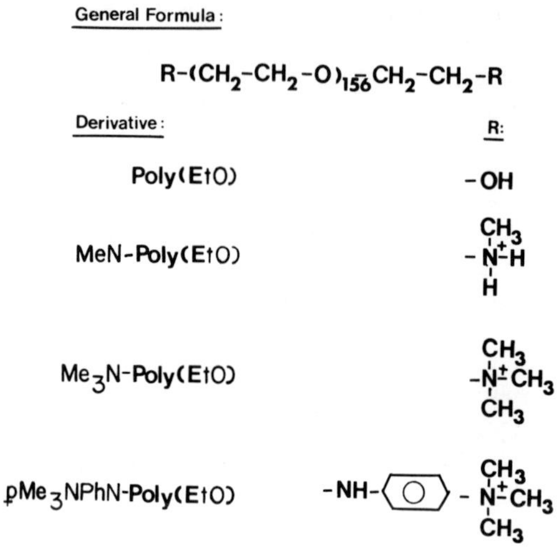

Fig. 1. Structure of poly(ethylene oxide) and derivatives. Abbreviations: MeN-poly(EtO), *bis* α,ω methylaminopoly(ethylene oxide) poly(EtO), poly(ethylene oxide); p-Me_3NPhN-poly(EtO), α,ω p-trimethylammonium(phenylamino)-poly(ethylene oxide); Me_3N-poly(EtO), *bis* α,ω trimethylammoniumpoly(ethylene oxide).

acetylcholine receptor (assayed by ^{125}I-labeled *Naja naja siamensis* α-toxin) was observed. Control of polymer, buffer and salt composition proved important; among the various systems tested, the system described in Fig. 2 gave the most favorable affinity partitioning effect. In the absence of poly(ethylene oxide) derivatives, more than 75% of the α-toxin binding activity partitioning into the bottom [dextran rich phase]. *bis*-α,ω-Methylamino-poly(EtO), a secondary amine derivative of poly(ethylene oxide), was the least potent in changing the distribution of membrane bound acetycholine receptor while *bis*- α, ω-trimethylaminopoly(EtO) and *bis*-α,ω-*p*-trimethylamino(phenylamino)-poly(EtO) proved considerably more potent. These results are consistent with known structure-activity relationships of cholinergic compounds. The authentic nature of the affinity partitioning effect shown here was further verified by the finding that the effects of *p*-Me_3PhN-poly(EtO) are blocked by saturating binding sites with unlabelled α-toxin (Fig. 2) or the addition of bisquaternary methonium compounds to the phase systems. At concentrations of α-toxin less than required to saturate binding sites, there was an apparent linear relationship between the fraction of sites occupied by α-toxin and the affinity partitioning effect observed.

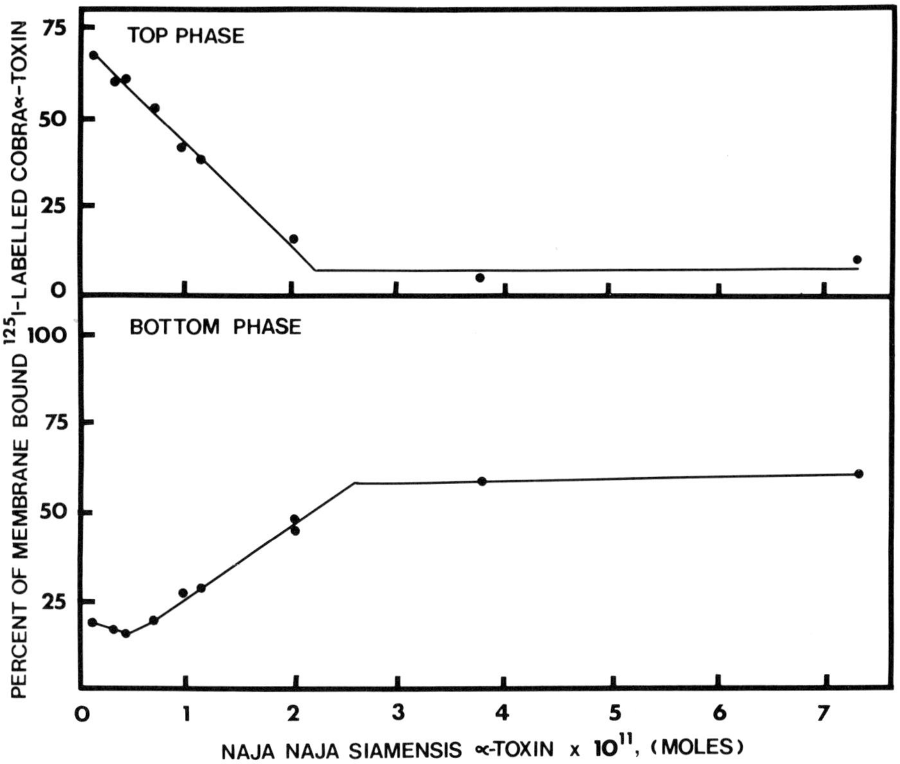

Fig. 2. Influence of α-toxin occupancy of acetylcholine receptor on the partitioning of membrane bound ^{125}I-labeled to win in a phase system containing p-ME$_3$PhNpoly(EtO). Purified membranes were incubated first with iodinated α-toxin for 1 hour. Then various amounts of unlabeled α-toxin were added giving the indicated moles of α-toxin. The incubation mixture was diluted into the phase system. The final phase system contained 4.32% (w/w) Dextran T-500, 3.59% (w/w) poly (EO) 6,000, 24.75 mmol of NaCl and 4.75 mmol Na phosphate pH 7.4 per kg system. A fraction of the unsubstituted poly (EO) was replaced with p-Me$_3$PhN-poly(EtO) to give a final concentration of substituted ligand of 1.8×10^{-4} M. The α-toxin binding activity in each phasing syringe was 2×10^{-11} moles on the basis of the filter toxin binding assay (19). In the absence of p-Me$_3$PhN-poly(EtO), incubation with excess unlabeled toxin has a negligible influence on the distribution of receptor enriched membranes. In this system free toxin has a partition coefficient of 0.97.

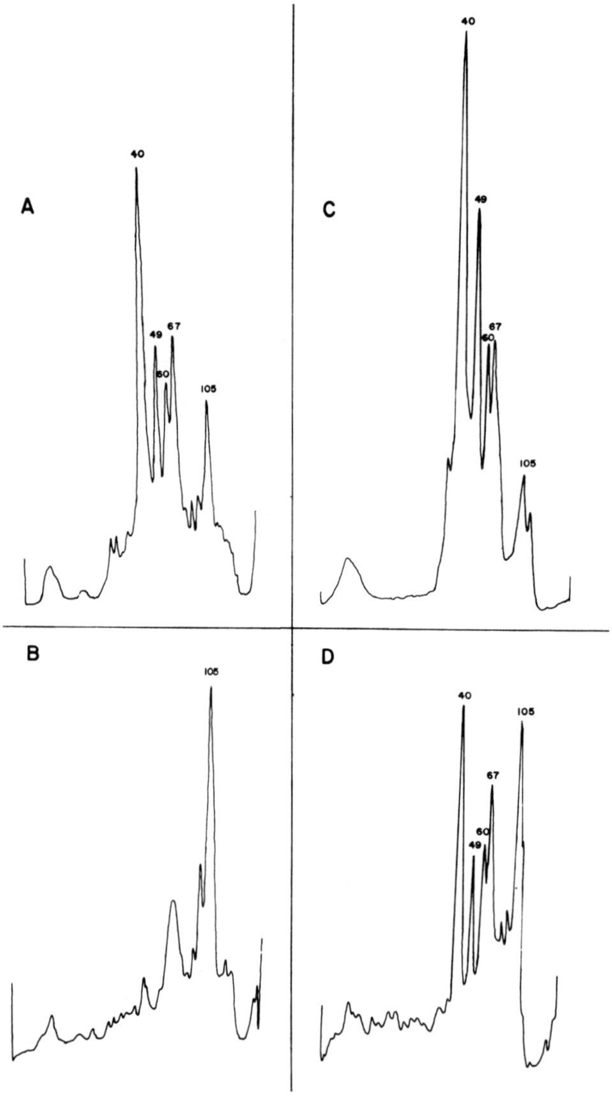

Fig. 3. Polyacrylamide gel electrophoresis in the presence of sodium dodecyl sulfate ($NaDodSO_4$) of isolated membrane fragments and the solubilized cholinergic receptor from electric organs of Torpedo californica. The individual fractions were solubilized with 1% $NaDodSO_4$ in the presence of 1% mercaptoethanol and then subjected to electrophoresis in 7.5% gels according to the prodecure of Weber et al. (24). The profiles shown are Coomassie blue staining densities where the left margin is the migration position of the bromophenol blue tracking dye and the right margin represents the top of the gel. A. Top poly(EO) rich phase of membranes subjected to affinity partitioning following differential and density gradient

(Fig. 3, Cont'd.)

procedures (specific activity - 3.6 nmoles of bound α-toxin/mg of protein) B. Bottom (dextran) phase from the corresponding phase separations C. Receptor solubilized and purified by method of Schmidt and Raftery (20). D. Membrane fraction after differential and density gradient procedures but prior to affinity partitioning (specific activity 2.2 nmoles bound α-toxin/mg of protein).

TABLE I

Purification of Receptor Enriched Membranes from the Electric Organ of *Torpedo californica**

		nmoles of bound toxin	mg of protein	S.A.(nmoles of toxin/mg of protein)
I.	Crude Membranes	16.80		0.19
II.	Differential Centrifugation	14.36	43.5	0.33
III.	Density Gradient Centrifugation	4.08	1.44	2.83
IV.	Affinity Partitioning			
	a) Top Phase	2.33	0.51	4.57
	b) Bottom Phase	0.64	0.65	0.98

*Steps I, II and III followed the procedure of Reed *et al.* (16). As shown by Reed *et al.*, two peaks of toxin binding activity were evident upon density gradient centrifugation. The high density peak (45% sucrose) contained most of the observable particulate matter and a low specific activity for toxin binding. The center region of a lower density (33% sucrose), which contained material of high specific activity, was employed in the subsequent affinity partitioning step. The portion removed amounted to 42% of the total receptor activity and 73% of the amount in the low density peak. It was dialyzed against the buffer used in the partitioning system. The final composition of phase system is the same as described in Fig. 2; however, no α-toxin was added before purification steps. Toxin binding was measured according to the procedures of Schmidt and Raftery (19). Membrane fragments were separated from phase solutions by centrifugation as described previously (10).

The utility of affinity partitioning must rest upon its effectiveness in purifying receptor-rich membranes from a hetergeneous mixture. A crude membrane preparation was subjected to a single affinity partitioning extraction in the presence of p-Me_3PhN-poly (EtO). The purified membranes were collected from the top (poly (EtO)-rich) by centrifugation. The specific activity of the purified membranes was 7-fold greater in α-toxin binding activity while adenosine triphosphatase and acetylcholinesterase were depleted 30-fold and 5-fold respectively (10). The affinity partitioning extraction requires thirty minutes or less. Polymer solutions were removed from purified membranes by centrifugation (8). The final specific activity of α-toxin binding activity was 2500 pmol per mg protein--in the range of purity reported using classical centrifugation methods (16). Further purification was achieved by classical centrifugation methods followed by an affinity partitioning extraction (Table I). The high purity of the resulting receptor-rich membrane preparation was verified by electron microscopy after negative staining with neutral phosphotungstic acid. In the highly purified preparations 60-80% of the membrane fragments contained closely packed rosettes, which are characteristic of the nicotinic acetylcholine receptor (4, 13). The protein profile of the purified membranes is remarkably similar to that of the solubilized receptor purified by affinity chromatography (Fig. 3). Affinity partitioning may be successfully applied to a variety of receptor systems. With the currently available polymer derivatives its utility may be limited to the purification of membrane fragments containing relatively high densities of binding sites. However, in the present application poly(ethylene oxide) derivatives of relatively low affinity for the nicotinic receptor were employed. By using derivatives of higher affinity for the receptor, lower densities of membrane binding sites may be required. Also, it is likely that the affinity partitioning effect may be amplified by the use of higher molecular weight poly(ethylene oxide) derivatives.

BIBLIOGRAPHY

1. ALBERTSSON, P. A., Partition of Cell Particles and Macromolecules, 2nd Ed., John Wiley and Sons, New York (1971).
2. ALBERTSSON, P. A., Biochemistry 12 (1973) 2525.
3. ALBERTSSON, P. A., Methods in Enzymology 31 (1974) 761.
4. CARTAUD, J., BENEDETTI, L., COHEN, J. B., MEUNIER, J. C. and CHANGEUX, J. P., FEBS Lett. 33 (1973) 109.
5. COHEN, J. B., WEBER, M., and CHANGEUX, J. P., Mol. Pharmacol. 10 (1974) 904.
6. DUGUID, J. R. and RAFTERY, M. A. Biochemistry 12 (1973) 3593.
7. FEWTRELL, C. M. S. and RANG, H. P. in Drug Receptors: A Symposium (H. P. Rang ed.) Macmillan Press, London (1973).
8. FLANAGAN, S. D., Ph.D. Thesis, University of California at San Diego (1974).

9. FLANAGAN, S. D. and BARONDES, S. H., J. Biol. Chem. 250 (1975) 1484.
10. FLANAGAN, S. D., TAYLOR, P. and BARONDES, S. H. , Nature 254 (1975) 441.
11. FLANAGAN, S. D., BARONDES, S. H. and TAYLOR, P. , J. Biol. Chem., (1975) in press.
12. JOHANSSON, G., HARTMAN, A. and ALBERTSSON, P. A., Eur. J. Biochem. 33 (1973) 379.
13. NICHOL, A. and POTTER, L. T., Brain Res. 57 (1973) 508.
14. O'CARRA, P. , Biochem. Soc. Trans. 2 (1974) 1289.
15. OLSEN, R. W., MEUNIER, J. P., and CHANGEUX, J. P., FEBS Lett. 28 (1972) 96.
16. REED, K., VANDLEN, R., BODE, J., DUGUID, J. and RAFTERY, M. A., Arch. Biochem. Biophys. 167 (1975) 138.
17. REITHERMAN, R., FLANAGAN, S. D. and BARONDES, S. H., Biochim. Biophys. Acta 297 (1973) 193.
18. RUTISHAUSER, U., D'ESUTACHIO, P. and EDELMAN, G. M., Proc. Nat. Acad. Sci. USA 68 (1973) 2153.
19. SCHMIDT, J. and RAFTERY, M. A., Anal. Biochem. 52 (1973) 349.
20. SCHMIDT, J. and RAFTERY, M. A., Biochemistry 12 (1973) 852.
21. SHANBHAG, V. P. and JOHANSSON, G., Biochem. Biophys. Res. Comm. 61 (1974) 1141.
22. TAKERKART, G., SEGARD, E. and MONSIGNY, M., FEBS Lett. 42 (1974) 218.
23. WALTER, H., Methods in Enzymology 32 (1974) 637.
24. WEBER, K., PRINGLE, J. and OSBORN, M., Methods in Enzymology 26 (1972) 3.

BIOCHEMICAL AND MOLECULAR CHARACTERISTICS OF β-ADRENERGIC RECEPTOR BINDING SITES

Lee E. LIMBIRD and Robert J. LEFKOWITZ

Department of Medicine, Duke University Medical Center
Durham, North Carolina 27706 (USA)

The availability of radiolabeled hormones, drugs and analogs which retain their characteristic biological activity and selectivity has provided a valuable tool for the study of the interaction of these agents with their target cell receptors. Despite the major advances in the investigation of receptors for polypeptide (14, 37) and steroid hormones (13) and nicotinic cholinergic drugs (11) using these techniques, the confident identification of the β-adrenergic receptors for catecholamines has remained until recently an elusive goal. Initial studies employed tritium labeled β-adrenergic agonists to identify binding sites in membrane fractions containing catecholamine-sensitive adenylate cyclase. However, some of the characteristics of binding observed with these agents diverged from which might be expected of the physiological β-adrenergic receptor (3, 17, 18). Perhaps the difficulties in these studies can be attributed to: 1) the multiple potential binding sites in addition to the adenylate cyclase coupled β-adrenergic receptor for labeled native catecholamines, 2) the relatively low affinity for catacholamines in *in vitro* membrane preparations ($K_D \cong 10^{-6}M$), 3) the unimpressive specific radioactivity of available tritiated ligands ($\cong$ 1 Ci/mmol), or 4) the absence of tissue preparations which provided a high concentration of β-adrenergic receptors relative to other constituents. Even in the case of the high affinity β-adrenergic antagonist propranolol, which would not likely interact with degradative enzymes and uptake mechanisms directed at catecholamines, binding characteristics indicated that this ligand binds disproportionately to non β-adrenergic receptor sites (40), possibly those responsible for the so-called "anaesthetic" properties of this drug.

The preparation of potent β-adrenergic antagonists labeled to high specific radioactivity has, however, resulted in the identification of membrane binding sites which appear to possess all of the essential characteristics expected of the physiologically relevant β-adrenergic receptors. β-Adrenergic agents appear to elicit most, if not all, of their physiologic effects through the stimulation of adenylate cyclase and synthesis of cAMP (35). This close relationship between the β-adrenergic receptor and adenylate cyclase has provided a valuable tool for studying this catecholamine sensitive system at the biochemical level, particularly in developing criteria for binding characteristics which are consistent with interaction with adenylate cyclase-coupled β-adrenergic receptors.

(-)[^{3}H]Alprenolol as a Radioactive Tracer of β-Adrenergic Receptor Binding

In our laboratory, (-)[^{3}H]alprenolol, a potent competitive β-adrenergic antagonist (17-33 Ci/mole, has been employed as a radioactive tracer for binding to the β-adrenergic receptor. Although (-)[^{3}H]alprenolol binding has been studied in a number of mammalian tissues (1, 42)[1], the most extensive characterization of the binding phenomenon has been carried out with membranes derived from frog erythrocytes since they possess an adenylate cyclase system extremely sensitive to catecholamines and provide a homogeneous and reproducible source of easily purified membrane preparations (19, 27, 28).

Documentation that specific (-)[^{3}H]alprenolol binding represents interaction with the adenylate cyclase coupled β-adrenergic receptors has been presented (1, 19, 27, 28, 42)[1] and reviewed (20, 22) in considerable detail elsewhere. Briefly, the ability of 60 β-adrenergic agonists and antagonists to compete for the (-)[^{3}H]alprenolol binding sites in frog erythrocyte membranes directly parallels their ability to interact with the adenylate cyclase system. The marked stereoselectivity for (-)isomers characteristic of the physiologic β-adrenergic receptor is also apparent in the competition of stereoisomers for (-)[^{3}H]alprenolol binding. The rapid rate of association and dissociation of (-)[^{3}H] alprenolol is also in accord with the rapid rates of onset and offset of the antagonist effects of alprenolol on the erythrocyte membrane adenylate cyclase. The binding of (-) [^{3}H] alprenolol to frog erythrocyte membranes is also saturable, which is consistent with the existence of a finite number of receptor sites.

The confident identification of β-adrenergic receptors using (-)[^{3}H]alprenolol has thus provided a valuable experimental tool to investigate the biochemical and molecular characteristics of binding

[1] Williams, L. T., Jarett, L., and Lefkowitz, R. J., submitted for publication.

TABLE I

EFFECTS OF ENZYMES AND DENATURANTS ON SPECIFIC (-)[^{3}H]ALPRENOLOL BINDING

Membrane preparations were preincubated in the presence of the indicated concentrations of enzymes and urea at 25° for 10 to 15 min. Controls were preincubated under identical conditions without enzymes. Aliquots were then transferred to the usual incubation medium for assay of specific binding. Concentrations of trypsin greater than 5 μg/ml caused membrane preparations to congeal and therefore could not be reliably tested. Values shown represent the mean of duplicate determinations in at least two experiments.

Enzyme	Concentration μg/ml	(-)[^{3}H]Alprenolol Bound % Control
Trypsin	5	31
	2	36
	1	47
	0.5	61
Phospholipase A (bee venom)	10	6
Phospholipase A (*Vipera russelli*)	10	5
Phospholipase C (*Clostridium Welchii*)	100	54
Phospholipase D (cabbage)	100	76
Combined Enzymatic Treatment		
Trypsin	0.5	71
Phospholipase A (bee venom)	0.5	64
Trypsin + Phospholipase A	0.5 + 0.5	16
Denaturant	**Concentration, M**	
Urea	5	32
	1	73
	0.1	106

TABLE II

EFFECT OF SPECIFIC RESIDUE - DIRECTED REAGENTS ON SPECIFIC (-)[^{3}H]ALPRENOLOL BINDING

Membranes preincubated with each reagent in 75 mM Tris, 25 mM $MgCl_2$ buffer at the indicated pH for 15 minutes at 25°. Controls were incubated under identical pH and solvent conditions. At the end of the preincubation period aliquots of the membrane preparations were added to the usual binding assay incubation medium. Data presented represent the mean of duplicate determinations ± SE for 2-6 separate experiments.

Reagent	Concentration mM	[^{3}H] (-)Alprenolol Bound % Control	Incubation Conditions
1 Ethyl-3-(3-dimethylamino propyl) carbodiimide	50 20	29 ± 3 41 ± 9	aqueous, pH 7.4
1 Ethyl -3-(3-dimethylaminopropyl) carbodiimide + glycine-o-methyl ester	20 50	29 ± 5	aqueous, pH 7.14
2,4,6-Trinitrobenzene sulfonic acid	10	20 ± 12	aqueous, pH 8.1
Phenylmethylsulfonyl fluoride	10	61 ± 12	1% benzene, pH 7.4
2-Hydroxy-5-nitrobenzyl bromide	5	5 ± 11	1% acetone, pH 7.4
2-Methoxy-5-nitrobenzyl bromide	5	45 ± 3	1% acetone, pH 7.4
N-Acetylimidazole	10	84 ± 9	aqueous, pH 7.4
Iodoacetamide	5	93 ± 4	aqueous, pH 8.1
5,5'-Dithiobis (2-nitrobenzoic acid)	0.1	108 ± 0	aqueous, pH 8.1
p-Chloromercuribenzoate	5	97 ± 4	aqueous, pH 8.1

to these receptors and, in addition, their relationship with the adenylate cyclase system.

Effects of Membrane Perturbation on β-Adrenergic Receptor Binding and Adenylate Cylase Activity

Exposure of frog erythrocyte membranes to a number of degradative enzymes and other so-called membrane perturbants has revealed a number of membrane constituents which appear crucial for (-)[^{3}H]alprenolol binding. Decreased radioligand binding in membranes exposed to proteases, urea (Table I), or temperatures greater than 45° for 15 minutes indicates the participation of protein membrane constituents in β-adrenergic receptor binding. The possible contribution of particular amino acid residues to the binding phenomenon was evaluated by treating the frog erythrocyte membranes with several protein-modifying reagents. Table II summarizes the reagents and incubation conditions employed and their effects on specific (-)[^{3}H]alprenolol binding. Dramatic reduction of specific ratioligand binding was observed after interaction of the membranes with 2,4,6-trinitrobenzene sulfonic acid (TNBS), a reagent used to modify lysine ε-amino groups. (2). This reagent is not eliciting its effects by combining with the secondary amino group of (-)[^{3}H]alprenolol or unlabeled propranolol during the binding assay, since extensive washing of the membranes or addition of excess lysine after TNBS treatment did not alter the extent of binding reduction. A water soluble carbodiimide reagent which can react with protein carboxyl groups (12) also markedly reduces (-)[^{3}H]alprenolol binding. Since carbodiimide is also capable of condensing carboxyl and amino residues to form amido linkages, it is possible that the observed decline in binding is due to the interaction of this reagent with crucial amino groups and only fortuitously with exposed carboxyl groups. Phenylmethylsulfonyl fluoride (PMSF), an agent which modifies reactive serine residues in proteolytic enzymes (29, 30, 41), reduces (-)[^{3}H]alprenolol binding even when five fold excess concentrations of serine are added prior to the binding assay as a reagent scavenger to prevent possible modification of the β-adrenergic agents. Treatment of the membranes with the nitrobenzyl bromide tryptophan-modifying reagents also caused diminished (-)[^{3}H]alprenolol binding. In addition to modifying tryptophan, these compounds can react with free sulfhydryl groups. However, since relatively drastic concentrations of the sulfhydryl directed reagents iodoacetamide, 5,5'-dithiobis (2-nitrobenzoic acid) (DTNB) and p-chloromercuribenzoate (PCMB) (36) did not significantly alter specific binding, the effect of the nitrobenzyl bromides is likely due to a modification of tryptophan residues in the frog erythrocyte membranes. N-Acetylimidazole, a reagent capable of acetylating the phenolic residues of tyrosine (34), only marginally reduces specific binding. Until the receptor molecule is purified to homogeneity, it cannot be unequivocally verified that the effects of these reagents are due to a specific modification of an amino acid residue located at the

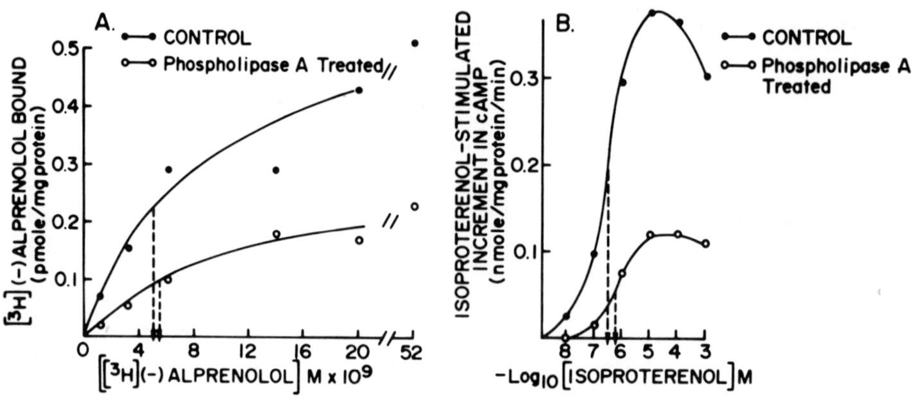

Fig. 1. A: Effect of phospholipase A treatment of purified frog erythrocyte membranes on specific (-)[^{3}H]alprenolol binding. ("Specific" binding will be used throughout to refer to that binding of (-)[^{3}H]alprenolol which is competed for by 10^{-5} M unlabeled (-)alprenolol). Membranes, prepared as described previously (13), were preincubated for 10 minutes at 25° in the presence ("treated") or absence ("control") of 1 μg/ml bee venom phospholipase A. The arrow (▼) indicates the concentration of (-)[^{3}H]alprenolol which half maximally saturates the sites in each preparation. B: Effects of phospholipase A treatment of erythrocyte membranes on isoproterenol-stumulated adenylate cyclase. "Treated" and "control" membrane preparations identical to those studied in Fig. 1A were assayed for adenylate cyclase responsiveness to catecholamines by methods described previously (13). The arrow (▼) indicates the concentration of isoproterenol which half maximally stimulates enzyme activity.

hormone binding site rather than to residues on adjacent membrane constituents responsible for maintaining a crucial architecture around the receptor molecule. Interpretation of the exact site of reagent interaction necessarily awaits complete purification of the β-adrenergic receptor. However, the requirement for unaltered carboxyl, amino hydroxyl and tryptophanyl residues for optimal binding in solubilized β-adrenergic receptor preparations(3)[2] suggests that reagents directed at these residues are likely binding to the receptor molecule itself rather than to adjacent membrane components which may no longer be juxtaposed in solubilized preparations. It is no surprise that both hydrophobic and hydrophilic residues are involved in the binding phenomenon, since β-adrenergic agonists and antagonists possess both hydrophobic (catechol moiety) and hydro-

---

[2] Caron, M. G. and Lefkowitz, R. J., submitted for publication.

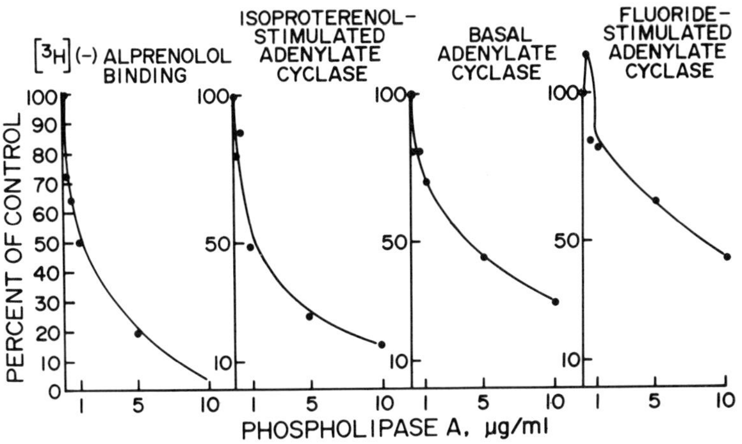

Fig. 2. Concentration dependence of phospholipase A treatment of erythrocyte membranes on specific (-)[^{3}H]alprenolol binding and adenylate cyclase activities. The source of the enzyme was *Vipera russelli*. Membranes were preincubated at 25° for 10 minutes in the absence or presence of the indicated concentrations of enzyme. Aliquots were then removed and assayed simultaneously under comparable incubation conditions for binding and enzyme activities.

philic (ethanolamine moiety) structural regions.

Treatment of frog erythrocyte membranes with phospholipases markedly diminishes (-)[^{3}H]alprenolol binding, indicating that lipid components of the frog erythrocyte membranes are also crucial for β-adrenergic receptor function. This decreased binding is not due to solubilization of intact receptors concomitant with phospholipase digestion, since assay of the soluble phase by equilibrium dialysis and Sephadex G50 chromatographic techniques after centrifugation of the incubate at 104,000x g revealed no binding activity. The possibility exists that phospholipase treatment results in simultaneous solubilization and degradation of the receptor, still indicating a crucial role for membrane lipids in β-adrenergic receptor function. Fig. 1A indicates that the decrease in (-)[^{3}H]alprenolol binding observed in membranes pretreated with phospholipase A is due to a decrease in the number of binding sites with insignificant alteration in the affinity of the remaining sites. The expected consequence of binding site degradation, a decrease in the sensitivity of adenylate cyclase to catecholamine stimulation, was observed and is shown in Fig. 1B. Figure 2 demonstrates that the effects of phospholipase A on the receptor and enzyme components of this catecholamine sensitive system are dose dependent. Digestion of erythrocyte membranes with 0.1 - 10 μg/ml phospholipase A reduced basal and fluoride-stimulated adenylate cyclase activities and caused an even more drastic decrease in isoproternol-stimulated catalysis. The

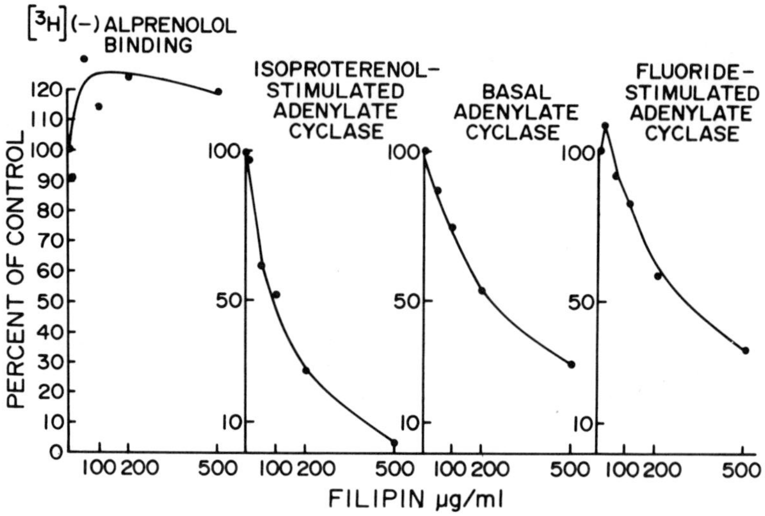

Fig. 3. Effect of exposure of frog erythrocyte membranes to filipin on (-)[^{3}H]alprenolol binding and adenylate cyclase activities. Filipin was suspended with membranes for 10 minutes at 25° by swirling the aqueous membrane preparation in a round bottom flask previously coated with filipin under a N_2 stream. Values shown represent the means of duplicate determinations from five separate experiments.

marked decline in catecholamine sensitivity of adenylate cyclase directly paralleled the reduction in (-)[^{3}H]alprenolol binding in the same preparations. Comparable dose dependent profiles were obtained when frog erythrocyte membranes were exposed to lysolecithin and the polyene antibiotic, amphotericin B, which is also known to interact with the lipidic regions of biological membranes.

The pattern of effects elicited by the polyene antibiotic, filipin, is distinct, however. As seen in Fig. 3., although filipin also produces a greater decline in catecholamine sensitive enzyme activity than in basal activity, this occurs in the absence of any decline in receptor binding. These findings suggest that filipin "uncouples" catecholamine stimulation of adenylate cyclase by perturbing processes distal to β-adrenergic receptor binding. Puchwein *et al*. have demonstrated that filipin interacts with membrane cholesterol to form a filipin-cholesterol complex in a comparable model system, the pigeon erythrocyte membrane (32). In addition, these investigators reported an "uncoupling" of the catecholamine sensitive adenylate cyclase system based on a decline in hormone sensitive catalysis in the absence of altered (±)[^{3}H]epinephrine and (±)[^{3}H]isoproterenol binding. Since (-)[^{3}H]alprenolol binding provides a more confident identification of β-adrenergic receptor binding, our studies confirm that filipin is capable of uncoupling the catecholamine-sensitive adenylate cyclase system and suggest that an

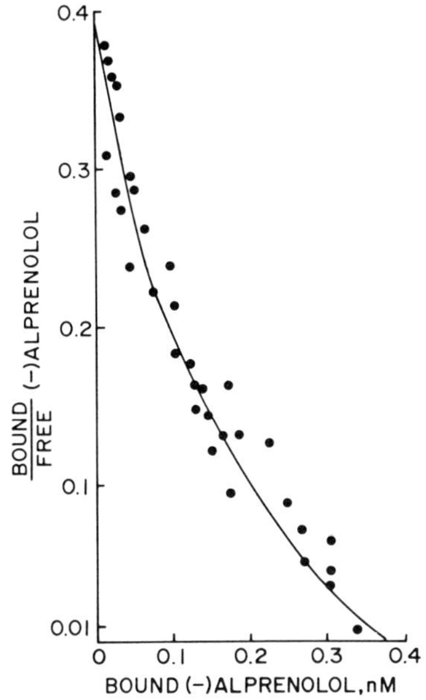

Fig. 4. A: Scatchard plot of specific binding of (-)[^{3}H]alprenolol to purified frog erythrocyte membranes. Radioligand was incubated with membranes for 10 min at 37°, pH 7.42, in the absence and presence of unlabeled hormone. Separation of membrane-bound from free (-)[^{3}H]alprenolol was accomplished by vacuum filtration through Whatman GF/C glass fiber filters which were then washed one time with ice cold buffer 75 mM Tris, 25 mM $MgCl_2$, pH 7.42 at 4° (36).

intact membrane matrix is essential for optimal catalytic response to hormone stimulation. The disproportionate decrease in isoproterenol-stimulated adenylate cyclase activity caused by treatment of the membranes with phospholipase A, lysolecithin and amphotericin B is also consistent with this hypothesis. However, the concomitant reduction in (-)[^{3}H]alprenolol binding caused by these agents prevents an assessment of the extent to which the decline in hormone sensitive adenylate cyclase activity is due to alterations at the level of receptor binding or to an interruption of "signal transmission."

Although both adenylate cyclase activity and β-adrenergic receptor binding are diminished by membrane exposure to proteases, phospholipases and non-degradative lipid-perturbing agents, a number of the properties of (-)[^{3}H]alprenolol binding sites are clearly distinguishable from adenylate cyclase. For example, receptor binding is stable to temperatures as high as 45° for 15 minutes, whereas adenylate cyclase activity is 50% decreased after exposure to 45° for 5 minutes (23).

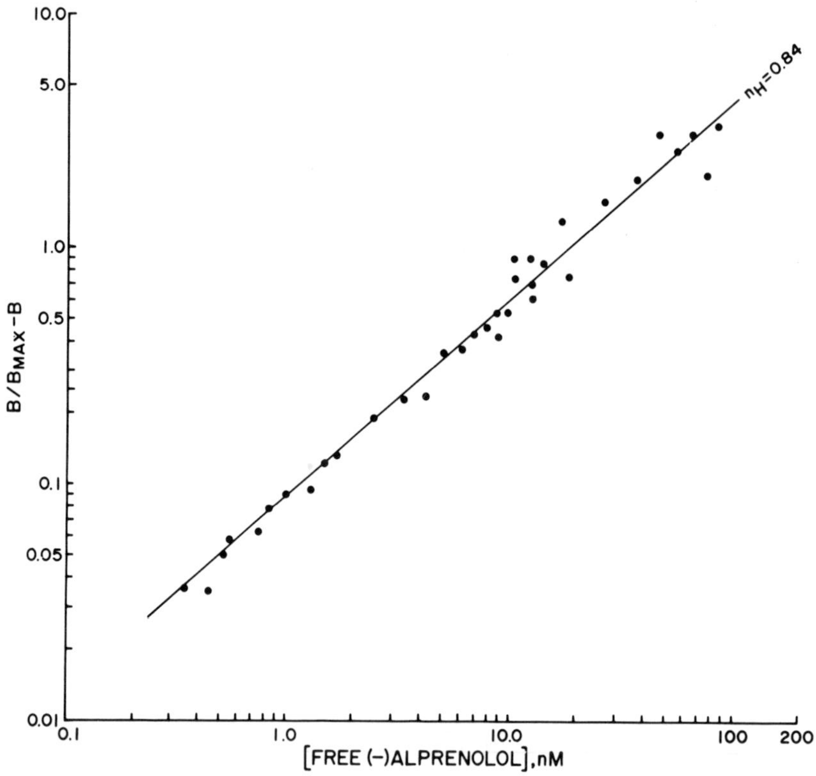

Fig. 4. B: Hill Plot of specific binding of (-)[^{3}H]alprenolol to purified frog erythrocyte membranes. Data was obtained under the experimental conditions described for Fig. 4A.

Monovalent and divalent cations are neither required nor are inhibitory to (-)[^{3}H]alprenolol binding. In contrast, the adenylate cyclase activity requires the presence of divalent cations (7), ATP-Mg++ being the physiologic substrate, and is inhibited by Ca^{++} (31, 39).

Negatively Cooperative Interactions Among the β-Adrenergic Receptor Sites

We have demonstrated that β-adrenergic receptor sites in frog erythrocyte membranes exhibit negatively cooperative site-size interactions (25). That negative cooperativity among these receptors might exist was suggested by steady state binding area. Saturation of the receptors with (-)alprenolol progresses from 10% to 90% saturation over a greater than 81-fold range, Scatchard plots are curvilinear with upward concavity (Fig. 4A), and Hill plots consistently display slopes, n_H, less than 1.0 (Fig. 4B). When certain theoretical considerations are met (4, 36, 38), these observations indicate the existence of multiple orders of binding sites with discrete affinities,

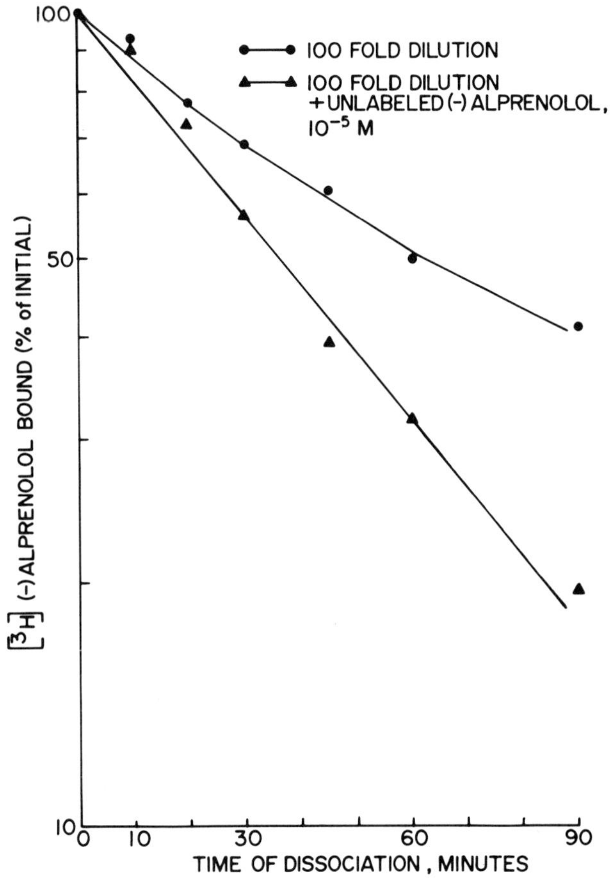

Fig. 5. Enhancement of (-)[^{3}H]alprenolol dissociation from frog erythrocyte membranes in the presence of unlabeled (-)alprenolol. (-)[^{3}H]alprenolol, 5 x 10^{-9} M, was incubated with purified frog erythrocyte membranes (12 mg/ml) for 45 minutes at 10° after which 100 μl aliquots were transferred to a series of tubes that contained 9.9 ml of buffer in the presence or absence of unlabeled (-)alprenolol, 10^{-5} M. At the intervals shown, two tubes from each set were vacuum filtered through Whatman GF/C glass fiber filters, the filters washed one time with 5 ml ice cold buffer, and the radioactivity in each filter counted. The radioactivity bound to the membranes, expressed as a percentage of the radioactivity at t=0, is plotted as a function of the time elapsed after dilution of the membranes (t=0 refers to the sampling time immediately after 1:100 dilution).

negatively cooperative interactions among the binding sites, or both. Negative cooperativity among the β-adrenergic receptors would mean that as the proportion of receptors occupied by β-adrenergic agents is increased, the overall affinity of all the receptor sites is decreased. Steady state binding data alone cannot distinguish between the two possibilities of heterogeneous binding sites or cooperative interactions among homogeneous binding sites. However, a study of radioligand dissociation kinetics has been demonstrated to directly determine the existence of negative cooperativity (5, 6).

This approach exploits the difference in affinity which exists in a negatively cooperative system when only a minority of the sites are occupied as compared with complete saturation of the sites. To study the dissociation kinetics under these two situations (i.e. fractional occupancy and full occupancy), membrane preparations are first incubated with concentrations of (-)[^{3}H]alprenolol which occupy only a fraction of the β-adrenergic receptors. When steady state binding is attained, the dissociation of radioligand is studied under two experimental conditions: 1) with dilution of receptors sufficient to prevent detectable rebinding of the dissociated tracer, and 2) with identical dilution in the presence of excess concentrations of unlabeled drug. If the filling of the empty sites by excess unlabeled (-)alprenolol does not alter the affinity of the sites already occupied by (-)[^{3}H]alprenolol, then the rate of radioligand dissociation under both experimental conditions will be identical. If, however, negatively cooperative interactions among the receptors exist, saturation of the sites with labeled ligand will decrease the affinity of all the sites. This decreased affinity will be reflected by an enhanced rate of (-)[^{3}H] alprenolol dissociation. The acceleration of (-)[^{3}H]alprenolol dissociation in the presence of excess unlabeled (-)alprenolol is precisely what we observed in the frog erythrocyte system, as shown in Fig. 5. The absence of detectable rebinding of radioligand has been documented by two experimental approaches and described in detail previously (25). Since the accelerated (-)[^{3}H]alprenolol dissociation in the presence of excess unlabeled ligand is therefore not due to the ability of the unlabeled (-)alprenolol to compete for radioligand rebinding, we interpret the observations represented in Fig. 5 to reflect the existence of negatively cooperative interactions among the β-adrenergic receptor sites.

A number of β-adrenergic agents were evaluated for their ability to induce this cooperative effect. As seen in Fig. 6, both β-adrenergic agonists and antagonists are capable of accelerating (-)[^{3}H] alprenolol dissociation. The efficacy of β-adrenergic agents in producing this phenomenon is directly related to their affinity for the β-adrenergic receptors rather than their intrinsic activity in the catecholamine-sensitive adenylate cyclase system. For example, the full agonist (-)isoproterenol has a $K_{S_{0.5}}$[3] calculated from competition

[3]$K_{S_{0.5}}$ refers to the concentration of β-adrenergic agent which half-maximally saturates the (-)[^{3}H]alprenolol binding sites. In a non-cooperative system where the affinity of the receptor is a constant descriptor, $K_{S_{0.5}}$ would be equal to the equilibrium dissociation constant, K_D.

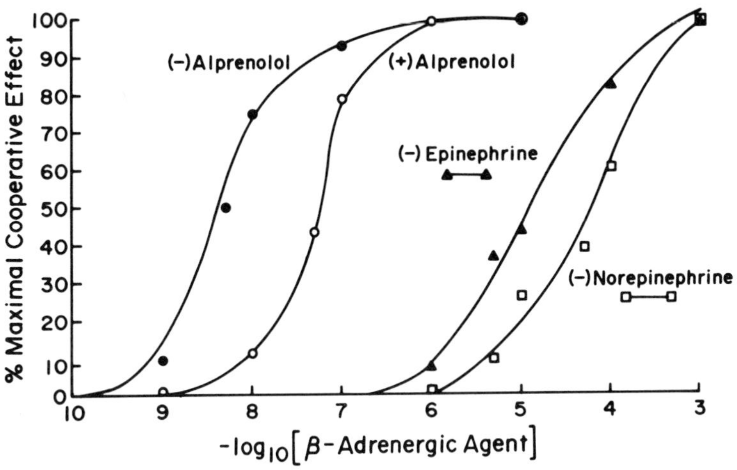

Fig. 6. Concentration dependent effects of (-)alprenolol, (+)alprenolol, (-)epinephrine and (-)norepinephrine on the dissociation of (-)[^{3}H]alprenolol from purified frog erythrocyte membranes. (-)[^{3}H] alprenolol was incubated with the membranes for 45 minutes at 10°. Membranes were diluted 1: 100 into fresh buffer in the absence or presence of varying concentrations of unlabeled β-adrenergic agents. After 45 minutes at 10°, the membranes were filtered as described in the legend to Fig. 5. The differences in the (-)[^{3}H]alprenolol bound to the membranes with "dilution" and "dilution + unlabeled agent" are expressed as % of the maximal cooperative effect, i.e. the maximal difference observed with "dilution + unlabeled (-)alprenolol."

binding studies of 4 x 10^{-7} M. (-)Soterenol, a pharmacologic agent of comparable affinity ($K_{S_{0.5}}$ = 1.2 x 10^{-6} M) but only 28% as effective as isoproterenol in stimulating adenylate cyclase, was equiactive in accelerating the dissociation of (-)[^{3}H]alprenolol, as seen in Fig. 7. Competitive antagonists (e.g. alprenolol), which possess no intrinsic activity in stimulating adenylate cyclase, are also able to induce negative cooperativity in a manner which parallels their receptor affinity. These observations indicate that the ability to fully activate adenylate cyclase is not necessarily directly related either to affinity for the β-adrenergic receptor or to potency in inducing negatively cooperative site-site interactions. It is also important to note that the steroselectivity for (-)isomers which is characteristic of the physiologic β-adrenergic receptor and the (-)[^{3}H]alprenolol binding sites is also apparent in the ability to elicit site-site interactions among the receptor sites. This is demonstrated for the antagonist alprenolol in Fig. 6 and for the β-adrenergic agonist isoproterenol in Fig. 8.

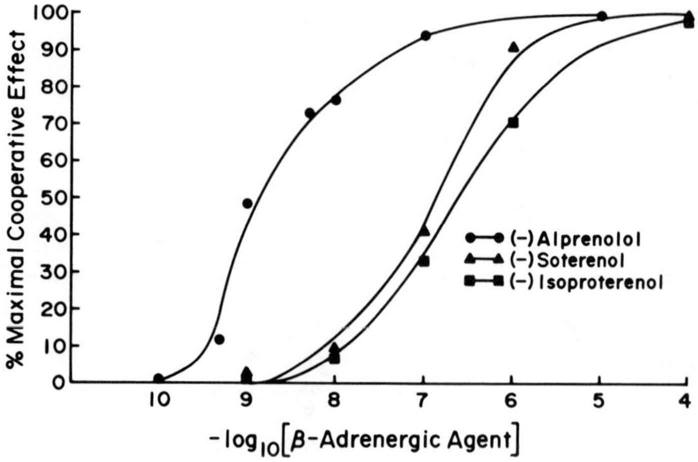

Fig. 7. Concentration dependent effects of (-)alprenolol, (-) isoproterenol and (-)soterenol on the dissociation of (-)[^{3}H]alprenolol from purified frog erythrocyte membranes. The experimental design was identical to that described in the legend to Fig. 6.

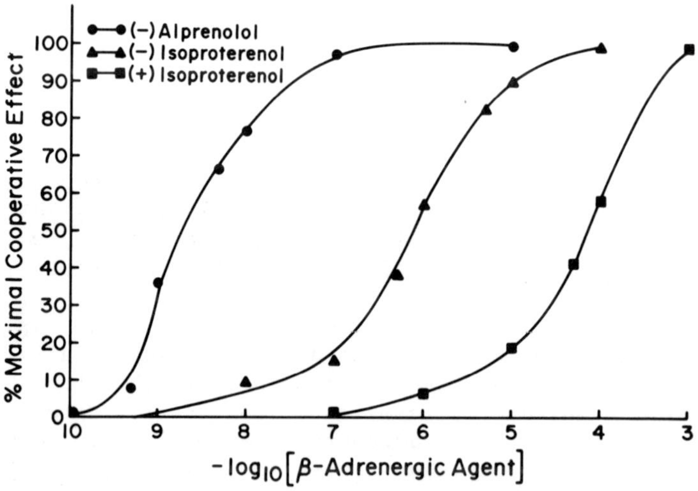

Fig. 8. Comparison of the concentration dependent effects of the stereoisomers of isoproterenol on (-)[^{3}H]alprenolol dissociation from frog erythrocyte membranes. The experimental design was identical to that described in the legend to Fig. 6.

The molecular phenomena responsible for the observed negative cooperativity are still speculative. The ability to accelerate (-)[^{3}H]alprenolol dissociation in the presence of excess unlabeled ligand is a temperature dependent process in the frog erythrocyte and is not detected at 4°.[4] This observation is consistent with a number of putative mechanisms for inducing site-site interactions, including conformational changes in the receptor molecules or aggregation of receptor sites in a manner analogous to known cooperative phenomena within other molecules (10, 16, 26). It is also possible that ligand binding induces mobility and clustering of receptor sites within the biological membrane (24), or that metabolically derived chemical energy plays a role in eliciting or modulating the cooperative effect.

The existence of negative cooperativity among hormone and drug receptor sites appears to be a widespread, though not universal, phenomenon. The acceleration of dissociation by unlabeled hormone has been demonstrated with the binding of insulin (5, 6), nerve growth factor (8, 9) and TSH (15) to their specific receptors. In addition, curvilinear Scatchard plots have been reported for a number of other hormone and drug receptors and thus represent potential cooperative systems. The existence of negative cooperativity among hormone and drug receptors may be of significant physiologic importance. Negative cooperativity provides for enhanced sensitivity to low concentrations of drug and a buffering against acutely elevated levels of these agents. It will be of considerable interest to establish to what extent cooperativity at the level of hormone and drug binding is reflected in the characteristic physiologic response produced by these agents. In the catecholamine responsive adenylate cyclase system, for example, the negative cooperativity among the β-adrenergic receptors might be transmitted with fidelity to the activation of adenylate cyclase in a tightly coupled receptor-enzyme mechanism. On the other hand, site-site interactions among the β-adrenergic receptors might provide a threshold or trigger mechanism for the activation of adenylate cyclase by catecholamines, which might then display apparent positive cooperativity at the level of catalysis and generation of cAMP. It is apparent that the potential importance of initial receptor binding phenomena in modulating the subsequent regulatory events elicited by hormones and drugs requires further investigation.

SUMMARY

The availability of a radioligand for the confident identification of the physiologic β-adrenergic receptors has provided a valu-

[4]Limbird, L. E. and Lefkowitz, R. J., Manuscript in preparation.

able tool for investigating the biochemical and molecular characteristics of this drug-receptor interaction. It has previously been documented that the binding of (-)[^{3}H]alprenolol to purified frog erythrocyte membrane preparations possesses all of the essential characteristics expected of binding to the adenylate cyclase coupled β-adrenergic receptors. Perturbation of either the protein or lipid constituents of the membrane abolishes receptor binding in a dose-dependent fashion and both hydrophobic and hydrophilic membrane resides appear to be crucial for hormone recognition and binding. A number of biochemical characteristics suggest that the β-adrenergic receptor and the adenylate cyclase enzyme are distinct molecular entities and that an intact lipid matrix is essential for the "coupling" of receptor binding to enzyme activation. Binding of β-adrenergic agents to the receptor induces negatively cooperative interactions among the binding sites. These site-site interactions occur within a physiologically relevant range of β-adrenergic agent concentrations and provide for both exquisite sensitivity to low levels of β-adrenergic agents and a protective mechanism against acutely elevated levels of these drugs. Still to be elucidated are the molecular mechanisms for these negatively cooperative interactions among the β-adrenergic receptors and their overall regulatory function in the catecholamine-sensitive adenylate cyclase systems of mammalian tissues.

REFERENCES

1. ALEXANDER, R. W., WILLIAMS, L. T., and LEFKOWITZ, R. J., Proc. Nat. Acad. Sci. USA 72 (1975) 1564.
2. BURTON, P. M., and JOSSE, J., J. Biol. Chem. 245 (1970) 4358.
3. CUATRECASAS, P., TELL, G. P. E., SICA, V., PARIKH, I., and CHANG, K. J., Nature 247 (1974) 92.
4. DE MEYTS, P., Methods in Receptor Research (M. Blecher ed.) M. Dekker Publishing Co., N. Y. in press.
5. DE MEYTS, P., Roth, J., NEVILLE, D. M., GAVIN, J. R. and LESNIAK, M. A., Biochem. Biophys. Res. Commun. 55 (1974) 154.
6. DE MEYTS, P., BIANCO, A. R., and ROTH, J., J. Biol. Chem. (1975), in press.
7. DRUMMOND, G. I., SEVERSON, D. L., and DUNCAN, L., J. Biol. Chem. 246 (1971) 4166.
8. FRAZIER, W. A., BOYD, L. F., PULLIAM, M. W., SZUTOWICZ, A. and BRADSHAW, R. A., J. Biol. Chem. 249 (1974) 5918.
9. FRAZIER, W. A., BOYD, L. F., and BRADSHAW, R. A., J. Biol. Chem. 249 (1974) 5513.
10. FRIEDEN, C., in The Regulation of Enzyme Activity and Allosteric Interactions. Universitets forlaget, Oslo and Academic Press, New York (1968) 59.
11. HALL, Z. W., Annual Review of Biochemistry 41 (1972) 925.
12. HOARE, D. E., J. Biol. Chem. 242 (1967) 2447.
13. JENSEN, G. V., and DE SOMBRE, E. R., Ann. Rev. Biochem. 41 (1972) 203

14. KAHN, C. R. , Methods in Membrane Biology, Vol. 3. (E. D. KORN, ed.) Plenum Press, N. Y., 1975, in press.
15. KOHN, L., and WINAND, R. , Molecular Aspects of Membrane Phenomena (Kaback, H. R., Radda, G., Schwyzer R., eds) Springer-Verlag, Heidelberg, (1976) in press.
16. KOSHLAND, D. E., NEMETHY, G., and FILMER, D. , Biochemistry 5 (1966) 365.
17. LEFKOWITZ, R. J. and HABER, E., Proc. Nat. Acad. Sci. USA 68 (1971) 1773.
18. LEFKOWITZ, R. J. , Biochemical Pharmacology 23 (1974) 2069.
19. LEFKOWITZ, R. J., MUKHERJEE, C., COVERSTONE, M. and CARON, M. G., Biochem. Biophys. Res. Commun. 69 (1974) 703.
20. LEFKOWITZ, R. J., LIMBIRD, L. E., MUKHERJEE, C. and CARON, M. G., B. B. A., Biomembrane Reviews (1975) in press.
21. LEFKOWITZ, R. J., MUKHERJEE, C., LIMBIRD, L. E., CARON, M. G., WILLIAMS, L. T., ALEXANDER, R. W., MICKEY, J. V., and TATE, R., Recent Progress in Hormone Research (1975) in press.
22. LEFKOWITZ, R. J., Methods in Receptor Research (M. Blecher ed.) M. Dekker Publishing Co., N. Y. (1975) in press.
23. LEFKOWITZ, R. J., and CARON, M. G., J. Biol. Chem. 250 (1975) 4418.
24. LEVITZKI, A. J., J. Theoret. Biol. 44 (1974) 367.
25. LIMBIRD, L. E., DE MEYTS, P., and LEFKOWITZ, R. J., Biochem. Biophys. Res. Commun. 64 (1975) 1160.
26. MONOD, J., WYMAN, J., and CHANGEUX, J. P., J. Mol. Biol. 12 (1965) 88.
27. MUKHERJEE, C., CARON, M. G., COVERSTONE, M. and LEFKOWITZ, R. J., J. Biol. Chem. 250 (1975) 4869.
28. MUKHERJEE, C., CARON, M. G., MULLIKIN, D., and LEFKOWITZ, R. J., Molecular Pharmacology (1975) in press.
29. NEET, K. E., and KOSHLAND, D. E., Proc. Nat. Acad. Sci. USA 56 (1966) 1606.
30. NEET, K. E., NANCI, A., and KOSHLAND, D. E., J. Biol. Chem. 243 (1968) 6392.
31. POHL, S. L., BIRNBAUMER, L., and RODBELL, M., J. Biol. Chem. 246 (1971) 1849.
32. PUCHWEIN, G., PFEUFFER, T., and HELMREICH, E. J. M., J. Biol. Chem. 249 (1974) 3232.
33. RIORDAN, J. F., and VALLEE, B. L., Methods in Enzymology XI (1967) 541.
34. RIORDAN, J. F., and VALLEE, B. L., Methods in Enzymology XI (1967) 570.
35. ROBISON, G. A., BUTCHER, R. W. and SUTHERLAND, E. W., Cyclic AMP, Academic Press, New York (1971), p. 17.
36. RODBARD, D., Receptors for Reproductive Hormones. (B. W. O'Malley and A. R. Means eds.) Plenum Publishing Co., N. Y. (1973) 289.
37. ROTH, J., Metabolism 22 (1973) 1059.
38. SCATCHARD, G., Ann. N. Y. Acad. Sci. 51 (1949) 660.
39. STEER, M. L., and LEVITZKI, A., J. Biol. Chem. 250 (1975) 2080.
40. VATNER, D. and LEFKOWITZ, R. J., Molecular Pharmacology 10 (1974) 450.

41. WEINER, H., WHITE, W. N., HOARE, D. G. and KOSHLAND, D. E., J. Amer. Chem. Soc. 88 (1969) 3851.

42. WILLIAMS, L. T., SNYDEPMAN, R., and LEFKOWITZ, R. J., J. Clin. Invest (1975) in press.

CHARACTERIZATION OF THE β-ADRENERGIC RECEPTOR AND THE REGULATORY CONTROL OF ADENYLATE CYCLASE

Alexander LEVITZKI

Department of Biological Chemistry, The Hebrew University of Jerusalem, Jerusalem (Israel)

Numerous tissues possess β-adrenergic receptors coupled to adenylate cyclase. The binding of *l*-catecholamines to the β-adrenergic receptor is specific, rapid and reversible (5, 16). The extent of receptor occupancy is controlled by the expression

$$K_D = \frac{(R)(H)}{(RH)} \quad [1]$$

where K_D is the receptor-hormone dissociation constant, (R) the concentration of free receptor, (H) the concentration of free hormone and (RH) the concentration of the receptor-hormone complex. The affinity of *l*-catecholamines to the β-receptor as measured by the dose response curve of adenylate cyclase is between 5×10^{-7} M to 10^{-5} M depending on the activating *l*-catecholamine ligand. Since the receptor concentration accessible experimentally rarely exceeds 5×10^{-9} M, attempts to probe the β-receptor by measuring ^{3}H-catacholamine binding were bound to be unsuccessful (6-8). Indeed, it was demonstrated that the ligand binding specificity did not match the specificity of the β-receptor as defined pharmacologically and biochemically (3, 4). Recent experiments (4, 9) have indeed shown that most of the binding signal is due to non-specific catecholamine binding to non-receptor binding sites. Therefore it became necessary to develop a reliable binding assay for the purpose of direct measurement of ligand binding to the receptor. Such an assay was first developed using ^{3}H-propranolol as the probing ligand (9). This drug binds to the β-receptor with a dissociation constant of 2.5×10^{-9} M thus allowing the probing of experimentally accessible receptor concentration. Indeed, it was demonstrated that the affinity of propranolol and the stereospecificity of binding (1, 9) match the values obtained from the capacity of the drug to inhibit specifically β-receptor stimulated adenylate cyclase. Recently (2) it has been demonstrated that the drug p-iodohydroxyphenylpindolol binds with an affinity of $\sim 10^{-11}$ to 10^{-10} M to β-receptors thus allowing to probe receptor concentrations in that concentration range (10^{-11} - 10^{-10} M), a situation often encountered in tissue culture work. This drug therefore allows the probing of a receptor concentration too low to be analyzed by ^{3}H-propranolol. At low receptor concentration the specific radioactivity attainable using ^{3}H is insufficient to probe the receptor and thus the radioactive ^{125}I-iodohydroxyphenyl pindolol is employed (2). Once direct binding measurements of β-receptor active ligands has become possible the detailed study of the β-receptor adenylate cyclase interrelationships has become feasible (12, 13, 15, 17, 19).

Ligand Specificity of the β-Receptor

The capacity of ligands to displace ^{3}H-propranolol from the β-receptor can be used to measure their affinity towards the receptor (Table I). The formula [2] is used to calculate the affinity of a ligand

$$K'_D = \frac{D_{0.5}}{\frac{L^*}{K'_R} - 1} \qquad [2]$$

from the concentration of the compound required to displace 50% of the specifically bound counts. K'_D is the dissociation constant for the displacing ligand, $D_{0.5}$ the concentration of the ligand required for 50% displacement, K'_R the dissociation constant for the radioactive drug used (in this case [^{3}H]propranolol). Equation [2] simplifies to.

$$K'_D = \frac{D_{0.5}}{L^*/K_{R'}} \qquad [3]$$

under conditions which most of these experiments were conducted. The affinity of ligand to the β-receptor can also be measured by measuring the capacity of a β-receptor active ligand to stimulate the enzyme adenylate cyclase coupled to the receptor. However, partial agonists possessing low efficacy as well as low affinity induce low cyclase activity. Furthermore, the low affinity of these ligands toward the receptor forces one to use high ligand concentrations which modulate in a non-specific fashion the response of adenylate cyclase. Thus accurate dose-response curves cannot be obtained by measuring adenylate cyclase activity. It is however possible to measure the affinity of partial agonists towards the β-receptor by a perturbation method where an adenylate cyclase dose response as a function of full agonist concentration is measured in the presence of increasing concentrations of partial agonist. The data is then analyzed according to equation [4].

TABLE I

THE LIGAND SPECIFICITY OF β-RECEPTORS

Ligand	Maximal Velocity Induced in Adenylate Cyclase pmoles cAMP/mg/min	β-Receptor - Ligand Dissociation Constant μM	
		From Adenylate Cyclase Activation	From Binding Measurements
l-isoproterenol	160 ± 15	0.50 ± 0.05	0.21 ± 0.02[a]
d*l*-isoproterenol	160 ± 15	1.00 ± 0.10	0.25 ± 0.02[b]
l-epinephrine	160 ± 15	6.0 ± 1.0	2.0 ± 0.4
d-epinephrine	1 ± 1[c]	No Effect[d]	No Binding[d]
l-norepinephrine	160 ± 0.5	6.0 ± 0.5	2.6 ± 0.5
D-norepinephrine	1 ± 1[c]	No Effect[d]	No Binding[d]
dopamine	16 ± 1	22 ± 2	24 ± 3
l-phenylephrine	16 ± 1	61 ± 5	50 ± 5
l-ephedrine	0.0	450 ± 50	370 ± 50
tyramine	0.0	1000 ± 200	410 ± 50
d*l*-metanephrine	0.0	800 ± 150	500 ± 100
2-isopropyl-ethanolamine	0.0	No Effect[d]	No binding[d]
l-propranolol	0.0	0.0012 ± 0.0002	0.0012 ± 0.0002
d-propranolol	1 ± 1	No Blocking[e]	No Binding[e]

[a] Data from (1).

[b] From equilibrium dialysis studies using d*l*-[^{3}H]isoproternol in the absence of propranolol and in the presence of propranolol (9).

[c] The basal adenylate cyclase activity in the absence of hormone or F^- is between zero and 1 pmoles cAMP/min/mg.

[d] Concentrations checked were up to 5 x 10^{-4} M.

[e] Concentrations checked were up to 2-5-fold higher than the concentration of *l*-propranolol giving the maximal effect.

$$V_{obs} = \frac{K_2L_2V_2}{1 + K_2V_2} + \frac{K_1}{1+K_2L_2}(V_1-V_{obs})L_1 \qquad [4]$$

where V_{obs} is the observed rate, L_1 the concentration of the full agonist, V_1 the maximal adenylate cyclase activity induced by L_1; L_2 the concentration of the partial agonist and V_2 the maximal velocity induced by the partial agonist. K_1 and K_2 are the intrinsic affinities (association constants) of L_1 and L_2 towards the β-receptor, respectively. The slope of equation [4] is given by:

$$\text{Slope} = \frac{K_1}{1+K_2L_2} \qquad [5]$$

and therefore

$$\frac{1}{\text{slope}} = \frac{1}{K_1} + \frac{K_2}{K_1}L_2 \qquad [6]$$

Thus from a family of dose response curves with respect to L_1 at different L_2's one can obtain from equations [4] through [6] the values of K_1, K_2, V_1 and V_2. In Table I, a summary of β-receptor ligand affinity constants as measured by the displacement technique as well as by the kinetic method is given. It can be easily seen that the affinity constants measured by the two methods compare fairly well.

Limitations of the Propranolol Binding Technique

Since the *l*-propranolol binds to the β-receptor with an affinity constant of 1.5×10^{-9} M its use is limited to receptor-containing preparations where the receptor concentration can be as high as 10^{-9} M to 10^{-8} M. In systems, such as cultured cells where the experimentally accessible receptor concentration is well below this range, high affinity β-blockers must be used. An extremely high affinity β-blocker has been introduced by Aurbach (2) and is being used extremely successfully to measure β-receptor content in a number of sources.

GppNHp as the Adenylate Cyclase Regulator

Adenylate cyclase activity induced by hormone alone is only a fraction of the maximal potential adenylate cyclase activity(Table II). In the presence of the ATPase resistant GTP analog GppNHp the enzyme is stimulated to its highest activity form. The synergistic action of a hormone and guanyl nucleotide triphosphate were originally

TABLE II

ADENYLATE CYCLASE SPECIFIC ACTIVITY AND ITS TURNOVER NUMBER

Ligand Present[a]	Specific Activity pmoles cAMP/min/mg	k_{cat}[b] min^{-1}
none	1 ± 1	1 ± 1
l-Isoproterenol (10^{-4} M)	160 ± 10	160 ± 10
Gpp(NH)p (10^{-4} M)	70 ± 10 [c]	70 ± 10
l-Isoproterenol + Gpp(NH)p	1400 ± 100 [c]	1400 ± 100
Fluoride (10 mM)	500 ± 50	500 ± 50
B. liquifaciens (pure)	30,030 x 10^3	1401[d]

[a] Saturating concentrations.

[b] Every milligram of ghost membrane protein binds 1.0 ± 0.1 pmoles [^{3}H]propranolol. Thus, if the receptor:enzyme ratio is 1:1, one can calculate k_{cat} per receptor molecule by dividing the specific activity by the number of pmole receptor per milligram membrane protein.

[c] The synergistic allosteric effects of Gpp(NH)p and *l*-catecholamines on the turkey erythrocyte adenylate cyclase will be described in detail elsewhere.

[d] This value is calculated by multiplying the value for the specific activity by the subunit molecular weight (30.03 x 10^{-3}moles cAMP/ min/ gm protein x 46,000).

demonstrated and analyzed by Rodbell and co-workers (12, 13, 15, 17-20, 22) in the glucagon-activated adenylate cyclase in fat cells and in liver cells. Subsequent work (14, 21, 22) revealed that GppNHp also stimulates β-receptor-activated adenylate cyclase. The interrelationships between GppNHp and the β-receptor as found in our laboratory (11)[1] can be summarized as follows:

(1) The total number of tight binding sites for GppNHp is independent of the presence of the agonist (such as *l*-epinephrine) as measured using [^{3}H]GppNHp.[2]

(2) The total number of [^{3}H]propranolol sites and the affinity towards the antagonist are independent of whether the guanyl nucleotide regulatory site is occupied.[3]

(3) Binding of GppNHp alone is incapable of inducing cyclase activity.

(4) The conversion of the low specific activity species to the high specific activity only takes place in the presence of hormone once the guanyl nucleotide site is occupied. The highest turnover number for the conversion of the low activity form to the high activity form is about 1-2 min^{-1} when both the guanyl regulatory site and the β-receptor are fully occupied.

(5) The highly active form of the enzyme, once obtained, is stable for many hours and retains its full activity even when excess free GppNHp and hormone are removed. When analyzed, using [^{3}H]GppNHp, it is found that tightly bound nucleotide is retained in the washed enzyme preparation.

(6) Propranolol is capable to compete with *l*-epinephrine for the β-receptor, inhibiting the effect of the β-agonist in affecting the conversion of the low activity cyclase to the high activity cyclase (Fig. 1).

(7) The permanently activated enzyme species possessing the highest activity (and GppNHp at its tight sites) in the absence of free GppNHp is inhibited to 50% of its maximal activity in the presence of saturating β-agonist (Fig. 2). This somewhat surprising inhibitory effect induced by the hormone is fully reversible (Table III) and upon removal of the hormone the original activity is restored.

---

[1] Sevilla, N., Steer, M. L. and Levitzki, A., submitted for publication.

[2] Levitzki, A., unpublished experiment.

[3] Sevilla, N. and Levitzki, A., unpublished experiment.

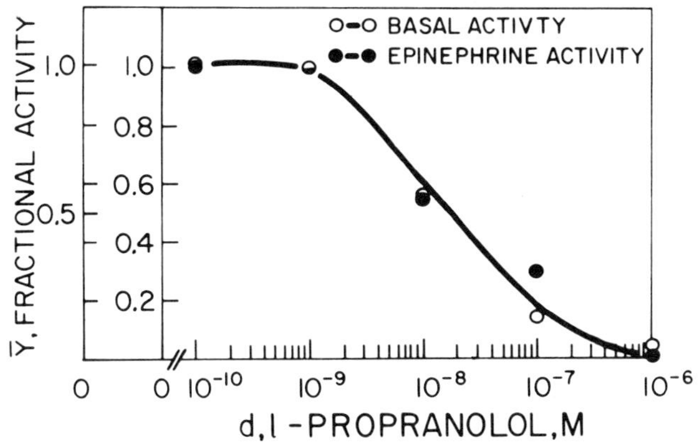

Fig. 1. The inhibition of adenylate cyclase activation by propranolol. One milligram of membrane protein was incubated with *l*-epinephrine (10^{-7} M) and Gpp(NH)p (10^{-4} M), 1.0 mM $MgCl_2$, 1 mM EDTA, Tris-HCl pH 7.4 (0.05 M) at 37°, for 40 min. Different concentrations of propranolol were added to the incubation mixture. After the incubation period the enzyme was thoroughly washed and assayed in the absence of added ligands. The maximal activity obtained in the absence of added propranolol to the preincubation mixture was 1250 pmoles of cAMP/mg/min. The same enzyme had a specific activity of 500 pmole cAMP/mg/min in the presence of epinephrine. At these low epinephrine concentrations one can therefore demonstrate that the β-blocker propranolol is capable of inhibiting the interaction of the β-receptor with epinephrine. Propranolol, however, is unable to inhibit an enzyme species already activated.

Also, by using [^{3}H]GppNHp it can be demonstrated that this partial inhibition is not due to the loss of tightly bound hormone.[3] The hormone induced antagonism displays a saturation curve with respect to agonist with an apparent affinity towards the agonist which is 4-fold higher than the affinity displayed against the agonist in the activation process of adenylate cyclase in the absence of GppNHp (Fig. 2).

(8) The partial hormone inhibition can be overcome by three independent mechanisms:

(a) removal of the hormone β-agonist from the β-receptor.

(b) keeping the β-agonist off the receptor by an antagonist such as propranolol. Propranolol occupying the receptor is incapable of inducing partial inhibition (Fig. 3).

TABLE III

REVERSIBLE NATURE OF THE PARTIAL INHIBITION

BY HORMONE OF THE Gpp(NH)p ACTIVATED ENZYME

Preincubation	Washings	Assay		
		Basal	10^{-4} M *l*-epinephrine	10^{-4} M epinephrine 10^{-4} M Gpp(NH)p
Gpp(NH)p+*l*-epinephrine	Buffer	1096±50	599±40	1100±50
Gpp(NH)p + *l*-epinephrine	10^{-4} M *l*-epinephrine	528±35[a]	528±35	-
Gpp(NH)p + *l*-epinephrine	10^{-4} M *l*-epinephrine then buffer	1130±50	649±65	-

[a]Assay in the presence of 10^{-4} M *l*-epinephrine.

(c) Addition of 10^{-5} M free GppNHp to the enzyme. The intricate hormone-GppNHp interrelationships as depicted above can be summarized in the scheme represented in Fig. 4.

It is not clear whether the reactivation of species HR'E" G by the addition of free GppNHp occurs via the decrease of agonist affinity to form the fully active species R"E"GG or reactivation of the enzyme without any effect on the β-receptor thus forming the species HR'E"GG as represented in the figure. It is however clear that the GppNHp activated enzyme has an altered β-receptor only with respect to agonist binding. The binding of an antagonist is unaltered as is seen both kinetically (Figure 3) and from direct [^{3}H]propranolol binding to the β-receptor.

In conclusion, the effect of GppNHp and hormone on adenylate cyclase is synergistic since the simultaneous presence of the two ligands is required to bring about the full expression of the enzyme.

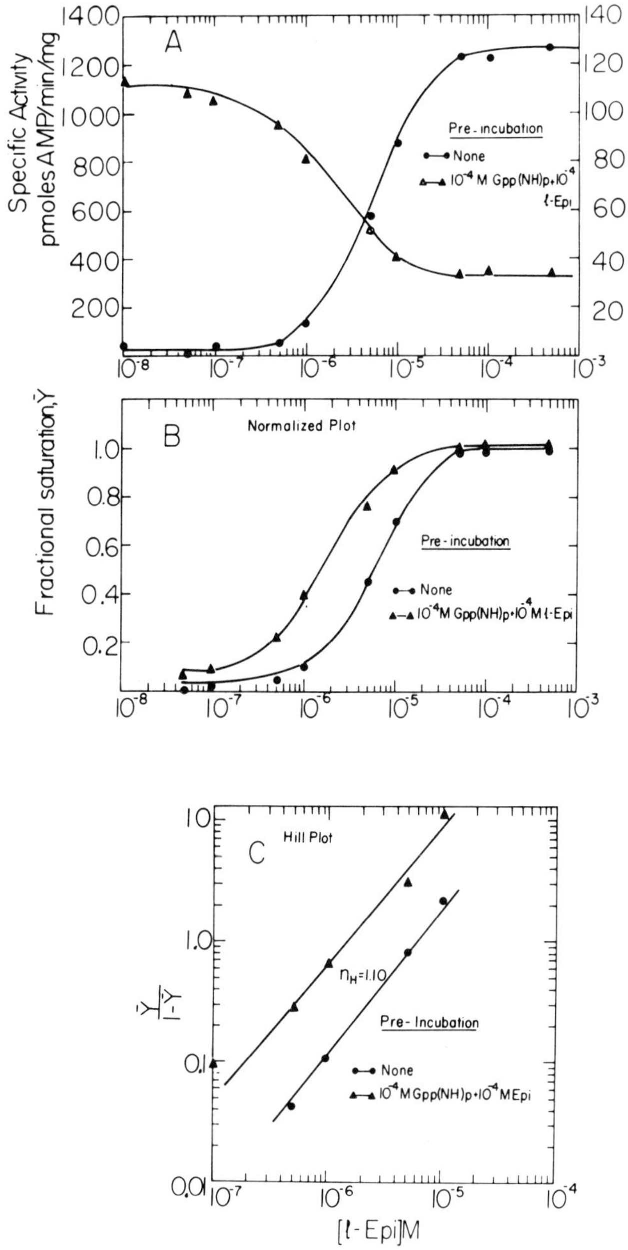

Fig. 2. The affinity of the β-receptor-cyclase complex towards *l*-epinephrine in the permanently active state. Plasma membranes were incubated for 40 min at 37° with Gpp(NH)p (10^{-4} M), or Gpp(NH)p(10^{-4} M) and *l*-epinephrine (10^{-4} M) 0.8 mM $MgCl_2$, washed throughly (dilution factor 1:10^8) at the end of preincubation, and assayed for adenylate cyclase activity as a function of *l*-epinephrine concentration in assay. Absolute values given (A), fractional activity (B), Hill plot of the above results (C).

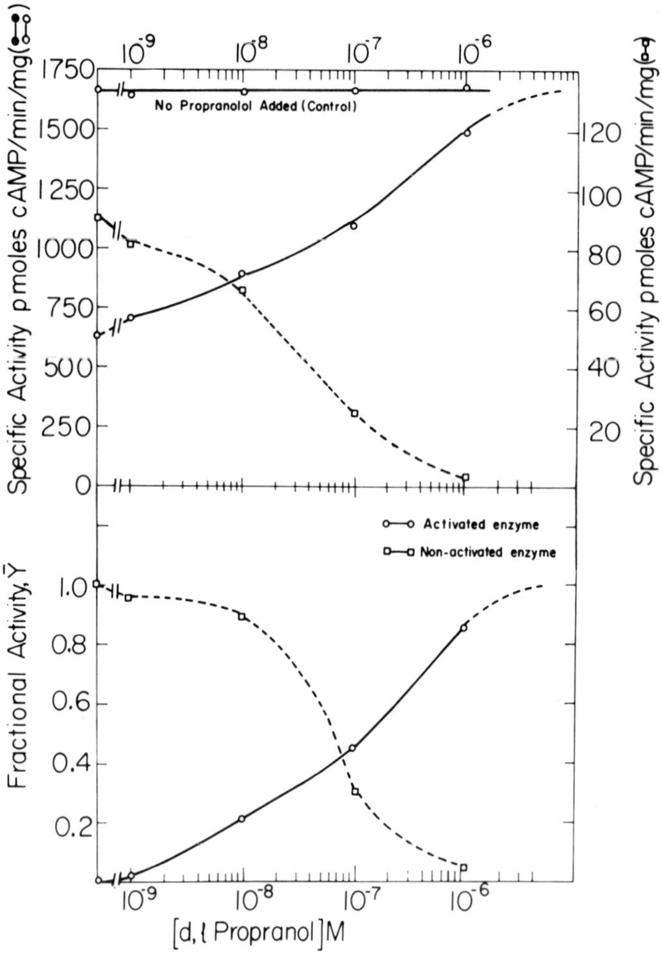

Fig. 3. The inhibition of the partial hormone Gpp(NH)p antagonism by propranolol. Plasma membranes were pretreated with Gpp(NH)p 10^{-4} M + *l*-epinephrine 10^{-4} M for 40 min at 37°; then thoroughly washed and assayed for adenylate cyclase as a function of propranolol concentration in the assay in the absence and in the presence of 5×10^{-5} M *l*-epinephrine.

The phenomenon of synergistic activation of regulatory enzymes by two ligands is well documented.

The action of GppNHp *in vitro* probably reflects the action of GTP *in vivo* in the whole cell. The presence of only a small GTP effect *in vitro* probably due to the presence of very highly active

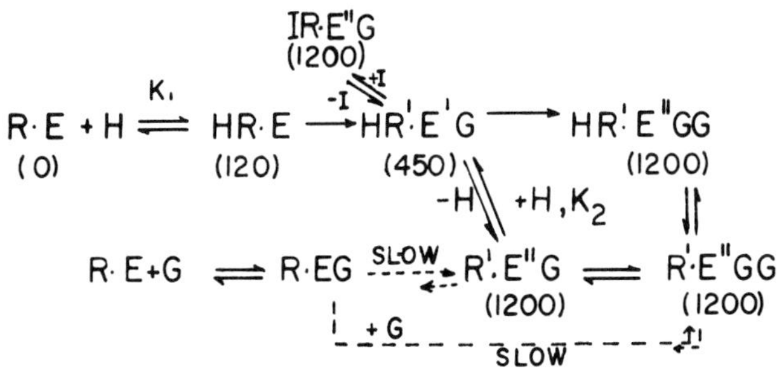

Fig. 4. The synergistic activation of the adenylate cyclase by *l*-catecholamines and Gpp(NH)p. The receptor enzyme complex R·E can combine reversibly with the *l*-catecholamine HR·E which possesses the turnover number of 120 min^{-1} (9). This value may reflect the k_{cat} of the system in which the regulatory site is occupied by GTP or partly by GTP and partly by GDP (see text). In the presence of Gpp(NH)p (G) the enzyme has a k_{cat} value of 450 min^{-1} if it occupies only the tight regulatory sites and at the same time the β-receptor is bound with a β-agonist (H) such as *l*-epinephrine. In the absence of β-agonist or in the presence of the β-blocker propranolol (16) the enzyme is fully active (k_{cat} = 1400 min^{-1}) when only the tight sites are filled. In the presence of hormone the enzyme will be fully active (k_{cat} = 1400 min^{-1}), only when Gpp(NH)p is present in excess thus probably saturating a low affinity site which counteracts the partial Gpp(NH)p hormone antagonism. The receptor affinity towards the β-agonist (H) is increased as is revealed by the dose response curve of hormone inhibition of pre-activated enzyme binding Gpp(NH)p only at its tight site (Fig. 2). The enzyme can be activated slowly by Gpp(NH)p alone in the absence of hormone by a different pathway (dotted lines) provided, for example, by excess free Mg^{2+} in the preincubation medium.

ATPase in the membrane fragments. Work is continuing in our laboratory to establish the extent of the GTP effects as compared to GppNHp effects.

Ca^{2+} as a Negative V_{max}-Effector

Ca^{2+} inhibits all form of adenylate cyclase in a cooperative fashion (Figure 5). The Ca^{2+} effect is a pure V_{max} effect since Ca^{2+} does not have any effect on the kinetic parameters with respect to the substrate ATP or the effect of Mg^{2+} (23). Furthermore, Ca^{2+} can also be demonstrated on whole cells (24) provided a Ca^{2+} ionophore A-23187 is incorporated into the erythrocyte membrane (Table IV) Recently we were able to demonstrate also that external Ca^{2+} inhibits adenylate cyclase in hormone responsive resealed ghosts[4] when the

[4] Steer, M. L., and Levitzki, A., manuscript in preparation.

TABLE IV

The Effect of Ca^{2+} on Epinephrine Stimulated cAMP Formation in Intact Cells

Assays were performed in the presence of $1x10^{-10}$ *l*-epinephrine after 2 hours preincubation with the additives described.

Additive	cAMP Formation (% of control)
none	100[a,b]
Ca^{2+} (0.1 mM)	100 ± 10
Ca^{2+} (1.0 mM)	100 ± 10
Ca^{2+} (50 mM)	100 ± 10
Ionophore (0 g/ml)	30 ± 10
Ionophore + EGTA ($3x10^{-4}$ M)	63 ± 10
Ionophore + Ca^{2+} (5 mM)	10 ± 3
Ionophore + Ca^{2+} (50 mM)	1 ± 1

[a] cAMP formation in the absence of epinephrine (basal activity) was found to be less than 5 pmoles/$5x10^8$ cells/min.

[b] cAMP formation in control cells stimulated with epinephrine was found to be 600±50 pmoles/$5x10^8$ cells/min.

Ca^{2+} ionophore is incorporated into the resealed ghost membrane.

CONCLUSION

The ligands which determine the degree of adenylate activity are the β-agonist *l*-epinephrine, guanine nucleoside triphosphate and Ca^{2+}. The hormone and the guanyl nucleotide act in a synergistic fashion to activate the enzyme adenylate cyclase. Ca^{2+} is a regulatory inhibitor which inhibits the catalytic machinery of the enzyme without affecting either agonist binding to the receptor or the regulatory site for the guanyl nucleotide.

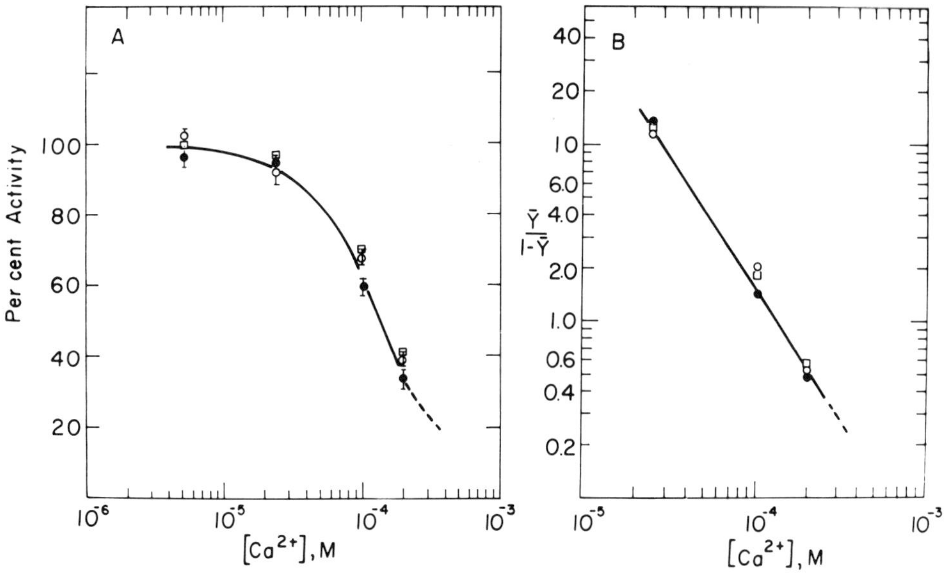

Fig. 5. Inhibition of adenylate cyclase activity by Ca^{++} treated or preheated with Gpp(NH)p and *l*-epinephrine. Ghost membranes were preincubated in the presence of Gpp(NH)p (10^{-4} M) *l*-epinephrine (10^{-4} M) and 0.9 mM $CaCl_2$ for 40 min at 37°. After this period the membranes were washed thoroughly and assayed for adenylate cyclase activity as a function of Ca^{++} concentration in the absence (o - o), or presence (● - ●) of *l*-epinephrine ($5x10^{-5}$ M). The maximal activity of the pre-activated enzyme was 1100 pmole/mg/min in the absence of epinephrine and 410 in the presence of epinephrine. The Ca^{++} inhibition of the hormone stimulated activity in the non-activated state (specific activity 100 ± pmole/min/mg) was measured in parallel ([]—[]).

REFERENCES

1. ATLAS, D., STEER, M. L. and LEVITZKI, A., Proc. Nat. Acad. Sci. USA 71 (1974) 4246.
2. AURBACH, G. D., FEDAK, S. A., WOODWARD, C. J., PALMER, J. S., MAUSER, D. and TROXLER, F., Science 186 (1974) 1223.
3. CUATRECASAS, P., Ann. Rev. Biochem. 43 (1974) 169.
4. CUATRECASAS, P., TELL, G. P. E., SICA, V. and PARIKH, I., Nature 247 (1974) 92.
5. FURCHGOTT, R. F., in Handbook of Experimental Pharmacology 33: Catecholamines (H. Blaschko and E. Muscholl eds.) Springer Verlag, New York, (1972) 283.
6. LEFKOWITZ, R. J., N. Eng. J. Med. 288 (1973) 1061.

7. LEFKOWITZ, R. J., HABER, E. and O'HARA, D., Proc. Nat. Acad. Sci. USA 69 (1972) 2828.
8. LEFKOWITZ, R. J., SHARP, G. W. G. and HABER, E., J. Biol. Chem. 284 (1973) 342.
9. LEVITZKI, A., ATLAS, D. and STEER, M. L., Proc. Nat. Acad. Sci. USA 71 (1974) 2773.
10. LEVITZKI, A., SEVILLA, N., ATLAS, D. and STEER, M. L., J. Biol. Chem. (1975) in press.
11. LEVITZKI, A., SEVILLA, N. and STEER, M. L., J. Supramolec. Structure (1975) in press.
12. LIN, M. C., SALOMON, Y., RENDELL, M. and RODBELL, M., J. Biol. Chem. 250 (1975) 4246.
13. LONDOS, C., SALOMON, Y., LIN, M. C., HARWOOD, J. P., SCHRAMM, M., WOLFF, J. and RODBELL, M., Proc. Nat. Acad. Sci. USA 71 (1974) 3087.
14. PFEUFFER, T. and HELMREICH, E. J. M., J. Biol. Chem. 250 (1975) 867.
15. RENDELL, M., SALOMON, Y., LIN, M. C., RODBELL, M. and BERMAN, M., J. Biol. Chem. 250 (1975) 4253.
16. ROBINSON, G. A., BUTCHER, R. W. and SUTHERLAND, E. W., Cyclic AMP, Academic Press, New York (1971).
17. RODBELL, M., BIRNBAUMER, L., POHL, S. L. and KRANS, H. M. J., J. Biol. Chem. 246 (1971) 1877.
18. RODBELL, M., LIN, M. C. and SALOMON, Y., J. Biol. Chem. 249 (1974) 59.
19. RODBELL, M., KRANS, H. M. J., POHL, S. L. and BIRNBAUMER, L., J. Biol. Chem. 246 (1971) 1872.
20. SALOMON, Y., LIN, M. C., LONDOS, C., RENDELL, M. and RODBELL, M., J. Biol. Chem. 250 (1975) 4233.
21. SCHRAMM, M. and RODBELL, M., J. Biol. Chem. 250 (1975) 2232.
22. SPIEGEL, A. M. and AURBACH, G. D., J. Biol. Chem. 249 (1974) 7630.
23. STEER, M. L. and LEVITZKI, A., J. Biol. Chem. 250 (1975) 2080.
24. STEER, M. L. and LEVITZKI, A., Arch. Biochem. Biophys. 167 (1975) 371.

CHEMOTAXIS IN BACTERIA

Julius ADLER

Departments of Biochemistry and Genetics
University of Wisconsin-Madison
Madison, Wisconsin 53706 (USA)

Bacterial chemotaxis, the movement of motile bacteria toward or away from chemicals, was discovered nearly a century ago by Engelmann (43) and Pfeffer (70, 71). The subject was actively studied for about fifty years, but then there were very few reports until quite recently. For reviews of the literature up to about 1960, see Berg (23), Weibull (90) and Ziegler (92). The present review will restrict itself to the recent work on chemotaxis in *Escherichia coli* and *Salmonella typhimurium*, some of which is also covered in Berg's review (23).

Motile bacteria are attracted by certain chemicals and repelled by others: this is positive and negative chemotaxis. Chemotaxis can be dissected by means of the following questions:

1. How do individual bacteria move in a gradient of attractant or repellent?

2. How do bacteria detect the chemicals?

3. How is the sensory information communicated to the flagella?

4. How do bacterial flagella produce motion?

5. How do flagella respond to the sensory information in order to bring about the appropriate change in direction?

6. In the case of multiple or conflicting sensory data, how is the information integrated?

DEMONSTRATION AND MEASUREMENT OF CHEMOTAXIS IN BACTERIA

For much of the work it was necessary to first develop conditions for obtaining optimum motility and chemotaxis in defined media (1-3, 7); up to the time of this work, studies had been carried out in complex media such as tryptone, peptone, or meat extract. Also, the earlier work was largely of a subjective nature, so it was necessary to find quantitative methods of demonstrating chemotaxis.

(a) Plate method. For positive chemotaxis, a Petri dish containing metabolizable attractant, salts needed for growth, and soft agar (a low enough concentration that the bacteria can swim) is inoculated in the center with the bacteria. As the bacteria grow, they consume the local supply of attractant, thus creating a gradient, which they follow to form a ring of bacteria surrounding the inoculum (1). For negative chemotaxis, a plug of hard agar containing repellent is planted into a Petri dish containing soft agar and bacteria concentrated enough to be visibly turbid; the bacteria soon vacate the area around the plug (86). By searching in the area of the plate traversed by wild-type bacteria, one can isolate mutants in positive or negative chemotaxis (for example: 2, 6, 47, 48, 86).

(b) Capillary method. In the 1880's, Pfeffer observed bacterial chemotaxis by inserting a capillary containing a solution test chemical into a bacterial suspension and then looking microscopically for accumulation of bacteria at the mouth of and inside the capillary (positive chemotaxis) or movement of bacterial away from the capillary (negative chemotaxis) (70, 71). For positive chemotaxis this has been converted into an objective, quantitative assay by measuring the number of bacterial accumulating inside a capillary containing attractant solution (2, 3). For negative chemotaxis repellent in the capillary decreases the number that will enter (86); alternatively repellent is placed with the bacteria, none in the capillary, and then the number of bacteria fleeing into the capillary for "refuge" is measured (86). Unlike in the plate method, where bacteria make the gradient of attractant by metabolizing the chemical, here the experimenter provides the gradient; hence nonmetabolizable chemicals can be studied.

(c) Defined gradients. Quantitative analysis of bacterial migration has been achieved by making defined gradients of attractant (35) or repellent (85), and then determining the distribution of bacteria in the gradient by measuring scattering of a laser beam by the bacteria. The method allows the experimenter to vary the shape of the gradient.

(d) Tumbling frequency. A change in the bacterium's tumbling frequency in response to a chemical gradient, described next, is also to be regarded as a demonstration and a measurement of chemotaxis.

THE MOVEMENT OF INDIVIDUAL BACTERIA IN A GRADIENT

The motion of bacteria can of course be observed microscopically by eye, or recorded by microcinematography, or followed as tracks that form on photographic film after time exposure (42, 87). Owing to the very rapid movement of bacteria, however, significant progress was not made until the invention of an automatic tracking microscope, which allowed objective, quantitative, and much faster observations (21). A slower, manual tracking microscope has also been used.[1] A combination of these methods has led to the following conclusions.

In the absence of a stimulus (i.e., no gradient of attractant or repellent present, or else a constant, uniform concentration) a bacterium such as E. coli or S. typhimurium swims in a smooth, straight line for a number of seconds--a "run", then it thrashes around for a fraction of a second--a "tumble" (or abruptly changes its direction--a "twiddle"); and then it again swims in a straight line, but in a new, randomly chosen direction (25). (A tumble is probably a series of very brief runs and twiddles.)

Compared to this unstimulated state, cells tumble less frequently--i.e., they swim in longer runs--when they encounter increasing concentration of attractant (25, 62) and they tumble more frequently when the concentration decreases (62). For repellents, the opposite is true: bacterial encountering an increasing concentration tumble more often, while a decreasing concentration suppresses tumbling (85). (See Figure 1). (Much smaller concentration changes are needed to bring about suppression of tumbling than stimulation of tumbling (25, 57)).

All this applies not only to spatial gradients (for example, a higher concentration of chemical to the right than to the left, but even to temporal gradients (a higher concentration of chemical now than earlier). This important discovery that bacteria can be stimulated by temporal gradients of chemicals was made by mixing bacteria quickly with increasing or decreasing concentration of attractant (62) or repellent (85), then immediately observing the alteration of tumbling frequency. After a short while (depending on the extent and the direction of the concentration change), the tumbling frequency returns to the unstimulated estate (62, 85). A different way to provide temporal gradients is to destroy or synthesize an attractant enzymatically; as the concentration of attractant changes, the tumbling frequency is measured (30). (For a history of the use of temporal stimulation in the study of bacterial behavior, see the introduction to ref. 30). The fact that bacteria can "remember" that there is a different concentration now than before has led to

---

[1] Lovely, P., Macnab, R., Dahlquist, F. W. and Koshland, D. E., Jr., unpublished experiments.

the proposal that bacteria have a kind of "memory" (57, 62).

(The possibility that a bacterium in a spatial gradient compares the concentration at each end of its cell has not been ruled out, but it is not necessary to invoke it now, and in addition, the concentration difference at the two ends would be too small to conceivably be effective (57, 62)).

These crucial studies (25, 30, 62, 85) point to the regulation of tumbling frequency as a central feature of chemotaxis. The results are summarized in Figure 1.

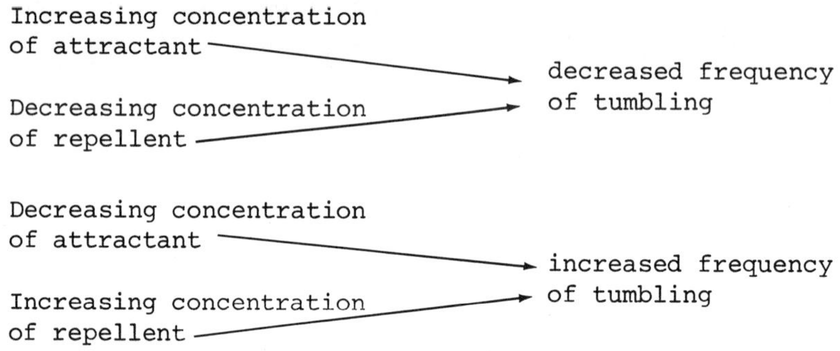

Fig. 1. Effect of change of concentration of chemical on tumbling frequency.

By varying the tumbling frequency in this manner, the bacteria, in a "biased random walk" (57),migrate toward attractants and away from repellents: motion in a favorable direction is prolonged, and motion in an unfavorable direction is terminated.

(Bacteria that have one or more flagella located at the pole ("polar flagellation," as for example in *Spirillum* or *Pseudomonas*) back up instead of tumbling (31, 83). Even bacteria that have flagella distributed all over ("peritrichous flagellation," as in *E. coli* or *S. typhirmurium*) will go back and forth instead of tumbling if the medium is sufficiently viscous (88).

THE DETECTION OF CHEMICALS BY BACTERIA: CHEMORECEPTORS

What is Detected?

Until 1969 it was not known if bacteria are capable of detecting the attractants themselves, or if instead they measure some product of metabolism of the attractants, for example ATP. The latter idea was eliminated, and the former established, by the following results (2). (a) Some chemicals that are extensively metabolized are not attractants. This includes chemicals that are the first products in the metabolism of chemicals that do attract. (b) Some chemicals that are essentially nonmetabolized attract bacteria: nonmetabolizable analogs of metabolizable attractants attract bacteria, and mutants blocked in the metabolism of an attractant are still attracted to it. (c) Chemicals attract bacteria even in the presence of a metabolizable, non-attracting chemical. (d) Attractants that are closely related in structure compete with each other but not with structurally unrelated attractants. (e) Mutants lacking the detection mechanism, but normal in metabolism, can be isolated. (f) Transport of a chemical into the cells is neither sufficient nor necessary for it to attract.

Thus it was established that bacteria can sense attractants _per se_: these cells are equipped with sensory devices--"chemoreceptors"--that measure changes in concentration of certain chemicals and report the changes to the flagella (2). It is a characteristic feature of this and many other sensory functions that when the stimulus intensity changes, there is a response for a brief period only, i.e. this response is _transient_ (62, 85). In contrast, all previously described responses of bacteria to changes in concentration of a chemical presist as long as the new construction is maintained. For example, when the concentration of lactose is increased (over a certain range), there is a persisting increase in the rate of transport of lactose, or the rate of metabolism of lactose, or the rate of synthesis of β-galactosidase.

Since metabolism of the attractants is not involved in sensing them (21), the mechanism of positive chemotaxis does not rely upon the attractant's value to the cell. Similarly, negative chemotaxis is not mediated by the harmful effects of a repellent (86): (a) Repellents are detected at concentrations too low to be harmful. (b) Not all harmful chemicals are repellents. (c) Not all repellents are harmful. Nevertheless, the survival value of chemotaxis must lie in bringing the bacteria into a nutritious environment (the attractants might signal the presence of other undetected nutrients) and away from a noxious one.

The Number of Different Chemoreceptors

For both positive and negative chemotaxis the following criteria have been used to divide the chemicals into chemoreceptor classes. (a) For a number of chemoreceptors, mutants lacking the corresponding taxis--"specifically nonchemotactic mutants"--have been isolated (2, 5, 6, 46, 86). (b) Competition experiments: chemical A, pressent in high enough concentration to saturate its chemoreceptor, will completely block the response to B if the two are detected by the same chemoreceptor but not if they are detected by different chemoreceptors (2, 10, 20, 46, 64, 75, 85, 86). (c) Many of the chemoreceptors are inducible, each being separately induced by a chemical it can detect (2, 6).

Altogether about 12 chemoreceptors have been identified so far for positive chemotaxis in E. coli, and about 10 for negative chemotaxis in E. coli. It should be emphasized that the evidence for each of them is not equally strong. S. typhimurium, insofar as its repertoire has been investigated, shows some of the same responses as E. coli (9, 10, 35, 62, 85).

Nature of the Chemoreceptors

Protein components of some of the chemoreceptors have been identified by a combination of biochemical and genetic techniques. Each chemoreceptor, it is believed, has a protein that recognizes the chemicals detected by that chemoreceptor--a "recognition component" or "binding protein." Wherever this protein has been identified, it has also been shown to function in a transport system for which the attractants of the chemoreceptor class are substrates. Yet both the transport and chemotaxis systems have other, independent components, and transport is not required for chemotaxis. These relationships are diagrammed in Figure 2.

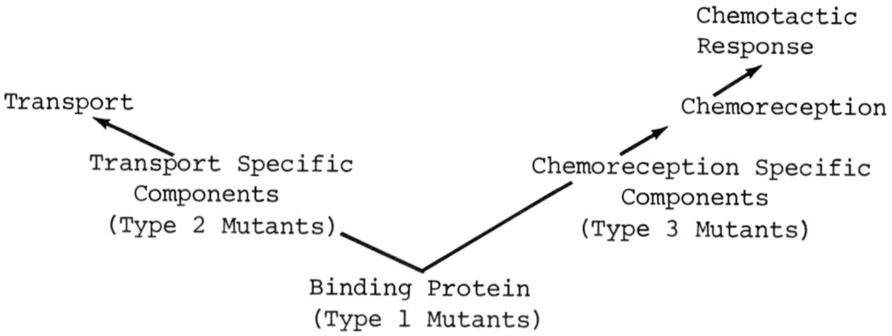

Fig. 2. Relation between chemoreception and transport.

Transport and chemotaxis are thus very closely related; but not all substances that are transported, or for which there are binding proteins, are attractants or repellents (6, 86).

The first binding protein shown to be required (47, 54) for chemoreception was the galactose binding protein (11). This protein is known to function in the β-methylgalactoside transport system (27, 28, 54), one of several by which D-galactose enters the E. coli cell (78). This is one of the proteins released from the cell envelope of bacteria--presumably from the periplasmic space, the region between the cytoplasmic membrane and the cell well--by an osmotic shock procedure (49).

The evidence that the galactose binding protein serves as the recognition component for the galactose chemoreceptor is the following: (a) mutants (Type 1 in Figure 2) lacking binding protein activity lack the corresponding taxis (47), and they are also defective in the corresponding transport (27). Following reversion of a point mutation in the structural gene for the binding protein (28); there was recovery of the chemotactic response[2] and of the ability to bind and transport galactose (28). (b) For a series of analogs, the ability of the analog to inhibit taxis towards galactose is directly correlated to its strength of binding to the protein (47). (c) The threshold concentration and the saturating concentration for chemotaxis toward galactose and its analogs are consistent with the values expected from the dissociation constant of the binding protein (65). (d) Osmotically shocked bacteria exhibit a greatly reduced taxis towards galactose, while taxis towards some other attractants is little affected (47). (e) Galactose taxis could be restored by mixing the shocked bacteria with concentrated binding protein (47), but this phenomenon requires further investigation and confirmation.

Binding activities for maltose and ribose were revealed by a survey for binding proteins, released by osmotic shock, which might function for chemoreceptors other than galactose (47). The binding protein for maltose has now been purified (55), and mutants lacking it fail to carry out maltose taxis as well as being defective in the transport of maltose (46, 55). That for ribose has been purified from S. typhimurium and has been shown to serve the ribose chemoreceptor by criteria (a) to (c) above (9, 10).

Mutants of Type 2 (Figure 2) are defective in a transport system but not in a chemoreceptor, even though the two share a common binding protein. Thus, certain components of the transport system, and the process of transport itself, are not required for chemotaxis (at least for certain chemoreceptors). This has been studied (2, 47, 67). Two genes of Type 2 were found for the β-methylgalactoside transport system for galactose (67). Some of the mutations in these genes abolished transport without affecting chemotaxis; other mutations in these genes affected chemotaxis as well (67). Such chemotactic defects may reflect interactions, direct or indirect, that these components normally have with the chemoreception machinery, or some kind of unusual interaction of the mutated component with the binding protein which would hinter its normal function in chemoreception. Two genes whose products are involved in the transport for maltose (51) can be mutated without affecting taxis toward that sugar (6, 75).

Mutants of Type 3 presumably have defects in gene products--"signallers" (67)--which signal information from the binding protein to the rest of the chemotaxis machinery, without having a role in

---

[2] Gay, M., unpublished observations.

the transport mechanism. Such mutants, defective only for the galactose chemoreceptor, or jointly for the galactose and ribose chemoreceptors, are known (67). In that connection it is interesting that there is a mutant in the binding protein gene, by the criterion of complementation, that binds and transports galactose normally but fails to carry out galactose taxis, presumably because this binding protein is altered at a site for interaction with the Type 3 gene product (67). Conversely, some mutations mapping in the gene for the maltose binding protein affect transport but not maltose binding or chemotaxis (46). Nothing is known about the biochemistry or location of Type 3 gene products.

While the binding proteins mentioned above can be removed from the cell envelope by osmotic shock, other binding proteins exist that are tightly bound to the cytoplasmic membrane. Examples of such are the enzymes II of the phosphotransferase system, a phosphoenolpyruvate-dependent mechanism for the transport of certain sugars (58, 74). A number of sugar chemoreceptors utilize enzymes II as recognition components: for example, the glucose and mannose chemoreceptors are serviced by glucose enzyme II and mannose enzyme II, respectively (5). In these cases, enzyme I and HPr (a phosphate-carrier protein) of the phosphotransferase system (74) are also required for optimal chemotaxis (5). This could mean either that phosphorylation and transport of the sugars are required for chemotaxis in these cases; or that enzyme I and HPr must be present for interaction of enzyme II with subsequent chemoreception components; or, as seems most likely, that the enzyme II binds sugars more effectively after it has been phosphorylated by phosphoenolpyruvate under the influence of enzyme I and HPr. The phosphorylated sugars, the products of the phosphotransferase system, are not attractants, even when they can be transported by a hexose phosphate transport system (5, 6); this rules out the idea that the phosphotransferase system is required to transport and phosphorylate the sugars so that they will be available to an internal chemoreceptor, and indicate instead that interaction of the sugar specifically with the phosphotransferase system somehow leads to chemotaxis (5). Certainly it is not the metabolism of the phosphorylated sugars that brings about chemotaxis: several cases of non-metabolizable phosphorylated sugars are known, yet the corresponding free sugars are attractants (5).

Bacteria detect changes over time in the concentration of attractant or repellent (30, 62, 85), and experiments with whole cells indicate that it is the time rate of change of the fractions of the binding protein occupied by ligand that the chemotactic machinery appears to detect (30, 65). How this is achieved remains unknown. A conformational change has been shown to occur when ligand (galactose) interacts with its purified binding protein (29, 77), and possibly this change is sensed by the next component in the system, but nothing is known about this linkage.

COMMUNICATION OF SENSORY INFORMATION FROM CHEMO-RECEPTORS TO THE FLAGELLA

Somehow the chemoreceptors must signal to the flagella that a change in concentration of chemical has been encountered. The nature of this system of transmitting information to the flagella is entirely unknown, but several mechanisms have been suggested (2, 40, 62).

(a) The membrane potential alters, either increasing or decreasing for attractants, with the opposite effect for repellents. The change propagates along the cell membrane to the base of the flagellum. Causing the change in membrane potential would be a change in the rate of influx or efflux of some ions(s) when the concentration of attractant or repellent is changed.

(b) The level of a low molecular weight transmitter changes, increasing or decreasing with attractant or repellent. The transmitter diffuses to the base of the flagella. Calculations (23) indicates that diffusion of a substance of low molecular weight is much too slow to account for the practically synthronous reversal of flagella at the two ends of *Spirillum volutans*, which occurs in response to chemotactic stimuli (31, 83). Thus for this organism, at least, a change in membrane potential appears to be the more likely of the two mechanisms.

Although the binding protein of chemoreceptors is probably distributed all around the cell, since it is shared with transport, possibly only those protein molecules at the base of the flagellum serve for chemoreception. In that case, communication between the chemoreceptors and flagella could be less elaborate, taking place by means of direct protein-protein interaction.

Several tools are available for exploring the transmission system. One is the study of mutants that may be defective in this system; these are the "generally nonchemotactic mutants," strains unable (fully or partly) to respond to any attractant or repellent (2, 17, 86). Some of these mutants swim smoothly, never tumbling (17, 60, 68, while others--"tumbling mutants"--tumble most of the time (12, 19, 68, 88). Genetic studies (15, 16, 69) have revealed that the generally nonchemotactic mutants map in four genes (68, 69). One of these gene products must be located in the flagellum, presumably at the base, since some mutations lead to motile, nonchemotactic cells while other mutations in the same gene lead to the absence of flagella (79). The location of the other three gene products is unknown. The function of the four gene products is also unknown, but it has been suggested (68) that they play a role in the generation and control of tumbling at the level of a "twiddle generator" (25).

A second tool comes from the discovery that methionine is required for chemotaxis (4), perhaps at the level of the transmission system. Without methionine, chemotactically wild-type bacteria do not carry out chemotaxis (3, 4, 13, 19) or tumble (4, 19, 82). This is not the case for tumbling mutants (19), unless they are first "aged" in the absence of methionine (82), presumably to remove a store of methionine or a product formed from it. There is evidence that methionine functions via S-adenosylmethionine (13, 14, 19, 59, 82).

We (56) have now identified a protein in the cytoplasmic membrane of *E. coli* that contains a metabolically labile methyl group donated by methionine. The involvement of this protein methylation reaction in chemotaxis is indicated by four lines of evidence (56): 1) the methylation reaction is altered in some generally nonchemotactic mutants, and is coreverted with the chemotaxis defects. 2) The methylation level of the protein is affected by chemotactic stimuli. 3) The methyl group of the protein is derived from methionine and is metabolically labile, in accord with the known fact that chemotaxis requires a continuous supply of methionine. 4) Methylation is abnormal in mutants having defective or missing flagella. (See ref. 56 for documentation.)

THE FUNCTIONING OF FLAGELLA TO PRODUCE BACTERIAL MOTION

For reviews of bacterial flagella and how they function, see refs. 18, 39, 41, 52, 66, 81 and 90.

The genetics of synthesis of bacterial flagella is being vigorously pursued in *E. coli* and *S. typhimurium* (50, 88, 91). It is consistent with such a complex structure that at least twenty genes are required for the assembly and function of an *E. coli* flagellum (50) and many of these are homologous to those described in *Salmonella* (88, 91).

For many years it was considered that bacterial flagella work either by means of a wave that propagates down the flagellum, as is known to be the case for eucaryotic flagella, or else by rotating as rigid or semi-rigid helices (for a review of the history, see ref. 24). Recently it was argued from existing evidence that the latter view is correct (24), and this was firmly established by the following experiment (80). *E. coli* cells with only one flagellum (obtained by growth on D-glucose, a catabolite repressor of flagella synthesis (7)) were tethered to a glass slide by means of antibody to the filaments. (The antibody, of course, reacts with the filament and just happens to stick to glass.) Now that the filament is no longer free to rotate, the cell instead rotates, usually counterclockwise, sometimes clockwise (80). By use of such tethered cells, the dynamics of the flagellar motor were then characterized (22).

The energy for this rotation comes from the intermediate of oxidative phosphorylation (the proton gradient in the Mitchell hypothesis), not from ATP directly (59, 84)--unlike in the case of eucaryotic flagella or muscle, and this is true for both counterclockwise and clockwise rotation (59).

In S. typhimurium light having the action spectrum of flavins brings about tumbling, and this might in some way be caused by interruption of the energy flow from electron transport (63).

It is now possible to isolate from bacteria "intact" flagella, i.e., flagella with the basal structure still attached (36-38). There is the helical filament, a hook and a rod. In the case of E. coli four rings are mounted on the rod (36), while flagella from gram positive bacteria have only the two inner rings (36, 38). For E. coli it has been established that the outer ring is attached to the outer membrane and the inner ring to the cytoplasmic membrane (37). The basal body thus a) anchors the flagellum into the cell envelope, b) provides contact with the cytoplasmic membrane, the place where the energy originates, and c) very likely constitutes the motor (or a part of it) that drives the rotation.

THE RESPONSE OF FLAGELLA TO SENSORY INFORMATION

Addition of attractants to E. coli cells, tethered to glass by means of antibody to flagella, causes counterclockwise rotation of the cells as viewed from above (60). (Were the flagellum free to rotate, this would correspond to clockwise rotation of the flagellum and swimming toward the observer, as viewed from above. But since a convention of physics demands that the direction of rotation be defined as the object is viewed moving away from the observer, the defined direction of the flagellar rotation is counterclockwise.) On the other hand, addition of repellents causes clockwise rotation of the cells (60). These responses last for a short while, depending on the strength of the stimulus; then the rotation returns to the unstimulated state--mostly counterclockwise (60).

Mutants of E. coli that swim smoothly and never tumble, rotate always counterclockwise, while mutants that almost always tumble, rotate mostly clockwise (60).

From these results, and from the prior knowledge that increase of attractant concentration causes smooth swimming (i.e., suppressed tumbling) (25, 30, 62) while addition of repellents causes tumbling (85), it was concluded that smooth swimming results from counterclockwise rotation of flagella and tumbling from clockwise rotation (60).

When there are several flagella originating from various places around the cell, as in E. coli or S. typhimurium, the flagella function together as a bundle propelling the bacterium from behind (72, 73, 84). Apparently the bundle of flagella survives counterclockwise rotation of the individual flagella, to bring about smooth swimming (no tumbling), but comes apart as a result of clockwise rotation of individual flagella to produce tumbling. That tumbling occurs concomitantly with a flying apart of the flagellar bundle has actually been observed by means of light of such high intensity that individual flagella could be seen (63).

Presumably less than a second of clockwise rotation can bring about a tumble, and the long periods of clockwise rotation reported (60) result from the use of unnaturally large repellent stimuli. (The corresponding statement can be made for the large attractant stimuli used.) Some kind of recovery process is required for return to the unstimulated tumbling frequency. The mechanism of recovery is as yet unknown, but it appears that methionine is somehow involved (19, 82).

The information developed so far is summarized in Figure 3.

Chemical Gradient → Chemoreceptor → Change in Signal → Counterclockwise → Suppression of Rotation of Flagella → Tumbling → Recovery (Smooth Swimming)

Chemical Gradient → Chemoreceptor → Change in Signal → Clockwise Rotation of Flagella → Tumbling → Recovery

Fig. 3. Summary scheme of chemotaxis

The reversal frequency of the flagellum of a pseudomonad is also altered by gradients of attractant or repellent, and reversal of flagellar rotation can explain the "backing up" of polarly flagellated bacteria (83).

INTEGRATION OF MULTIPLE SENSORY DATA BY BACTERIA

Bacteria are capable of integrating multiple sensory inputs, apparently by algebraically adding the stimuli (85). For example, the response to a decrease in concentration of repellent could be overcome by superimposing a decrease in concentration of attractant (85). Whether bacteria will "decide" on attraction or repulsion in a "conflict" situation (a capillary containing both attractant and repellent) depends on the relative effective concentration of the two chemicals, i.e., how far each is present above its threshold concentration (8, 71). The mechanism for summing the opposing signals is unknown.

ROLE OF THE CYTOPLASMIC MEMBRANE

There is increasing evidence that the cytoplasmic membrane plays a crucial role in chemotaxis. (a) Some of the binding proteins that serve in chemoreception--the enzymes II of the phosphotransferase system (5)--are firmly bound to the cytoplasmic membrane (58, 74). Binding proteins that can be released by osmotic shock are located in the periplasmic space (49), perhaps loosely in contact with the cytoplasmic membrane. (b) The base of the flagellum has a ring that is embedded in the cytoplasmic membrane (50). (c) The energy source for motility comes from oxidative phosphorylation (59, 84), a process that along with electron transport is membrane-associated (45). (d) Chemotaxis, but not motility, is unusually highly dependent on temperature, which suggested a requirement for fluidity in the membrane lipids (3). This requirement for a fluid membrane was actually established by measuring the temperature dependence of chemotaxis in an unsaturated fatty acid auxotroph that had various fatty acids incorporated (61). (e) A number of reagents (for example, ether or chloroform) that affect membrane properties inhibit chemotaxis at concentrations that do not inhibit motility (31, 32, 44, 76). (f) The methyl-accepting chemotaxis protein (56) is located in the cytoplasmic membrane.

This involvement of the cytoplasmic membrane in chemotaxis, especially the location there of the chemoreceptors, the flagella, and the methyl-accepting chemotaxis protein, makes the membrane potential hypothesis for transmission of information from chemoreceptors to flagella plausible, but of course by no means proves it.

UNANSWERED QUESTIONS

While the broad outlines of bacterial chemotaxis have perhaps been sketched, the biochemical mechanisms involved remain to be elucidated: How do chemoreceptors work? By what means do they communicate with the flagella? What is the mechanism that drives the motor for rotating the flagella? What is the mechanism of the gear that shifts the direction of flagellar rotation? How does the cell recover from the stimulus? How are multiple sensory data processed? What are the functions of the cytoplasmic membrane in chemotaxis?

With regard to the protein methylation reaction recently discovered (56) the following questions remain to be answered:

1. What is the nature of the methylation system (reactants, products, methylation and demethylation enzymes involved, defects in various types of mutants)?

2. How do sensory receptors bring about a change in methylation?

3. How does a change in methylation bring about a change in the direction of rotation of the flagella?

RELATION OF BACTERIAL CHEMOTAXIS TO BEHAVIORAL BIOLOGY AND NEUROBIOLOGY

The inheritance of behavior and its underlying biochemical mechanisms are nowhere more amenable to genetic and biochemical investigation than in the bacteria. From the earliest time of studies of bacterial behavior (26, 53, 70, 71, 89) to the present (1, 2, 33, 34, 40, 57, 65) people have hoped that this relatively simple system can tell us something about the mechanisms of behavior of animals and man. Certainly striking similarities exist between sensory reception in bacteria and in higher organisms (33-35, 57, 65).

REFERENCES

1. ADLER, J., Science 153 (1966) 708.
2. ADLER, J., Science 166 (1969) 1588.
3. ADLER, J., J. Gen. Microbiol. 74 (1973) 77.
4. ADLER, J. and DAHL, M. M., J. Gen. Microbiol. 46 (1967) 161.
5. ADLER, J. and EPSTEIN, W., Proc. Nat. Acad. Sci. USA 71 (1974) 2895.
6. ADLER, J., HAZELBAUER, G. L. and DAHL, M. M., J. Bacteriol. 115 (1973) 824.
7. ADLER, J. and TEMPLETON, B., J. Gen. Microbiol. 46 (1967) 175.
8. ADLER, J. and TSO, W.-W., Science 184 (1974) 1292.
9. AKSAMIT, R. and KOSHLAND, D. E., Jr., Biochem. Biophys. Res. Commun. 48 (1972) 1348.
10. ASKAMIT, R. R. and KOSHLAND, D. E., Jr., Biochemistry. 13 (1974) 4473.
11. ANRAKU, Y., J. Biol. Chem. 243 (1968) 3116.
12. ARMSTRONG, J. B., Ph.D. Thesis, University of Wisconsin-Madison (1968) 43, 85.
13. ARMSTRONG, J. B., Can. J. Microbiol. 18 (1972) 591.
14. ARMSTRONG, J. B., Can. J. Microbiol. 18 (1972) 1695.
15. ARMSTRONG, J. B. and ADLER, J., Genetics 61 (1969) 61.
16. ARMSTRONG, J. B. and ADLER, J., J. Bacteriol. 97 (1969) 156.
17. ARMSTRONG, J. B., ADLER, J. and DAHL, M. M., J. Bacteriol. 93 (1967) 390.
18. ASAKURA, S., Advan. Biophys. 1 (1970) 99.
19. ASWAD, D. and KOSHLAND, D. E., Jr., J. Bacteriol. 118 (1974) 640.
20. BARACCHINI, O. and SHERRIS, J. C., J. Pathol. Bacteriol. 77 (1959) 565.
21. BERG, H. C., Rev. Sci. Instruments 42 (1971) 868.
22. BERG, H. C., Nature 249 (1974) 77.

23. BERG, H. C., Ann. Rev. Biophysics Bioengineering (1975) in press.
24. BERG, H. C. and ANDERSON, R. A., Nature 245 (1973) 380.
25. BERG, H. C. and BROWN, D. A., Nature 239 (1972) 500.
26. BINET, A., The Psychic Life of Micro-organisms. Open Court Chicago. (Translated from the French.) (1889) iv.
27. BOOS, W., Eur. J. Biochem. 10 (1969) 66.
28. BOOS, W., J. Biol. Chem. 247 (1972) 5414.
29. BOOS, W., GORDON, A. S., HALL, R. E. and PRICE, H. D., J. Biol. Chem. 247 (1972) 917.
30. BROWN, D. A. and BERG, H. C., Proc. Nat. Acad. Sci. USA 71 (1974) 1388.
31. CARAWAY, B. H. and KRIEG, N. R., Can. J. Microbiol. 18 (1972) 1749.
32. CHET, I., FOGEL, S. and MITCHELL, R., J. Bacteriol. 106 (1971) 863.
33. CLAYTON, R. K., Arch. Mikrobiol. 19 (1953) 141.
34. CLAYTON, R. K., Encyclopedia of Plant Physiology. (W. Ruhland ed.) Springer-Verlag., Berlin, 17-I: (1959) 371.
35. DAHLQUIST, F. W., LOVELY, P. and KOSHLAND, D. E., Jr., Nature New Biol. 236 (1972) 120.
36. DE PAMPHILIS, M. L. and ADLER, J., J. Bacteriol. 105 (1971) 384.
37. DE PAMPHILIS, M. L. and ADLER, J., J. Bacteriol. 105 (1971) 396.
38. DIMMITT, K. and SIMON, M., J. Bacteriol. 105 (1971) 369.
39. DOETSCH, R. N., CRC Critical Revs. Microbiol. 1 (1971) 73.
40. DOETSCH, R. N., J. Theor. Biol. 35 (1972) 55.
41. DOETSCH, R. N. and HAGEAGE, G. J., Biol. Revs. 43 (1968) 317.
42. DRYL, S., Bull. de l'Acad. Polon. des Sci. 6 (1958) 429.
43. ENGELMANN, T. W., Pfluger's Archiv für die gesammte Physiologie 25 (1881) 285.
44. FAUST, M. A. and DOETSCH, R. N., Can. J. Microbiol. 17 (1971) 191.
45. HAROLD, F. M., Bacteriol. Rev. 36 (1972) 172.
46. HAZELBAUER, G. L., J. Bacteriol. (1975) in press.
47. HAZELBAUER, G. L. and ADLER, J., Nature New Biol. 230 (1971) 101.
48. HAZELBAUER, G. L., MESIBOV, R. E. and ADLER, J., Proc. Nat. Acad. Sci. USA 64 (1969) 1300.
49. HAPPEL, L. A., Science 156 (1967) 1451.
50. HILEM, M., SILVERMAN, M. and SIMON, M., J. of Supramolecular Structure 2 (1974) in press.
51. HOFNUNG, M., HATFIELD, D. and SCHWARTZ, M., J. Bacteriol. 117 (1974) 40.
52. IINO, T., Bacteriol. Revs. 33 (1969) 454.
53. JENNINGS, H. S., Behavior of the Lower Organisms (1906) (Republished by Indiana University Press, Bloomington, 1962).
54. KALCKAR, H. M., Science 174 (1971) 557.
55. KELLERMAN, O. and SZMELMAN, S., Eur. J. Biochem. 47 (1974) 139.

56. KORT, E. N., GOY, M. F., LARSEN, S. H., ADLER, J., Proc. Nat. Acad. Sci. USA (1975) in press.
57. KOSHLAND, D. E., Jr., FEBS Letters 40 (1974) S3.
58. KUNDIG, W., ROSEMAN, S., J. Biol. Chem. 246 (1971) 1407.
59. LARSEN, S. H., ADLER, J., GARGUS, J. J. and HOGG, R. W., Proc. Nat. Acad. Sci. USA 71 (1974) 1239.
60. LARSEN, S. H., READER, R. W., KORT, E. N., TSO, W.-W. and ADLER, J., Nature 249 (1974) 74.
61. LOFGREN, K. W. and FOX, C. F., J. Bacteriol. 118 (1974) 1181.
62. MACNAB, R. M. and KOSHLAND, D. E., Jr., Proc. Nat. Acad. Sci. USA 69 (1972) 2509.
63. MACNAB, R. M. and KOSHLAND, D. E., Jr., J. Mol. Biol. 84 (1974) 399.
64. MESIBOV, R. and ADLER, J., J. Bacteriol. 112 (1972) 315.
65. MESIBOV, R., ORDAL, G. W. and ADLER, J., J. Gen. Physiol. 62 (1973) 203.
66. NEWTON, B. A. and KERRIDGE, D., Symp. Soc. Gen. Microbiol. 15 (1965) 220.
67. ORDAL, G. W. and ADLER, J., J. Bacteriol. 117 (1974) 517.
68. PARKINSON, J. S., Nature 252 (1974) 317.
69. PARKINSON, J. S., J. Bacteriol. (1975) in press.
70. PFEFFER, W., Untersuch. Botan. Inst. Tübingen 1 (1884) 363.
71. PFEFFER, W., Untersuch. Botan. Inst. Tübingen 2 (1888) 582.
72. PIJPER, A., Ergebn. Mikrobiol. Immunforsch. Exp. Ther. 30 (1957) 37.
72. PIJPER, A. and NUNN, A. J., J. Roy. Microscop. Soc. 69 (1949) 138.
74. ROSEMAN, S., Metabolic Pathways (3rd Edition) Vol. VI, (L. E. Hokin ed.) Academic Press, New York (1972) 41.
75. ROTHERT, W., Flora 88 (1901) 371.
76. ROTHERT, W., Jahrb. Wiss. Bot. 39 (1904) 1.
77. ROTMAN, B. and ELLIS, J. H., Jr., J. Bacteriol. 111 (1972) 791.
78. ROTMAN, B., GANESAN, A. K. and GUZMAN, R., J. Mol. Biol. 36 (1968) 247.
79. SILVERMAN, M. and SIMON, M., J. Bacteriol. 116 (1973) 114.
80. SILVERMAN, M. and SIMON, M., Nature 249 (1974) 73.
81. SMITH, R. W. and KOFFLER, H., Advances in Microbiol. Physiology (A. H. Rose ed.) Academic Press, London 6 (1971) 219.
82. SPRINGER, M. S., KORT, E. M., LARSON, S. H., ORDALL, G. W., READER, R. W. and ADLER, J., Proc. Nat. Acad. Sci. USA (1975) in press.
83. TAYLOR, B. L. and KOSHLAND, D. E., Jr., J. Bacteriol. 119 (1974) 640.
84. THIPAYATHASANA, P. and VALENTINE, R. C., Bioch. Biophys. Acta 347 (1974) 464.
85. TSANG, N., MACNAB, R. and KOSHLAND, D. E., Jr., Science 181 (1973) 60.
86. TSO, W.-W. and ADLER, J., J. Bacteriol. 118 (1974) 560.

87. VAITUZIS, Z. and DOETSCH, R. N., Applied Microbiol. 17 (1969) 584.
88. VARY, P. S. and STOCKER, B. A. D., Genetics 73 (1973) 229.
89. VERWORN, M., Psycho-Physiologische Protisten-Studien. Fischer, Jena (1889).
90. WEIBULL, C. The Bacteria, (I. C. Gunsalus and R. Y. Stanier eds.) Academic Press, New York 1 (1960) 153.
91. YAMAGUCHI, S., IINO, T., HORIGUCHI, T. and OHTA, K., J. Gen. Microbiol. 70 (1972) 59.
92. ZIEGLER, H., Encyclopedia of Plant Physiology, (W. Ruhland ed.) Springer-Verlag, Berlin, 17-II (1962) 484.

CYCLIC AMP RECEPTORS AT THE SURFACE OF AGGREGATING *DICTYOSTELIUM DISCOIDEUM* CELLS

Dieter MALCHOW*, David ROBINSON*, and Fritz ECKSTEIN†

Biozentrum der Universitat Basel,* Basel (Switzerland)
and Max Planck Institut für Experimentalle Medizin,†
Göttingen (West Germany)

Two types of cell-surface sites of the cellular slime mold *Dictyostelium discoideum* are known to react with extracellular cAMP: a cAMP-phosphodiesterase (9, 12, 14) and cAMP-binding sites (6, 7, 10, 13). The latter can be detected by using phosphodiesterase inhibitors such as cGMP (10) or dithiothreitol (6). Both cAMP-phosphodiesterase and cAMP-binding sites are developmentally regulated exhibiting their maximal activity during aggregation (9, 11). Also, the specificity of binding correlates well with the chemotactic activity of cAMP analogs (8, 11) thus indicating that cAMP-binding to cell-surface receptors is the first step leading to the chemotactic response (1) and to signal relaying (15).

A Mutant Partially Defective in Cell-bound Phosphodiesterase

By screening mutants which might be defective in cell-bound phosphosdiesterase, we found a non-aggregating mutant (Wag-6) which was partially impaired in the developmental regulation of cell-surface sites (Table I). Maximal cAMP-binding in the mutant was reduced by 50% and phosphodiesterase activity by 80% as compared to the wild type. Most importantly, binding was easily detectable in the absence of cGMP which can be explained at least in part by the relatively low activity of the phosphodiesterase at the cell-surface.

Interaction or Multiplicity of Binding Sites

Scatchard plots of cAMP-binding to Wag-6 cells were concave indicating multiple binding sites or negative cooperativity at 23°. The asymptotic K_a's were 0.2 and 10 μM. At 4-6° the Scatchard plot was convex suggesting a temperature dependent transition of the

TABLE I

Developmental regulation of cell-surface sites in an aggregateless mutant Wag-6 and in the wild type. Methods were as described (11) except that cAMP-binding to Wag-6 cells was measured in absence of cGMP. Unless otherwise stated experiments were performed at 23° and (^{3}H) cAMP-binding was measured for 5 sec.

Strain	cAMP-Hydrolysis (nmol per 10^7 cells per min)		cAMP-Binding (%)	
	growth phase	aggregation competence*	growth phase	aggregation competence
Ax-2	0.10	1.8	10	60
Wag-6	0.14	0.3	8	32

* In case of the agg- mutant Wag-6 the parameters were analyzed 8-10 hours after washing free of nutrient.

binding characteristics (4). Green and Newell also found different cAMP-binding sites or multiplicity of binding in the wild type using dithiothreitol to inhibit phosphodiesterase activity. Measuring at 4-6° they found a higher range of apparent affinity constants (6).

An Attractant Which is a Poor Substrate for the cAMP Phosphodiesterase

Adenosine -3', 5' cyclic-phosphorothioate (cAMPS) (2) induced a chemotactic response and a half-maximal change in light-scattering measurements (3) at about 10 times the concentration that cAMP does (Gerisch, personal communication). In agreement with the contention that cAMP receptors mediate the chemotactic response it was found that cAMPS competed well for cAMP-binding. By contrast, the thio-analog was a poor substrate for cell-bound phosphodiesterase: 30 μM cAMPS was needed to achieve a 50% inhibition of cAMP-hydrolysis (Table II). As indicated in Table II the use of cGMP and cAMPS allows one to distinguish cAMP receptor sites from phosphodiesterase at the cell-surface; enzyme and receptor display just opposite specificities towards these nucleotides.

TABLE II

Cell-bound Phosphodiesterase and cAMP-binding Sites Display Opposite Specificities towards cGMP and cAMPS

	Chemotactic activity (8) M	Half maximal fast-response in light scattering (3) M	Cell-bound phosphodi-esterase $K_{0.5}$	50% Inhibition of cAMP-binding M
cAMP	10^{-8}-10^{-9} (8)	$3x10^{-9}$ (3)	$5x10^{-7}$M (12)	$1\text{-}2x10^{-7}$ (11)
cAMPS	10^{-8} *	10^{-8} *	$3x10^{-4}$M **	$5x10^{-6}$ †
cGMP	$10^{-5}10^{-6}$ (8)	>$3x10^{-6}$ (3)	$6x10^{-6}$M (12)	>>$5x10^{-4}$ ‡ (10)

*not determined exactly

**50% inhibition of cell-bound phosphodiesterase which was measured at 75 mM cAMP

†mean of three independent experiments performed with Wag-6 cells. In one experiment an enhancement of cAMP-binding at lower cGMP concentrations occurred.

‡no inhibition by cGMP was observed up to that concentration. However, in presence of dithiothreitol (6) 400 μM cGMP inhibited cAMP-binding by 50%

cGMP Actions

As described above, cAMP-binding to aggregating cells can be measured in the presence of excess cGMP (50 - 500 μM) (10). Binding was found to be transient and the rate of decrease of bound cAMP concentration was similar to the time course of hydrolysis (11). Therefore we assumed that a primary effect of cGMP was to inhibit cAMP-hydrolysis. However, the following experiments indicate further functions of cGMP as Ax-2 cells were preincubated with 500 μM cGMP and cAMP-binding (10 nM) was assayed at different time intervals (Fig. 1). Inhibition of binding occurred, which was maximal at 2 min. If Wag-6 cells were used, inhibition was also observed but it proceeded at a slower rate (Fig. 2).

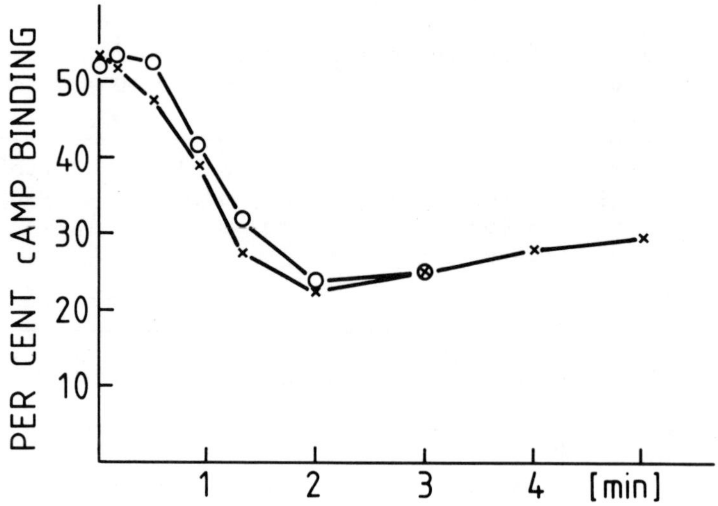

Fig. 1. Binding of cAMP to Ax-2 cells as a function of the time of preincubation with cGMP. Two independent experiments are shown. cAMP and cGMP concentrations were 10 nM and 500 μM respectively.

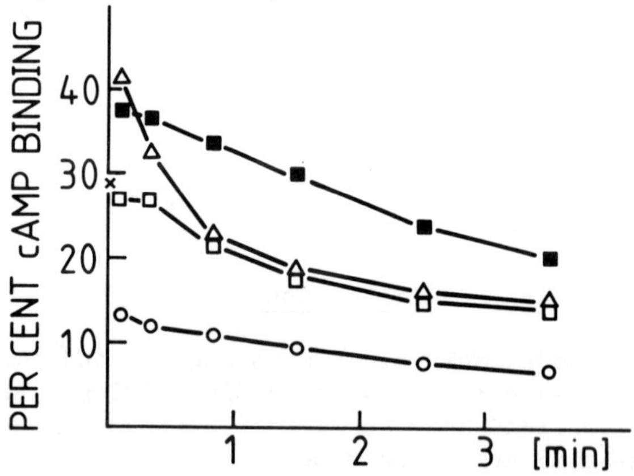

Fig. 2. Binding of cAMP to Wag-6 cells as a function of the time of preincubation with cGMP and/or cAMPS. ■ preincubation with 500 μM cGMP, Δ with 0.7 μM cAMPS, O with 10 μM cAMPS and □ with 500 μM cGMP and 10 μM cAMPS, x represents binding of 10 nM [^{3}H] cAMP without any addition.

It was shown (5) that cGMP at $5x10^{-4}M$ concentration induces a release of radioactive cAMP by aggregation competent amoebae. This release was maximal at 180 s. Therefore we can assume that at least part of the inhibition seen above is due to competition of cAMP released by the cells. The slow decrease of cAMP-binding may also reflect in part a desensitization of receptors corresponding to a refractory period which occurs during signal relaying (15).

Preincubation of Wag-6 cells with 10 μM cAMPS almost immediately inhibited cAMP-binding. Occasionally, at lower cAMPS concentrations (0.7 μM) a transient enhancement of binding was observed (Fig. 2).

To find out if cAMPS and cGMP compete for the same sites, we preincubated Wag-6 amoebae simultaneously with 10 μM cAMPS and 500 μM cGMP (Fig. 2). Interestingly cGMP seemed to antagonize the inhibitory effect of the thioanalog. This effect of cGMP is not due to inhibition of phosphodiesterase activity since the latter is low in that mutant and cAMP-binding was measured for 5 sec only.

The action of cGMP therefore seems to be at least threefold:

1. It competes for cAMP-hydrolysis
2. Since it is an attractant, though weak, it should bind to a site which causes a chemotactic response (chemotactic response site(s)). This binding also results in a release of cAMP by the cells which could lead to a slow inhibition of cAMP-binding.
3. It binds to a regulatory site thereby increasing the number of cAMP-binding sites, the affinity or both. Thus the inhibitory effect of cAMPS is immediately antagonized. Likewise cAMP-binding seems not to be inhibited but rather enhanced at high cGMP concentrations.

The suggestion of distinct chemotactic "response" sites and a regulatory site for cyclic nucleotide-binding is suggested by the following observations: 500 μM cGMP did not inhibit cAMP-binding by itself but did so in the presence of dithiothreitol. Similarly Green and Newell (6) found a 50% inhibition of cAMP-binding at 400 μM cGMP in the presence of dithiothreitol. Using the latter they found more than one binding site or negative cooperativity as discussed above. These sites showed the same relative affinities to cAMP and cGMP indicating that in the presence of dithiothreitol the enhancement of cAMP-binding by cGMP is inhibited. Therefore a regulatory cyclic nucleotide-binding site seems to be present in addition to what is found in the presence of dithiothreitol.

We do not know whether this regulatory site is located at the cell surface or inside the cell although the fast action of cGMP on cAMPS inhibition of cAMP-binding (Fig. 2) would tend to favor a cell surface location.

ACKNOWLEDGEMENTS

We thank Dr. J. M. Ashworth, Colchester, for the mutant Wag-6 and Dr. G. Gerisch for advice and support. Our work was supported by the Deutsche Forschungsgemeinschaft, the Stiftung Volkswagenwerk and the Swiss National Fund.

REFERENCES

1. BONNER, J. T., BARKLEY, D. S., HALL, E. M., KONIJN, T. M., MASON, J. W., O'KEEFE, G. and WOLFE, P. B., Develop. Biol. 20 (1969) 72.
2. ECKSTEIN, F., SIMONSON, L.-P. and BAER, H. P., Biochemistry 13 (1974) 3806.
3. GERISCH, G. and HESS, B., Proc. Nat. Acad. Sci. USA 71 (1974) 2118.
4. GERISCH, G., MALCHOW, D., HUESGEN, A., NANJUNDIAH, V., ROOS, W. and WICK, U., Proc. ICN-UCLA Symposium on Develop. Biol. (1975) in press.
5. GERISCH, G. and MALCHOW, D., (as cited from W. Roos) Adv. Cyclic Nucleot. Res. (1967) in press.
6. GREEN, A. A. and NEWELL, P. C., Cell 6 (1975) 129.
7. HENDERSON, E., J. Biol. Chem. 250 (1975) 4730.
8. KONIJN, T. M., Adv. Cyclic Nucleotide Res. 1, (1972) 17.
9. MALCHOW, D., NÄGELE, B., SCHWARZ, H. and GERISCH, G., Eur. J. Biochem. 28 (1972) 136.
10. MALCHOW, D. and GERISCH, G., Biochem. Biophys. Res. Com. 55 (1973) 200.
11. MALCHOW, D. and GERISCH, G., Proc. Nat. Acad. Sci. USA 71 (1974) 2423.
12. MALCHOW, D., FUCHILA, J. and NANJUNDIAH, V., Biochim. Biophys. Acta 385 (1975) 421.
13. MATO, J. M. and KONIJN, T. M., Biochim. Biophys. Acta 385 (1975) 173.
14. PANNBACKER, R. G. and BRAVARD, L. J., Science 175 (1972) 1014.
15. SHAFFER, B. M., Adv. Morphogenesis 2 (1962) 109.

CYCLIC AMP RECEPTOR ACTIVITY IN DEVELOPING CELLS OF DICTYOSTELIUM DISCOIDEUM

Christopher D. TOWN

Imperial Cancer Research Fund
Mill Hill Laboratories
London (UK)

During the period of interphase which precedes chemotactic aggregation in *Dictyostelium discoideum*, a number of new components appear on the cell surface. These include 1) cyclic AMP (cAMP) binding activity (receptors) (8, 11) 2) cAMP phosphodiesterase activity, (10) and 3) a new class of sites mediating cell-cell adhesion (1), which have been designated contact-sites A.

Recently it has been shown that cAMP signals are involved not only in the chemotactic aggregation process *per se* but also in the differentiation of cells to aggregation-competence (4, 6). I have been interested in the possibility that extracellular cAMP signals are present in the multicellular aggregates at late stages and may be involved in the control of cell movement, cell differentiation and pattern formation. To investigate this I have begun to isolate mutants with altered cAMP receptor activity with a view to following their fate in the developmental cycle.

This communication describes some new aspects of the temporal regulation of receptor activity in wild-type and some characteristics of one such receptor mutant.

Materials and Methods

Dictyostelium discoideum strain V-12 M2 (kindly supplied by Dr. G. Gerisch) was used for all the investigation. Conditions for growth and development in shaken suspension were as described by Henderson (8). For development on a surface, either millipore filters on support pads or 1% agar were used, both containing the same phosphate buffer.

Binding of cAMP to cells was measured in one of two ways: 1) as ^{3}H cAMP removed from the supernatant was first described by Malchow and Gerisch (11) and slightly modified by Henderson (8) 2) using the same reaction mix but measuring directly the ^{3}H cAMP bound to the cell pellet after centrifugation through a layer of silicone oil. The second technique is more sensitive and permits the characterization of low levels of binding activity, such as are found in vegetative cells and certain mutants. Details will be published shortly.

cAMP phosphodiesterase activity was measured in freeze-thawed cell lysates by the method of Brooker *et al.* (3) as described by Henderson (8).

The ability of cells to form EDTA-stable agglomerates under controlled shear conditions was used as an indication of the development of contact-sites A.

The mutants were isolated following treatment with N-methyl-N-nitro-nitrosoguanidine. Mutagenised cells which did not enter aggregates were plated and gave rise to some aggregation-defective plaques. These cells were tested for their ability to respond chemotactically to cAMP, and non-responders were assayed for the development of cAMP binding and cAMP phosphodiesterase activity. Mutant 1.1 developed no cAMP binding but approximately normal cAMP phosphodiesterase activity during starvation.

Characterization of cAMP binding to vegetative cells using the pellet assay

Figure 1 shows the binding of cAMP to vegetative cells as function of cAMP concentration for wild-type and mutant 1.1.

The data are shown on double logarithmic plots; the straight lines which best fit the data at high cAMP concentration for wild-type (Fig. 1a), and at all cAMP concentrations for mutant 1.1 (Fig. 1b) have slopes of 1.04. A slope of exactly 1.0 would indicate that the cAMP bound is a linear function of the free cAMP and does not saturate at high cAMP concentrations. Within the accuracy of the experiments this appears to be the case for mutant 1.1 at all cAMP concentrations, and for wild-type cells above $\sim 10^{-6}$ M cAMP. The magnitude of this non-saturable component appears to be very similar in the two cases. At cAMP concentrations less than 10^{-6} M, the data points for wild-type cells lie above the line, suggesting the presence of a small saturable component superimposed on the non-saturating binding.

Thus it appears that there are a small number of saturable cAMP binding sites already present on wild-type cells in the vegetative

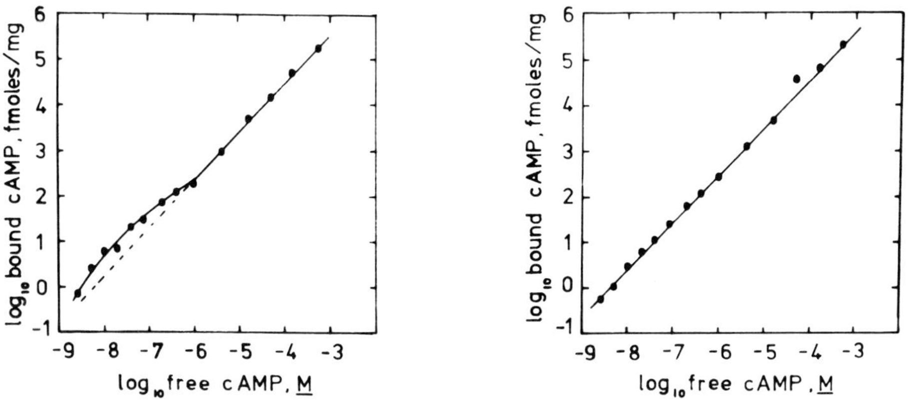

Fig. 1. Binding of cAMP to vegetative cells of D. discoideum a) wild-type cells b) mutant 1.1. Exponentially growing cells at ∿ 4 x 10^6/ml were washed three times and suspended in assay buffer (12.5 mM MES, 1mM azide, 60 mM KCl, pH 6.0) at 4°. They were then incubated at 4° for 8-12 min with ^{3}H cAMP, cGMP (5 x 10^{-4}M) and ^{14}C inulin (0.2 μCi/ml) before centrifugation through silicone oil and determination of the radioactivity in the cell pellet. The ^{14}C inulin radioactivity was used to correct for the unbound ^{3}H cAMP in the extracellular water in the cell pellet.

phase (roughly 700 sites/cell assuming the same K_d as during development). These sites are apparently absent in mutant 1.1.

Demonstration of cAMP receptor activity on cells developing on a solid surface at the aggregation and culmination stages

Previous studies have shown the appearance of receptor-like cAMP binding activity on cells developing in suspension (8, 11). Binding activity increases within a few hours of starvation and reaches a plateau value at between 6 hours (11) and 12 hours (8). Cells developing in suspension never undergo the normal morphogenetic sequence because agitation hinders cell contact formation and more significantly because the agglomerates are never presented with an air-water interface, do not form tips, and express none of the later functions (5, 13). If extracellular cAMP signals acting on the membrane-bound cAMP receptors are important in the late stages, it should be possible to demonstrate cAMP receptor activity on cells from late aggregates undergoing the normal morphogenetic sequence on a solid substrate.

To do this cells were taken after either 9-1/2 hours (late aggregation) or 19-1/2 hours (start of culmination) development on phosphate buffered 1% agar and their cAMP binding activity charac-

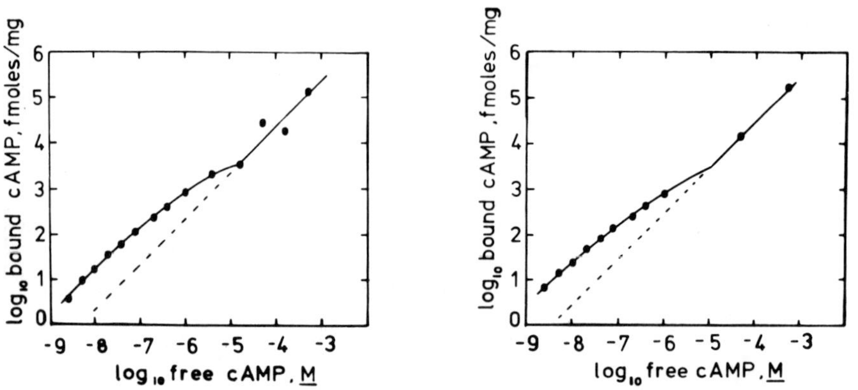

Fig. 2. Characterization of cAMP binding to cells derived from multicellular aggregates after (a) 9-1/2 hours development and (b) 19-1/2 hours development. Aggregates were washed off the agar and dissociated by moderate pipetting during 3 washes with cold assay buffer to give a predominantly single cell suspension. Binding was assayed in cell pellets as described for Fig. 1. The straight lines, broken at lower cAMP concentrations, have slopes of 1.0 and are assumed to represent the non-saturable vegetative component of binding.

terized as a function of cAMP concentration. The results are shown in Figure 2. It can be seen that at both time points suitable binding is present superimposed on the non-saturable vegetative background.

In the data shown, cAMP binding (per mg of protein) is slightly higher at 19-1/2 than 9-1/2 hours. In a number of similar experiments, binding activity reached its maximum by 9-10 hours, and this average value was the same as that for the 19-1/2 cells shown here. Since the protein content of cells at 19-1/2 hours is about two thirds of that at 9-1/2 hours, the number of receptors per cell is less at the later time. This binding activity (i.e. ∿ 23 fmoles/mg protein at 10^{-8} M cAMP) is about 20-fold lower than the corresponding value observed for cells developing in suspension (8, 11 and unpublished data). Peak cell-associated phosphodiesterase activities for cells of this strain developing on agar are also 5-10-fold less than for cells developing in suspension. In this, and other situations, there is a degree of coordination between expression of cAMP binding and phosphodiesterase activities.

Scatchard analysis (14) of the binding data reveals a complex curve suggesting either negative cooperativity, or the existence of multiple classes of cAMP binding sites, or both. Scatchard plots of the binding data for 9-10 hour and 19-1/2 hour cells have similar shapes indicating that there are no major qualitative changes in the spectrum of cAMP receptors between the early and late stages.

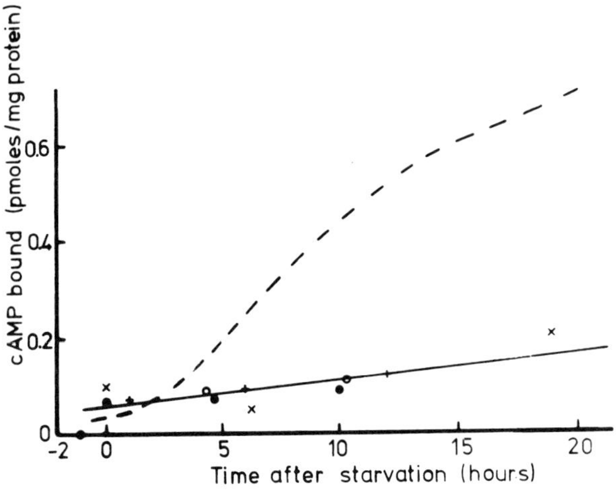

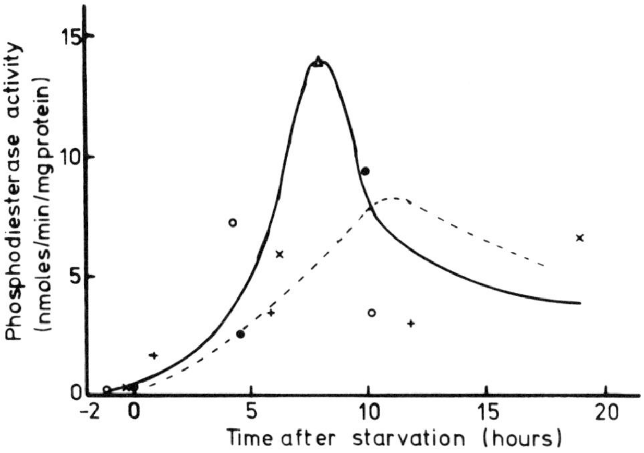

Fig. 3. Development of cAMP binding and cAMP phosphodiesterase activity in mutant 1.1 during starvation in suspension. Cells were shaken at 10^7 cells/ml and at various times samples were taken and assayed for (a) cAMP binding (using the supernatant assay) and (b) cAMP phosphodiesterase activities. The broken lines show comparable data for wild-type cells determined in the same laboratory by Dr. Ellen Henderson (8). Data from independent experiments are shown by different symbols.

Curved Scatchard plots have also been reported by Green and Newell (7) for cAMP binding to Ax-2 cells developing on filters and were interpreted by them as indicating the existence of two classes of binding sites. While this may be true, some negative cooperativity is to be expected when a negatively-charged molecule binds to a negatively-charged cell surface (see e. g. 12). A rough estimate of the high affinity part of the data suggests a K_d of 1-2 x 10^{-7} M and about 5,000 binding sites per cell.

Characterization of Receptor Mutant 1.1.

Figure 3 shows the changes in cAMP binding activity and cAMP phosphodiesterase activity during development of 1.1 in suspension. Also shown in the figure are binding and phosphodiesterase activities for M2 wild-type developing under identical conditions. The data for wild-type were collected by Dr. Ellen Henderson (8) while we were both working in the same laboratory in the University of Edinburgh and are strictly comparable as controls.

It can be seen that in the mutant 1.1 there is little or no increase in cAMP binding activity while the appearance and disappearance of phosphodiesterase follows a course similar to wild-type. The small increase in cAMP bound per mg of protein in 1.1 probably reflects protein degradation. When binding in 1.1 at $\sim$ 12 hours starvation was characterized as a function of cAMP concentration (using the supernatant assay), no difference was found between vegetative or 12 hours developing cells of 1.1 nor between vegetative cells of 1.1 and of wild-type. However, the supernatant assay is not sufficiently sensitive to demonstrate the small putative "receptor-component" seen in Fig. 1.

This mutant 1.1 appears to be completely defective in the developmental appearance of cAMP binding activity while its cell-associated phosphodiesterase appears approximately normally both in time and amount.

During the characterization of cAMP binding and PDE activities it was observed that, even at late stages, suspensions of 1.1 failed to form the tight agglomerates characteristic of wild-type. Using a roller tube apparatus as described by Gerisch (5) and the light scattering criterion E/E_o (1), it was determined that there is no developmental change in the ability of 1.1 to form EDTA-stable agglomerates, even after 24-30 hours in suspension. The mutant also forms mixed agglomerates with wild-type very poorly (see below). Thus it appears that 1.1 is unable to make any component of contact-sites A.

Development of Mutant 1.1 Alone and in Mixed Agglomerates with Wild-Type

On growth plates, mutant 1.1 is almost always aggregation-defective, the cells sometimes forming loose mounds or ripples on moist plates. On rare occasions a few minute fruiting bodies are seen on very old growth plates 1-2 weeks after starvation. When cells are washed and transferred to agar or millipore filters at normal densities (i.e. $\sim$ confluent), no development is seen.

In order to by-pass the chemotactic aggregation phase, multicellular agglomerates were formed of 1.1 either alone or in mixtures with wild-type cells. A spontaneous acriflavin-resistant derivative of 1.1 was used for these experiments. Because 1.1 is unable to form the EDTA-stable contacts, agglomerates containing the mutant cells are rather shear-sensitive. Very low shear speeds were used to form the agglomerates which were then handled as little as possible. Mixtures of either vegetative or eight hour starved cells were gently shaken for eight to twenty-four hours. After agglomerates had formed, they were gently transferred to phosphate-buffered 1% agar. A sample of the agglomerates was dissociated to approximately single cells and plated with and without acriflavin to determine the proportion of acriflavin-resistant mutant cells which had entered the agglomerates. The remaining agglomerates were allowed to develop tips and fruit. The spores were collected, heated to 45° for 30 min (to permit only spores to survive) and then plated with and without acriflavin to determine the proportion of mutant to wild-type spores in the mature fruiting bodies.

The results of 6 such experiments with three independent cultures have been variable. On all three occasions, a 50:50 mixture of mutant and wild-type cells give agglomerates containing 10-20% mutant cells. After allowing for this reduced proportion, the results show that the mutant forms spores with a reduced efficiency compared with the wild-type ranging from about 10^{-4} to 10^{-2} ($<10^{-5}$, 10^{-4}, 8×10^{-5}, 3×10^{-4}, 2×10^{-2}), with the exception of one experiment in which little or no reduction in mutant spore frequency was observed. In this same experiment (and in no others) a small fraction ($\lesssim$ 1%) of the agglomerates formed from 1.1 alone developed tips and subsequently fruited. The fruiting bodies arising from the mixed agglomerates frequently had a mound of cells around their base. Few, if any, amoeboid cells were found in the spore head.

It is difficult to draw any firm conclusions from these experiments. In general, the receptorless mutant cells differentiate into spores with an efficiency much less than that of wild-type cells, either alone or in association with wild-type. However, it is not possible to say whether this is due specifically to the receptorless character at the time of spore formation, to the absence of contact-sites A or some other receptor-dependent function, or even to the presence of other mutations in the receptorless strain.

Stalk Cell Induction by cAMP

Bonner (2) reported that a proportion of amoebae of strain NC4 developed into large vacuolated stalk-like cells when plated in the presence of 10^{-3} M cAMP at cell densities that were too low to permit aggregation in controls without cAMP. I have repeated these experiments with strain V-12 M2. When the cells are plated on 10^{-3} M cAMP at 2 x 10^4 cells/cm^2, a density at which only a few minute fruits are formed in the controls, 80% or more of the amoebae are converted into stalk-like cells as defined by morphology and staining with Calcofluor (2). On the cAMP plates, about half the cells are found in small clumps, and the frequency of stalk cells is significantly higher in the clumps than in the single cell population. As the plating density is reduced, the yield of stalk cells decreases progressively reaching a plateau value of about 5%. At these low densities (3 x 10^2 to 2 x 10^3 cells/cm^2) there is no cell clumping; all the stalk cells are isolated single cells. It seems therefore that cAMP can induce differentiation of stalk cells in the absence of any aggregation, but that the efficiency of conversion is greatly enhanced by cell-cell interaction.

It was of interest to determine whether cAMP can induce stalk cell formation under conditions that permit aggregation but normally block the expression of late developmental functions. I have examined this question under two conditions: 1) with cells plated on agar under a layer of salt solution and prevented by a sheet of cellophane from making contact with the air-water interface (5, 13) if they float off the agar; 2) with cells plated on agar in the normal way but overlaid directly with a sheet of cellophane or agar. This latter treatment may block late development by constraining the cells in two dimensions (Gross and Peacey, unpublished). In either case, cells retain their amoeboid shape indefinitely in the absence of cAMP. In its presence, they are converted into stalk cells with close to 100% efficiency.

The concentration of cAMP required to induce the formation of stalk-like cells is several orders of magnitude greater than that which promotes the acquisition of aggregation-competence when applied periodically during interphase (4, 6). It may well be that "read-out" of both pre- and post-aggregative developmental programmes requires the elvation of intracellular cAMP levels (9), and that this can be achieved either by activation of the membrane receptor system, or by continuous leakage of cAMP into the cells from a sufficiently high external source. It is not clear at what point the programs leading to stalk versus spore cell differentiation diverge, and why cAMP preferentially induces the appearance of stalk-like cells.

I have examined stalk cell induction in the receptorless mutant 1.1 and in a number of other aggregation-defective mutants of V-12 which are less well characterized. The results are shown in Table I. No stalk cells are induced in 1.1 (nor several of the other aggregation-defective mutants). This behavior may be due to the absence of the cAMP receptor, but more probably results from failure of the mutant cells to express functions required for the induction process.

TABLE I

Strain	Phenotype	Response to 10^{-3} M cAMP in Stalk-Cell Test.
V-12 M2	Wild Type	+++
1.1	Aggregateless, no receptor normal PDE	-
4.2.10	Delayed aggregation, delayed receptor appearance	-
2.5.1	Aggregateless	-
4.4.34	Aggregateless, occasional minute fruits	+++
4.1.16	Very delayed aggregation, minute fruits	+
4.3.10	Delayed aggregation	+++
4.3.11	Delayed aggregation	+++

SUMMARY

Several lines of evidence published recently (4, 6) suggest that extra-cellular cyclic AMP signals are involved in initiating some of the early steps of differentiation of vegetative cells to aggregation-competence. The data presented here suggest that there are small numbers of cAMP receptors present already on vegetative cells, which could mediate these early cAMP effects.

In cells which are allowed to develop into multicellular aggregates and undergo morphogenesis on a solid surface, much less binding activity is detected (Fig. 2) than on cells shaken in suspension. A similar observation has already been made for strain Ax-2 by Henderson (8), and Green and Newell (7) although in the latter case, this difference was less marked. Neither of these works demonstrated the existence of cAMP receptor-like binding to cells at a late stage of development. When cells of strain V-12 M2 are dissociated from 19-1/2 hour culminating aggregates and the binding activity characterized, approximately the same binding activity as at 9-1/2 hours is observed. The binding activities at 9-1/2 and 19-1/2 hours show similar saturation kinetics, suggesting that there are no major qualitative changes in cAMP receptors on the cell-surface. Thus cAMP receptors, quantitatively and qualitatively similar to those which appear during aggregation, are present on cells at the late stages of development.

In an effort to elucidate the possible role of the cAMP receptors at the late stages of development, mutants with altered cAMP receptor activity (but hopefully otherwise normal) were sought. One mutant (1.1) has been described here. Cyclic AMP receptor activity seems to be completely absent from this mutant both during growth and development, contact-sites A are not found but cell-associated cAMP phosphodiesterase appears about normally. The properties of this mutant are consistent with the idea that the appearance of contact-sites A is dependent on cAMP receptor activity (6) but suggest that the appearance of phosphodiesterase is not.

The ability of the receptorless mutant to undergo terminal differentiation into spore or stalk cells has been investigated in two ways. It has been asked whether the mutant, when artificially agglomerated, will form spores either alone or in mosaics with wild-type cells. The results show that it does so, but with an efficiency several orders of magnitude less than wild-type. It has also been asked whether the mutant can be induced to form stalk cells by cAMP, as can wild-type cells. The answer is that it can not. Both these could be due to the absence of the cAMP receptor at the time of final differentiation. However, it seems more likely that the inability of mutant 1.1 to differentiate results from the failure of the mutant cells to express other functions required for differentiation. This failure may be due to the lack of receptor at earlier times.

ACKNOWLEDGEMENTS

I should like to thank Dr. Julian Gross for his constant encouragement and helpful advice, and Dr. Ellen Henderson for an introduction to the receptor and phosphodiesterase assays. This research was initiated in the Department of Molecular Biology, University of Edinburgh where I was supported by a Beit Memorial Fellowship for Medical Research.

REFERENCES

1. BEUG, H., KATZ, F. E. and GERISCH, G., J. Cell Biol. 56 (1973) 647.
2. BONNER, J. T., Proc. Nat. Acad. Sci. USA 65 (1970) 110.
3. BROOKER, G., THOMAS, L. J., JR. and APPLEMAN, M. M., Biochemistry 7 (1968) 4177.
4. DARMON, M., BRACHET, P. and PEREIRA DA SILVA, L., Proc. Nat. Acad. Sci. USA (1975) in press.
5. GERISCH, G., Roux' Archiv für Entwicklungsmechanik 152 (1960) 632.
6. GERISCH, G., FROMM, H., HUESGEN, A. and WICK, U., Nature 255 (1975) 547.
7. GREEN, A. A. and NEWELL, P. C., Cell (1975) in press.

8. HENDERSON, E. J., J. Biol. Chem. 250 (1975) 4730.
9. KLEIN, C. and BRACHET, P., Nature 254 (1975) 432.
10. MALCHOW, D., NÄGELE, B., SCHWARTZ, H. and GERISCH, G., Eur. J. Biochem. 28 (1972) 136.
11. MALCHOW, D. and GERISCH, G., Proc. Nat. Acad. Sci. USA 71 (1974) 2423.
12. MC LAUGHLIN, S. and EISENBERG, M., Ann. Rev. Biophys. Bioeng. (1975) in press.
13. NEWELL, P. C., Biochemical Society Transactions 1 (1973) 1025.
14. SCATCHARD, G., Ann. N. Y. Acad. Sci. 51 (1949) 660.

PROSTAGLANDIN $F_{2\alpha}$ RECEPTORS IN CORPORA LUTEA

William S. POWELL+, S. HAMMARSTRÖM, U. KYLDÉN and Bengt SAMUELSSON

Department of Chemistry, Karolinska Institutet
Stockholm (Sweden)

The luteolytic effect of prostaglandin $F_{2\alpha}$ ($PGF_{2\alpha}$) was first reported in rats by Pharriss and Wyngarden in 1969 (18). Since that time this prostaglandin has been found to have a similar effect in many mammalian species, including monkeys, but not humans (16). Prostaglandin $F_{2\alpha}$ has been shown to have a physiological role as a luteolytic hormone in the sheep (17). It is produced in high concentrations by the uterus at the end of the estrus cycle and is transferred from the uteroovarian vein to the ovarian artery by a countercurrent mechanism (17).

The mechanism of the luteolytic effect of $PGF_{2\alpha}$ is not known with certainty. Pharris and Wyngarden (18) originally postulated that it may act by constricting the ovarian vein but subsequent experiments have not confirmed this (4, 7) and it has been shown that $PGF_{2\alpha}$ does not affect the capillary blood flow in the corpus luteum (11). Another possibility is that $PGF_{2\alpha}$ could cause luteolysis by increasing the fragility of the lysosomes of the corpus luteum as it does in rat liver (27). This would be in agreement with earlier studies which indicated that the fragility of lysosomal membranes was increased towards the end of the estrus cycle (10). It is difficult to know whether this was a cause or a consequence of cell death, however. Prostaglandin $F_{2\alpha}$ has been shown to counteract the effects of luteotrophic hormones on the corpus luteum. Behrman and coworkers (3) reported that administration of $PGF_{2\alpha}$ to rats caused a reduction of cholesterol ester synthetase activity

---

+Present address: Endocrine Laboratory, Royal Victoria Hospital, Montreal, Quebec, Canada in the ovary, resulting in a large decrease in the concentration of cholesterol esters available for progesterone synthesis. In another study (26) this prostaglandin was found to induce the activity of 20α-hydroxysteroid dehydrogenase, and enzyme which converts progesterone to 20α-hydroxypregn-4-en-3-one, which is inactive. The effects of these enzymes could be explained by an effect of $PFG_{2\alpha}$ on the concentration of luteotrophic hormone receptors. Administration of $PGF_{2\alpha}$ to rats about 24 hours before sacrifice caused a large decrease in the concentration of LH receptors in the corpus luteum without having a significant effect on the dissociation constant for the hormone-receptor interaction (14). Similar results were obtained for an estrogen receptor in rabbit corpora lutea (15).

Irrespective of the precise mechanism of action of $PGF_{2\alpha}$, it seemed probable that its effect would be mediated by its interaction with a receptor in the corpus luteum. These experiments were designed to determine whether or not such a receptor existed, and, if so, to investigate some of its properties and its subcellular localization.

Occurrence of $PGF_{2\alpha}$ Receptors in Ovine, Bovine and Human Corpora Lutea

Initial experiments with ovine (19) and later with bovine (20) corpora lutea indicated that $[9\beta\text{-}^3H]PGF_{2\alpha}$ (1 Ci/nmol) was indeed specifically bound by particulate fractions from these tissues. Since the properties of these two receptors were found to be very similar, they will be discussed together. The bound and free prostaglandins could be conveniently separated by chromatography at 4° on small Sephadex columns. Although this procedure took several minutes the results obtained were quite reproducible since the $PGF_{2\alpha}$ receptor complex did not dissociate under the conditions employed for the separation. Thin-layer chromatography of the bound and free $PGF_{2\alpha}$ showed that no metabolism of this compound occurred during the incubations.

$PGF_{2\alpha}$ was not bound by parts of the ovary other than the corpus luteum. The binding of this compound was unaffected by large excesses of various non-prostaglandin fatty acids, steroids or nucleotides or by LH or FSH. It was, however, inhibited by preincubation with proteolytic agents, sulfhydryl reagents or phospholipase A. The optimum pH for the $PGF_{2\alpha}$-receptor interaction was 6.3, although under these conditions the non-specific binding was rather high. The amount of $PGF_{2\alpha}$ bound was directly proportional to the protein concentration of the incubation mixtures.

Figure 1 shows the time courses for the association and dissociation reactions between $PGF_{2\alpha}$ and the receptor in bovine corpora lutea at various temperatures. When the association reaction was

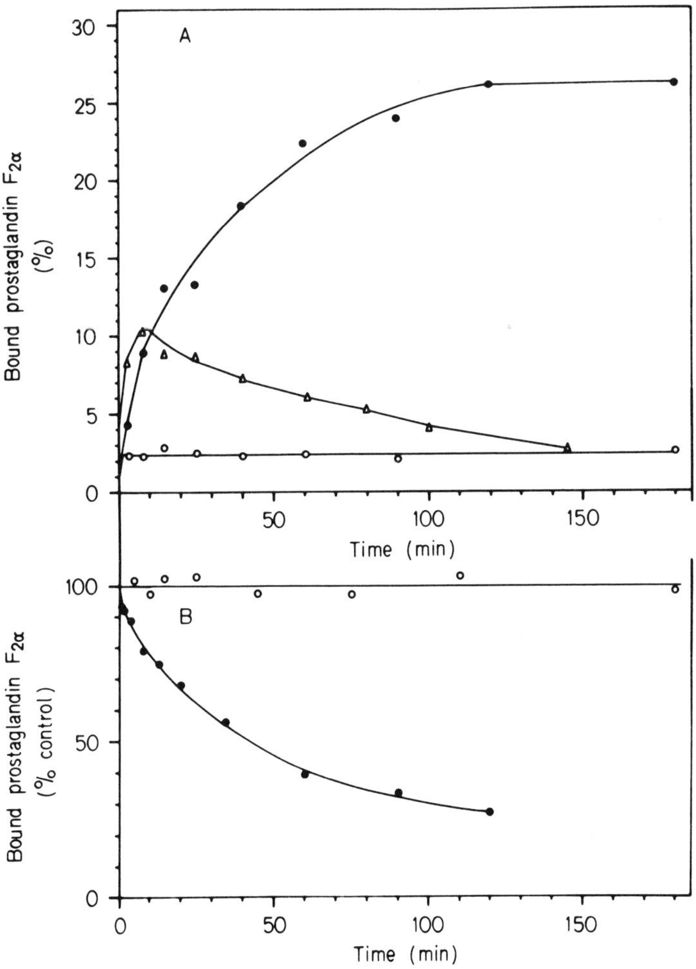

Fig. 1. (TOP) Time courses for the association of [9β-^{3}H] $PGF_{2\alpha}$ with a particulate fraction from bovine corpora lutea at 23° (●) and 37° (Δ). The nonspecific binding (O) was identical at both temperatures. (BOTTOM) Time courses for the dissociation reaction after incubation of a [9β-^{3}H] $PGF_{2\alpha}$-bovine corpus luteum receptor complex with a 500-fold excess of unlabeled $PGF_{2\alpha}$ at 0° (O) and 23° (●). Reproduced with permission from ref. 20.

followed at 23° (Fig. 1A) there was no decline in the concentration of bound $PGF_{2\alpha}$ after equilibrium had been reached, indicating that the receptor was reasonably stable under these conditions. At 37°, however, although the reaction was initially more rapid, the concentration of bound $PGF_{2\alpha}$ began to decline after 8 minutes, indicating that degradation of the receptor was taking place. When the data from the time course of the reaction at 23° was plotted accord-

TABLE I

Dissociation constants, luteolytic potencies in hamsters and relative rates of dehydrogenation by 15-hydroxyprostaglandin dehydrogenase for some prostaglandins, prostaglandin metabolities and prostaglandin analogues.

Compound	K_d (ovine corpra lutea) (μM)	K_d (bovine corpora lutea) (μM)	Relative ED100 for luteolysis	Relative rate of metabolism
Naturally occurring prostaglandins				
PGA_2	>260	140		
PGB_2	>260	150		
PGD_2		0.46		
PGE_1	34	38		
PGE_2	2.7	2.4	2.5	2.04
PGE_3		16		
$PGF_{1\alpha}$	2.0	2.0		
$PGF_{2\alpha}$	0.1	0.05	1.0	1.0
$PGF_{3\alpha}$		0.17		
Prostaglandin metabolities				
13, 14-dihydro$PGF_{2\alpha}$	0.23	0.20		
13,14-dihydro-15-oxo$PGF_{2\alpha}$	11	14	10	0
15-oxo-$PGF_{2\alpha}$		9		
1, 9 , 11 , 15(S)-tetrahydroprosta-5-cis-13-trans-diene	38	5.4		
2-nor$PGF_{2\alpha}$		2.6	1.6	0.22
$PGF_{2\beta}$	7.3			

TABLE I (continued)

Compound	K_d (ovine corpora lutea) (μM)	K_d (bovine corpora lutea) (μM)	Relative ED100 for luteolysis	Relative rate pf metabolism
13,14-didehydroPGF$_{2\alpha}$		0.21	0.12	0
ent-13,14-didehydroPGF$_{2\alpha}$		1300	15	0
ent-13,14-didehydro-15-epiPGF$_{2\alpha}$		6.3	4.0	0
15-methylPGF$_{2\alpha}$	0.18	0.063	0.1	0
15-O-acetylPGF$_{2\alpha}$		0.46	2.5	0
PGF$_{2\alpha}$ 15-methyl ether		0.06	0.25	0
16,16-dimethylPGF$_{2\alpha}$	0.29	0.085	0.1	0
16-(p-fluorophenoxy)-17,18,19,20-tetranorPGF$_{2\alpha}$		0.37[a]	<0.01[b]	
16-(m-chlorophenoxy)-17,18,19,20-tetranorPGF$_{2\alpha}$		0.02[a]	<0.01[b]	
16-(m-trifluoromethylphenoxy)-17,18,19,20-tetranorPGF$_{2\alpha}$		0.046[a]	<0.01[b]	
17-phenyl-18,19,20-trinorPGF$_{2\alpha}$	0.05	0.027	0.01	0.34
15-methyl-17-phenyl-18,19,20-trinorPGF$_{2\alpha}$		0.028	0.02	0
20-ethylPGF$_{2\alpha}$		0.12	0.25	0.34
PGE$_2$ methyl ester		4.1	2.5	1.7
16,16-dimethylPGE$_{2\alpha}$		8.8	0.5	0
17-phenyl-18,19,20-trinorPGE$_{2\alpha}$		1.7	0.5	0
15-methylPGE$_2$ methyl ester		7.5	0.1	0
15-methyl-15-epiPGE$_2$ methyl ester		88	2.5	0
rac-7-oxa-13-prostynoic acid		460	antagonist	0

[a]W. S. Powell, S. Hammarström and B. Samuelsson, unpublished work.
[b]Taken from Ref. 5.

ing to the second order rate equation, the initial slope of the resulting curve gave a value of 7500 liters/mole/sec for the rate constant for the association reaction. This value was independent of the concentration of $PGF_{2\alpha}$ employed. To investigate the dissociation of the $PGF_{2\alpha}$-receptor complex, tritium-labeled $PGF_{2\alpha}$ was incubated with the particulate fraction at 23° until equilibrium had been reached. At this time a 500-fold excess of unlabeled PGF_2 was added and the concentration of bound tritium labelled $PGF_{2\alpha}$ was determined at various time intervals. At 0° there was no detectable dissociation of the complex after 3 hours, whereas at 23°, the half-life of the complex was about 40 min. When the data from the time course at 23° was plotted according to the first order rate equation a value of 2.1×10^{-4} sec^{-1} was obtained for the rate constant for the dissociation reaction. The ratio of the rate constants for the dissociation and association reactions gave a value of 28 nM for the equilibrium dissociation constant.

Experiments performed after equilibrium had been reached, in which the concentration of tritium-labeled $PGF_{2\alpha}$ was varied, enabled the determination of the dissociation constant (K_d) from Scatchard plots (cf. Fig. 2). The dissociation constant determined in this way was 50 nM for bovine corpora lutea, in agreement with the value obtained from kinetic data. The concentration of receptor sites in crude bovine corpus luteum homogenates was about 0.8 pmoles/mg protein. The K_d for the interaction of $PGF_{2\alpha}$ with a particulate fraction from ovine corpora lutea was 100 nM.

Preliminary experiments with human corpora lutea (15, 22) suggest that a $PGF_{2\alpha}$ receptor also exists in this tissue. The dissociation constant for the interaction of $PGF_{2\alpha}$ with a particulate fraction from this tissue was 50 nM, close to those observed for ovine and bovine corpora lutea, but the concentration of binding sites was about 10-fold lower than in the latter two species.

Specificities of the Ovine and Bovine Corpus Luteum Receptors

Table I gives the dissociation constants for the interactions between the ovine and bovine corpus luteum receptors and some prostaglandins (19, 20), prostaglandin metabolities (19, 20) and prostaglandin analogues (21). The luteolytic potencies of some of these compounds as well as their relative rates of metabolism by 15-hydroxyprostaglandin dehydrogenase are also given (21). From these data it is apparent that the specificities of the ovine and bovine receptors are almost identical. Fig. 2 shows a typical series of Scatchard plots of $[9\beta-^3H]PGF_{2\alpha}$-recepter interactions in the absence and in the presence of several prostaglandin analogs. All four lines have a common intercept with the abscissa, indicating that these compounds inhibit the binding of $PGF_{2\alpha}$ in a competitive manner. This was true for all of the prostaglandins and prostaglandin-like compounds listed in Table I.

TABLE II

Effects of various structural modifications on the dissociation constant for prostaglandin - bovine corpus luteum receptor interactions. All modifications refer to a single change in the structure of $PGF_{2\alpha}$ unless otherwise indicated.

Structural Modification	Increase in K_d
CO_2H → CH_2OH	108
CH_2CO_2H → CO_2H	52
5,6-CH=CH → 5,6,-CH_2CH_2	40
9α-OH → 9β-OH[a]	146
→ 9-oxo	54
11α-OH → 11-oxo	9.2
13,14-CH=CH → 13,14-CH_2CH_2	4
→ 13,14-C≡C	4.2
15-OH → 15-oxo	180
→ 15-epi-OH[b]	12
→ 15-OAc	9.2
→ 15-OMe	1.2
15-H → 15-Me	1.3
15-pentyl → 15-(1,1-dimethylpentyl)	1.7
16-butyl → 16-(*m*-chlorophenoxy)	0.40
→ 16-(1-butenyl)	3.4
17-propyl → 17-phenyl	0.54
20-H → 20-ethyl	2.4
Inversion of all asymmetric centers (carbons 8, 9, 11, 12 and 15)[c]	6200
Inversion of all asymmetric centers except carbon-15[c]	30

[a] Ovine corpora lutea

[b] The reference compound is 15-methylPGE_2 methyl ester

[c] The reference compound is 13,14-didehydro$PGF_{2\alpha}$

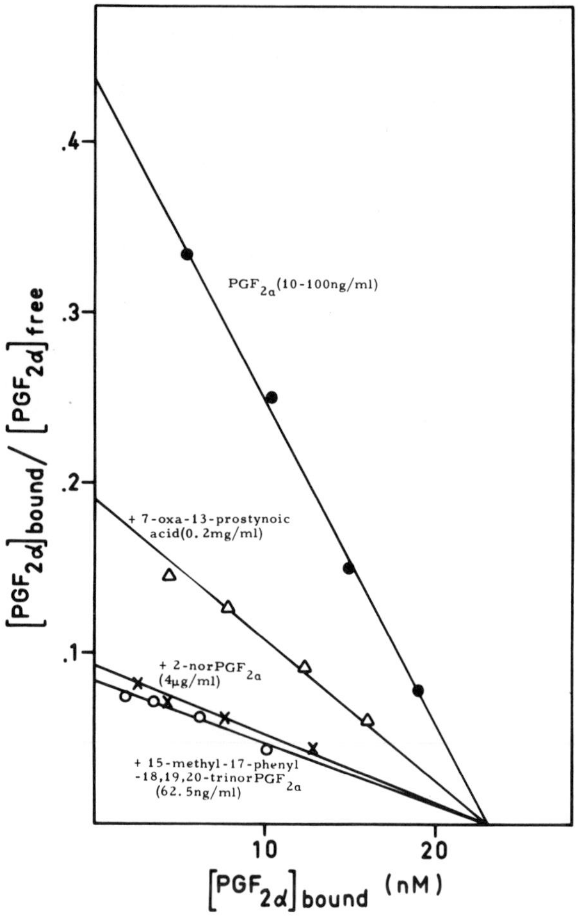

Fig. 2. Scatchard plot for the binding of [9β-^{3}H]$PGF_{2\alpha}$ to a particulate fraction from bovine corpora lutea in the presence and absence of some synthetic prostaglandin analogues. Adapted from Ref. 21 with permission.

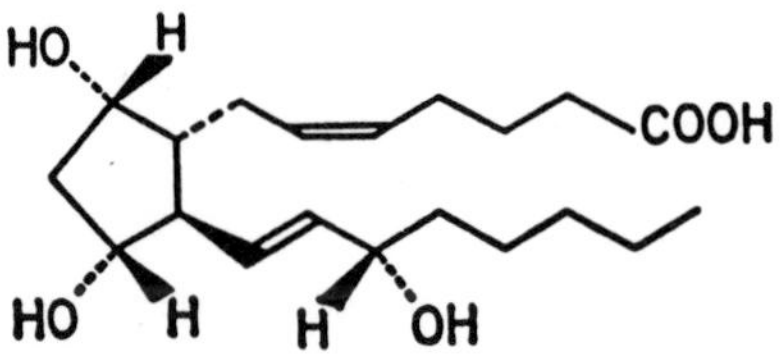

Fig. 3 Structure of $PGF_{2\alpha}$.

The results of the specificity studies involving the bovine corpus luteum receptor are summarized in Table II, which shows the effects of various structural modifications on affinity for the receptor. Reduction of the carboxylic acid group of $PGF_{2\alpha}$ (Fig. 3) to a hydroxymethyl group increases the dissociation constant by over 100 times, indicating that it plays an important role in the $PGF_{2\alpha}$ receptor interaction. The length of the carboxylic acid side chain is also crucial, since shortening it by one carbon unit increased the K_d 52-fold. The 5,6-double bond of $PGF_{2\alpha}$ is well recognized by the receptor, since the K_d of this prostaglandin is 40 times lower than that of $PGF_{1\alpha}$. This effect could be due either to a direct interaction between the double bond and the receptor or to the lengthening of the carboxylic acid side chain which occurs upon saturation of the _cis_-double bond. The 9α-hydroxyl group of $PGF_{2\alpha}$ is about as important as the carboxylic acid group for binding to the receptor, since its oxidation to an oxo group, as in PGE_1, increased the K_d by over 50-fold, and inversion of the stereochemistry about the 9-carbon atom caused an increase of about 150-fold. Oxidation of the 11α-hydroxyl group had a much smaller effect on the K_d than oxidation of the 9α-hydroxyl group since the affinity of PGD_2 for the receptor was about 6 times greater than that of $PGE_{2\alpha}$ and only 9 times lower than that of $PGF_{2\alpha}$. Hydrogenation or dehydrogenation of the 13,14-double bond had a rather small effect on the K_d. Modifications of the 15-hydroxyl group had quite variable effects, however. Its oxidation to an oxo group caused a dramatic increase in the K_d. Inversion of the stereochemistry about the 15-carbon atom caused a 12-fold increase in the K_d (15-methyl-15-epi-PGE_2 _vs_ methylPGE_2), but since this modification was in the PGE series it is less meaningful. Acetylation of the 15-hydroxyl group caused a 9-fold reduction in affinity for the receptor, whereas its methylation had almost no effect.

Various modifications of the terminal pentyl group of $PGF_{2\alpha}$ had relatively small effects on affinity for the receptor, and in some cases even resulted in increased affinity. The presence of two methyl groups at carbon-16 or an ethyl group at carbon-20 doubled the K_d whereas a double bond between carbons 17 and 18 ($PGF_{3\alpha}$) caused a 3.4 fold increase. The presence of an aryl group in the terminal part of the alkyl side chain increased affinity for the receptor. The $PGF_{2\alpha}$ analogue, 16-(_m_-chlorophenoxy)-17, 18, 19, 20-tetranor$PGF_{2\alpha}$, for example, had 2.5 times the affinity of $PGF_{2\alpha}$ for the receptor and 17-phenyl-18, 19, 20-trinor $PGF_{2\alpha}$ had almost twice the affinity.

Inversion of all the asymmetric centers of 13,14-didehydro$PGF_{2\alpha}$ caused an increase in the K_d of 6200-fold, whereas inversion of all the centers except carbon-15 caused only a 30-fold increase. The reason for the dramatic effect of reinverting only one of the 5 asymmetric centers is not clear. In this connection it has been reported that ent-15-15-epi-PGA_1 and ent-15-epiPGE_1 were more potent than their natural enantiomers on some smooth muscle preparations (23).

From these experiments it can be concluded that all the polar groups of $PGF_{2\alpha}$ (i.e. the carboxylic acid group and the 3-hydroxyl groups) participate in the $PGF_{2\alpha}$-receptor interaction. This interaction is rather sensitive to modifications in the carboxylic acid side chain which affect the distance of the carboxyl group from the hydroxyl substituents. A direct interaction between the 5, 6-double bond and the receptor is also possible. The affinity of prostaglandins for the receptor is not greatly affected by modifications of the alkyl side chain not directly affecting the 15-hydroxyl group.

Comparison between receptor affinity and luteolytic effect

If the $PGF_{2\alpha}$ receptor is responsible for luteolysis, then there should be a correlation between the specificities of the receptor and the luteolytic potencies of different prostaglandins and prostaglandin analogues. The antifertility potencies in hamsters of a number of prostaglandins and prostaglandin analogues have been determined and compared to the K_d's for the interactions of these compounds with the bovine corpus luteum $PGF_{2\alpha}$ receptor (21) (see Table I). The potencies of these compounds in vivo, however, depend not only on their affinities for the receptor but also on other factors such as their rate of metabolism. The rates at which these prostaglandins were dehydrogenated by 15-hydroxyprostaglandin dehydrogenase were therefore also determined (21) (see Table I), since this is the first step in the metabolism of the primary prostaglandins (25).

From the in vitro binding experiments it could be predicted that those analogues which had dissociation constants lower than that of $PGF_{2\alpha}$ (i.e. the two 17-phenyl-18, 19, 20-trinor$PGF_{2\alpha}$ derivatives and the three 16-aryloxy-17, 18, 19, 20-tetranor$PGF_{2\alpha}$ derivatives) would be more potent luteolytic agents than $PGF_{2\alpha}$ itself. In addition other compounds which had K_d's in the range of $PGE_{2\alpha}$ and in which metabolism by 15-hydroxyprostaglandin dehydrogenase had been completely blocked in some way would also be expected to be more potent than $PGF_{2\alpha}$ in producing luteolysis. This group would include the two 15-methyl$PGF_{2\alpha}$ derivatives, 16, 16-dimethyl$PGF_{2\alpha}$, $PGF_{2\alpha}$ 15-methyl ether and 13, 16-didehydro$PGF_{2\alpha}$. One might also expect 20-ethyl$PGF_{2\alpha}$ to be more potent than $PGF_{2\alpha}$ in vivo since it is metabolized 3 times less readily than the parent compound but has only 2.4-fold less affinity for the receptor.

The predictions are borne out quite well in practice. The most active analogues in vivo are those which contain an aromatic ring in the alkyl side chain or those in which metabolism by 15-hydroxyprostaglandin dehydrogenase is completely blocked by a methyl group at carbon-15. All of these compounds are at least 50-100 times more potent than $PGF_{2\alpha}$ in vivo (see Table I). 16, 16-Dimethyl$PGF_{2\alpha}$, $PGF_{2\alpha}$ 15-methyl ether, 13,14-didehydro$PGF_{2\alpha}$ and 20-ethyl$PGE_{2\alpha}$ are all considerably more potent than $PGF_{2\alpha}$ as luteolytic agents. It is apparent, however, that factors other than receptor affinity and metabolism by 15-hydroxyprostaglandin dehydrogenase are important in determining the luteolytic potencies of these prostaglandin analogues. The affinity for the receptor of 17-phenyl-18, 19, 20-trinor-$PGF_{2\alpha}$ is about twice that of $PGF_{2\alpha}$ and it is metabolized 3 times more slowly by 15-hydroxyprostaglandin dehydrogenase. Yet it is 100 times more potent than $PGF_{2\alpha}$ in vivo. Similarly, although none of the $PGE_{2\alpha}$analogues are as active as their $PGF_{2\alpha}$ counterparts, as would be predicted from the building experiments, some of them (i.e. 15-methyl$PGE_{2\alpha}$methyl ester, 16, 16-dimethyl$PGE_{2\alpha}$and 17-phenyl-18,19,20-trinor PGE_2) are more potent than $PGF_{2\alpha}$ in vivo, in spite of the fact that they have considerably lower affinities for the receptor. This may be partially explained in the cases of the two former compounds by the fact that they are not substrates for 15-hydroxyprostaglandin dehydrogenase. This explanation is not valid for 17-phenyl-18,19,20-trinor$PGE_{2\alpha}$, however, since this compound is metabolized more rapidly than $PGF_{2\alpha}$. The reasons for these discrepancies are not known at present. It can only be postulated that there may be differences in the rates at which the different prostaglandins reach the receptor in the corpus luteum subcutaneous deposition in the hamster or that there may be some species differences in the specificities of the receptor for $PGF_{2\alpha}$ in bovine and hamster corpora lutea.

The experiments with synthetic prostaglandin analogues show that it is possible to make considerable modifications in the prostaglandin alkyl side chain and still retain, or even enhance, the affinity of the analogue for the corpus luteum receptor. In this way metabolism of the analogue by 15-hydroxyprostaglandin dehydrogenase can be blocked, resulting in a greatly enhanced luteolytic potency. The presence of an aromatic ring in the alkyl side chain of $PGF_{2\alpha}$ increases both the affinity for the receptor and the luteolytic potency.

Subcellular Localization of the $PGF_{2\alpha}$ Receptor in Bovine Corpora Lutea

The localization of the receptor within the corpus luteum cell was investigated by examining the distribution of the receptor in various fractions obtained by differential and sucrose gradient centrifugation of bovine corpus luteum homogenates. This was compared with the distributions in these fractions of marker enzymes known to be localized in different subcellular components.[1]

[1] Powell, W. S., Hammarström, S. and Samuelsson, B., submitted for publication.

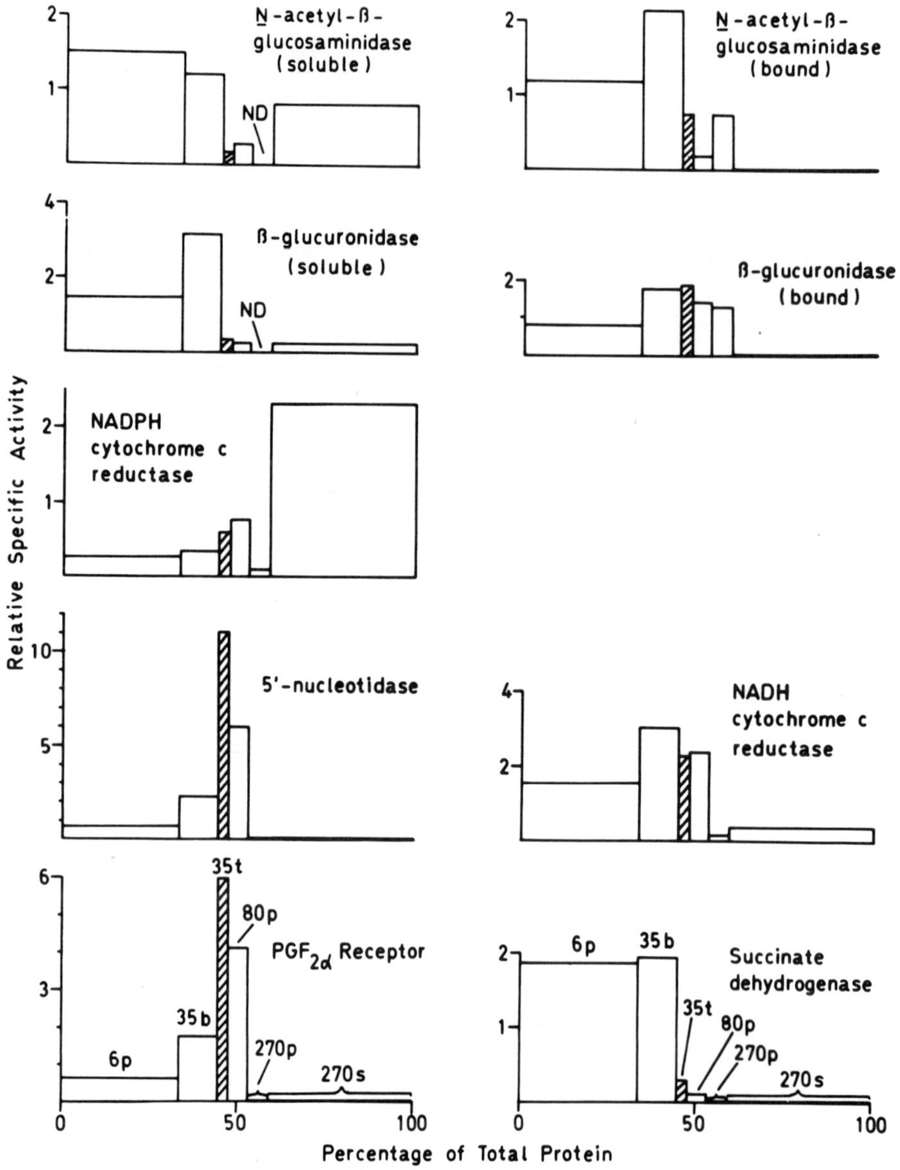

Fig. 4. Distributions of marker enzymes and a $PGF_{2\alpha}$ receptor after differential centrifugation of a bovine corpus luteum homogenate. Abscissa: the amount of protein in each fraction as a percentage of the total protein. Ordinate: purification of each constituant with respect to its activity in the homogenate. 6p 6000 x g pellet; 35b, bottom layer of 35,000 x g pellet; 35t, top layer of 35,000 x g pellet; 80p, 80,000 x g pellet; 270p, 270,000 x g pellet; 270s, 270,000 x g supernatant.[1]

Figure 4 shows the results of an experiment in which a homogenate from bovine corpora lutea was centrifuged successively at 6000xg, 35,000xg, 80,000xg and 270,000xg. The pellet obtained upon centrifugation of the 6000xg supernatant at 35,000xg had two layers which were separated and designated 35b for the bottom layer and 35t for the top layer. On the ordinate is given the purification of the constituent with respect to its activity in the homogenate and on the abscissa, the percentage of the total protein present in each of the fractions.

The distribution of the $PGF_{2\alpha}$ receptor in the various fractions resembled very closely that of 5'-nucleotidase, a marker enzyme for plasma membranes (9), suggesting that the receptor is localized on the plasma membranes of the corpus luteum cells. The receptor and 5'nucleotidase were purified maximally in fraction 35t, the purification of the former being 6-fold and of the latter, 11-fold. None of the other marker enzymes was purified maximally in this fraction with the exception of particle bound β-glucuronidase (see below), which was rather evenly distributed throughout all the fractions. Succinate dehydrogenase, a marker enzyme for mitochondria (24), was concentrated in the 6000xg pellet and in the bottom layer of the 35,000xg pellet. The particle-bound activity of NADPH cytochrome c reductase, a marker enzyme for the endoplasmic reticulum (1), was purified to the greatest extent in the 80,000xg pellet. Most of the activity of this enzyme in bovine corpus luteum homogenates, however, is in the 270,000xg supernatant, which is not the case with liver (1). It is possible that this could be a non-specific activity involving some other enzyme. The activities of β-glucuronidase and N-acetyl-β-glucosaminidase which could be released by freeze-thawing, and are designated as being soluble in Fig. 4, are indicative of the presence of lysosomes (1). The"soluble" fractions of both of these enzymes were purified to some extent in the 6000xg pellet and in fraction 35b, but had very low activities in fraction 35t. About half the activities of each of these two enzymes could not be released by freeze-thawing, but remained particle bound. This may be due to absorption (6), or, in the case of β-glucuronidase, to its presence in the endoplasmic reticulum as is the case with liver (1).

Further purification of the receptor present in fraction 35t by sucrose gradient centrifugation was not possible, mainly because of the difficulty to separate plasma membrane vesicles from endoplasmic reticulum vesicles. It was possible to remove these contaminants to a greater extent at the expense of lower yield and greater mitochondrial contamination by using lower centrifugal forces for the differential centrifugation step (Fig. 5, left hand side). The homogenate was centrifuged at 1000xg, 10,000xg and 80,000xg. The top and bottom layers of the 10,000xg pellet were separated and designated 10t and 10b. The receptor and 5'-nucleotidase were purified maximally in fractions 10t and 80p (the 80,000xg

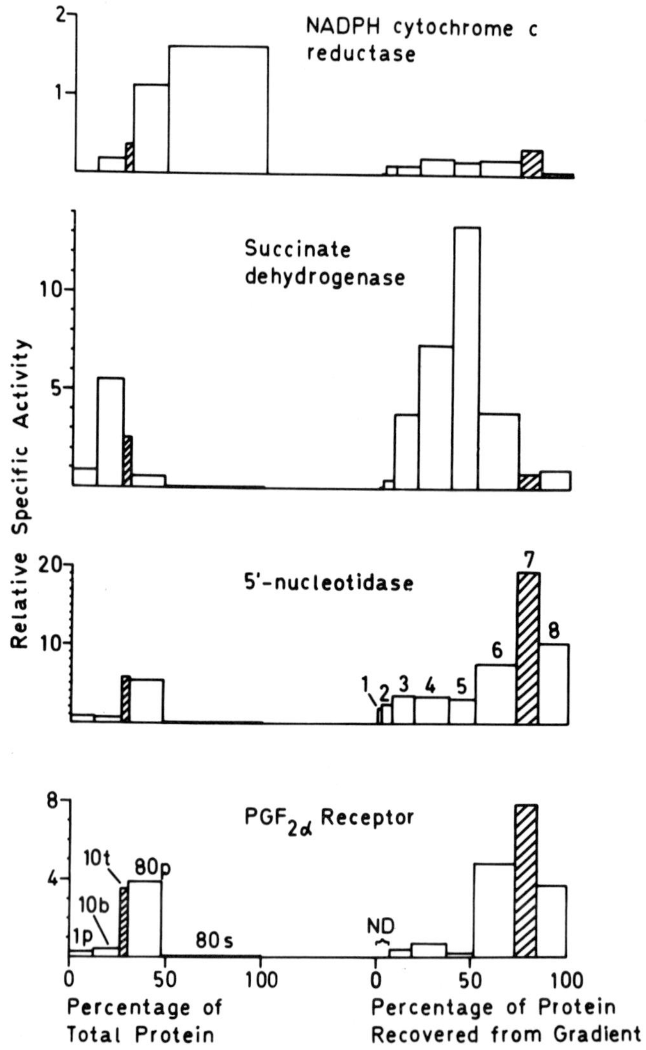

Fig. 5. The left side shows the distributions of some constituents after differential centrifugation of bovine corpus luteum homogenates. 1p, 1000xg pellet; 10b, bottom layer of 10,000xg pellet; 10t, top layer of 10,000xg pellet; 80p, 80,000xg pellet; 80s, 80,000xg supernatant. The right side shows the distributions of some constituents after sucrose density gradient centrifugation of fraction 10t. The numbers refer to the fractions indicated in Fig. 6. Reproduced with permission.[1]

pellet). Succinate dehydrogenase was purified to the greatest extent in fraction 10b, although a significant amount was present in fraction 10t. Particulate NADPH cytochrome c reductase was purified maximally in fraction 80p.

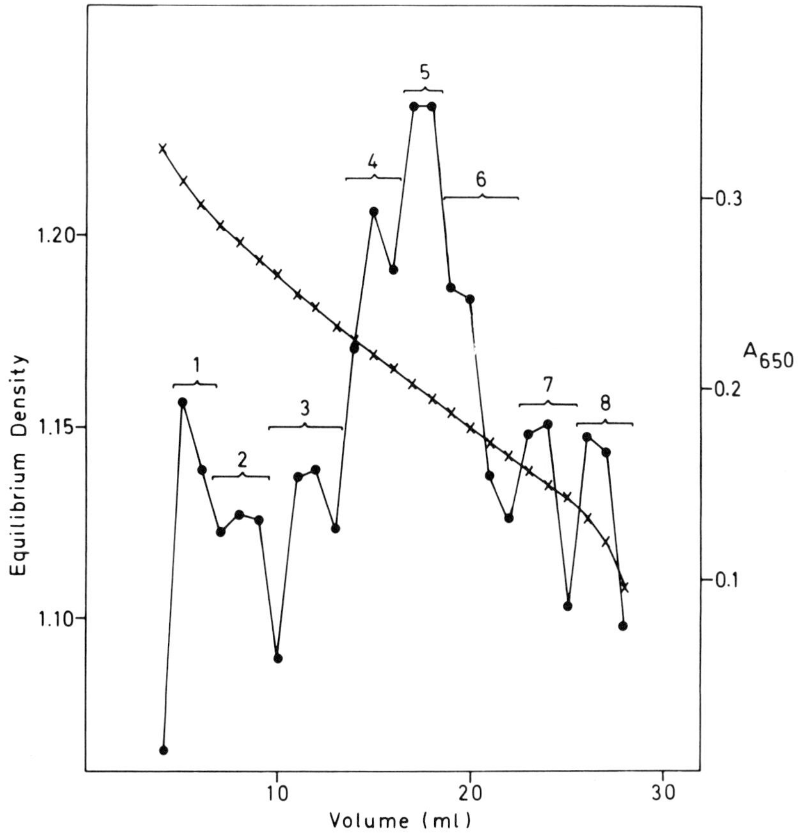

Fig. 6. Sucrose density gradient centrifugation of fraction 10t. The absorbance at 650nm (O) and the sucrose density (x) were determined for each fraction. The various 1 ml fractions were combined as indicated to give 8 larger fractions. Reproduced with permission.[1]

Fraction 10t was applied to a continuous gradient of sucrose and centrifuged for 10 hours at 96,000xg. One ml fractions were collected and their absorbances at 650 nm and their densities were determined (Fig. 6). The fractions were pooled as indicated in Fig. 6. The right side of Fig. 5 shows the distributions of the receptor and some marker enzymes in the fractions collected after sucrose gradient centrifugation. Succinate dehydrogenase was purified 4-fold with respect to the homogenate in fraction 5, which had a mean density of 1.160 g/ml. The $PGF_{2\alpha}$ receptor and 5'-nucleotidase were purified 8- and 20-fold, respectively, in fraction 7, which had a mean density of 1.135 g/ml, within the range (1.12-1.14) found for the light subfraction of liver plasma membranes (2, 8, 12). NADPH cytochrome c reductase was also purified to some extent in this fraction, but succinate dehydrogenase was almost completely absent.

These results are all consistent with the localization of the $PGF_{2\alpha}$ receptor on the plasma membranes of the bovine corpus luteum cells. This receptor is not present in appreciable amounts in the mitochondria, lysosomes or endoplasmic reticulum.

Solubilization of the $PGF_{2\alpha}$ Receptor Complex

The $PGF_{2\alpha}$-receptor complex has been solubilized with a variety of detergents (13). Although these detergents inhibited the binding of $PGF_{2\alpha}$ to the receptor, once the hormone-receptor complex was formed it was stable in the presence of most of the detergents. Treatment of the $PGF_{2\alpha}$-particulate fraction complex with sodium deoxycholate (0.5%) for 30 min at 0° completely solubilized the complex, whereas Triton X-100 (0.5%) and Nonidet P-40 (0.5%) solubilized about 70-80%. Incubation with sodium dodecyl sulphate resulted in a total loss of $PGF_{2\alpha}$ binding, however. The solubilized $PGF_{2\alpha}$-receptor complex was quite stable at 0°, since about 70% still remained after incubation at this termperature for one week.

Fig. 7 shows the results of chromatography of the sodium deoxycholate solubilized complex on Sepharose 6B in the presence of 0.5% sodium deoxycholate. When the initial incubation of the particulate fraction with tritium-labeled $PGF_{2\alpha}$ took place in the presence of an excess of the unlabeled prostaglandin, the first peak, but no other peaks, was absent, indicating that it was due to the $PGF_{2\alpha}$ receptor complex. The largest peak had the same elution volume as tritium-labeled $PGF_{2\alpha}$ and therefore represented the free $PGF_{2\alpha}$. Similar results were obtained when the $PGF_{2\alpha}$-receptor complex was solubilized with Triton X-100, although its elution volume changed somewhat, indicating that the size of the hormone-receptor-detergent complex was not the same with this detergent.

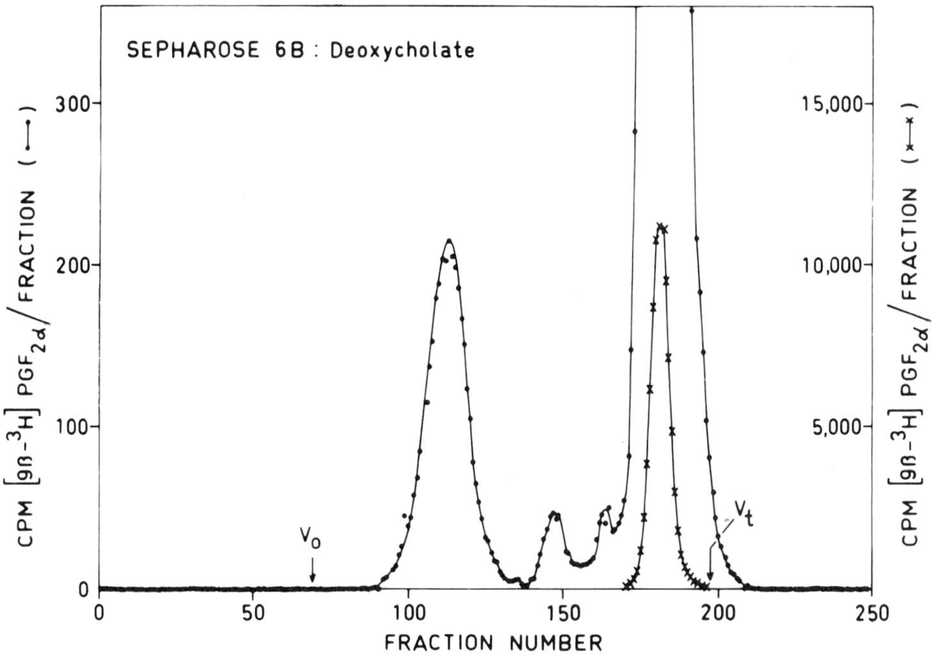

Fig. 7. Chromatography of the sodium-deoxycholate (0.5% solubilized [9β-^{3}H] PGF$_{2\alpha}$-receptor complex on Sepharose 6B in the presence of 0.5% sodium deoxycholate. Reproduced with permission from Ref. 13.

ACKNOWLEDGEMENT

This work was supported by a grant from the World Health Organization. W. S. P. is the holder of a fellowship from the Medical Research Council of Canada.

REFERENCES

1. AMAR-COSTESEC, A., BEAUFAY, H., WIBO, M., THINÈS-SEMPOUX, D., FEYTMANS, E., ROBBI, M. and BERTHET, J., J. Cell. Biol. 61 (1974) 20.
2. BEAUFAY, H., AMAR-COSTESEC, A., THINÈS-SEMPOUX, D., WIBO, M. ROBBI, M. and BERTHET, J., J. Cell Biol. 61 (1974) 213.
3. BEHRMAN, H. R., MACDONALD, G. J. and GREEP, R. O., Lipids 6 (1971) 791.

4. BEHRMAN, H. R., YOSHINAGA, K. and CREEP, R., Ann. N. Y. Acad. Sci. 180 (1971) 426.
5. BINDER, D., BOWLER, J., BROWN, E. D., CROSSLEY, N. S., HUTTEN, J. SENIOR, M., SLATER, L., WILKINSON, P. and WRIGHT, N. C. A., Prostaglandins 6 (1974) 87.
6. BOWERS, W. E. and DE DUVE, C., J. Cell. Biol. 32 (1967) 339.
7. CHAMLEY, W. A., BUCKMASTER, J. M., CAIN, M. D., CERINI, J., CERINI, M. E., CUMMING, I. A.and GODING, J. R., J. Endocr. 55 (1972) 253.
8. COLEMAN, R., MICHELL, R. H., FINEAN, J. B. and HAWTHORNE, J. N., Biochim. Biophys. Acta 135 (1967) 573.
9. DE PIERRE, J. W. and KARNOVSKY, M. L., J. Cell Biol. 56 (1973) 275.
10. DINGLE, J. T., HAY, M. F. and MOOR, R. M., J. Endocr. 40 (1968) 325.
11. EINER-JENSEN, N.and MC CRACKEN, J. A., in: Proc. Int. Conf. on Prostaglandins, Florence, Raven Press, New York (1975) in press.
12. EVANS, W. H., Biochem. J. 116 (1970) 833.
13. HAMMARSTRÖM, S., KYLDÉN, U., POWELL, W. S. and SAMUELSSON, B., FEBS Lett. 50 (1975) 306.
14. HICHENS, M., GRINWICH, D. L. and BEHRMAN, H. R., Prostaglandins 7 (1974) 449.
15. JACOBSON, H., OBRAMS, G. I. and SWARTZ, D. P., Fertil. Steril. 25 (1974) 293.
16. KIRTON, K. T., GUTKNECHT, G. D., BERGSTROM, K. K., WYNGARDEN, L. J. and FORBES, A. D., J. Reprod. Med. 9 (1972) 266.
17. MC CRACKEN, J. A., CARLSON, J. C., GLEW, M. E., GODING, J. R., BAIRD, D. T., GREÉN, K. and SAMUELSSON, B., Nature New Biol. 238 (1969) 129.
18. PHARRISS,B. B., and WYNGARDEN, L. J., Proc. Soc. Exp. Biol. Med. 130 (1969) 92.
19. POWELL, W. S., HAMMARSTRÖM, S. and SAMUELSSON, B., Eur. J. Biochem. 41 (1974) 103.
20. POWELL, W. S., HAMMARSTRÖM, S. and SAMUELSSON, B., Eur. J. Biochem. 56 (1975) 73.
21. POWELL, W. S., HAMMARSTRÖM, S., SAMUELSSON, B., MILLER, W. L., SUN, F. F., FRIED, J., LIN, C. H. and JARABAK, J., Eur. J. Biochem. (1975) in press.
22. POWELL, W. S., HAMMARSTRÖM, S., SAMUELSSON, B. and SJOBERG, B., Lancet 1 (1974) 1120.
23. RAMWELL, P. W., SHAW, J. E., COREY, E. J. and ANDERSON, N., Nature 221 (1969) 1251.
24. REID, E. in: Subcellular Components: Preparation and Fractionation. (G. D. Birnie ed.) Butterworths, London (1972) 93.
25. SAMUELSSON, B., GRANSTRÖM, E., GREÉN, K., HAMBURG, M. and HAMMARSTRÖM, S., Ann. Rev. Biochem. 44 (1975) 669.
26. STRAUSS, J. F. and STAMBAUGH, R. L., Prostaglandins 5 (1974) 73.
27. WEINER, R. and KALEY, G., Nature New Biol. 236 (1972) 46.

PARTICIPANTS IN THE

ADVANCED STUDY INSTITUTE

SURFACE MEMBRANE RECEPTORS:

INTERFACE BETWEEN CELLS AND ENVIRONMENT

Mark Achtman
Max-Planck-Institut
fur Molekulare Genetik
Ihnestr 63-73
Berlin 33 Dahlem
W. Germany

Julius Adler
Dept. of Biochemistry
University of Wisconsin
Madison, WIS. 53706

Richard R. Almon
Dept. of Biology
Univ. of California, San Diego
LaJolla, CA. 92037

Gilbert Ashwell
Lab. of Biochem. and Metabolism
National Inst. of Health
Bethesda, MD. 20014

Clinton E. Ballou
Dept. of Biochemistry
University of California
Berkeley, CA. 94720

Samuel Barondes
Dept. of Psychiatry
School of Medicine
Univ. of California, San Diego
LaJolla, CA. 92037

Luisa Bertolini
Lab. di Biologia Cellulare
Via G. Romagnosi 18A
Roma, Italy

Friedrich J. Bonhoeffer
Max-Planck-Institut
fur Virusforschung
Spemann Str. 35
Tubingen, W. Germany

Ralph A. Bradshaw
Dept. of Biological Chemistry
Washington University
School of Medicine
St. Louis, MO. 63110

Max Burger
Biozentrum
Universitat Basel
Dept. of Biochemistry
Basel, Switzerland

Jean P. Christophe
Dept. of Biochemistry
Univ. Libre de Bruxelles
Bruxelles 1000, Belgium

Elmon L. Coe
Dept. of Biochemistry
The Medical and Dental Schools
Northwestern University
Chicago, ILL. 60611

Pierre DeMeyts
Diabetes Branch
NIAMDD
National Institutes of Health
Bethesda, MD. 20014

Alex N. Eberle
E.T.H.
Institute of Molecular Biology and Biophysics
Honggenberg
8049 Zurich, Switzerland

Steven D. Flanagan
Friedrich Meischer Institut
P.O.B 273
Basel 4002, Switzerland

William A. Frazier
Dept. of Biological Chemistry
Washington University
School of Medicine
St. Louis, MO. 63110

Gunther Gerisch
Biozentrum Der Universitat Basel
Klingelbergstr 70
Basel 4056, Switzerland

Mary Jane Gething
Imperial Cancer Research Fund
Lincolns Inn Fields
London WC2A 3PX, England

Luis Glaser
Dept. of Biological Chemistry
Washington University
School of Medicine
St. Louis, MO. 63110

Beat E. Glatthaar
Hofmann-LaRoche
Basel, Switzerland

Sina Gökce
Inst. Univ. Med. Faculty
Dept. of Biophysics
University of Istanbul
Istanbul, Turkey

Rosanne Goodman
Biozentrum Der Universitat Basel
Basel 4056, Switzerland

David I. Gottlieb
Dept. of Anatomy
Washington University
School of Medicine
St. Louis, MO. 63110

Allan R. Gould
Dept. of Botany
University of Edinburgh
Mayfield Road
Edinburgh, Scotland

Hans H. Grünhagen
Institut Pasteur
75024 Paris, France

Bernard Haber
Marine Biomedical Institute
UTMB
Galveston, TEX. 77550

Harvey Herschman
Lab. of Nuclear Medicine and Radiation Biology
UCLA
Los Angeles, CA. 90024

Ruth A. Hogue-Angeletti
Neuropathology-School of Med.
University of Pennsylvania
Philadelphia, PA. 19111

Magnus Höök
Medicinsk-Kemiska Institutionen
Veterinar Hogskolan
Biomedicinska Centrum
Uppsala, Sweden

Louis L. Houston
Dept. of Biochemistry
University of Kansas
Lawrence, KAN. 66045

James E. Jumblatt
Biozentrum Der Universitat Basel
Basel 4056, Switzerland

Torbjorn Karlsson
Medical & Physiological Chemistry
Uppsala University
Biomedicum
Uppsala, Sweden

Carolyn King
St. Cross College
Oxford University
Oxford OX1 3TU, England

Claudette Klein
Dept. de Biologie Moleculaire
Institut Pasteur
25 Rue de Docteur Roux
Paris, France

Kam Hung Leung
Dept. of Biochemistry
Wing Hall
Cornell University
Ithaca, N.Y. 14853

Alex Levitzki
Dept. of Biological Chemistry
Hebrew University
Jerusalem, Israel

Lee E. Limbird
Dept. of Medicine
Duke University Medical Center
Durham, N.C. 27710

Anthony R. Limbrick
Biophysics Department
University College, London
Gower Street
London, WCIE 6BT, England

Ingrid C. Ljungstedt
Inst. of Pharmaceutical Biochem.
University of Uppsala
Uppsala, Sweden

Dieter Malchow
Biozentrum Der Universitat Basel
Basel 4056, Switzerland

Ronald Merrell
Dept. of Surgery
Washington University
School of Medicine
St. Louis, MO. 63110

Michel Monsigny
Centre de Biophysique
Moleculaire
45045 Orleans, France

David L. Nelson
Dept. of Biochemistry
University of Wisconsin
Madison, WIS. 53706

Hugh Niall
Howard Florey Inst.
University of Melbourne
Parkville, Victoria, Australia

Sjur Olsnes
Norsk Hydro's Institute for
Cancer Research
Norwegian Radium Hospital
Montebello, Oslo 3
Norway

Roland Pochet
Institut de Recherche
Interdisciplinaire
Univ. Libre de Bruxelles
Boulevard de Waterloo 115
Bruxelles, Belgium

William S. Powell
Dept. of Chemistry
Karolinska Institutet
Stockholm 60, Sweden

Joav M. Prives
Neurobiology Section
Biochemistry Department
Weizmann Institute
Rehovot, Israel

Morris W. Pulliam
Dept. of Biological Chemistry
Washington University
School of Medicine
St. Louis, MO. 63110

Efraim Racker
Section of Biochemistry,
Molecular and Cell Biology
Wing Hall
Cornell University
Ithaca, N.Y. 14853

Brian D. Read
Dept. of Chemistry
Chelsea College
Manersa Road
London SW 3 6LX, England

Gerald R. Reeck
Dept. of Biochemistry
Kansas State University
Manhattan, KAN. 66506

Clay Reilly
Lehrstuhl für Immunbiologie der
Universität Freiburg
Stefan-Meier Str. 8
78 Freiburg, W. Germany

Roberto Revoltella
Lab. di Biologia Cellulare
Via Romagnosi 18A
Roma, Italy

Jesse Roth
Diabetes Branch
NIAMDD
National Institute of Health
Bethesda, MD. 20014

Zehra Sayers
Guy's Hospital
Medical School
Physics Dept.
London Bridge SE1
England

Werner Schlegel
Institute of Molecular Biology
and Biophysics
E.T.H.
Honggerberg
8049 Zurich, Switzerland

Susanna Schlegel-Haueter
Institut of Biochemistry II
E.T.H.
Universitat Str. 16
8006 Zurich, Switzerland

Uli Schwarz
Max-Planck-Institut fur
Virusforschung
Spemann Str. 35
D-74 Tubingen, W. Germany

Robert Silverman
Dept. of Biological Chemistry
Washington University
School of Medicine
St. Louis, MO. 63110

Lee Simon
Institute of Cancer Research
Fox Chase
Philadelphia, PA. 19111

S. J. Singer
Dept. of Biology
Univ. of California, San Diego
LaJolla, CA. 92037

Chariclia I. Stassinopoulou
N.R.C. "Democritos"
Aghia Paraskevi Attikis
Athens, Greece

Klaus M. Stockel
Biozentrum Der Universitat Basel
Basel 4056, Switzerland

Andrzej Szutowicz
Dept. of Clinical Biochemistry
Institute of Pathology
Medical Academy
Gdansk, Poland

Christopher D. Town
Imperial Cancer Research Fund
Mill Hill Laboratories
Burton Hole Lane
London NW7 1AD, England

W. E. van Heyningen
St. Cross College
Oxford University
Oxford OX1 3TU
England

Dido Varoucha
N.R.C. "Democritos"
Aghia Paraskevi
Athens, Greece

Alida Vreugdenhil
Dept. of Oral Biochemistry
Vrije Universiteit
P. O. Box 7161
Amsterdam, Netherlands

Michael Young
Lab. Physical Biochemistry
Harvard Medical School
Massachusetts General Hospital
Boston, MASS. 02114

Jesse M. Zimmerman
University Institute for
Medical Microbiology
Geneva, Switzerland

Christine Zioudrou
N.R.C. "Democritos"
Aghia Paraskevi
Athens, Greece

SUBJECT INDEX

1

M